PROGRESS IN CLINICAL AND BIOLOGICAL RESEARCH

For information regarding previous volumes in this series, please contact the publisher.

LIMB DEVELOPMENT AND REGENERATION PART A

ORGANIZING COMMITTEE
Third International Conference on Limb Morphogenesis and Regeneration

Arnold I. Caplan

> *Professor of Biology and Anatomy, Case Western Reserve University, Cleveland, Ohio*

John F. Fallon

> *Professor of Anatomy, University of Wisconsin School of Medicine, Madison, Wisconsin*

Paul F. Goetinck

> *Professor of Animal Genetics and Genetics and Cell Biology, University of Connecticut, Storrs, Connecticut*

Robert O. Kelley

> *Professor and Chairman of Anatomy, The University of New Mexico School of Medicine, Albuquerque, New Mexico*

Jeffrey A. MacCabe

> *Associate Professor of Zoology, University of Tennessee, Knoxville, Tennessee*

LIMB DEVELOPMENT AND REGENERATION

Part A

Proceedings of the Third International Conference on Limb Morphogenesis and Regeneration University of Connecticut, Storrs, June 27–July 2, 1982

Editors

John F. Fallon
Professor of Anatomy
University of Wisconsin
School of Medicine
Madison, Wisconsin

Arnold I. Caplan
Professor of Biology and Anatomy
Case Western Reserve University
Cleveland, Ohio

Alan R. Liss, Inc. • New York

QL
950
.7
.I57
1982
V.1

Address all Inquiries to the Publisher
Alan R. Liss, Inc., 150 Fifth Avenue, New York, NY 10011

Copyright © 1983 Alan R. Liss, Inc.

Printed in the United States of America

Library of Congress Cataloging in Publication Data

International Conference on Limb Morphogenesis and
    Regeneration (3rd : 1982 : University of Connecticut)
    Limb development and regeneration.

    (Progress in clinical and biological research ;
v. 110)
    Vol. 2 edited by Robert O. Kelley, Paul F. Goetinck,
and Jeffrey A. MacCabe.
    Includes bibliographies and indexes.
    1. Extremities (Anatomy) — Regeneration — Congresses.
2. Morphogenesis — Congresses. I. Fallon, John F.
II. Caplan, Arnold I.
III. Title. IV. Series. [DNLM: 1. Extremities — Growth
and development — Congresses. 2. Extremities — Physiology
— Congresses. 3. Regeneration — Congresses. W1 PR668E
v.110 / WE 800 I59 1982L]
QL950.7.I57   1982   596'.031   82-20391
ISBN 0-8451-0110-2 (set)
ISBN 0-8451-0170-6 (v. 1)
ISBN 0-8451-0171-4 (v. 2)

to the memory of
Professor Edgar Zwilling
1913–1971

Participants in the Third International Conference on Limb Morphogenesis and Regeneration, held at the University of Connecticut, Storrs, June 27–July 2, 1982.

Contents of Part A

Contents of Part B

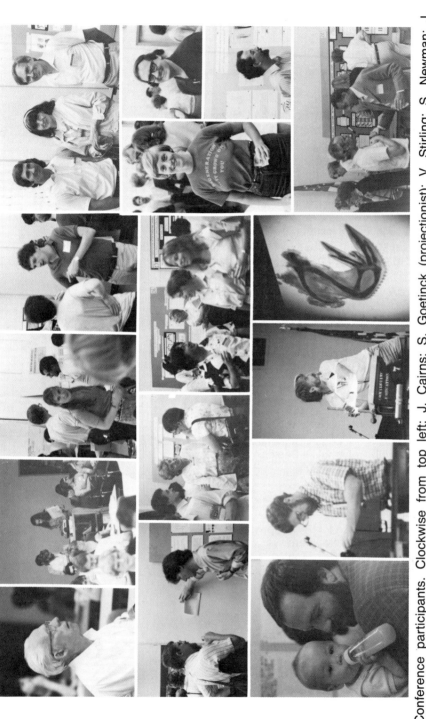

Conference participants. Clockwise from top left: J. Cairns; S. Goetinck (projectionist); V. Stirling; S. Newman; J. Pennypacker, U. Abbott and P. Goetinck; D. Dayton; V. Hascall; A. Caplan and M. Michael; the duplicated limb; D. Summerbell; J. MacCabe; C. Ordahl and offspring; D. Heinegård.

Center group, from left: M. Maden and J. McCredie; K. Muneoka, M. Runner, C. Rollman-Dinsmore and N. O'Rourke; C. Olsen.

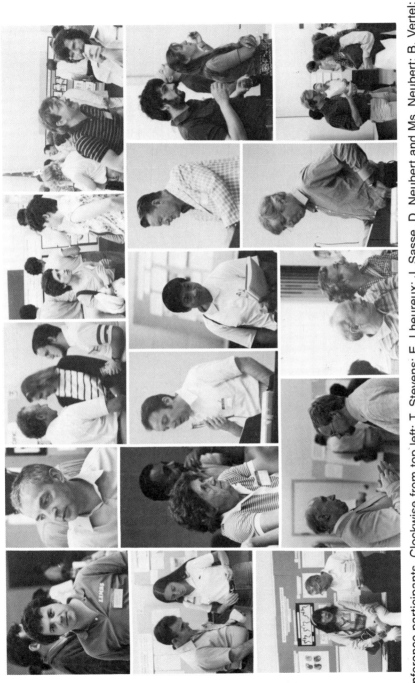

Conference participants. Clockwise from top left: T. Stevens; E. Lheureux; J. Sasse, D. Neubert and Ms. Neubert; B. Vertel; T. Linsenmayer and J. Slack; N. Holder and V. Stirling; R. Kelley and L. Iten; L. Wolpert; L. Rosenberg and D. Fischman; D. Ede and H.J. Jacob; U. Abbott and M. Runner; K. Sparks and P. McKeown-Longo.
Center group, from left: Mrs. E. Zwilling and M. Michael; J. Fallon; M. Yasuda; J.W. Saunders, Jr.

Participants and Contributors

Ursula K. Abbott [A:13]
University of California, Davis, Davis CA 95616

Thomas R. Adams [B:349]
T.H. Morgan School of Biological Sciences, University of Kentucky, Lexington, KY 40506

Loulwah Al-Ghaith [A:195]
Department of Biology as Applied to Medicine, Middlesex Hospital Medical School, London W1P 6DB, United Kingdom

Rodolfo Amprino [A:155]
Institute of Human Anatomy, University of Bari, 70124 Bari, Italy

C.W. Archer [A:267]
Department of Biology as Applied to Medicine, The Middlesex Hospital Medical School, London W1P 6DB, United Kingdom

John R. Baker [B:17]
University of Alabama in Birmingham, Birmingham, AL 35294

Mark Ballow [B:113]
Department of Pediatrics, University of Connecticut Health Center, Farmington, CT 06032

J. Biehl [B:271]
Department of Anatomy, Medical School of the University of Pennsylvania, Philadelphia, PA 19104

David E. Birk [A:245]
Department of Pathology, UMDNJ-Rutgers Medical School, Piscataway, NJ 08854

Philip H. Bonner [B:349]
T.H. Morgan School of Biological Sciences, University of Kentucky, Lexington, KY 40506

Fred L. Bookstein [A:525]
Center for Human Growth and Development, The University of Michigan, Ann Arbor, MI 48109

Richard B. Borgens [A:597]
Institute for Medical Research, San Jose, CA 95128

Edward G. Buss [B:85]
Department of Poultry Science, Pennsylvania State University, University Park, PA 16802

Jo Ann Cameron [A:491]
Department of Anatomical Sciences, University of Illinois, Urbana, IL 61801

Arnold I. Caplan [A:143, B:229, 379]
Department of Biology, Developmental Biology Center, Case Western Reserve University, Cleveland, OH 44106

The boldface number in brackets following each contributor's name indicates the opening page of that author's article in either Part A or Part B of these proceedings.

Bruce M. Carlson [A:433]
Departments of Anatomy and Biological Sciences, University of Michigan, Ann Arbor, MI 48109

Jill L. Carrington [A:33]
Department of Anatomy, University of Wisconsin, Madison, WI 53706

David A. Carrino [B:379]
Department of Biology, Case Western Reserve University, Cleveland, OH 44106

Bruce Caterson [B:17]
University of Alabama in Birmingham, Birmingham, AL 35294

Matthias Chiquet [B:359]
Department of Embryology, Carnegie Institution of Washington, Baltimore, MD 21210

H. Choi [B:67]
Department of Orthopedic Research, Montefiore Hospital and Medical Center, Bronx, NY 10467

Bodo Christ [B:281,313,333]
Institute of Anatomy, Ruhr-University Bochum, D-4630 Bochum, Federal Republic of Germany

James E. Christner [B:17]
University of Alabama in Birmingham, Birmingham, AL 35294

Sandra B. Conlon [B:417]
Department of Medicine/Oncology, Stanford School of Medicine, Stanford, CA 94305

Thomas G. Connelly [A:525]
Department of Anatomy and Cell Biology, The University of Michigan, Ann Arbor, MI 48109

Thomas Cooper [B:391]
Department of Anatomy, Temple University School of Medicine, Philadelphia, PA 19140

Michael T. Crow [B:417]
Department of Medicine/Oncology, Stanford School of Medicine, Stanford, CA 94305

Ann Dannenberg [B:85]
Department of Animal Genetics, University of Connecticut, Storrs, CT 06268

Charles E. Dinsmore [A:577]
Department of Anatomy, Rush Medical College, Chicago, IL 60612

Albert Dorfman [B:175]
Department of Pediatrics, The University of Chicago, Chicago, IL 60637

D.A. Ede [A:45]
Department of Zoology, University of Glasgow, Glasgow G12 9LU, United Kingdom

William A. Elmer [A:355]
Biology Department, Emory University, Atlanta, GA 30322

John F. Fallon [A:33,119]
Department of Anatomy, University of Wisconsin, Madison, WI 53706

Douglas M. Fambrough [B:359]
Department of Embryology, Carnegie Institution of Washington, Baltimore, MD 21210

Marc Y. Fiszman [B:401]
Department of Molecular Biology, Pasteur Institute, Paris Cedex 15, France

John M. Fitch [B:369]
Department of Medicine, Harvard Medical School of Massachusetts General Hospital, Boston, MA 02114

Jeanne M. Frederick [A:33]
Cullen Eye Institute, Baylor College of Medicine, Houston, TX 77030

Mary T. Gasseling [A:67]
Department of Biological Sciences, State University of New York, Albany, NY 12222

Jacqueline Géraudie [A:289]
Equipe de recherche "Formations squelettiques," Laboratoire d'Anatomie comparée, Université de Paris VII, 75251 Paris Cedex 05, France

Morton Globus [A:513]
Department of Biology, University of Waterloo, Waterloo, Ontario, Canada N2L 3G1

Suresh C. Goel [A:175]
Department of Zoology, University of Poona, Poona 411 007, India

Paul F. Goetinck [B:85,113]
Department of Animal Genetics, University of Connecticut, Storrs, CT 06268

John M. Graham, Jr. [A:413]
Department of Maternal and Child Health, Dartmouth Medical School, Hanover, NH 03755

M. Grim [B:333]
Department of Anatomy, Faculty of Medicine, Charles University Prague, 12800 Praha 2, Czechoslovakia

Brian K. Hall [B:323]
Department of Biology, Dalhousie University, Halifax, Nova Scotia, Canada B3H 4J1

Fiona Harvey [A:109]
National Institute for Medical Research, London NW7 1AA, United Kingdom

Vincent C. Hascall [B:3]
Laboratory of Biochemistry, National Institute of Dental Research, National Institutes of Health, Bethesda, MD 20205

John R. Hassell [B:105]
Laboratory of Developmental Biology and Anomalies, National Institute of Dental Research, National Institutes of Health, Bethesda, MD 20205

Stephen Hauschka [B:303]
Department of Biochemistry, University of Washington, Seattle, WA 98195

L.L. Hearson [A:587]
Department of Biology, Wabash College, Crawfordsville, IN 47933

Dick Heinegård [B:35]
Department of Physiological Chemistry, University of Lund, S-220 07 Lund, Sweden

A. Tyl Hewitt [B:25,105]
Laboratory of Developmental Biology and Anomalies, National Institute of
Dental Research, National Institutes of Health, Bethesda, MD 20205

Stuart M. Heywood [B:409]
Biological Sciences Group, Genetics and Cell Biology Section, University of
Connecticut, Storrs, CT 06268

J. R. Hinchliffe [A:131]
Department of Zoology, University College of Wales, Aberystwyth, Wales,
Dyfed SY23 3DA, United Kingdom

Kenneth S. Hirsch [A:423]
Eli Lilly and Co., Greenfield Laboratories, Greenfield, IN 46140

Nigel Holder [A:477]
Department of Anatomy, King's College University of London, London WC2R
2LS, United Kingdom

Margaret Hollyday [A:183]
Department of Pharmacological and Physiological Sciences, The University
of Chicago, Chicago, IL 60637

Lewis B. Holmes [A:311,317]
Department of Pediatrics, Embryology-Teratology Unit, Massachusetts
General Hospital, Boston, MA 02114

Howard Holtzer [B:159,271]
Department of Anatomy, Medical School, University of Pennsylvania,
Philadelphia, PA 19104

S. Holtzer [B:271]
Department of Anatomy, Medical School, University of Pennsylvania,
Philadelphia, PA 19104

Lawrence S. Honig [A:57,99]
Laboratory for Developmental Biology, University of Southern California, Los
Angeles, CA 90007

Marcia Honig [A:207]
Department of Biology, Yale University, New Haven, CT 06511

David R. Hootnick [A:327]
Department of Orthopedic Surgery, SUNY Upstate Medical Center, Syracuse,
NY 13210

Elizabeth A. Horigan [B:105]
Laboratory of Developmental Biology and Anomalies, National Institute of
Dental Research, National Institutes of Health, Bethesda, MD 20205

Amata Hornbruch [B:343]
Department of Biology as Applied to Medicine, The Middlesex Hospital
Medical School, London W1P 6DB, United Kingdom

Laurie E. Iten [A:77]
Department of Biological Sciences, Purdue University, West Lafayette, IN
47907

Heinz Jürgen Jacob [B:281,313,333]
Institute of Anatomy, Ruhr-University Bochum, D-4630 Bochum, Federal
Republic of Germany

Monika Jacob [B:281,313]
Institute of Anatomy, Ruhr-University Bochum, D-4630 Bochum, Federal
Republic of Germany

David M. Jargiello [A:143]
Department of Biology, Developmental Biology Center, Case Western
Reserve University, Cleveland, OH 44106

Janice A. Jerdan [B:105]
Laboratory of Developmental Biology and Anomalies, National Institute of
Dental Research, National Institutes of Health, Bethesda, MD 20205

Thomas Johnson [B:67]
Department of Orthopedic Research, Montefiore Hospital and Medical
Center, Bronx, NY 10467

Arnold J. Kahn [B:239]
Department of Biomedical Sciences, Washington University School of Dental
Medicine, St. Louis, MO 63110

Gary H. Karpen [A:609]
Department of Genetics, University of Washington, Seattle, WA 98195

Robert O. Kelley [A:119]
Department of Anatomy, University of New Mexico School of Medicine,
Albuquerque, NM 87131

Terry Kenny-Mobbs [B:323]
Department of Biology, Dalhousie University, Halifax, Nova Scotia, Canada
B3H 4J1

Agnes Kesik [A:513]
Department of Biology, University of Waterloo, Waterloo, Ontario, Canada
N2L 3G1

Akbar Khan [A:195]
Department of Anatomy, King's College University of London, London WC2R
2LS, United Kingdom

Madeleine A. Kieny [B:293]
Laboratoire de Zoologie et Biologie animale, Université Scientifique et
Médicale, Grenoble 38041, France

D.M. Kochhar [B:203]
Department of Anatomy, Jefferson Medical College, Philadelphia, PA 19107

Robert A. Kosher [A:279]
Department of Anatomy, University of Connecticut Health Center,
Farmington, CT 06032

Dean Kravis [B:175]
Department of Pediatrics, The University of Chicago, Chicago, IL 60637

Ralf Krowke [A:387]
Institut für Toxikologie und Embryopharmakologie der Freien Universität
Berlin, D-1000 Berlin 33, Federal Republic of Germany

Marilyn Krukowski [B:239]
Department of Biology, Washington University, St. Louis, MO 63130

Gerald Kuncio [B:391]
Department of Anatomy, Temple University School of Medicine, Philadelphia,
PA 19140

Thomas E. Kwasigroch [A:335]
Department of Anatomy, Quillen-Dishner College of Medicine, East
Tennessee State University, Johnson City, TN 37614

Alan H. Lamb [A:227]
Department of Pathology, University of Western Australia, Nedlands 60009,
Western Australia

Karen M. Lamb [B:45]
Department of Anatomy, Cell Biology Division, New Jersey Medical School,
Newark, NJ 07103

Lynn T. Landmesser [A:207]
Department of Biology, Yale University, New Haven, CT 06511

Mark E. Lanser [A:33]
Department of Anatomy, University of Wisconsin, Madison, WI 53706

Bernard Lassalle [A:547]
Laboratoire de Morphogenèse animale, Université des Sciences et
Techniques de Lille, 59655 Villeneuve d'Ascq Cedex, France

C. W. Leal [A:237]
Department of Zoology, University of Tennessee, Knoxville, TN 37996

K.W. Leal [A:237]
Department of Zoology, University of Tennessee, Knoxville, TN 37996

Claire M. Leonard [A:251]
Department of Anatomy, New York Medical College, Valhalla, NY 10595

E. Mark Levinsohn [A:327]
Department of Radiology, SUNY Upstate Medical Center, Syracuse, NY
13210

Julian Lewis [A:195]
Department of Anatomy, King's College London WC2R 2LS, United Kingdom

Emile Lheureux [A:455]
Laboratoire de Morphogenèse animale, Université des Sciences et
Techniques de Lille, 59655 Villeneuve d'Ascq Cedex, France

Thomas F. Linsenmayer [B:369]
Department of Anatomy and Medicine, Harvard Medical School and
Massachusetts General Hospital, Boston, MA 02114

Jeffrey A. MacCabe [A:237]
Department of Zoology, University of Tennesse, Knoxville, TN 37996

Thomas McCarthy [B:409]
Biological Sciences Group, Genetics and Cell Biology Section, The University
of Connecticut, Storrs, CT 06268

Janet McCredie [A:399]
Department of Radiology, University of Sydney, New South Wales, Australia
2006

M.E. McGinnis [A:587]
Department of Biological Sciences, Purdue University, West Lafayette, IN
47907

John C. McLachlan [B:343]
Department of Zoology, University of Oxford, Oxford OX1 3PS, United
Kingdom

Malcolm Maden [A:445]
Division of Developmental Biology, National Institute for Medical Research, London NW7 1AA, United Kingdom

Robert J. Majeska [B:249]
Department of Oral Biology, University of Connecticut Health Center, School of Dental Medicine, Farmington, CT 06032

J. David Malone [B:239]
Department of Internal Medicine, Jewish Hospital of St. Louis, St. Louis, MO 63110

Klaus von der Mark [B:159]
Max-Planck-Institut für Biochemie, 8033 Martinsried bei München, Federal Republic of Germany

George R. Martin [B:25,105]
Laboratory of Developmental Biology and Anomalies, National Institute of Dental Research, National Institutes of Health, Bethesda, MD 20205

Mary Martini [B:229]
Biology Department, Case Western Reserve University, Cleveland, OH 44106

Pauline Mayne [B:125]
Department of Anatomy, University of Alabama in Birmingham, Birmingham, AL 35294

Richard Mayne [B:125,369]
Department of Anatomy, University of Alabama in Birmingham, Birmingham, AL 35294

Anthony L. Mescher [A:501]
Anatomy Section, Medical Sciences Program, Indiana University School of Medicine, Bloomington, IN 47405

Guy Milton [A:513]
Department of Biology, University of Waterloo, Waterloo, Ontario, Canada N2L 3G1

Jay E. Mittenthal [A:619]
Department of Anatomical Sciences, University of Illinois, Urbana, IL 61801

Didier Montarras [B:401]
Department of Molecular Biology, Pasteur Institute, Paris Cedex 15, France

Barbara Mroczkowski [B:409]
Biological Sciences Group, Genetics and Cell Biology Section, The University of Connecticut, Storrs, CT 06268

Helen Muir [B:55]
Department of Biochemistry, Kennedy Institute of Rheumatology, Hammersmith, London W6 7DW, United Kingdom

Ken Muneoka [A:77]
Developmental Biology Center, University of California, Irvine, Irvine, CA 92717

Douglas J. Murphy [A:77]
Department of Biological Sciences, Purdue University, West Lafayette, IN 47907

Harukazu Nakamura [A:301]
Department of Anatomy, Hiroshima University School of Medicine, Hiroshima 734, Japan

Mark A. Nathanson [B:215]
Department of Anatomy, New Jersey Medical School, Newark, NJ 07103

Diether Neubert [A:387]
Institut für Toxikologie und Embryopharmakologie der Frein Universität Berlin, D-1000 Berlin 33, Federal Republic of Germany

Daniel A. Neufeld [A:407]
Department of Anatomy, University of South Dakota School of Medicine, Vermillion, SD 57069

Stuart A. Newman [A:251]
Department of Anatomy, New York Medical College, Valhalla, NY 10595

Yoshifumi Ninomiya [B:183]
Department of Biochemistry, UMDNJ-Rutgers Medical School, Piscataway, NJ 08854

Michael J. O'Donovan [A:207]
Department of Biology, Yale University, New Haven, CT 06511

Bjorn Reino Olsen [B:183]
Department of Biochemistry, UMDNJ-Rutgers Medical School, Piscataway, NJ 08854

Cherie L. Olsen [A:537]
Department of Zoology, The Ohio State University, Columbus, OH 43210

Pamela S. Olson [B:417]
Department of Medicine/Oncology, Stanford School of Medicine, Stanford, CA 94305

Charles P. Ordahl [B:391]
Department of Anatomy, Temple University School of Medicine, Philadelphia, PA 19140

Philip Osdoby [B:229]
Department of Biomedical Sciences, School of Dentistry, Washington University, St. Louis, MO 63110

Maurizio Pacifici [B:159,271]
Department of Anatomy, Medical School, University of Pennsylvania, Philadelphia, PA 19104

David S. Packard, Jr. [A:327]
Department of Anatomy, SUNY Upstate Medical Center, Syracuse, NY 13210

Subhash Pal [B:67]
Department of Orthopedic Research, Montefiore Hospital and Medical Center, Bronx, NY 10467

Mats Paulsson [B:35]
Department of Physiological Chemistry, University of Lund, S-220 07 Lund, Sweden

R. Payette [B:271]
Department of Anatomy, Medical School, University of Pennsylvania, Philadelphia PA 19104

Isaac Peng [B:391]
Department of Anatomy, Temple University School of Medicine, Philadelphia, PA 19140

John D. Penner [B:203]
Department of Anatomy, Jefferson Medical College, Philadelphia, PA 19107

John P. Pennypacker [B:167]
Department of Zoology, University of Vermont, Burlington, VT 05405

I. Pidoux [B:67]
Joint Diseases Research Laboratories, Shriners Hospital for Crippled
Children, Montreal, Quebec, Canada H3G 1A6

A. Robin Poole [B:67]
Joint Diseases Research Laboratories, Shriners Hospital for Crippled
Children, Montreal, Quebec, Canada H3G 1A6

A.H. Reddi [B:261]
Bone Cell Biology Section, Laboratory of Biological Structure, National
Institute of Dental Research, National Institutes of Health, Bethesda, MD
20205

Charles A. Reese [B:125]
Department of Anatomy, University of Alabama in Birmingham, Birmingham,
AL 35294

Agnes Reiner [B:67]
Joint Disease Research Laboratories, Shriners Hospital for Crippled Children,
Montreal, Quebec, Canada H3G 1A6

Susan Reynolds [A:477]
Department of Anatomy, King's College University of London, London, WC2R
2LS, United Kingdom

Gideon A. Rodan [B:249]
Department of Oral Biology, University of Connecticut Health Center, School
of Dental Medicine, Farmington, CT 06032

P. Rooney [A:267]
Department of Biology as Applied to Medicine, The Middlesex Hospital
Medical School, London W1P 6DB, United Kingdom

Marcel André Rooze [A:365]
Laboratory of Human Anatomy and Embryology, Faculty of Medicine, Free
University of Brussels, B-1000 Brussels, Belgium

Lawrence C. Rosenberg [B:67]
Department of Orthopedic Surgery, Albert Einstein College of Medicine,
Montefiore Hospital and Medical Center, Bronx, NY 10467

Peter Roughley [B:67]
Joint Diseases Research Laboratories, Shriners Hospital for Crippled
Children, Montreal, Quebec, Canada H3G 1A6

Meredith N. Runner [A:345]
Department of Molecular, Cellular and Developmental Biology, University of
Colorado, Boulder, CO 80309

Richard Rutz [B:303]
Department of Biochemistry, University of Washington, Seattle, WA 98195

Linda Sandell [B:175]
Department of Biochemistry, Rush-Presbyterian-St. Lukes Medical Center,
Chicago, IL 60612

J.D. Sandy [B:55]
Department of Biochemistry, Kennedy Institute of Rheumatology, Hammersmith, London W6 7DW, United Kingdom

Joachim Sasse [B:159,271]
Department of Anatomy, Medical School, University of Pennsylvania, Philadelphia, PA 19104

John W. Saunders, Jr. [A:67]
Department of Biological Sciences, State University of New York, Albany, NY 12222

Claire M. Schreiner [A:423]
Division of Pathologic Embryology, Children's Hospital Research Foundation, Cincinnati, OH 45229

Gerold Schubiger [A:609]
Department of Zoology, University of Washington, Seattle, WA 98195

Nancy B. Schwartz [B:97]
Department of Pediatrics and Biochemistry, The University of Chicago, Chicago, IL 60637

William J. Scott [A:423]
Division of Pathologic Embryology, Children's Hospital Research Foundation, Cincinnati, OH 45229

Robert L. Searls [A:165]
Department of Biology, Temple University, Philadelphia, PA 19122

Robert E. Seegmiller [B:193]
Department of Zoology, Division of Genetics and Developmental Biology, Brigham Young University, Provo, UT 84602

Maryam Shamslahidjani [A:45]
Department of Biology, National University of Iran, Teheran, Iran

Val C. Sheffield [B:175]
Department of Pediatrics, The University of Chicago, Chicago, IL 60637

Thomas H. Shepard [A:377]
Central Laboratory for Human Embryology, University of Washington School of Medicine, Seattle, WA 98195

Kohei Shiota [A:377]
Central Laboratory for Human Embryology, University of Washington School of Medicine, Seattle, WA 98195

Allan M. Showalter [B:183]
Department of Biochemistry, UMDNJ-Rutgers Medical School, Piscataway, NJ 08854

Frederick H. Silver [A:245]
Department of Pathology, UMDNJ-Rutgers Medical School, Piscataway, NJ 08854

Michael H. Silver [B:25]
Department of Laboratory Medicine and Pathology, University of Minnesota Medical Center, Minneapolis, MN 55455

B. Kay Simandl [A:33]
Department of Anatomy, University of Wisconsin, Madison, WI 53706

Richard G. Skalko [A:335]
Department of Anatomy, Quillen-Dishner College of Medicine, East Tennessee State University, Johnson City, TN 37614

Jonathan M.W. Slack [A:557]
Department of Experimental Morphology, Imperial Cancer Research Fund, London NW7 1AD, United Kingdom

J.C. Smith [A:57]
Imperial Cancer Research Fund, Mill Hill Laboratories, London NW7 1AD, United Kingdom

Michael Solursh [B:139]
Department of Zoology, University of Iowa, Iowa City, IA 52242

Yngve Sommarin [B:35]
Department of Physiological Chemistry, University of Lund, S-220 07 Lund, Sweden

Kenneth Sparks [B:113]
Department of Animal Genetics, University of Connecticut, Storrs, CT 06268

Trent D. Stephens [A:3]
Department of Biology, Idaho State University, Pocatello, ID 83209

R. Victoria Stirling [A:217]
Department of Developmental Biology, National Institute for Medical Research, London NW7 1AA, United Kingdom

Frank E. Stockdale [B:417]
Department of Medicine/Oncology, Stanford School of Medicine, Stanford, CA 94305

David L. Stocum [A:467]
Department of Genetics and Development, University of Illinois, Urbana IL 61801

Dennis Summerbell [A:109,217]
Division of Developmental Biology, National Institute for Medical Research, London NW7 1AA, United Kingdom

Gavin Swanson [A:195]
Department of Anatomy, King's College University of London, London WC2R 2LS, United Kingdom

Lih-Heng Tang [B:67]
Department of Orthopedic Research, Montefiore Hospital and Medical Center, Bronx, NY 10467

Patrick W. Tank [A:565]
Department of Anatomy, University of Arkansas for Medical Sciences, Little Rock, AR 72205

Roy A. Tassava [A:537]
Department of Zoology, The Ohio State University, Columbus, OH 43210

Steven L. Teitelbaum [B:239]
Department of Pathology, Jewish Hospital of St. Louis, St. Louis, MO 63110

C. Tickle [A:89]
Department of Biology as Applied to Medicine, The Middlesex Hospital Medical School, London W1P 6DB, United Kingdom

Madeleine Toutant [B:401]
Groupe de Recherches de Biologie et de Pathologie neuromusculaires, CNRS, Paris 75005, France

David P. Treece [A:537]
Department of Zoology, The Ohio State University, Columbus, OH 43210

Robert L. Trelstad [A:245]
Department of Pathology, UMDNJ-Rutgers Medical School, Piscataway, NJ 08854

William B. Upholt [B:175]
Department of Pediatrics, The University of Chicago, Chicago, IL 60637

Joseph W. Vanable, Jr. [A:587]
Department of Biological Sciences, Purdue University, West Lafayette, IN 47907

Hugh H. Varner [B:25,105]
Laboratory of Developmental Biology and Anomalies, National Institute of Dental Research, National Institutes of Health, Bethesda, MD 20205

N.S. Vasan [B:45]
Department of Anatomy, Cell Biology Division, New Jersey Medical School, Newark, NJ 07103

Barbara M. Vertel [B:149]
Department of Biology, Syracuse University, Syracuse, NY 13210

Swani Vethamany-Globus [A:513]
Department of Biology, University of Waterloo, Waterloo, Ontario, Canada N2L 3G1

F. Wachtler [B:281]
Institute of Histology and Embryology, University of Vienna, A-1090 Wien, Austria

Rachel Warga [A:619]
Department of Biology, University of Oregon, Eugene, OR 97403

Cynthia C. Williams [B:125]
Department of Anatomy, University of Alabama in Birmingham, Birmingham, AL 35294

L. Wolpert [A:267]
Department of Biology as Applied to Medicine, The Middlesex Hospital Medical School, London W1P 6DB, United Kingdom

J. Thomas Wright [A:355]
Biology Department, Emory University, Atlanta, GA 30322

B.L. Yallup [A:131]
Department of Zoology, University College of Wales, Aberystwyth, Wales, Dyfed SY23 3DA, United Kingdom

Mineo Yasuda [A:301]
Department of Anatomy, Hiroshima University School of Medicine, Hiroshima 734, Japan

David J. Zaleske [A:317]
Department of Orthopedics, Pediatric Orthopedic Unit, Massachusetts General Hospital, Boston, MA 02114

Preface

The developing vertebrate limb continues to be an excellent model for investigation of morphogenesis, cytodifferentiation, and regeneration. Considerable theoretical and experimental attention has been devoted by investigators around the world to defining the nature of a series of complex morphogenetic interactions that ultimately lead to the development of a definitive structure. These studies have ranged through multiple levels of analysis, from the organismal to the molecular, and attempts are now underway to integrate this vast accumulation of information into meaningful hypotheses that can be tested and that will unify our understanding of limb morphogenesis. In addition, it is hoped that further understanding of basic events in normal limb development will aid in the generation of experimental protocols that will clarify problems of abnormal development and function.

Prior to the meeting on which these volumes are based, investigators interested in the developing limb had participated in two international meetings. The first was held in Grenoble, France, in 1972 and the second in Glasgow, Scotland, in 1976. The major impact of the Grenoble meeting was that it brought together, for the first time, experimentalists in the field with those who were building theoretical models to account for experimental observations. As a result of that meeting, the theoretical constructs generated experimental avenues leading to exploration of fundamental aspects of limb development. The second meeting in Glasgow, Scotland, served two important functions. First, the meeting served as a forum for the consolidation of both theoretical and experimental information that had been generated over the preceding four years. In addition, it drew attention to the importance of directing efforts towards the cellular and molecular aspects of experimentation in this field.

The third international conference was held at the University of Connecticut, Storrs campus, from June 27 through July 2, 1982. The Organizing Committee sought to continue the established emphasis in the field of limb development by bringing together investigators with the latest information in the various experimental aspects of limb development. The meeting provided a unique forum in which focused discussions and interchange of ideas took place between scientists with diverse experimental and theoretical backgrounds, including experimental morphology, cellular biology, and molecular biology.

As an aid to this exchange of ideas and methodological approaches, the Organizing Committee planned two entire sessions in workshop format. In one workshop, the molecular and cellular biologists were requested to emphasize the methodological aspects of their work during a presentation of their recent experimental findings for the explicit benefit of individuals who emphasize the morphological and clinical approaches to the study of the limb. It was anticipated that this type of presentation would demonstrate that molecular techniques of recombinant DNA, monoclonal antibody production, and protein characterization can be successfully applied to the study of the limb at the organized tissue level. The second workshop was designed to analyze and compare current models used to explain limb development and regeneration. Again, the intent was to continue the development of new avenues of communication between theoretical and experimental biologists. This successful combination of workshop format with regular sessions greatly facilitated verbal interchange between audience and speakers, with the result that technological and conceptual strengths of all participants merged and created new experimental approaches.

The Organizing Committee is especially grateful to Mr. John Farling, Assistant Director of Conferences, Institutes and Administrative Services, and Ms. Deborah McSweeney and Ms. Danitza Nall in the School of Extended and Continuing Education at the University of Connecticut for their professional management of the conference. In addition acknowledgment is extended to Ms. Jean O'Neill at the University of New Mexico School of Medicine and to Ms. Sue Leonard and Ms. B. Kay Simandl at the University of Wisconsin School of Medicine for their assistance in the final preparation of these proceedings for publication.

Finally, special acknowledgment is given to the SmithKline Corporation (Smith, Kline and French Laboratories) for financial contributions that defrayed most of the expenses necessitated by a meeting of this size and scope. It is doubtful that the conference would have been possible without this grant assistance. The Organizing Committee also acknowledges support from the National Science Foundation, the International Society for Developmental Biology, the Ortho Pharmaceutical Corporation, and the National Institute of Child Health and Human Development, National Institutes of Health. Collectively, these agencies assured a productive and successful conference.

Considerable effort was expended by members of the Organizing Committee to plan and successfully implement the meeting that formed the basis for these volumes. We share equal responsibility for credit as well as criticism.

The Organizing Committee

SECTION ONE
CELL AND TISSUE INTERACTIONS

Limb Development and Regeneration
Part A, pages 3–12
© **1983 Alan R. Liss, Inc., 150 Fifth Avenue, New York, NY 10011**

PARAMETERS ESTABLISHING THE LOCATION AND NATURE OF THE
TETRAPOD LIMB

Trent D. Stephens, Ph.D.

Department of Biology
Idaho State University
Pocatello, Idaho   83209

INTRODUCTION

The majority of the research concerning the nature of
limb morphogenesis has involved the use of already developing limb buds while the pre-limb events have yet to be
thoroughly explored. My research interests are directed
toward discovering the parameters involved in determining
the location and size of the initial limb field.

In order to discover the determining factors involved
in establishing limb location it is important to understand
the nature of the limb field in the pre-limb and early limb
bud embryo. In 1759 Caspar Friedrich Wolff described a
ridge of tissue which extended along the flank of pre-limb
chick embryos. He observed that the tissue at the end of
each ridge condenses and then extends outward to form the
extremities. Wilhelm His named the structure the Wolffian
ridge in 1868 in honor of its discoverer. Knowledge of the
importance of the Wolffian ridge has nearly slipped into
oblivion (Stephens, 1982) and as a result, modern theories
of limb morphogenesis ignore the concept of a lateral or
Wolffian ridge precursor to the limb. The concept of
isolated limb fields has replaced that of a continuous
totipotent lateral plate.

Recent experimental evidence suggests that the early
lateral plate exhibits limb forming potential along its
entire length. Searls and Janners (1971) demonstrated that
limb buds form as a result of decreased proliferative
activity in the flank rather than increased activity

in the buds. Such an observation is consistant with a
lateral ridge precursor to the limbs but was not what was
expected on the basis of modern limb theory. Balinsky
(1957) and Crosby (1967) have both demonstrated that the
non-limb region of the lateral plate (central lateral ridge)
has significant limb forming potential. Stephens et al.
(1980) suggested that the central flank region may be
inhibited from rapid proliferation and limb formation by
the production of extracellular matrix.

The purpose of this paper is to present a model of
limb location determinants which takes the lateral ridge
into consideration and accounts for limb positioning in a
variety of chordates. A discussion of the apparent metamerism of the tetrapod limb will also be presented with its
possible relation to the model.

MODEL OF DETERMINANTS OF LIMB POSITION

There are distinct topographic variations along the
surface of any chordate embryo from very early phases of
development. Such variations are a result of both underlying predispositions in the embryo's own environment, such
as the massive yolk, and forces created by the embryo itself, such as the expansion of the neural tube. The model
proposes that although the extremities of various animals
are located at widely divergent locations in relation to
the head and somite number, there is a highly conservative
relationship maintained between the location of the limb
and the extent of the yolk sac.

Establishment of Tissue Ridges

During the growth of an embryo there tends to be a
disproportionate increase in surface area of the ectoderm
as compared to that of the underlying mesoderm. As a
result cells begin to pile up along frontiers where one
region of ectodermal expansion meets another (Geddes, 1912).
At the simplest, ectodermal sheets expand away from the
lateral bo y wall and accumulate along dorsal and ventral
ridges where they encounter the expanding edge of the ectoderm from the opposite side of the embryo (Fig 1A). Such
a pattern of expansion will give rise to a body shape
similar to what is seen in an amphibian tadpole's tail.

The next, more complex type of body form may result from an
increase in mass of the ventral regions of the developing
embryo. Such a·configuration may result in three regions
of expansion and three intervening ridges (Fig 1B). Such a
body form can be seen in *Amphioxis* (Fig 2) and many of the
primitive *Acanthodian* fishes (Fig 3; now extinct). If expansion of the body is more or less even, expansive quad-
rants may be proposed with four intervening ridges (Fig 1C).
Such body form may be seen in the posterior body regions of
some elasmobranches (Fig 4) and is the hypothetical form of
the prognathostomal ancestor of the tetrapods (Fig 5). The
ridges of that body form would overlay the four interior
septa present in vertebrates (Fig 6; Hyman, 1942). In those
regions of the body where neural tube expansion is excessive
or where influence of a large yolk sac may be seen, the dorsal and ventral ridges are reduced and ridges form in the
epimere-hypomere junction (Fig 1D).

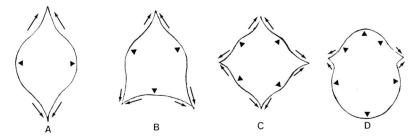

Fig. 1. Proposed types of body expansion and ridge formation. A. Simplest form with two expansion points and two
ridges. B. Three quadrant form with three ridges. C. Four
quadrant form with four ridges. D. Form with large yolk
ventrally and neural tube expansion dorsally so that the
ridges form at the point of junction.

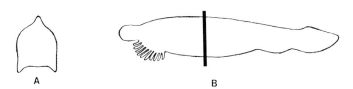

Fig. 2. A. Cross section of *Amphioxis* taken from the level
indicated by the line in B which is a lateral view.

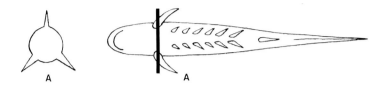

Fig. 3. A. Cross section of an *Acanthodian* taken from the level indicated by the line in B which is a lateral view.

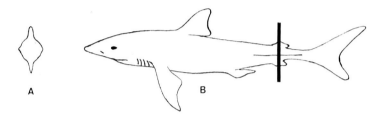

Fig. 4. A. Cross section of an elasmobranch taken from the level indicated by the line in B, a lateral view.

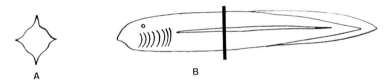

Fig. 5. A. Cross section of a prognathostome taken from the level indicated by the line in B, a lateral view.

Epimeric expansion may cause cells to pile up along the dorsal line of the embryo as the expanding frontiers meet from opposite sides of the body. Expansion of the epimere in the ventral direction has a somewhat different result. As the expanding ectoderm of the epimere reaches the hypomere it encounters both the epimeric-hypomeric fossa and the resistance of the yolk sac. Cell accumulation occurs along the fossa resulting in formation of the Wolffian or lateral ridge. Once the downward expanding ectoderm passes onto the hypomere there is little resistance until it meets the expanding counterpart from the other side of the body and cell accumulation occurs along the ventral line.

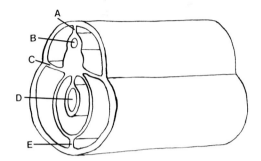

Fig. 6. Diagram of a section of vertebrate body form. A. Dorsal septum, B. Neural tube, C. Lateral septum, D. Gut, E. Ventral septum.

It is predicted that in all vertebrates the lateral ridge will overlay the horizontal skeletogenous septum and will develop between the epimere and hypomere (Fig 6C). A given animal is not confined to develop by only one mode of body expansion. Combinations of expansion and ridge pattern can be seen in many animals (Fig 7).

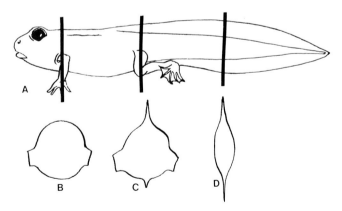

Fig. 7. A. Frog tadpole with three lines indicating the points at which the cross sections B, C, and D were taken.

Emergence of Limbs from the Lateral Ridges

In animals where lateral ridges are clearly visable, the limb buds develop as localized lateral extensions of those ridges. However, whether the animal exhibits a lateral ridge during development or not, it is proposed that the limbs always develop around the margin of the yolk sac and in most cases between the yolk sac and epimere. Examination of a large number and variety of embryos (both from illustrations in the literature and from personal observation) has revealed that in every case the limb buds develop along the margin of the yolk sac (Table 1). The limb buds were located along the fossa between the epimere and yolk sac in all but two of thirty-five species examined.

The actual position of the limbs in relation to the rest of the body is apparently dependent upon at least the following criteria: 1) presence or absence and extent of the subcephalic pocket (if no subcephalic pocket develops, as in fishes, the anterior limbs will be located at the base of the head); 2) length of the yolk sac in relation to body length (if the yolk sac intersects the epimere-hypomere junction at two points as in most vertebrates, limbs will form at each point but if the intersection is at only one point, as in many fishes, only one limb will form at that point and the other will form at the ventral margin of the yolk sac); 3) timing of bud initiation (the actual time during development when the buds appear in relation to relative changes in yolk sac - embryo size). The above criteria are partially illustrated in Fig 8.

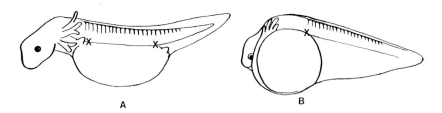

A   B

Fig. 8. A. *Necturus* embryo. The extreme points of contact of the yolk sac with the epimere-hypomere line are marked by an "X". The limb buds can be seen to the side of each "X". B. Lung fish embryo. The single point of contact of the yolk sac and epimere-hypomere line is marked by the "X".

Table 1
Location of Limb Buds in Relation to the Yolk Sac of Embryos

Animal	Yolk Shape	Limb Position Pect. / Pelv.		Source	
Port Jack. Shark	D	1	2	Smith,	1942
Dogfish Shark	D	1	2	Fischer,	1911
Afr. Lungfish	B	3	4	"	1909
Aust. Lungfish	B	3	4	"	1897
Flutemouth	C	8	–	Delsman,	1921
Skipper	A	10	–	"	1924
Dussumier	C	8	–	"	1925
Chirocentrus	C	8	9	"	"
Shad	A	7	–	"	1926
Herring	A	7	–	"	"
Lajang	B	3	–	"	"
Minnow	B	5	6	Balinsky,	1948
Cichlid	A	10	–	Fryer & Iles,	1972
"	B	3	11	PO	
Salmon	A	10	–	PO	
Necturus	B	3	4	Fischer,	1909
Salamander	B	3	4	"	1938
Lizard	D	1	2	"	1904
Eur. Lizard	D	1	2	Porter,	1972
Snapping Turtle	D	1	2	"	"
Leathery Turtle	D	1	2	Deraniyagala,	1939
Chick	D	1	2	PO	
Rat	D	1	2	PO	
Mouse	D	1	2	PO	
Rabbit	D	1	2	Fischer,	1904
Deer	D	1	2	"	"
Pig	D	1	2	PO	
Horse	D	1	2	PO	
Dog	D	1	2	PO	
Lemur	D	1	2	Fischer,	1907
Monkey	D	1	2	PO	
Human	D	1	2	PO	

PO: Personal Observation
Yolk shape: letters; Limb location:numbers

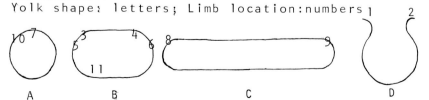

A   B   C   D

Limbs develop at either the point where the yolk sac contacts the epimere-hypomere junction or where the yolk sac contacts the ventro-lateral body margin. Limb development is inhibited in all other positions around the yolk sac. The mechanism by which the precise location of the limb bud is determined along the yolk sac margin is yet unclear and the nature of limb inhibition in other regions is not fully understood.

This model only accounts for the earliest, most formative development of the limb buds and the placement relative to the remainder of the body. Subsequent outgrowth is dependent upon the more specialized AER and its maintenance by the wing mesoderm. This model does not propose to account for the origin of either of those factors. The wingless mutant in chickens may be an example of what the factors described in this model are capable of by themselves without the added assistance of the MF and AER.

METAMERISM OF THE LIMB

The fin folds of primitive fishes which perhaps developed according to the simple mechanism described above (Fig 5) were apparently supported by short metameric rods. It is presumed that the skeletons of tetrapod limbs evolved from one or more of those rods. The most widely accepted hypothesis is that the humerus or femur evolved from only one of those metameric rods as a metapterygian stem (Fig 9A; Gregory, 1951). Stephens and McNulty (1981) demonstrated that placement of foil barriers either parallel or perpendicular to the long axis of chick embryos between the somites and lateral plate could elicit the development of four separate regions of the chick humerus. Foil barriers placed adjacent to specific somite levels inhibited specific limb regions from developing. The humeri from the deficient wings could be combined into four separate groups according to their pattern of skeletal loss. These data suggest that the humerus develops as a result of the combination of four separate metameric elements. These data also suggest that the concept of a metapterygian stem, first introduced by Gegenbaur (1878) in relation to his erroneous gill arch theory and currently used in every textbook of comparative anatomy, should be superseded by the concept of metameric condensation of elements to form the humerus or femur (Fig 9B).

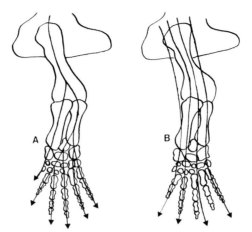

Fig. 9. Comparison of the concept of a metapterygian stem (A) and a pattern of metameric condensation (B) in the development of the tetrapod limb. The former is based on the archaic gill arch theory and the latter is based on recent experimental data.

In an attempt to determine the origin of the metamerism observed in the chick wing, Stephens and Strecker (unpublished data) placed foil barriers between the neural tube and somites (rather than lateral to the somites as in the study described above) so as to inhibit nerve entrance into the developing limb. The resulting limb skeletons were normal. These data confirmed earlier work of Hamburger (1938) and others in suggesting that nerves do not play an important role in limb skeletal formation. These data also suggest that nerves are not responsible for the establishment of a metameric pattern in the limb skeleton. It is possible that the foil barriers which affected limb development were inhibiting normal expansion of the epimeric ectoderm. It may be that cell accumulation in the epimere-hypomere junction accounts for both the formation of a lateral ridge and the position and metamerism of the limb bud itself. Experiments are currently being planned to test this hypothesis.

Balinsky BI (1948). On the development of specific characters in Cyprinid fishes. Proc Zool Soc London 118:335.

Balinsky BI (1957). New experiments on the mode of action
of the limb inductor. J Exp Zool 134:239

Crosby GM (1967). Developmental capabilities of the
lateral somatic mesoderm of early chick embryos.
Brandeis Univ. Thesis, Univ. Microfilms.

Delsman HC (1921, 1924, 1925, 1926). Fish eggs and larvae
from the Java sea. Extrait de Treubia 2:98, 5:408,
6:297, 8:199.

Deraniyagala EP (1939). "The Tetrapod Reptiles of Ceylon"
vol. 1, London: Dulau, p 67.

Fischer VG (1897, 1904, 1907, 1909, 1911, 1938). "Normen-
tafeln zur Entwicklung g. d. Wirbelthiere".

Fryer G and Iles, TD (1972). "The Cichlid Fishes of the
Great Lakes of Africa." Neptune City, NJ:TFH Pub., p 127.

Geddes AC (1912). The origin of the vertebrate limb. J
Anat 46:351.

Gegenbaur C (1878). "Elements of Comparative Anatomy."
Bell, FJ (trans), London: Macmillan pp 416-491.

Gregory WK (1951). "Evolution Emerging." New York:
Macmillan, vol 2 p 358.

Hamburger V (1938). Morphogenetic and axial self-differen-
tiation of transplanted limb primordia of 2-day chick
embryos. J Exp Zool 77: 379.

Hyman LH (1942). "Comparative Vertebrate Anatomy."
Chicago: Univ. Chicago Press, 2nd ed, p 100.

Porter KR (1972). "Herpatology" Philadelphia: Saunders,
pp 418 and 428.

Searls RL and Janners MY (1971). The initiation of limb
bud outgrowth in the embryonic chick. Devel Biol 24: 198.

Smith BG (1942). The heterodontid sharks. In Gudger, EW
(ed): "Bashford Dean Memorial Volume Archaiz Fishes."
New York: Am Mus Nat Hist, Art 8, plate II, no 26.

Stephens TD (1982). The Wolffian ridge: history of a mis-
conception. Isis 73:254.

Stephens TD and McNulty TR (1981). Evidence for a metameric pattern in the development of the chick humerus.
J Embryol exp Morph 61:191.

Stephens TD, Vasan NS, and Lash JW (1980). Extracellular
matrix synthesis in the chick embryo lateral plate
prior to and during limb outgrowth. J Embryol exp Morph
59:71.

Limb Development and Regeneration
Part A, pages 13–31
© 1983 Alan R. Liss, Inc., 150 Fifth Avenue, New York, NY 10011

GENETIC MODIFICATION OF LIMB MORPHOGENESIS

Ursula K. Abbott, Ph.D.

University of California, Davis
Davis, California 95616

The developing vertebrate limb is one of the most suitable systems for the study of pattern formation and its control. The many experimental advantages offered by the embryonic chick have made the developing avian limb the object of choice in such studies. Efforts to explore the genesis of limb patterning and its regulation have been limited only by the ingenuity of investigators in developing means of perturbing the system. Mutants altering limb pattern provide one such means and one that has been exploited less than its potential warrants. This is, perhaps, surprising in that so much of the recent advance in our understanding of pattern and its control in invertebrates has resulted from the ingenious manipulation of mutants.

Abnormal limbs may result from interference with a number of the processes involved in the establishment of normal pattern. Much of the work identifying these processes and forming the basis of current models of limb pattern formation, is based on various kinds of physical or chemical defect experiments (Saunders, 1977). The detection of mutants that alter postulated steps in the process establishes that such steps are under direct genetic control. An extensive and increasing number of single genes affecting limb pattern are available. Such genes, individually and collectively, can provide the means of identifying important gene-controlled functions.

That is to say, the identification of a gene
preventing or altering a particular developmental
event implies an important control point. In so
far as limb development is concerned, genes
affecting a wide variety of the steps postulated
in the control of limb morphogenesis are
available, including those affecting early limb
bud formation, apical ectodermal ridge (AER)
induction and maintenance, response of the
epithelium, cell death, circulation to the limb
bud, muscle and cartilage formation and mitotic
activity.

This discussion will focus on the first two
classes of limb mutants, the polydactylous and the
hypodactylous. First, some general features of
pattern development in limb mutants will be
briefly reviewed. Normally, limbs, either
embryonic or fully formed, are characterized by
very exact bilateral concordance in all features
of pattern. In general, right and left limbs are
identical in size, proportions and in digit
number. They differ in symmetry and,
superficially, in the fine detail of the
integument, i.e. lines on the palm or fingers in
man or scale pattern in birds. In a limb pattern
mutant the situation is very different. Here
discordance in expression is the most common
manifestation. Contralateral limbs may be
strikingly different, with different numbers of
bones present or, if the numbers are identical,
the size, shape or arrangement may differ.

Limb mutants may be either dominant, semi-
dominant, or recessive; autosomal or sex-linked
and, either viable (sometimes only in restricted
environments) or lethal during the embryonic
period. The majority of the limb mutants have
pleiotropic expressions, frequently involving the
bones of the skull, face or beak, and the spine.
Their expression may be altered by changing the
genetic background. Thus, outcrossing, inbreeding
and selection may alter the expressivity,
penetrance, dominance or degree of pleiotropy
manifest by a given mutation.

Mutants resulting in distal deficiencies of
the legs or wings or of both sets of appendages
include three forms of wingless (Waters and
Bywaters, 1943; Zwilling, 1956, 1974; Pease, 1962)
stumpy, ectrodactyly, coloboma (Abbott, 1967) and

limbless (Fallon, 1982). All are pleiotropic and all except limbless are variable in expression. The wingless phenotype has been ascribed to lack of maintenance activity and to an enlarged and precocious anterior necrotic zone (ANZ). Ectrodactyly leads to limb deficiencies only in the homozygous presence of scaleless (MacCabe and Abbott, 1966) and coloboma (Abbott, et. al., 1970) is characterized by tremendous variability in expression with the wings typically being more reduced than the legs.

Stumpy in turkeys is a semi-dominant gene. Heterozygotes lack one or two toes and die during the period just before hatching while homozygotes have a more severe expression, with rear limb development usually restricted to a femur. The wings are normal. An autosomal recessive stumpy gene in chickens differs from the turkey form. Stumpy is of special interest in that it is expressed as a defect in the formation and pattern of blood vessels within the three circulatory arcs of the embryo. The alterations in the ground plan of the circulatory vessels leads to abnormal blood flow and hemmorhagic pools of blood in different sites throughout the embryo. The curtailment of blood supply to posterior regions of the embryo leads to malformations in the leg. The complete interruption of the important circulatory channels eventually causes death of stumpy embryos by stage 26-27 (Plate 2 A).

Avian polydactylous mutants are of three general types. The first class, exemplified by the diplopod series ($dp^{1,2,3,4}$) (Abbott, 1967; MacCabe and Abbott, 1966; MacCabe and Abbott, 1974) have three of the normal digits, II, III and IV, and usually the hallux (digit 1). Additional digits with varying degrees of mirror image symmetry form pre-axially, but on the same plane. The wings form two to three extra digits pre-axial to the normal pollex. Reciprocal recombination grafts of normal and diplopod limb mesenchyme and epithelium have demonstrated that in this type of mutant the mesenchyme, through an increase in the length of the AER, is initially responsible for the polydactyly. Diplopod limb epithelium when combined with normal limb mesenchyme gives rise to a limb with the normal number of bones. The zone of polarizing activity

(ZPA) in these mutants shows normal activity and additional areas with polarizing activity have not been located in the anterior limb corner nor in midsections (MacCabe et. al. 1975).

The second general type of polydactylous mutants are those in which a multitude of digits are formed on both wings and legs. There have been three talpid mutants, all autosomal and recessive, reported in chickens (Cole, 1942; Abbott et. al., 1960; Goetinck and Abbott, 1964; Hinchliffe and Ede, 1967). Similar mutants have been found in other bird species, including the turkey.

We have worked primarily with talpid^2, which differs from the others in time of embryonic mortality. Most talpid^2 embryos survive until fourteen to sixteen days of incubation and some slightly longer. Recombination experiments have shown the mesenchyme to be involved in the development of an extensive AER, which leads to the formation of the broad wing and leg paddles seen in Plate 1 C. Cell death is reduced in talpid embryos leading to syndactyly in the fully developed limb, here illustrated by a talpid turkey embryo (Plate 1 D). Note that the entire limb pattern is disturbed and that it is generally not possible to identify specific digits. Talpid^2 embryos have short trunks, micromelia, reduced upper beaks and celosomia.

Eudiplopodia provides a third type of polydactyly. It appears to be particularly suitable for studies of pattern formation in the chick limb in that in this mutant both the epithelial response and limb polarity are disturbed (Goetinck and Abbott, 1964; Fraser and Abbott, 1971). Early limb development in eudiplopod embryos appears normal and the mutants are indistinguishable from their normal siblings until stage 20 or 21. At this time additional ectodermal ridge structures become evident on the dorsal surfaces of both pairs of limb buds. These structures are more extensive on the hind than on the fore limb buds. Mesenchyme beneath the accessory ridge of the leg buds proliferates, pre-cartilaginous condensations form and result in a second group of digits located dorsal to the normal complement. Proximal structures are not duplicated. Eudiplopod wing buds at early

stages, have more accessory ridge structures than
are eventually manifest in final wing expression,
presumably reflecting the more extensive cell
death characterizing wing than leg morphogenesis.
Typically only minor accessory dorsal structures
(knobs or bumps) are present and occasional
duplicate feather follicles. The supernumerary
toes of eudiplopod embryos have footpads on their
ventral surfaces and conically shaped nails.

Scanning electron microscope preparations of
eudiplopod limbs at several stages reveal the
complex relationships of the variety of dorsal
ridge structures formed (Plate 2 B&C). It is rare
for the accessory ridge to closely resemble normal
AER and right and left limb buds are usually not
identical. Plate 2 B&C exemplify the extensive
and discordant development of dorsal ectodermal
ridge in stage 26 leg (B top) and wing (B bottom),
and in stage 28, eudiplopod leg buds (C).

Studies of expressivity and its modification
have employed crosses of lines segregating for a
specific limb mutant to an inbred (virtually
isogenic) line with backcrossing to the inbred
line, so that after several generations, mutant
homozygotes have an increasingly constant (nearly
homozygous) genetic background. This process has
altered expression and decreased its variability
in several of the more advanced lines. Currently
several are in their fourth or fifth backcross
generation as exemplified by the eudiplopod embryo
in Plate 1 B).

Lines segregating for mutants with similar or
opposite phenotypic effects were crossed. From
these crosses we have produced a variety of
double, triple and quadruple carriers. These have
been mated in various combinations to provide
material for comparisons of expression in doubly
or triply homozygous embryos with that of those
homozygous for a single mutant gene. These have
included:

a) combinations of genes each acting
individually to increase digit numbers, such as
diplopodia^1 with diplopodia^4 or talpid^2;
diplopodia with eudiplopodia; talpid^2 with
eudiplopodia, etc.

b) combinations of genes both acting to
decrease limb development, i.e. wingless with
coloboma.

c) combinations of genes with opposite effects on limb expression, i.e. eudiplopodia or talpid with wingless or coloboma.

Some specific examples of such embryos are shown in Plates 3 and 4.

Genetic combination of eudiplopodia with any of the diplopod mutants reveal that the control of pre-axial extension of ridge is quite distinct from that resulting in development of ridge in the dorsal ectoderm. In these embryos the dorsal ridge derivatives also extend further pre-axially so that one observes a more or less complete duplication of the diplopod pattern in the legs of double homozygotes. Expression in the wings of double homozygotes is similar to that of diplopod embryos (Plate 3 A), reflecting the greater amount of cell death in wing than in leg sculpting. Thus, the wings possess three digits pre-axial to the normal pollex (similar to diplopod wings) and an occasional small bony protrusion or duplicate feather follicle as found in eudiplopod wings. Combinations of eudiplopodia, in which the epithelial response is abnormal, with any of the polydactylous limb mutants shown to alter the properties of limb mesenchyme appear to support the concept of a diffusable morphogen involved in ridge induction and maintenance. These combinations also point up the fact that distances within the developing limb must be considered in three dimensions. Thus, eudiplopod embryos have dorsal limb ectoderm that misinterprets its position and responds to the more extensive distribution, in the single-plane polydactylous mutants, of what Zwilling and Hansborough (1956) called apical ectodermal ridge maintenance factor. When eudiplopodia is combined with coloboma, the latter producing limbs with variable degrees of distal deficiency, the resulting pattern again reflects the formation of dorsal ridge but in this instance the extent of ridge in both planes and thus the numbers of both normal and supernumerary digits are reduced (Plate 3 D).

Several mutations have been reported to increase or alter pattern through effects on the amount of cell death occurring during limb morphogenesis. Thus, increased cell death in the ridge and sub-ridge mesenchyme has been observed in several of the hypodactylous mutants, including

two forms of wingless, limbless, and stumpy. The
lethal wingless mutants are also characterized by
extensive pleiotropy. In addition to the absence
or severe reduction of wing structures, the
majority lack most leg elements as well. They
also have edema and highly abnormal feathers (an
extreme form of clubbed down), Plate 1 A.

Both talpid^2 and talpid^3 (Hinchliffe and
Thorogood, 1974) embryos have been found to have
very reduced cell death or complete absence of
areas of normal cell death. Cairns (1977) has
reported that the anterior necrotic zone is
entirely absent in talpid^2 embryos, apparently
leading to the characteristic broader base of the
limb. That cell death is clearly more important
in the development of wing than leg pattern, is
dramatically shown in the talpid mutants (Plate 1
C) where the overall effect of the lack of cell
death results in wings very similar to legs in
pattern. That changes in the amount of cell death
and in particular the lack of an ANZ are important
in patterning is suggested also by our studies of
doubly homozygous embryos combining talpid^2 with
other limb mutants. In the combination of talpid^2
and eudiplopodia dorsal epithelium of both the
fore and the hind limb buds forms numerous ridge
structures and in the double mutant these persist
in both the leg and the wing and give rise to
large numbers of supernumerary digits.
Accordingly, in contrast to the situation in
eudiplopod embryos, where the leg but not the wing
is extensively duplicated, the double mutant has
fore and hind limbs affected to approximately the
same degree with eighteen to twenty digits per
limb (Plate 3 B). In the double mutant the lack
of cell death due to the talpid^2 phenotype
predominates, and in this case, permits a more
extreme expression of eudiplopodia in the wing.
It is notable that the plethora of digits formed
are talpid^2-like in pattern. This example clearly
demonstrates the important role of cell death in
limb sculpturing.

When talpid^2 is combined with any one of the
three available diplopod mutations, the pattern
resulting in both fore and hind limbs is
predominantly that of talpid^2. The wings of
double homozygotes are indistinguishable from
those of talpid^2 embryos while the leg digits

appear to show less syndactylism. However, it is clear that the effect of the talpid^2 genotype in drastically reducing cell death and possibly altering other aspects of cell behavior, predominate over those of the diplopod mutants.

The majority of embryos homozygous for wingless alone lack wings completely while the legs may be completely absent, be represented by small femoral structures only or, rarely, form femurs with an occasional digit. Limb buds form initially but extensive cell death in the ANZ and in subridge mesenchyme accompanied by degeneration of the AER halts distal limb development. The time of onset and the extent of necrosis is reflected in the degree of amputation. Embryos homozygous for both mutants initially form broad based limb buds similar to those of talpid^2. In the double mutants produced to date, the wing bud degenerates completely, and the leg develops only a partial femur. The base of this rudimentary limb is broad like talpid^2. The double homozygotes possess other characteristics of talpid^2, such as reduced upper beak, short trunk, and celosomia (Plate 3 C) and older embryos also show the extreme clubbed down characteristic of wingless^2.

Combinations of the contrasting mutants eudiplopodia and wingless^2 provided evidence for interaction and amelioration of both as well as for a possible attenuation of wingless expression in embryos also heterozygous for eudiplopodia. In this study, males doubly heterozygous for eudiplopodia and wingless^2 were individually mated with double and both types of single carriers, i.e.:

        eu + +wg   x   eu+ +wg
                   x   ++ +wg
                   x   eu+ ++

Accordingly progenies segregating for normal and three mutant classes (double homozygotes, eudiplopodia, wingless) and either wingless or eudiplopodia were produced. In addition to typical wingless (Plate 1 A, far right and left) and eudiplopod embryos (Plate 1 B), a number of the expressions illustrated in Plate 4 were obtained. Three of these embryos are from the eu+ +wg x eu+ +wg cross and the fourth from eu+ +wg x ++ +wg. All four are wingless but the expression in limb and skin and feather development is modified. Double homozygotes lack wings and have clubbed down. Several digits typical of eudiplopodia are present in more than one plane, together with conical and split toenails. In a second type encountered, spike-like structures with additional toes on either a different plane, or arising from the base of the limb, are found. These embryos also show areas of normal feathering, especially in association with the limb. One wingless embryo had legs with several normal appearing digits accompanied by one reduced digit and areas of abnormal scale formation. This specimen (Plate 4 C) also had the largest number of normal feathers seen in any wingless produced in matings of the wingless stock.

The current models offered to explain limb morphogenesis are difficult to reconcile with certain aspects of the process revealed by some of the limb mutants, in particular, some of the types of patterning in the ploydactylous mutant. Also, the various kinds of, and causes for cell death appear not to have been adequately reflected in such models. It is hoped that many investigators will increasingly make use of genetic materials in their work employing the many ingenious methodologies that have been presented at this Conference.

Abbott UK, Taylor LW, Abplanalp H (1960). Studies
with talpid$_2$, an embryonic lethal of the fowl.
J. Hered. 51:195-202.
Abbott UK (1967). "Avian Developmental Genetics"
In Methods in Developmental Biology (eds. FW
Wilt and NK Wessels). Thomas Y. Crowell Co.,
New York.
Abbott UK, Craig RM, Bennett EB (1970). Sex-
linked coloboma in the chicken. J. Hered.
61:95-102.
Cairns JM (1977). Growth of normal and talpid$_2$
chick wing buds: an experimental analysis. In
Vertebrate Limb and Somite Morphogenesis (eds.
D.A. Ede, J.R. Hinchliffe and M. Balls),
Cambridge University Press, Cambridge, pp. 123-
137.
Cole RK (1942). The "talpid lethal" in the
domestic fowl. J. Hered. 33:82-86.
Fallon JF (1982). Studies on a limbless mutant in
the chick embryo. This volume.
Fraser RA, Abbott UK (1971). Studies on limb
morphogenesis V. The expression of
eudiplopodia and its experimental modification.
J. Exp. Zool. 176:219-236.
Goetinck PF, Abbott UK (1964). Studies on limb
morphogenesis. I. Experiments with the
polydactylous mutant, talpid$_2$. J. Exp. Zool.
155:161-170.
Hamburger V, Hamilton HL (1951). A series of
normal stages in the development of the chick
embryo. J. Morph. 88:49-92.
Hinchliffe JR, Ede DA (1967). Limb development in
the polydactylous talpid$_3$ mutant of the fowl.
J. Embryol. Exp. Morphol. 17:385-404.
Hinchliffe JR, Thorogood PV (1974). Genetic
inhibition of mesenchymal cell death and the
development of form and skeletal pattern in the
limbs of talpid 3 (ta$_3$) mutant chick embryos.
J. Embryol. Exp. Morphol. 31:747-760.
MacCabe JA, Abbott UK (1966). Polarizing and
maintenance activities in two polydactylous
mutants of the fowl: diplopodia 1 and talpid 2.
J. Embryol. exp. Morph. 31:735-746.
MacCabe JA, Abbott UK (1974). Polarizing and
maintenance activities in two polydactylous
mutants of the fowl: diplopodia$_1$ and talpid$_2$.
J. Embryol. Exp. Morphol. 31:735-746.

MacCabe JA, MacCabe AB, Abbott UK, McCarrey JR (1975). Limb development in Diplopodia$_4$: a polydactylous mutation in the chicken. J. Exp. Zool. 191:383–394.

Pease MS (1962). Wingless poultry. J. Hered. 53:109

Saunders JW, Jr. (1977). The experimental analysis of chick limb bud development. In Vertebrate Limb and Somite Morphogenesis (eds. DA Ede, JR Hinchliffe and M Balls), Cambridge University Press, Cambridge, pp. 1–24.

Waters NF, Bywaters JH (1943). A lethal embryonic wing mutation in the domestic fowl. J. Hered. 34:213.

Zwilling E (1956). Interactions between limb bud ectoderm and mesoderm in the chick embryo. IV. Experiments with a wingless mutant. J. Exp. Zool. 132:241.

Zwilling E (1974). Effects of contact between mutant (wingless) limb buds and those of genetically normal chick embryos: confirmation of a hypothesis. Dev. Biol. 39:37–48.

Zwilling E, Hansborough LA (1956). Interaction between limb bud ectoderm and mesoderm in the chick embryo. III. Experiments with polydactylous limbs. J. Exp. Zool. 132:219–239.

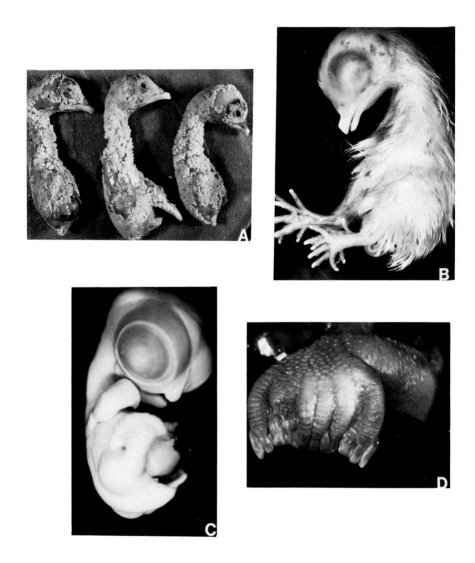

Plate 1

Plate 1. Expression in Wingless, Eudiplopod and Talpid Embryos.

A) Three 14-day wingless^2 embryos. Right and left lack wings and legs. Center has femoral stumps of different lengths. All three show the characteristic, highly abnormal, clubbed down. Bare areas due to extreme skin and down fragility.

B) Nineteen day eudiplopod embryo from the isogenic line. Nine digits, on different planes, per foot. Digits from dorsal epithelium have scales on ventral surface and conical toe nails. wings normal except for small extra finger.

C) Eight day talpid^2 embryo showing broad-based limb paddles. The neck and trunk are shortened and the embryo is celosomic.

D) Right leg from a twenty-four day talpid^1 turkey embryo. Note the eight syndactylous digits, eleven toenails, and the severe micromelia.

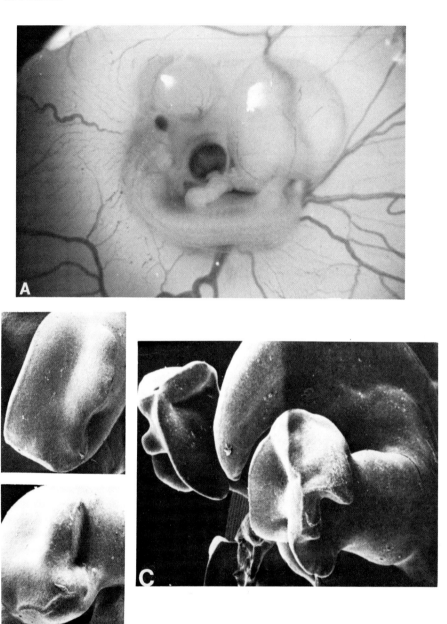

Plate 2

Plate 2.

A) A stumpy embryo of 5 3/4 days (stage 27) (Hamburger and Hamilton, 1951) photographed just prior to its death, illustrating reduced rear limb development and defects in the three embryonic circulatory systems: yolk sac, allantoic and that within the embryo. The reduced, thick-walled allantois and one of the characteristic pools of blood (base of brain) are evident, as is the lack of blood supply to the rear limb.

B and C) Extra AER formation in eudiplopod leg (top) and wing (bottom) buds at stage 26(B) and both right and left leg buds (stretched out on both sides of tail) at stage 28(C). Note variable shape and location of ridge formation.

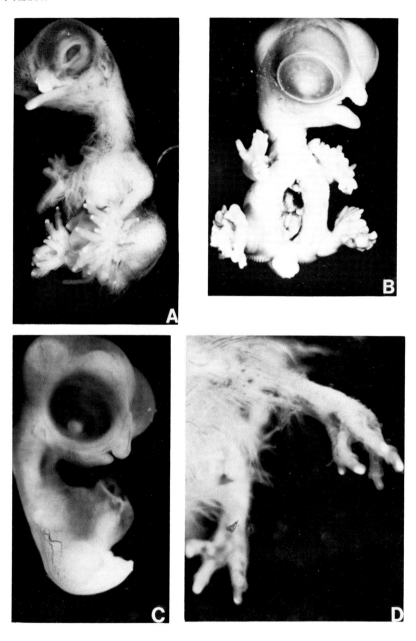

Plate 3

Plate 3. Expression in Double Mutants Affecting **Different Aspects of Limb** Differentiation.

A) Diplopodia[1] and Eudiplopodia (stage 37)
    Expression in the wings (3_1 extra wing fingers) characteristic of Diplopodia[1] as is the reduced upper beak and shortened trunk. The legs show the effect of both genes in an essentially additive fashion (i.e., the number of digits characteristic of diplopodia is approximately doubled) and 14-16 digits per foot are located on 2 or more planes. Eudiplopod expression is also shown in the conical and duplicated toenails of the extra digits and in the scale patterning.

B) Talpid[2] and Eudiplopodia (stage 35)
    Both legs and wings show extensive duplication of digits on 2 or more planes. There are 18-20 fused and syndactylous digits per limb. In this case, the lack of cell death characteristic of talpid[2] has prevented the removal of additional dorsal wing structures. The embryo also shows the short trunk, reduced upper beak, micromelia and celosomia characteristic of talpid but not eudiplopodia.

C) Talpid[2] and Wingless[2] (stage 30)
    Embryo showing vestigial wing stump. The breadth of the leg base is characteristic of talpid[2] as is the short trunk and celosomia. The extensive edema is typical of wingless embryos at this stage.

D) Eudiplopodia and Coloboma Limbs (stage 39)
    The effect of the coloboma mutation on the AER with subsequent reduction in limb distal development together with that of eudiplopodia in producing additional AER from dorsal ectoderm here results in a two plane duplication of fewer than the normal number of digits. Conical nails and scale location permit identification of the supernumerary digits.

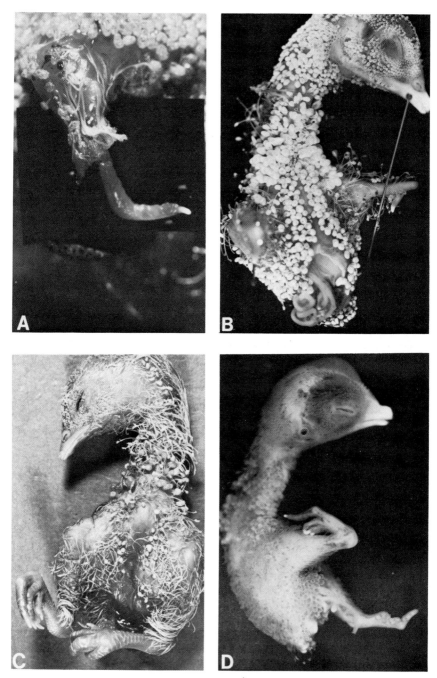

Plate 4

Plate 4. Expression in Wingless -- Eudiplopod Combinations

Four embryos produced in matings of eu+ wg^2+ x eu+ wg+ (A, B, C) and eu+ wg^2+ x ++ wg^2+ (D). All embryos lack wings and have variable degrees of limb reduction. All show characteristics of eudiplopodia, digits on different planes, conical toenails, and areas with normal feathering rather than clubbed down (especially C, but also evident in B and D). Specimen C is of special interest in that while the wings are absent, limb defects are confined to digit reduction. Specimen D is homozygous for wingless and, at best, heterozygous for eudiplopodia. It has, however, a two plane duplication with the more proximal toe showing a split toenail.

Limb Development and Regeneration
Part A, pages 33–43

STUDIES ON A LIMBLESS MUTANT IN THE CHICK EMBRYO

John F. Fallon, Jeanne M. Frederick, Jill L.
Carrington, Mark E. Lanser, B. Kay Simandl
Department of Anatomy
University of Wisconsin
Madison, Wisconsin 53706

OVERVIEW

 Reciprocal interactions between the limb bud mesoderm
and overlying ectoderm are required for normal avian limb
development (reviewed in Zwilling 1961; Saunders 1977).
For a relatively short period during development, mesoderm
from the limb regions of the somatopleure will cause the
formation, in competent ectoderm, of a pseudostratified
columnar epithelium at the tip of the forming bud (Kieny
1960; Dhouailly, Kieny 1972; Saunders, Reuss 1974). This
apical specialization is called the apical ectodermal
ridge. If the ridge is removed surgically, only the
proximal parts of the limb already determined in the
mesoderm will develop (Saunders 1948; Summerbell 1974;
Rowe, Fallon 1982). These experiments show that once the
apical ridge is induced by limb bud mesoderm, its presence
is required until all limb parts are determined. Further,
once removed the apical ridge cannot regenerate.

 Zwilling designed an experiment which tested whether
or not the apical ridge possessed specific limb-type (wing
vs. leg) information. Using enzymatic treatment, Zwilling
was able to separate the ectodermal and mesodermal
components of the limb bud. In controls when wing
components (or leg components) were simply recombined and
grafted to an appropriate site, normal limbs developed.
When wing bud mesoderm was combined with leg bud ectoderm,
the recombinant limb that developed was a wing. In the
reverse recombinant limb, a leg developed (see Zwilling
1961). Thus, specificity for fore- or hindlimb resides

within the mesoderm.

Rubin and Saunders (1972) made heterochronic limb recombinants, e.g., between stage 19 mesoderm and stage 24 ectoderm or vice versa. Had there been specific information about the proximal-distal limb axis emanating from the ridge, there would have been missing parts in the former case and additional parts in the latter case. However, normal limbs developed in both cases.

The experiments of Zwilling, and of Rubin and Saunders make it clear that the type of limb that will form and as well, the proximal-distal sequence of the laying down of limb parts are controlled by the mesoderm. The apical ridge is required for the realization of the potential of the limb bud mesoderm.

The recombinant limb technique has proved very useful in the analysis of limb mutants in the chick embryo (reviewed in Goetinck 1966). An example of such a mutant is $diplopodia_4$, where a more extensive apical ridge than normal develops anteriorly by the 4th day of incubation. This is correlated with increased outgrowth of the anterior wing bud and anterior polydactylism. Recombinant limbs made of the mutant and wild type limb components demonstrated that the mesoderm was the limb bud component affected by the mutation (MacCabe, MacCabe, Abbott, McCarrey 1975). Of the various limb mutants that have been tested previously in this way, all but one have proven to be the result of changes in the mesoderm. The exception is $eudiplopodia$ where, in mutant embryos, a second apical ridge develops dorsally on leg buds at about stage 24 and is associated with a dorsal supernumerary outgrowth. Recombinant limb experiments showed the mutant ectoderm formed the second ridge when combined with wild type mesoderm and the mutant phenotype was realized. The reverse experiment resulted in a normal limb (Goetinck 1964).

THE LIMBLESS MUTANT

Recently, we have been working with a line of chickens which carries a simple Mendelian autosomal recessive gene that, in the homozygous condition, yields a chick without fore- or hindlimbs. In the homozygous condition, there is

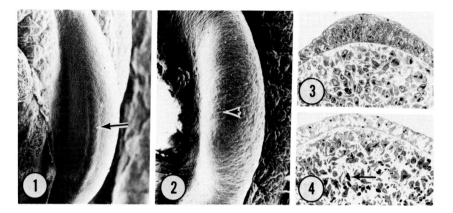

Figure 1. Scanning electron micrograph of stage 20 normal
  wing bud. A well formed apical ectodermal ridge is
  indicated by the arrow. (X75)
Figure 2. Scanning electron micrograph of stage 20
  limbless wing bud. Note absence of apical ectodermal
  ridge (arrowhead). (X75)
Figure 3. Cross section through the apex of a stage 20
  normal wing bud showing typical ridge. (X269)
Figure 4. Cross section through the apex of *limbless* wing
  bud; no ridge is present. Note dead cells in mesoderm
  (arrow). (X269)

100% penetrance. This is a lethal gene in the sense that
without legs the affected chick will not be able to break
out of the shell. When hatched by hand, we have kept the
animals alive for three weeks. Aside from the fact they
have no limbs, all other systems and their functions appear
normal. The mutant was first described by Prahlad, Skala,
Jones, and Briles (1979). Briles and Jones sent the entire
stock to Wisconsin for our use. These investigators as
well as Somes (1981), have agreed to call the mutant
limbless with the designation *ll*.

At the time for hatching, the homozygous *limbless*
embryo has no appendages. It is important to note that
limb buds do arise at the proper time (stage 17) in the
mutant embryos but disappear by stage 24. Scanning
electron microscopy and histological sections (Figures 1-4)
showed that the ectoderm failed to form an apical

ectodermal ridge. Limb bud development was grossly normal until late in stage 18 when the apical ectodermal ridge should appear as a morphological entity. Instead of the ridge developing, a characteristic saddle-shaped depression was seen and the mutant bud did not appear to grow as fast as wild type buds. Mesodermal cells began to die (Figure 4) where the depression was seen in the bud. Mesodermal cell death spread slowly throughout the limb bud so that by late stage 22 the mesoderm was a mass of necrotic cells.

RECOMBINANT LIMB ANALYSIS OF LIMBLESS

Using the techniques devised by Zwilling, we made recombinant limbs constructed of stage 19 and 20 normal and *limbless* limb bud components (see Figure 5). We found that a recombinant limb bud composed of the mesoderm from a *limbless* limb bud and wild type ectoderm produced a normal limb in at least 50% of the cases. This is comparable to controls of recombinant limbs made of wild type limb bud components. However, the results were different when the recombinant was made of wild type limb bud mesoderm and *limbless* limb bud ectoderm (cf. Figure 6). In all cases these recombinant limbs failed to develop.

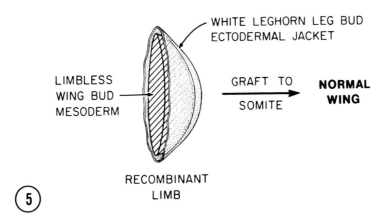

Figure 5. Drawing depicting recombinant limb bud of *limbless* mesoderm and normal ectoderm. A normal limb developed from such a recombinant.

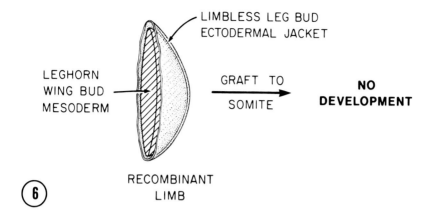

LIMBLESS LEG BUD
ECTODERMAL JACKET

LEGHORN
WING BUD
MESODERM

GRAFT TO
SOMITE

NO
DEVELOPMENT

6

RECOMBINANT
LIMB

Figure 6. Drawing depicting recombinant limb bud of normal
limb mesoderm and *limbless* ectoderm. No limb developed
from such a recombinant.

Our observations make it clear that during the limb
bud stages of development, the *limbless* ectoderm is
incapable of supporting growth of the normal limb bud
mesoderm. However, the *limbless* limb bud mesoderm is
capable of responding to the stimulus of wild type
ectodermal ridge and will form a normal limb. At the limb
bud stages the *limbless* gene affects the ectoderm.

A MODIFIED RECOMBINANT LIMB PROCEDURE

The question of whether or not the *limbless* mesoderm
can induce an apical ridge was not answered by the approach
described above. To answer this question, we surgically
removed the prospective wing bud ectoderm from possible
limbless embryos during stage 15. Non-mutant stage 15
flank ectoderm was then grafted over the bare mesoderm. In
order to properly keep track of the graft and to be sure no
mesoderm was grafted, quail flank ectoderm was used (see
Figure 7). Quail cells are distinguishable from chick
cells on the basis of large Feulgen positive heterochromatin
clumps in the quail cell nuclei (LeDouarin, Barq 1969).
The result of this experiment was that the *limbless*
mesoderm induced an apical ridge in the competent quail
ectoderm (Figure 8) and a normal limb developed at that
site, while the other three limb buds regressed (Figure 7).

These experiments clearly demonstrate the mesoderm is normal in the *limbless* mutant and the ectoderm is the limb bud component affected by the *limbless* gene.

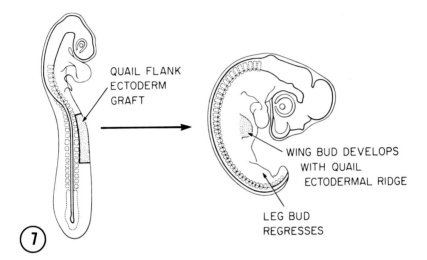

Figure 7. Drawing depicting graft of quail flank ectoderm to stage 15 *limbless* embryo which is shown on the left. On the right, the same embryo is drawn at about stage 23 showing a normal wing covered by quail ectoderm. The right leg bud is shown regressing.

POLARIZING ZONE

Grafts of polarizing zone from stages 19-21 embryos to the anterior wing border of a normal embryo resulted in good polarized duplications (Figure 9). Therefore, *limbless* has good polarizing activity. After stage 21, *limbless* polarizing zone will not cause duplications when grafted to the anterior border of a normal wing. The loss of activity is likely due to cell death of the limb bud mesoderm.

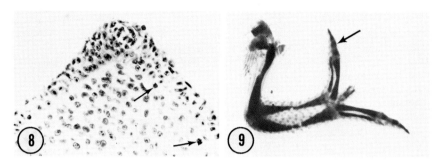

Figure 8. Cross section through stage 23 *limbless* limb bud with quail apical ridge. Stage 15 quail flank ectoderm was grafted to a stage 15 *limbless* host. The Limbless mesoderm induced the ridge in the quail ectoderm. Feulgen stain of the heterochromatin in the quail nuclei stands out from the pale chick mesoderm cells. Arrows point to dividing cells in the mesoderm. (X388)

Figure 9. Mirror image duplication that developed from normal wing after stage 20 *limbless* polarizing zone was grafted to the anterior border of a stage 20 normal wing bud. Arrow points to supernumerary digits 3 and 4.

CELL DEATH IN THE LIMBLESS LATERAL MOTOR COLUMN

During normal development, spinal motor neurons at the level of the chick wing and leg undergo massive degeneration between the 5th and 9th days of development (reviewed in Hollyday 1980). The brachial lateral motor column contains about 17,000 cells on the 6th day of incubation. Cell death begins in the column at least by the 8th day of incubation and continues so that by the 10th day, 71% (about 12,000 cells) survive. There is further attrition so that by hatching only 35% of the cells remain in the lateral motor column (about 6,000 cells) (Oppenheim, Majors-Willard 1978). Similarly, at the lumbosacral level of the lateral motor column, there is a maximum of 20,000 cells during the 5th day of incubation and this is reduced to 60% by the 10th day (12,000 cells) (Hamburger 1975).

If the periphery (wing or leg bud) is removed surgically, cell death occurs in the lateral motor column over the same interval, but appears to be accelerated when

compared to normal. By the 15th day, the number of
surviving neurons in the lumbosacral lateral motor column
approaches 10% of the population on the 5th day of
incubation (Oppenheim, Chu-Wang, Maderdrut 1978). Thus, it
has been argued that, at the very least, the survival of
the lateral motor column requires the presence of the
periphery. It has also been argued that the cell death
occurring naturally, with the periphery present, is due to
the fact that some neurons lose out in the competition for
limited sites in the periphery (see Hamburger 1975;
Oppenheim, Chu-Wang, Maderdrut 1978; Hollyday 1980, and as
well, Lamb this volume).

We have begun an analysis of the lack of the periphery
on the brachial lateral motor column in the *limbless* embryo
(Figures 10, 11). It is difficult to stage the *limbless*
embryo and initially the data have been collected on the
basis of the day of incubation. We are now staging the
embryos according to the procedures outlined in Hamburger
(1948, 1958).

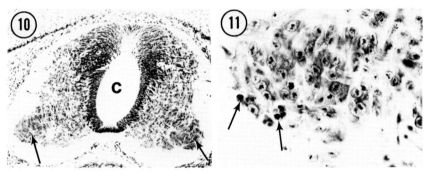

Figure 10. Cross section through brachial spinal cord of a
7-day *limbless* embryo. Central canal is indicated by the
C. The lateral motor columns are indicated by the
arrows. (X106)
Figure 11. Higher magnification of lateral motor column
shown on left of Figure 10. Arrows show degenerating
cells in the lateral motor column. (X500)

Very preliminary data are shown in Table 1. It can be
seen from these data that on the 6th day of incubation,
there are fewer cells in the *limbless* lateral motor column
when compared with the normal embryo. It is also clear

from Table 1 that cell death in the *limbless* brachial
lateral motor column is greater than in the normal embryo.
This results in 15% (1,698 cells) of the *limbless* cells
surviving on the 18th day of incubation compared with 44%
(6,811 cells) surviving in normal embryos. While these
numbers compare favorably with data published for surgical
removal of the periphery, the question remains whether the
gene may affect the central nervous system, which is an
ectodermal derivative. Experiments are in progress to make
observations on the lateral motor column of *limbless*
embryos that have had normal ectoderm grafted to the limb
region at stage 15. Such embryos will have a normal
forelimb bud on the right side at all stages of
development.

TABLE 1

Day of Development	6	9	10	12	18
# cells normal brachial LMC[1]	15,311(5) ±1,342	13,465(6) ±1,570	10,882(3) ±1,846	9,434(4) ±1,343	6,811(2)[44%] ±273
# cells *limbless* Brachial LMC[1]	11,585(5) ±2,133	3,624(5) ±100	3,466(1)	1,894(3) ±110	1,698(2)[15%] ±223

[1]LMC - Lateral Motor Column
Numbers in parentheses = embryos counted for that day
Numbers in brackets on day 18 = % cells surviving from day 6

CONCLUSION

We hope the use of this mutant will further
understanding of limb development. The data gathered so
far indicate that the gene affects the ability of the
ectoderm to form an apical ectodermal ridge. We have

initial data indicating that the *limbless* gene is not an allele of *wingless* or *talpid*[2]. Since *limbless* embryos form limb buds, it must be concluded that the apical ridge is not required for the initiation of limb bud outgrowth. However, the cells of the limb bud mesoderm require the inductive influence of the apical ridge by late stage 18, or early stage 19, or they will die. We are carrying the analyses reported here in a preliminary way to completion and will communicate them in complete form in the near future.

Acknowledgements

This research was supported by NSF Grant #PCM8205368 and NIH Grant #T32HD7118. We are grateful to Eugenie L. Boutin, Allen W. Clark, Mary Ellen McCarthy, William Todt and David B. Slautterback for constructive criticism of this manuscript. We thank Lucy Taylor for making the drawings and Sue Leonard for typing the manuscript.

REFERENCES

Dhouailly D, Kieny M (1972). The capacity of the flank somatic mesoderm of early bud embryos to participate in limb development. Devel Biol 28:162.
Goetinck PF (1964). Studies on limb morphogenesis. II. Experiments with the polydactylous mutant *eudiplopodia*. Devel Biol 10:71.
Goetinck PF (1966). Genetic aspects of skin and limb development. Curr Top Devel Biol 1:253.
Hamburger V (1948). The mitotic patterns in the spinal cord of the chick embryo and their relation to histogenetic processes. J Comp Neurol 88:221.
Hamburger V (1958). Regression versus peripheral control of differentiation in motor hypoplasia. Amer J Anat 102:365.
Hamburger V (1975). Cell death in the development of the lateral motor column of the chick. J Comp Neurol 160:535.
Hollyday M (1980). Motoneuron histogenesis and the development of limb innervation. Curr Top Devel Biol 15:181.

Kieny M (1960). Role inducteur du mesoderme dans la differentiation precoce du bourgeon de membre chez l'embryon de poulet. J Embryol Exp Morphol 8:457.

LeDouarin N, Barq G (1969). Sur l'utilization des cellules de la Caille japonaise comme "marquers biologiques" en embryologique experimentale. Cr hebd Seanc Acad Sci, Paris D269:1543.

MacCabe JA, MacCabe AB, Abbott UK, McCarrey JR (1975). Limb development in $diplopodia_4$: a polydactylous mutation in the chicken. J Exp Zool 191:383.

Oppenheim RW, Chu-Wang IW, Maderdrut JL (1978). Cell death of motorneurons in the chick embryo spinal cord. III. The differentiation of motorneurons prior to their induced degeneration following limb-bud removal. J Comp Neurol 177:87.

Oppenheim RW, Majors-Willard C (1978). Neuronal cell death in the brachial spinal cord of the chick is unrelated to the loss of polyneural innervation in wing muscle. Brain Res 154:148.

Prahlad KV, Skala G, Jones DG, Briles WE (1979). Limbless: A new genetic mutant in the chick. J Exp Zool 209:427.

Rowe DA, Fallon JF (1982). The proximodistal determination of skeletal parts in the developing chick leg. J Embryol Exp Morphol 68:1.

Rubin L and Saunders JW Jr (1974). Ectodermal-mesodermal interactions in the growth of limb buds in the chick embryo: constancy and temporal limits of the ectodermal induction. Devel Biol 28:94.

Saunders JW Jr (1948). The proximo-distal sequence of the origin of parts of the chick wing and the role of the ectoderm. J Exp Zool 108:363.

Saunders JW Jr (1977). Experimental analysis of chick limb development. In: DA Ede, JR Hinchliffe, M Balls (eds.): "Vertebrate Limb and Somite Morphogenesis", Cambridge: Cambridge University Press, p1.

Saunders JW Jr, Reuss CM (1974). Inductive and axial properties of prospective wing bud mesoderm in the chick embryo. Devel Biol 38:41.

Somes RG (1981). International Registry of Poultry Genetic Stocks. Storrs Agricultural Station, Storrs CT.

Summerbell D (1974). A quantitative analysis of the effects of excision of the AER from the chick limb bud. J Embryol Exp Morph 32:651.

Zwilling E (1961). Limb morphogenesis. Adv Morphog 1:301.

Limb Development and Regeneration
Part A, pages 45–55
© 1983 Alan R. Liss, Inc., 150 Fifth Avenue, New York, NY 10011

ECTODERM/ MESODERM RECOMBINATION, DISSOCIATION
AND CELL AGGREGATION STUDIES IN NORMAL AND TALPID[3]
MUTANT AVIAN EMBRYOS

D. A. Ede

Department of Zoology, University of Glasgow,
Developmental Biology Building,
124 Observatory Road, Glasgow G12 9LU, U.K.

and

Maryam Shamslahidjani

Department of Biology, National University of
Iran, Teheran, Iran

The ectodermal jacket and the mesodermal core of the limb
bud each contribute to its developing form, separately and by
their interactions. One method of investigating their respective
roles is to use the technique of reciprocal recombination of
these tissues between normal and mutant limb buds, first
developed by Zwilling (1956) in studies on a wingless mutant.
A more recent development of the technique by Zwilling (1964)
enables the cells of the limb mesoderm to be dissociated,
mixed and repacked in ectodermal jackets, giving information
about control mechanisms acting within the mesoderm. Here
we give a resumé of results which will be published in more
detail elsewhere. Details of the technical operations are
given by Shamslahidjani (1980).

1. LIMB DEVELOPMENT IN THE TALPID[3] MUTANT

The talpid[3] mutant has been described by Ede & Kelly (1964).

The limb bud is fan-shaped and produces a limb with the following characteristic features (illustrated for the leg in Fig. 1)

1. Polydactyly, with absence of A-P polarity. There are 5-8 digits, which cannot be identified as corresponding to particular digits of the normal limb; those at the anterior and posterior limits are longest and about equal; smaller digits often occur ectopically between the fork formed by two larger digits.

2. Digits are joined by interdigital webs.

3. Skeletal elements are partially or completely fused along the A-P axis, and to a lesser extent along the P-D axis.

4. With alcian blue, there are lightly-stained areas of cartilage, corresponding to regions where the chondrocytes are enlarged and there is consequently less intercellular matrix.

5. No ossification occurs.

2. RECOMBINATIONS OF ECTODERM WITH INTACT MESODERM

Because talpid[3] embryos do not generally survive to a stage when the results of the experiments can be assessed, the operated limbs of both normal and mutants have been cultured as grafts to normal wing buds to ensure their survival. The results reported here are based on a minimum of 6 surviving grafts for each experiment, assessed on the structure of the whole limb, stained for cartilage and bone and cleared. The illustrations are of experiments on legs, but the experiments have been repeated for the wings and the results are similar.

Talpid[3] limb mesenchyme cells behave differently in culture from normal cells (Ede & Flint, 1975) because they are not polarized and connect with neighbours by filopodia produced all over their surface. In the ectoderm the same cellular abnormality disturbs the normal palisade arrangement of the epidermal cells through a much closer bonding between cells of the basal layer. This may alter the mechanical properties of the ectodermal jacket and it is possible that this causes the talpid[3] modification of limb outgrowth and development.

Reciprocal recombinations of normal/talpid[3] ectoderm/

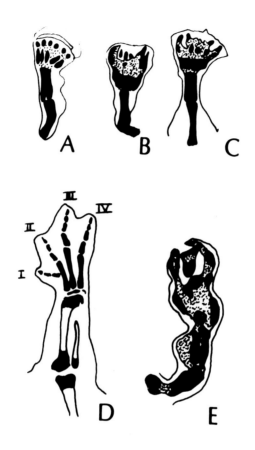

Fig 1

A,B,C. Right legs from representative ta³ embryos which
have survived to 12-12 d.
D.   Right leg of normal 10 d embryo.
E.   Right leg of talpid³ at 10 d, cultured as CAM graft
from st. 21-23.

Stain: Alcian Blue and Alizarin Red.
Key: Black = cartilage; stippled = less dense cartilage;
white = bone.

mesoderm show development according to the mesodermal genotype (Fig 2). When normal mesoderm is cultured in a talpid^3 jacket development is modified to the extent that the digits are bent back from the carpals or tarsals (Fig 2C). It appears that the mutant ectoderm has had no effect on the early growth and fundamental characteristics of the limb, but that it has restricted the extension of the digits, causing them to bend backwards.

3. RECOMBINATIONS WITH DISSOCIATED MESODERM

Dissociating mesoderm cells before repacking in the jackets also shows that talpid^3 characteristics go with the mesoderm (Fig 3B,C,D). With normal mesoderm, whatever jacket is used, though the P-D sequence of skeletal elements is formed, A-P polarity is lost (Fig 3A,C), so that in this respect the limbs are talpid-like. No bending back of the digits is found, presumably because the dissociated cells are more loosely packed in the jacket.

4. RECOMBINATIONS WITH MIXED MESODERMS

Dissociated cells from quail (Coturnix) or talpid^3 meso-derms were mixed (1:1) with normal fowl cells before repacking in normal jackets. In the resulting chimaeras (Fig 3E,F), quail/normal limbs were as dissociated normals. Talpid^3/normal limbs showed one normal feature - a tendency to separation of digits, but in other respects they were as talpid^3. Digital separation depends on interdigital cell death, which is absent in the mutant; in the chimaeras there must be sufficient dying normal cells to eliminate most of the web.

Autoradiographic studies showed that normal fowl/normal fowl and normal fowl/normal quail cells become uniformly mixed and undergo no subsequent segregation (Fig 4A). In the case of normal fowl/talpid^3 the talpid^3 cells mix uniformly at first but segregate later to form small clusters of cells embedded among normal cells (Fig 4B). These results are

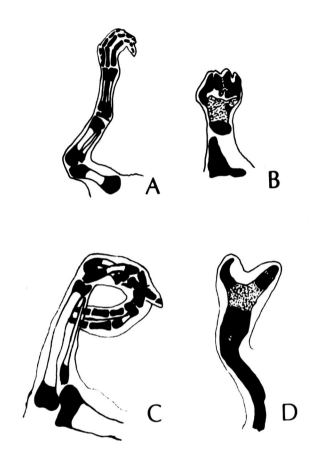

Fig 2

Ectoderm/intact mesoderm recombinations from legs of st. 21-22 embryos. The mesoderm is inserted into an ectodermal jacket, grafted to a normal st. 21-22 host wing and recovered at 10-11 d.

A. Normal mesoderm in normal ectoderm.
B. talpid³ mesoderm in talpid³ ectoderm.
C. Normal mesoderm in talpid³ ectoderm.
D. talpid³ mesoderm in normal ectoderm.

supported by experiments in which small blocks of labelled normal and talpid^3 limb bud tissue were inserted into normal limb buds. In the normal blocks the boundary between host and graft is not clear, and cells from the graft move out into host tissue (Fig 4C). In talpid^3 blocks the boundary remains very distinct, and mutant cells do not move out among normal cells. The imposition of talpid^3 characteristics on the chimaeric normal/talpid^3 limbs therefore occurs although the mutant cells will be distributed patchily in the developing bud.

5. RECOMBINATIONS WITH DISSOCIATED MESODERM AND ADDITION OF ZPA

If the A-P polarity in limb development depends on a morphogen gradient originating from cells of the zone of polarizing activity (ZPA) specifically located in the posterior mesenchyme, it is not surprising that when ZPA cells are scrambled among all the others polarity fails to develop in recombinations with even normal dissociated mesoderm. The question of the existence and activity of ZPA in talpid^3 was investigated by adding ZPA-region tissue from normal and talpid^3 limb buds in the appropriate location in dissociated mesoderm/ectoderm recombinations.

Fig 3

Ectoderm/dissociated mesoderm recombinations. Mesoderm from st. 19-20 legs was dissociated, mixed (sometimes) and reaggregated, then inserted into st. 22-23 ectodermal jackets and grafted to a normal st. 21-22 host wing and recovered at 10-11 d.
A.   Normal dissociated mesoderm in normal ectoderm.
B.   talpid^3 dissociated mesoderm in talpid^3 ectoderm.
C.   Normal dissociated mesoderm in talpid^3 ectoderm.
D.   talpid^3 dissociated mesoderm in normal ectoderm.
E.   Normal/quail dissociated mesoderm in normal ectoderm.
F.   Normal/talpid^3 dissociated mesoderm in normal ectoderm.

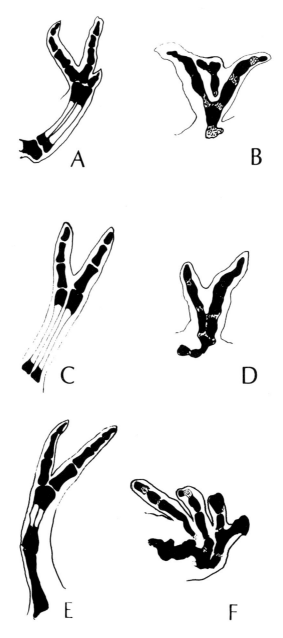

Figure 3

Normal ZPA added to normal dissociated mesoderm produced only a slight asymmetry in the A-P axis, but it did lead to a marked increase in overall limb outgrowth and digit production (Fig 5A). Cooke (1982) has also noted an activity of the ZPA, distinct from its polarizing role, as enhancing the rate of entry of cells into the cell cycle. Neither effect - polarizing or growth promoting - was produced when talpid[3] ZPA was added to normal or talpid[3] mesoderm (Fig 5C,D), suggesting that production of ZPA morphogen is absent or weak in the mutant. However, adding normal ZPA to talpid[3] mesoderm has no effect either, suggesting that the mutant cells are incapable of responding to the morphogen even when it is present. Similar results have been noted by MacCabe & Abbott (1974) in the case of the talpid[2] mutant.

These results tend to confirm the suggestion (Ede, 1982) that dual mechanisms operate in the control of pattern in the A-P axis in the pentadactyl limb bud. One is periodic, producing digits with no A-P polarized characteristics and which continues to operate in dissociated mesoderm. The other - which depends on a ZPA-morphogen gradient - gives the digits their specific characteristics, and fails to operate in dissociated normal mesoderm and in the talpid[3] mutant

---

Fig 4

Sectioned grafts of mixed (1:1) labelled (methyl-^{3}H thymidine)/unlabelled leg mesoderm cells, recovered 70 hours after grafting.
A,B. Set up as in Fig 3.
   A. Normal labelled/ Normal unlabelled in normal ectoderm.
   B. talpid[3] labelled/ Normal unlabelled in normal ectoderm.
C,D. Small blocks of leg mesoderm inserted into normal wing buds at st. 21-22.
   C. Normal block (labelled) in normal wing.
   D. talpid[3] block (labelled) in normal wing.
   Arrows indicate labelled cells which have moved out into unlabelled host tissue.

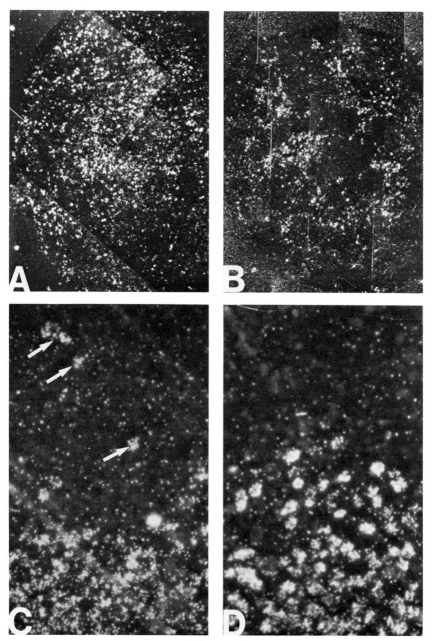

Figure 4

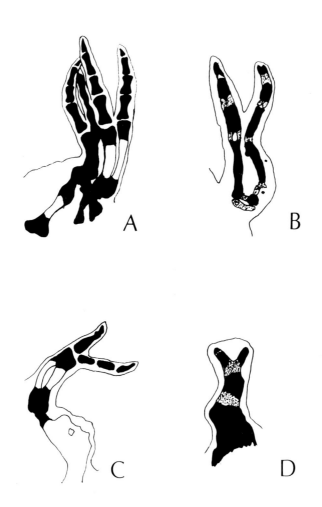

Fig 5
     Ectoderm/dissociated mesoderm recombinations, set up
as in Fig 3 but with addition of zone of polarizing activity
(ZPA) material to posterior border of mesoderm. The ectoderm-
al jacket is from a normal embryo.
A.   Normal dissociated mesoderm with normal ZPA.
B.   talpid[3] dissociated mesoderm with normal ZPA.
C.   Normal dissociated mesoderm with talpid[3] ZPA.
D.   talpid[3] dissociated mesoderm with talpid[3] ZPA.

even in intact embryos.

In the course of evolution it may be that a positional information gradient system has been superimposed on a more primitive prepattern system controlling development of the vertebrate limb. In the case of the talpid³ mutant, the gradient system fails.

REFERENCES

Cooke J (1982). Does wing bud epidermis mediate the growth promoting effect of ZPA tissue? Abstract British Society for Developmental Biology Programme 29 March-2 April 1982.

Ede DA (1982). Levels of complexity in limb-mesoderm cell culture systems. In Yeoman MM, Truman DES (eds): "Differentiation In Vitro" British Soc Cell Biol Symp 4, p 207.

Ede DA, Flint OP (1975). Intercellular adhesion and formation of aggregates in normal and ta³ mutant chick limb mesenchyme. J Cell Sci 18:97.

Ede DA, Kelly OK (1964). Developmental abnormalities in the trunk and limbs of the talpid³ mutants of fowl. J Embryol Exp Morphol 2:339.

MacCabe JA, Abbott UK (1974). Polarizing and maintenance activities in two polydactylous mutants of the fowl: diplopodia and talpid. J Embryol Exp Morphol 31:735.

Shamslahidjani, M (1980). Experimental studies on limb morphogenesis in normal and talpid³ mutant chick embryos. PhD Thesis University of Glasgow.

Zwilling E (1956). Interaction between limb bud ectoderm and mesoderm in the chick embryo. IV. Experiments with a wingless mutant. J Exp Zool 132:241.

Zwilling E (1964). Development of fragmented and of dissociated limb bud mesoderm. Develop Biol 9:20.

Limb Development and Regeneration
Part A, pages 57–65
© **1983 Alan R. Liss, Inc., 150 Fifth Avenue, New York, NY 10011**

GROWTH AND THE ORIGIN OF ADDITIONAL STRUCTURES IN
REDUPLICATED CHICK WINGS

J.C. Smith[1] and L.S. Honig[2]
[1] Imperial Cancer Research Fund, Mill Hill
Laboratories, London, NW7 1AD, U.K.
[2] Laboratory of Developmental Biology,
University of Southern California, Los Angeles,
CA 90007, U.S.A.

When an additional zone of polarizing activity (ZPA) is
grafted to the anterior margin of a developing chick wing
bud a mirror-image duplication is formed (see papers by
Fallon, Honig, MacCabe, Summerbell and Tickle in this volume).
This is most dramatic in the hand, where there may be a
complete doubling of structures (Figure 1); formation of
duplicated wings therefore involves both growth and pattern
formation. Until recently most attention has been directed
towards the specification of pattern (Tickle, Summerbell and
Wolpert, 1975; Summerbell and Tickle, 1977; Smith, Tickle
and Wolpert, 1978; Iten and Murphy, 1980) but Cooke and
Summerbell (1980, 1981), Summerbell (1981a) and Smith and
Wolpert (1981) have now been studying how this may be
linked with growth. The picture that emerges is not a
simple one. Cooke and Summerbell (1980) have demonstrated
that the ZPA will produce an increase in rate of entry into
S phase in responding cells within 8-9 hours, not enough
time to specify new structures, which takes at least 12-15
hours (Smith, 1980; Cooke and Summerbell, 1981). This
suggests that the two processes are not connected but
Smith and Wolpert (1981) have shown that interference with
growth after a ZPA graft will alter the pattern of digits
which results.

In this paper we analyze further the increase in growth
produced by a ZPA graft and attempt to discover whether this
is predominantly widespread or local. Our results also
enable us to determine the size of the rudiment of the newly-
specified structures compared with that of the primary axis.

 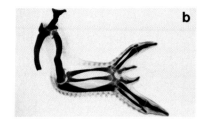

Figure 1. (a) A normal chick wing. (b) A reduplicated wing following a ZPA graft.

The technique we use is that of 'differential dilation' (Lewis, 1977). The left wing bud of a stage 20-21 chick embryo which has been uniformly labelled with tritiated thymidine is grafted to the stump of the right wing bud of an unlabelled host so as to reverse only the antero-posterior axis (Figure 2). This confronts anterior tissue of the grafted bud with the ZPA of the host and a reduplication develops (Saunders and Gasseling, 1968). The tissue that has expanded most in the reduplicated limb will have had its radioactive label correspondingly diluted; and similarly, the amount of label within a particular region will be proportional to the size of its rudiment at the time of the graft. By measuring the amount of radioactivity in different regions of reduplicated limbs we can deduce whether the ZPA brings about a local or a widespread increase in growth and also the size of the rudiment of the secondary axis.

METHODS

White Leghorn embryos were incubated at 36.5 - 38.0°C. At stages 17-18 (Hamburger and Hamilton, 1951) 30 - 40μCi of (methyl-^{3}H)-thymidine (Amersham, 45 Ci/mmol) was added to some eggs and labelled and unlabelled embryos were allowed to proceed to stages 20-21. Then, the left wing bud of the labelled embryos was removed, rinsed for 5 - 10 minutes in salts solution (usually containing 1 mM cold thymidine) and grafted to the stumps of the right wing buds of the unlabelled embryos to reverse the antero-posterior axis. Some right wing buds from labelled embryos were grafted in the correct orientation to unlabelled hosts as controls. The embryos were allowed to develop for a further 2 - 4 days and then the grafted limbs were photographed and fixed for autoradiography or drawn with a camera lucida and the distal 300 μm was dissected into anterior and posterior halves and

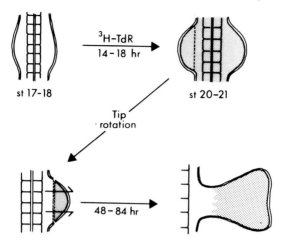

Figure 2.   The experimental design.

frozen in 100 mM NaCl, 10 mM EDTA, 10 mM tris pH 7.0 for
biochemical analysis. Here, and in further discussion,
'anterior' and 'posterior' refer to the axis of the grafted
limbs, not the axis with respect to the host embryo.

Autoradiography

    Limbs were embedded in Araldite and sectioned
perpendicular to the proximo-distal axis. Sections at the
widest part of the limb were stained with toluidine blue or
dipped in Kodak NTB2 emulsion diluted 1:1 with water. The
slides were exposed for 2 - 3 weeks, developed in D-19 and
mounted. The density of label over different parts of the
sections was estimated under dark field with the exposure
meter of a Zeiss Photomicroscope III. 30 - 40 fields were
measured for each limb.

Biochemical Analysis

    Pieces of limb tissue were thawed and disrupted by
brief sonication. Aliquots were precipitated by the
addition of trichloroacetic acid to 5.5% using bovine serum
albumin as a carrier. The precipitate was collected on a
GF/C filter, dried and liquid-scintillation counted. DNA was
determined by a Hoechst 33258 binding assay (Cesarone,
Bolognesi and Santi, 1979) using a Perkin-Elmer LS-3
Fluoresence Spectrometer and salmon sperm DNA as a standard.

RESULTS

Sections of limb buds fixed at stages 20 - 21, immediately after labelling, showed that incorporation of ^{3}H-thymidine was uniform along the antero-posterior axis. When labelled right wings were grafted to unlabelled hosts the resulting normal limbs, fixed 2 - 4 days later, were more heavily labelled posteriorly. The regions of heavy labelling corresponded to formation of cartilage condensations, which differentiate in a posterior-to-anterior sequence (Summerbell, 1976). However, this will have little effect on the pattern of labelling in reduplicated limbs (especially in the symmetrical limbs chosen for study here) because differentiation in the additional digits proceeds in an anterior-to-posterior sequence and lags only slightly behind the normal host digits (Smith, 1979; see Figure 3b).

Of 47 grafts of left wings to host embryos 33 survived and 6, the most symmetrical, were chosen for study. Four of these were examined by autoradiography. In all cases the anterior part of the limb was less heavily labelled than the posterior (Figure 3), indicating that it had undergone greater expansion; that is, it had arisen from a smaller rudiment. Photometric analysis showed that on average the density of label in the anterior portion of the limb was 57% of that in the posterior portion (the individual results were 58%, 43%, 58% and 67%), Thus, the anterior part of the limbs had expanded 1.8 times as much as the posterior; the rudiment of the anterior portion was 57% the size of that of the posterior.

These results were confirmed qualitatively by the biochemical experiments. In both limbs the specific activity of the DNA in the anterior portion of the limb was lower than in the posterior (Table 1). However, the figures suggest that the expansion of the anterior part of the limb is even greater than that suggested by autoradiography; in one case it had expanded 2.3 times as much as the posterior, in the other 5.5 times as much. To investigate this further serial reconstructions of these reduplicated wings are required.

DISCUSSION

The results presented here indicate that the ZPA preferentially stimulates growth in the anterior part of the limb bud and suggest that the rudiment of the additional structures is, at the time of the graft, smaller than the

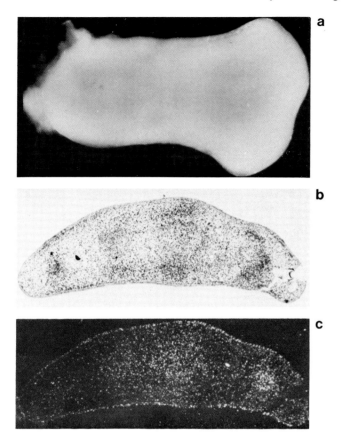

Figure 3. (a) Dorsal view of a reduplicated wing.
(b) Section perpendicular to the proximo-distal axis of the
limb in (a), at the widest part of the limb. Stained with
toluidine blue. (c) Autoradiograph of section adjacent to
that in (b), viewed in dark field. In both (b) and (c) the
anterior portion of the limb is to the left and in (c) this
is less heavily labelled than the posterior portion.

rudiment of the primary structures. According to the autoradiography results (which are the more consistent) the
anterior portion of the limb expands about 1.8 times as much
as the posterior portion and the anterior rudiment is about
57% the size of the posterior rudiment. From the fate maps
of Stark and Searls (1973) the normal antero-posterior
extent of the wing rudiment at stages 20 - 21 is about 350μm

Table 1 Specific activity of DNA in posterior and anterior halves of reduplicated wings

	DNA specific activity, cpm/µg		
Limb	Posterior	Anterior	Post/Ant
60	610	267	2.3
65	538	97	5.5

so this suggests that the ZPA specifies structures over a distance of about 200µm. This is in agreement with Honig's (1981) estimate but less than Summerbell's (1981a) figure of 300µm.

The technique we use cannot tell us when the anterior stimulation of growth occurs. Cooke and Summerbell (1980) observed an anterior stimulation of ^{3}H-thymidine labelling index 4 - 5 hours and 8 - 9 hours after a ZPA graft but this was masked by a posterior stimulation at 16 - 17 hours. Without detailed knowledge of cell cycle kinetics in the chick limb bud we cannot calculate whether this stimulation is sufficient to produce the anterior dilution in label we observe here. There is some indirect evidence that there is additional enhanced anterior growth 28 and 42 hours after a ZPA graft for if grafted limbs are treated with X-irradiation at these times the supernumerary structures are lost; more rapidly-growing structures are particularly sensitive to radiation (results of Hornbruch and Wolpert, reported in Smith and Wolpert, 1981).

What brings about the increase in growth and cell division ? Two views are expressed by Cooke and Summerbell (1980, 1981) and Summerbell (1981a). The first two papers suggest that the ZPA produces distinct growth and pattern specification signals whereas Summerbell believes that one signal can affect both processes. He puts forward a model in which responding cells measure the slope of a morphogen, compare it with the expected equilibrium slope and change their rate of growth to bring the two together. His model fits the double ZPA grafts used in his paper but would predict in a conventional ZPA graft that the labelling index in the posterior half of a limb would be depressed, not enhanced.

We would suggest that two types of growth occur in response to a ZPA graft. First, a widespread increase produced as suggested by Cooke and Summerbell (1980, 1981) and second the 'programmed growth' of the newly-specified anterior structures. Summerbell (1981a) has described how newly-specified structures grow, with an initial lag, a rapid burst of acceleration and then with a falling-off of growth. If this scheme applies to the normal antero-posterior axis of the limb then a model may be put forward which explains our results and predicts the kind of regulation described by Summerbell (1981b).

A hypothetical growth curve for the normal antero-posterior axis of the chick wing bud is shown in Figure 4.

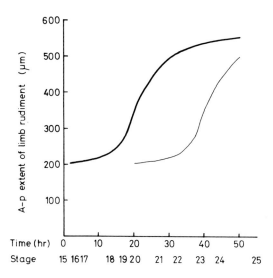

Figure 4. Suggested growth curve for the antero-posterior axis of the chick wing bud. Heavy line is the curve for the normal antero-posterior axis of the limb, beginning at 200μm at stage 15 and reaching 550μm by stage 25. Light line is the curve for newly-specified anterior structures after a ZPA graft at stage 20, when the normal antero-posterior axis is at 350μm. The newly-specified structures also begin at 200μm and reach 500μm by stage 25.

The characteristics of the curve agree with Summerbell's description above, and the antero-posterior axis extends 200μm at stage 15 and 550μm at stage 25, when the digits are specified (Summerbell, 1974). If a ZPA graft is made at stage 20 and the new structures grow according to the same schedule as those in the primary axis it may be seen that by stage 25 supernumerary structures have expanded 1.7 times as much as the posterior structures and they are 91% of the size. Thus we can explain the enhanced growth in the anterior portion of a limb bud after a ZPA graft and also the ability of the supernumerary structures to approach but not exceed the size of the posterior structures. It may also be seen that the earlier the graft the closer to the posterior structures in size will be the anterior structures, as Summerbell (1981b) has shown.

We thank Dr. Michael Sargent for use of the Fluorescene Spectrometer and Shirley Williams for help with photography.

REFERENCES

Cesarone CF, Bolonesi C, Santi L (1979) Improved micro-fluorometric DNA determination in biological material using 33258 Hoechst. Anal. Biochem. 100:188
Cooke J, Summerbell D (1980) Cell cycle and experimental pattern duplication in the chick wing during embryonic development. Nature 287:697
Cooke J, Summerbell D (1981) Control of growth related to pattern specification in chick wing-bud mesenchyme. J. Embryol. Exp. Morphol. 65 (supplement):169
Hamburger V, Hamilton HL (1951) A series of normal stages in the development of the chick embryo. J. Morphol. 88:49
Honig LS (1981) Positional signal transmission in the developing chick limb. Nature 291:72
Iten LE, Murphy DJ (1980) Pattern regulation in the embryonic chick limb: supernumerary limb formation with anterior (non-ZPA) limb bud tissue. Devl. Biol. 75:373
Lewis J (1977) Growth and determination in the developing limb. In Ede DA, Hinchcliffe JR, Balls M (eds): "Vertebrate Limb and Somite Morphogenesis", Cambridge: Cambridge University Press, p.215

Saunders, JW Jr, Gasseling, MT (1968) Ectodermal - mesenchymal interactions in the origin of limb symmetry. In Fleischmajer R, Billingham RE (eds): "Epithelial - Mesenchymal Interactions", Baltimore: Williams and Wilkins, p78.

Smith JC (1979) Studies of positional signalling along the antero-posterior axis of the developing chick limb. PhD thesis, University of London.

Smith JC (1980) The time required for positional signalling in the chick wing bud. J. Embryol. Exp. Morphol. 60:321

Smith JC, Tickle C, Wolpert L (1978) Attenuation of positional signalling in the chick limb by high doses of γ-radiation. Nature 272:612

Smith JC, Wolpert L (1981) Pattern formation along the anteroposterior axis of the chick wing : the increase in width following a polarizing region graft and the effect of X-irradiation. J. Embryol. Exp. Morphol. 63:127

Stark RJ, Searls RL (1973) A description of chick wing bud development and a model of limb morphogenesis. Dev. Biol. 33:138

Summerbell D (1974) A quantitative analysis of the effect of excision of the AER from the chick limb bud. J.Embryol. Exp. Morphol. 32:651

Summerbell D (1976) A descriptive study of the rate of elongation and differentiation of the skeleton of the developing chick wing. J. Embryol. Exp. Morphol. 35:241

Summerbell D (1981a) The control of growth and the development of pattern across the anteroposterior axis of the chick limb bud. J. Embryol. Exp. Morphol. 63:161

Summerbell D (1981b) Evidence of regulation of growth, size and pattern in the developing chick limb bud. J. Embryol. Exp. Morphol. 65 (supplement) : 129

Summerbell D, Tickle C (1977) Pattern formation along the antero-posterior axis of the chick limb bud. In Ede DA, Hinchliffe JR, Balls M (eds) : "Vertebrate Limb and Somite Morphogenesis", Cambridge: Cambridge University Press, p.41

Tickle C, Summerbell D, Wolpert L (1975) Positional signalling and the specification of digits in chick limb morphogenesis. Nature 254:199

Limb Development and Regeneration
Part A, pages 67–76
© **1983 Alan R. Liss, Inc., 150 Fifth Avenue, New York, NY 10011**

NEW INSIGHTS INTO THE PROBLEM OF PATTERN REGULATION IN THE
LIMB BUD OF THE CHICK EMBRYO

John W. Saunders, Ph.D., and Mary T. Gasseling, M.S.

Department of Biological Sciences
State University of New York, Albany
Albany, New York 12222

INTRODUCTION

In 1964, Gasseling and Saunders reported the presence
of supposedly unique properties in a mesodermal zone located
posteriorly in the limb buds of early avian embryos. Cells
from this zone, grafted to the apex of a wing bud or to its
anterior edge, induced the formation of supernumerary forelimb and digital parts, often of remarkable perfection,
from preaxial limb-bud tissues. Because the supernumerary
structures were oriented with their posterior parts facing
the graft, the source of the graft was later termed the
Zone of Polarizing Activity (ZPA).

Our findings, later reported in more detail (Saunders &
Gasseling, 1968), have been seminal to dozens of publications
from many laboratories, most reporting results of experiments
implicit in whose design, is that: the activity attributed
to the ZPA is unique to a restricted zone of the wing bud
or leg bud; this activity consists in the production of a
diffusible morphogen; the highest concentrations of morphogen determine limb parts of the most posterior positional
value on the anteroposterior axis of the limb.

At the Symposium on Vertebrate Limb and Somite Morphogenesis held in Glasgow in 1976, one of us (Saunders, 1977)
reported that the activity, hitherto regarded as the sole
property of the ZPA, can be exercised by a number of tissues of non-limb origin. This report, described as " . . .
most disturbing . . . " by the organizers of the Symposium

(Ede et al., 1977), has apparently disturbed very few, however, for it has had little impact on the way in which other investigators have designed experiments supposedly directed to the analysis of mechanisms that determine positional values along the anteroposterior axis of the limb bud. For this reason we have felt challenged to design test conditions under which non-limb tissues will elicit pattern regulation in chick limb buds as effectively as the ZPA and with a predictability of 100 percent.

We have met that challenge! How we have done so is the subject of this report. We hope that what we say here will challenge others to take a new and fresh look at the problems posed by pattern regulation in the embryonic avian limb.

MATERIALS AND METHODS

Host embryos of stage 18-21 were used. We gently teased the AER from subjacent mesoderm at the apex of the wing bud (Saunders, et al., 1959b) or from a preaxial portion of the bud and then inserted a test tissue fragment (500-1500 cells) between the ectoderm and mesoderm. Tissues tested are listed in Tables 1 and 2. Quail implants were used in most cases so as to provide nucleolar markers in sectioned, Feulgen-stained material. Most embryos were allowed to develop for 10 days after operation.

RESULTS

Apical Implants

Implants were placed subjacent to the AER at the level of somite 18. Effects of implants from various sources on the digital pattern of the wing are recorded in Table 1.

At the graft site, 24 hours after operation, all specimens except those receiving AWM implants showed a slight "dimpling" effect at the graft site as a result of diminution of the AER over the graft. All specimens receiving ALP implants and AWM implants developed a normal wing skeleton except in two cases in which the ALP implant suppressed formation of digit 2. These are included in Table 1 in the 234 category.

Table 1

Morphogenetic Effects of Various Mesodermal Tissues Implanted
   Subjacent to the AER at the Apex of the Chick Wing bud

Site	#	234	2334	3334	23334	23434	234334	234434	2344334
ALP	7	7							
AWM	14	14*							
ZPA	6		1					3	2
FM	5			3			1		1
TBM	12		3		1	1	3	1**	3

*Specimen sectioned; see Figure 1; **Specimen sectioned,
see Figure 3. ALP, anterior lateral plate; AWM, anterior
wing bud mesoderm; ZPA, "Zone of Polarizing Activity"; FM,
flank mesoderm; TBM, tail-bud mesoderm.

One specimen that received an implant of quail anterior
wing-bud mesoderm was sectioned. The skeletal pattern and
approximate distribution of quail cells in the wing are
shown in Figure 1.

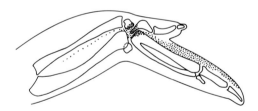

Figure 1. Semidiagrammatic reconstruction of the skeletal
pattern and distribution of quail anterior wing-bud mesoderm implanted beneath the AER apically in a chick host at
stage 20+. The host was fixed for sectioning 10 days after
operation. Stippled areas designate regions in which quail
cells predominate. They are found chiefly in soft tissues
but constitute parts of metacarpals 2 and 3 dorsally, and a
part of the radiale, also. Feather germs, not shown, are
largely composed of quail mesoderm on the anterior edge of
the hand.

In all specimens that received implants of ZPA, flank
or ventral tail-bud mesoderm, the wing buds showed extensive
pattern regulation. Table 1 shows the anteroposterior digital sequences generated during pattern regulation, but does

not show effects on the forearm skeleton. Sequences showing two successive supernumerary digits 3 represent cases in which a single metacarpal 3 articulates distally with two sets of phalanges or two partial supernumerary metacarpals 3 are present, each with a set of phalanges.

Flank implants showed little effect on the skeletal pattern of the forearm, regardless of host stage at the time of operation. Characteristically a normal ulna was formed and the radius developed normally at proximal levels, expanding to an ulna-like articulatory configuration distally. Both ZPA and tail-bud implants, however, brought about drastic modifications of the pattern of forearm bones. Forearms with 3 and 4 bones were found frequently.

Mesodermal implants from the tail-bud tip are incapable of eliciting the formation of supernumerary skeletal parts when implanted in the wing bud. Mesoderm from the ventral side of the tail bud (non-somite mesoderm), however, is equally as effective as mesoderm from the ZPA in bringing about the formation of supernumerary wing parts. Depending on the precise placement of the graft, after 48 hours the wing bud shows a very symmetrical dual outgrowth, on either side of a deep cleft, and a vascular pattern that is precisely duplicated distally (Figure 2).

a   **b**

Figure 2. Left wing bud, a, and right wing bud, b, of a chick embryo injected intravascularly with india ink 48 hours after receiving, at stage 19+, an implant of quail tail-bud mesoderm subjacent to the AER at the apex of the wing bud. Both wings shown in dorsal view.

One specimen, which received a quail implant of ventral tail bud (mesoderm) showed the digital order 234434. It was sectioned for analysis. A reconstruction of the limb, based on serial sections, shows that the chick ectodermal lining of the cleft between the two outgrowths is almost completely lined with quail cells. The latter are the principal components of the dermis and dermal portion of the feather germs, and of the skeletal parts bordering the cleft (Figure 3).

The skeletal parts bordering the cleft parallel to metacarpal 3 on each side have been designated as metacarpals 4 on the basis of their shape and relationship to metacarpals 3. Distally, however, abnormal phalanx-like elements; entirely composed of quail cells, are found. Three such elements are seen anteriorly, the distal one being capped with a claw, whose dermal component is of quail origin and whose ectodermal component is chick.

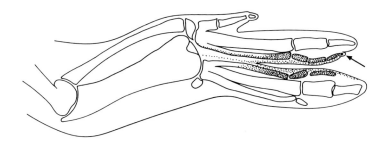

Figure 3. Semidiagrammatic reconstruction from serial sections of the skeletal pattern and the distribution of grafted tail-bud cells in the right wing of a chick embryo operated on at stage 21 and fixed for sectioning 10 days later. The quail implant was placed subjacent to the AER apically. Stippled areas indicate soft tissues and bone constituted predominantly of quail cells. Arrow indicates a claw-capped digit. Note the massive ulna.

Preaxial Implants

We have carried out a few experiments in which flank and tail implants have been placed subjacent to the AER of a normal limb bud preaxially. The results are presented in Table 2. They merit little comment other than that they confirm conclusions that may be derived from perusal of Table 1, namely that both flank and tail bud tissues, appropriately placed in association with AER, induce the formation of supernumerary limb parts.

An undergraduate student in our laboratory, Ron Sadler,

has been analyzing the effects of flank implants on the development of preaxial half-limb buds. Briefly, he has caused the formation of complete left limbs by placing the implants subjacent to the AER proximally on the half-limb bud and complete right limbs by placing the implants distally. The presence of the implant abolishes the pattern of massive necrosis that is regularly visited upon such half limbs within hours after amputation (Hinchliffe and Gumpel-Pinot, 1981).

Table 2

Morphogenetic Effects of Various Mesodermal Tissues Implanted Subjacent to the Preaxial AER of the Chick Wing Bud

Site[*]	#	3234	344334	243334	43234	432234
ZPA	2				1	1
FM	5	1	1	1	1	1
TMB	4	1			1	2

*Abbreviations as in Table 1.

DISCUSSION

Our results show unequivocally that tissues from morphogenetic fields other than the limb field, appropriately associated with the AER and sub-ridge limb mesoderm, can cause the formation of supernumerary limb digits from wing tissue and, furthermore, direct the polarity of those digits with respect to the anteroposterior skeletal axis. But, what do our findings tell us about the way in which the anteroposterior axial patterning of the limb is accomplished in normal development? Since 1975 (reviewed by Tickle, 1980), the predominant thesis has been that a concentration gradient of diffusible morphogen, emanating from the ZPA, specifies positional values with respect to the anteroposterior axis. We believe that our new evidence makes the popular view much less persuasive.

Host limb buds respond to sub-ridge implants of flank or tail tissue identically as they do to ZPA. Overlying AER thins, adjacent anterior ridge thickens, cells enter division more rapidly, (Cooke and Summerbell, 1981), the apex of the wing bud expands, necrosis in anterior half-

wings is diminished. It must follow from these observations
that, if the ZPA is the source of a "morphogen", then tail
cells and flank cells must either: 1) have been producing
the same morphogen and continue doing so in their new environment; 2) start producing the morphogen when implanted
in the wing bud; 3) cause adjacent wing bud tissues to generate the morphogen (cf. Tickle, et al., 1982). To our
minds, none of these alternatives is acceptable.

Iten (1980) excised triangular wedges of tissues comprising about one-third of the wing bud, including ZPA and
adjoining material destined to form the distalmost wing
parts. When the cut edges of the wound were juxtaposed and
pinned together, pattern regulation occurred and wings
showing a normal skeletal pattern usually resulted. In our
laboratory, Heinkel (1963), inspired by earlier studies of
Amprino and Camosso (1955), carried out similar experiments
involving the excision of large triangular wedges from
various levels of the wing periphery. When cut edges were
apposed and held together by pins or when the wound surfaces
spontaneously "zippered up", normal skeletal patterns usually arose. It is evident that in the experiments of Iten
and of Heinkel: 1) numerous positional values relative to
the anteroposterior axis were deleted by wedge removal;
2) the missing positional values were re-established when
their anterior and posterior bordering values were approximated. Distal outgrowth was made possible because the AER
rimmed the entire mesodermal mass after cut edges were apposed.

Results of hundreds of other experiments not reported
here show that non-limb tissue implants that induce supernumerary outgrowth must come from positions posterior to
the limb site. This reinforces the notion of Wolpert (1969)
that " . . . the same mechanisms that specify positional
information may be operative in different fields in the
same organism . . .". When the limb field is established,
its initial pattern of positional values is specified asy-
metrically, with a higher concentration of positional values
located posteriorly. Pattern regulation occurs when tissues
of the anterior portion of the field are confronted with
tissues of posterior positional value provided that the latter can establish an intimate relationship with the wing-
bud mesoderm and the AER. Thus, for example, fragments of
somites elicit supernumerary wing bud outgrowths much more

effectively if the somite cells are first dissociated and
then reaggregated centrifugally before being implanted.

We suggest that the ability of a tissue to elicit
pattern regulation in the limb bud is gradually lost as
differentiation proceeds. As the wing bud elongates, the
ability of grafts of its posterior tissue to elicit pattern
regulation in young wing buds is progressively relegated
to more distal, less differentiated regions (A.B. MacCabe,
et al., 1973; Amprino and Camosso, 1959). Grafts of reoriented wing-bud apices to the base of a wing bud severed
at the level of the body wall show twinning only if the
host embryo is younger than stage 25 (Saunders, et al.,
1959a).

In sum, the limb field undergoes pattern regulation
when perturbed by procedures that bring tissues of disparate positional value into apposition. Posterior positional information may be introduced into the limb field from
non-limb sources and be as effective in eliciting pattern
regulation as can the so-called ZPA. The elicitation of a
twinning pattern by non ZPA tissues, as with the ZPA involved both enhanced cell proliferation and pattern regulation. As Cooke (1982a) pointed out, the signal for acceleration of mitotic rate that follows the confrontation
of tissues of unlike positional value may not be the same
as that which initiates pattern respecification (see, also
Maden, 1982). Neither response, however, seems to require
that a diffusible morphogen be involved.

Wolpert, himself, observed that the " . . . almost
obsessive involvement . . ." of earlier investigators
" . . . with the process of induction and inducing substances: " . . . almost totally obscured the problem of
pattern formation . . . " (Wolpert, 1969, p.4). There
exists today an almost obsessive concern to demonstrate a
"morphogen" that issues from the ZPA, and we feel that this
concern has beclouded the problem of the origin and regulation of limb patterns. We suggest that it is time to
bid farewell to the concept of the "zone of polarizing
activity". Let the energy now directed to the search for
the polarizing "morphogen" be redirected to what occurs
when two tissues of unlike but still labile positional
value are apposed. As Cooke (1982b) has observed " . . .
the domain to be searched for the relevant effect is very
great". We suggest, as have many others (e.g., Bryant et

al., 1981), that short range cellular interactions characterize pattern regulation. These interactions may well
bring about changes on surface proteins, altered patterns of
cellular contacts and changes in the composition of intracellular matrix material that are propagated through the
limb field. If such changes can be delineated, perhaps
they will lead us further along the path to understanding
what positional information really is.

Acknowledgments

This research supported by grant RO1 HDO7390-01-08
from the National Institute of Child Health and Human Development. We gratefully acknowledge this support and the
skilled technical assistance of Paula DiGiuseppe and
Deborah Rannelone.

REFERENCES

Amprino R, Camosso M (1955). Ricerche sperimentali sulla
  morfogenese degli arte nel pollo. J Exp Zool 129:453.
Amprino R, Camosso M (1959). Observations sur les duplications expérimentales de la partie distale de l´ebauche de
  l'aile chez l'embryon de poulet. Arch Anat Micr Morph
  Exp 48:261.
Bryant, SV, French V, Bryant PJ, (1981). Distal regenera-
  tion and symmetry. Science 212:993.
Cooke J (1982a). The relation between scale and the com-
  pleteness of pattern in vertebrate embryogenesis: models
  and experiments. Am Zool 22:91.
Cooke, J (1982b). Vitamin A, limb patterns and the search
  for the positional code. Nature 296:603.
Cooke J, Summerbell D (1981). Control of growth related to
  pattern specification in chick wing-bud mesenchyme. J
  Embryol Exp Morphol, Suppl 65:169.
Ede DA, Hinchliffe JR, Balls M (eds) (1977). "Vertebrate
  Limb and Somite Morphogenesis",Cambridge: Cambridge
  University Press, p ix.
Gasseling MT, Saunders JW (1964). Effect of the "Posterior
  Necrotic Zone" of the early chick wing bud on the pattern
  and symmetry of limb outgrowth. Am Zool 4:303.
Heinkel DE (1965). An experimental analysis of the regu-
  latory capacity of the wing bud in the chick embryo.
  Master's Thesis, Marquette University.

Hinchliffe JR, Gumpel-Pinot M (1981). Control of maintenance and anteroposterior skeletal differentiation of the
anterior mesenchyme of the chick wing bud by its posterior
margin (the ZPA). J Embryol Exp Morphol 62:63.
Iten L (1980). Supernumerary limb structures with regenerated posterior chick wing bud tissue. J Exp Zool 213:327.
MacCabe AB, Gasseling MT, Saunders JW (1973). Spatiotemporal distribution of mechanisms that control outgrowth and
antero-posterior polarization of the limb bud in the chick
embryo. Mech Ageing Develop 2:1.
Maden M (1982). Vitamin A and pattern formation in the regenerating limb. Nature 295:672.
Saunders JW (1977). The experimental analysis of chick limb
development. In Ede DA, Hinchliffe JR, and Balls M (eds):
"Vertebrate Limb and Somite Morphogenesis", Cambridge:
Cambridge University Press, p 1.
Saunders, JW, Gasseling MT (1968). Ectodermal-mesenchymal
interactions in the origin of limb symmetry. In
Fleischmajer R, Billingham RE (eds): "Epithelial-Mesenchymal Interactions", Baltimore: Williams and Wilkins,
p 78.
Saunders, JW, Gasseling MT, Bartizal J (1959a). The distri-
bution of factors affecting the symmetry of skeletal parts
in the wing bud of the chick embryo. Anat Rec 133:332.
Saunders, JW, Gasseling MT, Cairns JM (1959b). The differentiation of prospective thigh mesoderm grafted beneath
the apical ectodermal ridge of the wing bud in the chick
embryo. Dev Biol 1:281.
Tickle C (1980). The polarizing region in limb development.
In Johnson MH (ed): "Development in Mammals", Vol 4,
Amsterdam: Elsevier/North-Holland, p 101.
Tickle C, Alberts B, Wolpert L, Lee J (1982). Local application of retinoic acid to the limb bond mimics the action
of the polarizing region. Nature 296:564.
Wolpert L (1969). Positional information and the spatial
pattern of cellular differentiation. J. Theoret Biol
25:1.

Limb Development and Regeneration
Part A, pages 77–88
© **1983 Alan R. Liss, Inc., 150 Fifth Avenue, New York, NY 10011**

DO CHICK LIMB BUD CELLS HAVE POSITIONAL INFORMATION?

Laurie E. Iten, Douglas J. Murphy and
Ken Muneoka*
Department of Biological Sciences, Purdue
University, West Lafayette, Indiana 47907
*Developmental Biology Center, University of
California, Irvine, California 92717

INTRODUCTION

In 1958, John Saunders, Jr. and his co-workers showed
that when the distal third of a right wing bud was reoriented 180° on its base juxtaposing graft/stump anterior
and posterior, as well as, dorsal and ventral cells,
supernumerary limbs or extra limb structures usually
formed. Depending on the angle of the cut severing the
wing bud tip from its base, they reported that either duplicate or triplicate distal limb structures resulted.
While they discussed predominately resulting duplicate
wings, we have studied the 180° rotation of a stage 21
wing bud tip on its stump and the resulting triplicate
wings (Iten, 1982; Iten and Javois, in preparation). Rotation of a stage 21 chick wing bud tip 180° on its base
usually results in a limb with extra forearm elements and
extra digits which represent three "hands" (Fig. 1). In
order to determine the handedness of the resulting structures, we examined the skeletal, integumentary, and muscle/
tendon patterns of resulting wings. Histological
examination of resulting limbs reveals a constant pattern
of numerous unidentifiable supernumerary forearm muscles.
However, the anterior-most and posterior-most forearm
skeletal elements can be identified as a radius and ulna
respectively. Distally the pattern of "hand" muscles and
tendons is considerably more organized and clearly indicates
the "handedness" of distal structures formed in all cases.
The tendons and muscle pattern indicates the rotated tip
gives rise to an upside-down right "hand", the anterior
supernumerary is also a right "hand", and the posterior

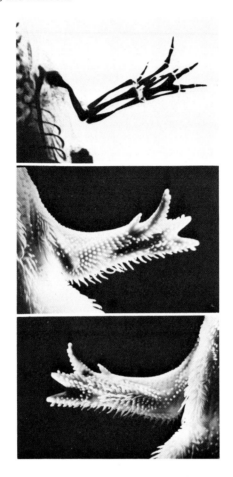

Fig. 1. Dorsal view of the skeletal pattern (top), dorsal integumentary pattern (middle), and ventral integumentary pattern (bottom) of a resulting wing after its tip had been rotated 180° at stage 21. This wing has two extra forearm skeletal elements and extra forearm muscles and triplicate "hands". The anterior to posterior digital sequence is 2, 3, 4, 3, 3, 4. The anterior-most "hand" (digits 2 and 3) has right asymmetry, the middle "hand" (digits 4 and 3) has upside-down right asymmetry, and the posterior-most "hand" (digits 3 and 4) has upside-down left asymmetry as determined from the skeletal, integumentary, and muscle pattern.

supernumerary is an upside-down left "hand". The anterior
to posterior sequence of digits and the overlying integument also substantiate this conclusion. By grafting rotated right quail wing bud tips onto right chick wing bud
stumps, we were also able to determine the contribution of
host (chick) stump cells and donor (quail) rotated tip
cells to the supernumerary limb structures that form. We
found that the progeny of the donor and host cells contribute to both the anterior and posterior supernumerary forearm and "hand" structures while the middle forearm and
"hand" structures are entirely of donor (rotated quail tip)
origin. In conclusion, 180° rotation of a stage 21 chick
wing bud tip on its stump results in the formation of
supernumerary limb structures and stump and rotated tip
cells contribute to the formation of extra limb structures.

    In the study presented here we show that after rota-
ting a stage 24 wing bud tip on its base, no supernumerary outgrowths are observed and only a wing with extra
muscles at the graft junction and sometimes with a duplicated digit results. These resulting wings were so strikingly different than those obtained when the same operation was done with stage 21 wing buds·that we decided to
examine whether or not stage 24 limb bud cells had lost
their ability to recognize normally nonadjacent neighbors
and respond by forming extra limb structures. In attempts
to answer this question, heterochronous 180° rotation of a
wing bud tip on a wing bud stump were performed. Transplanting and rotating a stage 24 wing bud tip on a stage 21
wing bud stump could result in extra forearm and hand structures that mainly came from the stump cells. The reciprocal transplantation operation could result in duplicated
forearm and/or hand structures that appeared to come from
the rotated stage 21 wing bud tip. These results suggest
to us that stage 21 and 24 wing bud cells have positional
information and that stage 21 cells have a greater capacity
to respond to being placed next to normally nonadjacent
neighbors than do stage 24 wing bud cells.

MATERIALS AND METHODS

    Embryos used in these experiments were from a White
Leghorn strain of chickens obtained from the Commercial
Chicks, Thorntown, Indiana or from a flock of Coturnix
coturnix japonica maintained at Purdue University, West
Lafayette, Indiana. Eggs were incubated at 38°C, and at

approximately 3 1/2 or 4 1/2 days incubation they were prepared for surgical manipulations. For chicken embryos, albumin was withdrawn to lower the yolk and embryo; a window was then sawed in the shell. Since quail embryos were only used for donor tissue, the embryo was exposed by simply opening the egg over the air cell. The extra-embryonic membranes surrounding the embryo were opened and tissue manipulations with the right wing bud were performed with sharpened tungsten needles and fine forceps. All surgical operations were recorded with a camera lucida drawing at the time they were performed and a drawing was made 1 and 2 days later. Resulting host embryos were sacrificed at approximately 12 days incubation.

In the experiments presented in this study, homo-chornous or heterochronous 180° rotation of a right wing bud tip on a right wing bud stump were performed. First, Hamburger and Hamilton (1951) stage 24 (length: width ratio of 1.0-0.8) right chick wing buds were severed from their base and rotated 180° about their proximodistal axis before being reattached to their base with tungsten microtacks, as illustrated in Figure 2. Next, the right wing bud tip of a stage 21 (l:w ratio of 2.3-2.9) and 24 chick embryos were exchanged and rotated 180° before being grafted to the heterochronous wing bud stump (Fig. 3). The same operations were also performed using donor quail wing buds.

All embryos resulting from operations with chick donor and host wing buds were fixed in Lillie's fixative and stained with Victoria Blue. The pattern of feather germs of resulting wings was recorded before the stained embryos were cleared in methyl salicylate and their wing skeletal pattern was analyzed. A number of resulting wings were embedded and serially cross sectioned to examine the muscle pattern and the contribution of donor and host limb bud cells to the wing structures formed. Sections of chimeric chick/quail wings were 7 µm thick and stained with the Feulgen nuclear reaction (Lillie, 1951) and sections of chick wings were 15 µm thick and stained with Mayer's Hemotoxylin and Eosin.

RESULTS

When a stage 21 wing bud was rotated 180° on its stump, three areas of limb bud outgrowth are typically

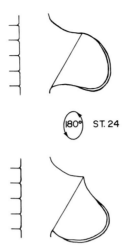

Fig. 2. Dorsal view outline of a stage 24 right wing bud before (top) and after (bottom) 180° rotation of the severed wing bud on its base.

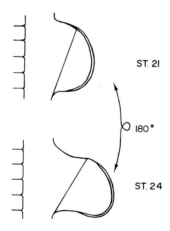

Fig. 3. Diagram illustrating the operation performed where the wing bud tips of right stage 21 and 24 were exchanged and rotated 180° before being grafted to the heterochronous wing bud stump.

observed one to two days after doing such an operation: one from the rotated tip, plus supernumerary anterior and posterior outgrowths; "triplicate" wings with extra forearm and hand structures resulted (Figure 1) (Iten, 1982; Iten and Javois, in preparation). However, when the same operation was performed with a stage 24 wing bud, no supernumerary limb bud outgrowths were observed. Of the 21 resulting wings obtained in this study where a stage 24 chick wing bud was rotated on its stump, upside down right wings developed and the dorsal-ventral integumentary asymmetry switches at the elbow of the wing and there are some duplicated secondary feather germs at the elbow (Fig. 4). Nine wings had an unidentifiable ectopic piece of cartilage at the elbow, four had a partially or completely duplicated digit 2, and the remainder had a normal complement of skeletal elements. Of the 10 wings sectioned, all had extra upperarm muscles at the elbow. Five resulting chimeric chick/quail wings from this category of operations were sectioned and Feulgen stained so that donor quail and host chick cells could be identified. The junction between donor and host cells was at the level of the elbow and the extra muscles and feather germs were also located at or near this graft junction and they were made up of both donor and host cells. None of the 11 resulting chimeric wings had a duplicated digit 2. Therefore, supernumerary outgrowths are seen after a stage 21 wing bud is rotated on its stump and extra forearm and hand structures result, whereas, when the same operation is performed with a stage 24 wing bud, no extra limb bud outgrowths are observed and most, if not all, supernumerary limb structures are restricted to the region of the graft junction.

When a right stage 21 chick wing bud tip was "tacked" to a right stage 24 stump after being rotated 180°, the rotated tip appears to broaden one to two days after the operation. In five of nine cases, a wing with a symmetric hand consisting of duplicated digits 3 and/or 4 resulted and four of these have a forearm with two ulnas or one complete ulna and the other element being radius-like proximally and ulna-like distally as determined from the muscle and integumentary patterns (Fig. 5). Four of the nine resulting wings have an unidentifiable piece of ectopic cartilage at the elbow (graft junction). Three of the nine wings only have one forearm skeletal element and it is an ulna.

Fig. 4. Dorsal view of the skeletal pattern (top), dorsal
integumentary pattern (middle), and ventral integumentary
pattern (bottom) of a resulting wing after its tip had
been rotated 180° at stage 24, as illustrated in Fig. 2.
This wing has a normal complement of skeletal elements
and the switch from a dorsal to ventral integumentary
pattern (and vice versa) is at the elbow. There are some
supernumerary secondary feather germs at the elbow.

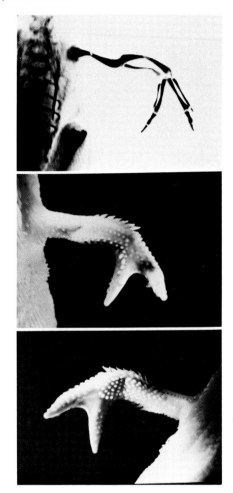

Fig. 5. Dorsal view of the skeletal pattern (top), dorsal integumentary pattern (middle), and ventral integumentary pattern (bottom) of a resulting wing after a right stage 21 chick wing bud tip was grafted to a right stage 24 chick stump after being rotated 180°. The forearm has one complete anterior ulna and the other element is radius-like proximally and ulna-like distally as determined from the muscle and integumentary patterns. The anterior to posterior digital sequence is 4, 3, 3, 4 and the anterior-most "hand" has upside-down right asymmetry and the posterior-most "hand" has upside-down left asymmetry.

A surprising result was obtained when a stage 21 quail right wing bud tip was grafted to a right stage 24 chick stump after being rotated 180°. The rotated quail tip did not appear to broaden and wings with a normal pattern of forearm and hand structures resulted in the 13 resulting wings. Of the three cases sectioned and Feulgen stained, all had extra muscles at the graft junction and they were made up of both donor and host cells. Distal to the graft junction, all limb structures were made up of quail cells and there were no extra muscles. Actually, two of these three limbs had only one skeletal element, an ulna, and they were missing some forearm muscles.

When a stage 24 right chick wing bud tip was grafted to a stage 21 right wing bud stump after being rotated 180°, a separate anterior supernumerary area of limb bud outgrowth was almost always observed. The rotated tip also appeared to broaden slightly in some cases. These same observations were made even when the rotated tip was from a donor quail embryo. Of the five cases obtained where donor and host wing buds are both chick tissue, all had at least a supernumerary anterior radius and often a digit 2 at the distal end of this extra radius (Fig. 6). One spectacular case had a supernumerary anterior radius, ulna and digits 2 and 3. An extra posterior digit 2 could also result (Fig. 6). Of the seven chimeric chick/quail wings, all appeared to have supernumerary anterior limb structures and three had an extra posterior digit 2 or 3. Five of these seven limbs were sectioned and where donor and host cells were located in these wings was determined from the Feulgen stained sections. The supernumerary anterior structures were made up almost entirely of host stump chick cells. Donor quail cells were seen at the base of this anterior supernumerary outgrowth and the super-numerary anterior structure was a radius, sometimes terminated with a digit 2. When a supernumerary posterior digit 2 or 3 formed, it was made up of donor and host cells. A strip of host chick cells could be seen the proximodistal length of the wing along its anatomical posterior edge. These cells appeared to be restricted to the dermis of the integument. When no supernumerary posterior structures formed, host cells were not found the entire length of the limb, but they did go distal to the graft junction. Extra muscles of donor and host origin were found at the graft junction.

Fig. 6. Dorsal view of the skeletal pattern (top), dorsal integumentary pattern (middle), and ventral integumentary pattern (bottom) of a resulting wing after a right stage 24 chick wing bud tip was grafted to a right stage 21 chick wing stump after being rotated 180°. There is an anterior supernumerary radius and digit 2 with right hand asymmetry. The other outgrowth formed an upside down right limb with an ulna and a partial distal radius terminated with a distally duplicated digit 2.

DISCUSSION

The ability of normally nonadjacent avian limb bud cells to interact and respond after being juxtaposed was studied to assess whether or not embryonic limb bud cells have positional information. Rotation of a stage 21 wing bud results in the formation of supernumerary limbs which are of both graft and host origin (Iten and Javois, in preparation). In contrast, rotation of a stage 24 wing bud results in the production of few supernumerary limb structures. Heterochronous 180° rotations between stage 21 and 24 wing buds result in the formation of super-numerary limb structures which appear to arise mostly from stage 21 wing bud tissue. These results suggest that stage 21 wing bud cells have the capacity to rec-ognize and respond to positional disparities while stage 24 wing bud cells can recognize, but cannot respond to positional disparities. These results indicate to us that during limb development, cells do not lose the ability to recognize or be recognized as a normally nonadjacent neighbor, but they do lose their ability to respond to a positional disparity.

The work reported here, along with previous work from our lab (Iten and Murphy, 1980; Iten et al, 1981; Iten, 1982; Javois and Iten, 1981, 1982; Javois et al, 1981), emphasized the fact that when normally nonadjacent cells of the chick wing bud are juxtaposed, supernumerary limb structures are formed. These results suggest that local cell-cell interactions are intimately involved with the development of the three-dimensional pattern of limb structures. Experiments directed at the level of the cell surface will help further elucidate the cellular and molecular nature of positional information. Significant advances have been made recently in examining cell surface properties using monoclonal antibodies to cell surface antigens. For example, a monoclonal antibody has been isolated which binds molecules distributed in a topo-graphic gradient in the avian retina (Trisler et al, 1981). These results suggest that the methodology is available and can be utilized with other developing systems to examine the role a cell's surface plays in the expression of its positional information.

ACKNOWLEDGEMENTS

This investigation was supported by a research grant awarded to L.E.I. by the National Science Foundation, PCM 81-10848. K.M. was supported by a National Institute of Health Training Grant, HD 07029, and was a recipient of a travel award from the University of California Chancellor's Patent Fund for Graduate Student Research. This is the first and perhaps the last publication from the Purdue Chapter of the Poets Club.

REFERENCES

Hamburger V, Hamilton HL (1951). A series of normal stages in the development of the chick embryo. J Morph 88:49.

Iten LE (1982). Pattern specification and pattern regulation in the embryonic chick limb bud. Amer Zool 22:117.

Iten LE, Murphy DJ (1980). Pattern regulation in the embryonic chick limb: supernumerary limb formation with anterior (non-ZPA) limb bud tissue. Develop Biol 75:373.

Iten LE, Murphy DJ, Javois LC (1981). Wing buds with three ZPA's. J Exp Zool 215:103.

Javois LC, Iten LE (1981). Position of origin of donor posterior chick wing bud tissue transplanted to an anterior host site determines the extra structures formed. Develop Biol 82:329.

Javois LC, Iten LE (1982). Supernumerary limb structures after juxtaposing dorsal and ventral chick wing bud cells. Develop Biol 90:127.

Javois LC, Iten LE, Murphy DJ (1981). Formation of supernumerary structures by the embryonic chick wing depends on the position and orientation of the graft in a host limb bud. Develop Biol 82:343.

Lillie RD (1951). Simplification of the manufacture of shift reagent for use in histochemical procedures. Stain Techol 26:163.

Saunders JW Jr, Gasseling MT, Gfeller MD (1958). Interactions of ectoderm and mesoderm in the origin of axial relationships in the wing of the fowl. J Exp Zool 137:39.

Trisler GD, Schneider MD, Nirenberg M (1981). A topographic gradient of molecules in retina can be used to identify neuron position. Proc Natl Acad Sci USA 73:2145.

Limb Development and Regeneration
Part A, pages 89–98
© 1983 Alan R. Liss, Inc., 150 Fifth Avenue, New York, NY 10011

POSITIONAL SIGNALLING BY RETINOIC ACID IN THE
DEVELOPING CHICK WING

C. Tickle

Department of Biology as Applied to Medicine

The Middlesex Hospital Medical School
Cleveland Street, London W1P 6DB, UK

The polarizing region, a small group of mesenchyme cells at
the posterior margin of the limb, acts as a signalling region
to specify the pattern of structures that develop across
the antero-posterior axis of the limb. This signalling
ability can be demonstrated by grafting a polarizing region
to an anterior position in a limb bud. Such a bud develops
to form a double set of structures along the antero-
posterior axis in mirror-image symmetry between the grafted
and host polarizing regions (Saunders, Gasseling, 1968:
Tickle, Summerbell, Wolpert, 1975). Thus the pattern of
digits in a wing to which a second polarizing region has
been grafted is, from anterior to posterior, $\underline{4}\ \underline{3}\ \underline{2}\ \underline{2}\ \underline{3}\ \underline{4}$,
instead of the normal pattern $\underline{2}\ \underline{3}\ \underline{4}$. There is little
cellular contribution to the additional digits from a viable
graft or none if the graft is irradiated so that the cells
cannot divide (Smith, 1979). The additional digits develop
from the host mesenchyme in response to the graft showing
that the polarizing region is indeed a signalling region
and that the action of the signal is to bring about a
change in the pattern of cellular differentiation.

The pattern of digits that develop between the grafted
and host polarizing regions depends on the distance between
the two polarizing regions (Tickle et al. 1975). When
the graft is made nearer the host polarizing region, instead
of a complete duplicate set of digits forming, the digit
pattern is still symmetrical but may be $\underline{4}\ \underline{3}\ \underline{2}\ \underline{3}\ \underline{4}$. With
grafts made still closer to the host polarizing region, the
digit pattern is $\underline{4}\ \underline{3}\ \underline{3}\ \underline{4}$ or even $\underline{4}\ \underline{3}\ \underline{4}$. In the latter
case, additional digits also form from the mesenchyme

anterior to the graft, the complete digit pattern being
2̲ 3̲ 4̲ 4̲ 3̲ 4̲.   Thus, there is a relationship between the
distance from the polarizing region and which structure
along the antero-posterior axis forms – digit 4̲ forming
close to the polarizing region, digit 3̲ a bit farther
away and digit 2̲ farther away still.

One way in which the polarizing region could signal
to the responding mesenchyme so that structures develop
according to their distance from the polarizing region
would be by producing a diffusible morphogen.   A
concentration gradient of morphogen would be established
across the antero-posterior axis of the bud and the
concentration at any point would provide information of
position relative to the polarizing region.   Thus, cells
close to the polarizing region would be exposed to a high
concentration of morphogen and participate in forming
posterior structures such as digit 4̲, cells slightly
farther away would be exposed to a lower concentration and
form digit 3̲, and cells, farther away still, would be
exposed to even lower concentration and form digit 2̲.
This idea can account quite well for the patterns of digits
that result following polarizing region grafts to different
positions along the antero-posterior axis.   An alternative
model based on intercalation of missing structures between
the grafted polarizing region and the adjacent host
mesenchyme can also account for these results (Iten,Murphy
1980).   However, the pattern of digits predicted by the
intercalation model is not obtained when two polarizing
regions are grafted to a bud.   The results instead confirm
that structures develop according to distance from the
polarizing region, consistent with the polarizing region
being a source of diffusible morphogen (Wolpert,Hornbruch
1981).

There are several other lines of evidence that are
consistent with the signal from the polarizing region being
a diffusible morphogen (reviewed, Tickle 1980).   For
example, progressive attenuation of the signal by grafting
smaller and smaller numbers of polarizing region cells
(Tickle 1981) or by treating the polarizing region before
grafting with increasing doses of γ-radiation (Smith,
Tickle, Wolpert 1978) leads to a successive sequence of
digit patterns in which first digit 4̲ is not specified, the
digit pattern being 3̲ 2̲ 2̲ 3̲ 4̲, then digit 3̲ is not specified,
the digit pattern being 2̲ 2̲ 3̲ 4̲ and finally no additional

digits are formed at all.   In contrast to this sequence of
loss of digits, when the bud is prevented from undergoing
the widening that normally follows a polarizing graft by
treating the grafted bud with X-irradiation, digit 2's are
not specified and in extreme cases the complete digit
pattern can be just 4 3 4 (Smith, Wolpert 1981).

However, direct evidence that the polarizing region
signals by producing a diffusible morphogen requires
identification of the morphogen and demonstration that cells
respond according to its concentration.   The assay to
identify substances with signalling activity is to apply
the substances locally to anterior positions of buds and
see whether additional digits are specified.   For local
application, the use of a range of inert carriers has been
explored in collaboration with B. Alberts.   For example,
to obtain slow release localized to the anterior of the bud
charged molecules would be carried on papers or beads with
the opposite charge.   Initial experiments in which
extracts of polarizing region cells and a number of
different chemicals such as hyaluronidase, thalidomide and
cAMP were locally applied to anterior positions in
developing wing buds on such carriers all gave negative
results - normal wings or ones with anterior defects
developed.   However, we have recently tested retinoic acid
because of its effects on cell communication (Pitts,Burk,
Murphy 1981) and cell differentiation (Strickland,
Mahdavi 1978).   To our surprise, as systemic application
of retinoic acid has been shown to produce limb defects in
mouse embryos (Kochhar 1977), we found that retinoic acid,
locally applied, acted as a signalling substance.
Positively charged papers soaked in retinoic acid grafted
to slits in the anterior margins of wing buds lead to the
development of additional digits (Tickle,Alberts,Wolpert,
Lee 1982).   Analysis of the data obtained so far suggests
that local application of retinoic acid closely mimics the
action of the normal signaller the polarizing region.   It
appears to be retinoic acid that is active and not some
minor contaminant as purified retinoic acid has the same
effect.

We will consider the similarities of the action of the
polarizing region and locally applied retinoic acid under
a number of headings.

1. Dose-dependent Response

   The response of the mesenchyme to the cells of the polarizing region is dose-dependent. This is shown by the pattern of digits that result when the polarizing region graft has been attenuated by high doses of radiation and by various metabolic inhibitors (Honig, Smith, Hornbruch, Wolpert 1981) and also when known numbers of polarizing region cells are grafted. Thus, in this latter case there is a quantitative relationship between the number of polarizing region cells grafted and the pattern of additional digits (Tickle 1981). This was shown by plating out polarizing region cells on to small pieces of plastic film and counting the number of adherent cells on each piece of film before grafting to anterior positions in host wing buds. The pattern of digits provides a measure of the polarizing activity of the graft - grafts with high signalling activity lead to digit 4 being specified next to the graft, grafts with slightly lower activity lead to specification of digit 3 next to the graft, and those with low activity lead to the development of an additional 2 only.

   Is the response of wing buds to grafts of paper soaked in different concentrations of retinoic acid similarly dose-dependent? The data obtained so far (Tickle et al. 1982) suggest that this may be so. Thus, with grafts of positively charged papers soaked in solutions of 5mg/ml retinoic acid, wings with an additional digit 2 only often result while grafts of papers soaked in higher concentrations (7.5 and 10mg/ml retinoic acid) lead to more complete duplications with digit 4 formed more frequently next to the graft.

2. Position of Graft and Digit Pattern

   The pattern of wing digits that develop following a polarizing region graft depends on the position along the antero-posterior axis of the wing bud margin to which the graft was made (as outlined earlier). As yet we have not completed a systematic investigation of the digit patterns that develop following grafts of retinoic acid impregnated paper to different positions along the antero-posterior axis of the limb bud margin. While the majority of the grafts have been made to anterior positions opposite somites 16 and the border between somites 16 and 17, a smaller number of grafts have been made to the centre of the wing bud margin opposite the border between somites 17 and 18 and to the posterior margin opposite somite 19. In this last position congruous with the polarizing region of the host bud, grafts of paper impregnated with 10mg/ml retinoic acid led, in nearly every case, to the development of a normal digit pattern (Fig.1a). This result is the same as that obtained when grafts of polarizing region cells are made to this position.

Grafts of polarizing region cells to the centre of the wing bud margin result in digit patterns 4 4 3 3 4 or 3 4 4 3 4. With positively charged paper soaked in 10mg/ml of retinoic acid grafted to this position, digit patterns of 3 3 4 and 3 4 4 have been obtained. However, in several cases, such grafts result instead in truncation of the developing wing at the elbow joint and no digits at all are specified. This result may be due to non-specific cytotoxic effects of retinoic acid. An alternative interpretation is that truncation is due to elevated signalling across the entire antero-posterior axis of the bud. Thus, when polarizing region cells are randomly distributed across the antero-posterior axis in buds reconstructed from disaggregated limb cells, then outgrowth of the limb is inhibited and no digits form (Crosby,Fallon 1975: Frederick, Fallon 1982).

The usual result of grafting polarizing cells opposite somite 16 and opposite the 16/17 border is the digit pattern 4 3 2 2 3 4 or 4 3 2 3 4. While several wings that developed following grafts of papers soaked in retinoic acid had these digit patterns and were indistinguishable from wings following grafts of polarizing regions (Fig.1b), there were some cases in which the digit pattern was markedly different despite digit 4 being formed next to the graft. For example, wings with complete digit patterns of 4 3 4 were also formed (Fig.1c). This digit pattern is only commonly formed following a polarizing region graft when the bud is X-irradiated to prevent widening. Preliminary observations of the shape of the bud at early times after grafts of retinoic acid impregnated papers show that the shape of the limb is variable and sometimes does not widen.

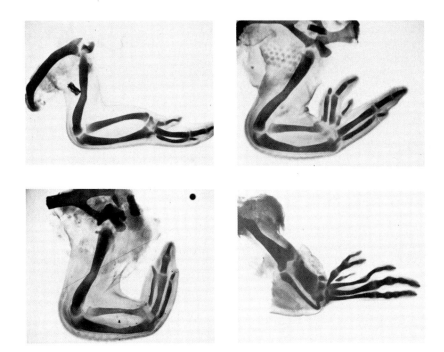

Figure 1.   Dorsal views of whole mounts of limbs that
developed following grafts of positively charged papers
impregnated with retinoic acid.   a. Wing that developed
after paper soaked in 10mg/ml retinoic acid had been grafted
at the posterior margin of the bud (stage 20) opposite
somite 19.   Digit pattern is 2 3 4.   The paper has ended
up halfway down the upper arm to the left of the humerus.
b.Wing that developed after paper soaked in 10mg/ml
retinoic acid had been grafted at the anterior margin of
the bud (stage 20) opposite somite 16.   Digit pattern is
4 3 2 3 4.   The paper has ended up in the shoulder.
c. Wing that developed after paper soaked in 10mg/ml
retinoic acid had been grafted at the anterior margin of
the bud (stage 20) opposite somite 16.   Digit pattern is
4 3 4.   The paper has ended up in the shoulder.   d. Leg
that developed after paper soaked in 10mg/ml retinoic acid
had been grafted to the anterior margin of the bud (stage
20) opposite somite 27.   Digit pattern is III II I II III IV

3. Universal Polarizing Signal

The signal from the polarizing region seems to be universal in all vertebrate limbs. Thus, the polarizing regions of the chick wing and leg are interchangeable and even the polarizing region from a mouse or hamster fore-limb bud can cause the formation of additional digits in the chick wing (Tickle, Shellswell, Crawley, Wolpert 1976: MacCabe, Parker 1976). We have therefore tested whether retinoic acid impregnated paper will affect the pattern of structures across the antero-posterior axis of the chick leg. The results show clearly that just as a wing polarizing region will cause the formation of additional toes so can grafts of paper soaked in 10mg/ml retinoic acid (Fig.1d).

4. Close Apposition with Apical Ridge Enhances Signalling

It is not clear whether the signal from the polarizing region acts directly on the responding mesenchyme or whether its effect is mediated by the apical ridge. The latter possibility is suggested by the more efficient signalling of polarizing region grafts made under an intact apical ridge. In addition, comparison of the estimates of the number of polarizing region cells required to produce each additional digit from grafts of cells on pieces of plastic and from grafts of pellets of polarizing region cells diluted with non-signalling cells shows that only the monolayer of cells in contact with the ridge is effective in positional signalling (Tickle 1981). Preliminary results show that, similarly, with grafts of retinoic acid impregnated papers, more efficient signalling is obtained when these are placed below the apical ridge rather than to slits cut through the apical ridge and the mesenchyme.

5. Duplicated Structures

Polarizing region grafts affect the development of other limb structures as well as the cartilage elements of the digits. For example, the forearm of the wing can be affected if the polarizing region graft is made to early buds before this limb segment has been completely laid down (Summerbell 1974) and the signal from the polarizing region also specifies the connective tissue sheaths of the muscle

(Shellswell, Wolpert 1977). The muscle patterns in wings
with duplicate sets of digits resulting from grafts of
retinoic acid impregnated papers are therefore being
investigated.

There are a number of possible ways in which the local
application of retinoic acid could act to bring about a
change in the pattern of cellular differentiation that
results in the development of additional digits. For
example, retinoic acid itself could be a morphogen and the
limb mesenchyme cells could be responding to a concentration
gradient of retinoic acid. A second possibility is that
retinoic acid acts locally to convert mesenchyme cells
adjacent to the implanted paper into signalling cells. Yet
another possibility is that the effect of retinoic acid is
mediated by the apical ridge. These possibilities are
currently being tested.

Retinoic acid has many well documented effects on cells
(reviewed Lotan 1980) and a number of vitamin A analogs are
available. These synthetic analogs have differing
properties as regards, for example, cellular toxicity,
control of differentiation and ability to bind to retinoic
acid binding proteins. The relative effectiveness of such
analogs may help to indicate which properties are important
for the signalling by retinoic acid.

The importance of the effects of retinoic acid on limb
development is that it is the first chemical that has been
found that brings about a change in the pattern of cellular
differentiation. Quite independent work by Maden (1982)
has shown that vitamin A analogs including retinoic acid can
also alter positional values in amphibian limb regeneration
blastemas. Thus it is clear that retinoic acid and several
other retinoids are indeed morphogenetically active chemicals
and this opens up exciting possibilities for exploring the
biochemical mechanisms of positional signalling.

This work was supported by the Medical Research Council,
U.K. I thank Professor L. Wolpert for reading the
manuscript and A. Crawley for help with the figures.

Crosby GM, Fallon JF (1975). Inhibitory effect on limb
    morphogenesis by cells of the polarizing zone
    coaggregated with pre- or postaxial wing bud mesoderm.
    Devl Biol 46:28

Frederick JM, Fallon JF (1982). The proportion and distribution of polarizing zone cells causing morphogenetic inhibition when coaggregated with anterior half wing mesoderm in recombinant limbs. J Embryol exp Morph 67:13.

Honig LS, Smith JC, Hornbruch A, Wolpert L (1981). Effects of biochemical inhibitors on positional signalling in the chick limb bud. J Embryol exp Morph 62:203.

Iten LE, Murphy DJ (1980). Pattern regulation in the embryonic chick limb: supernumerary limb formation with anterior (non-ZPA) limb bud tissue. Devl Biol 75:373.

Kochhar DM (1977). Cellular basis of congenital limb deformity induced in mice by vitamin A. In Bergsma D, Lenz W (eds): "Morphogenesis and Malformation of the Limb," The National Foundation - March of Dimes Birth Defects Vol.XIII No.1, New York: Alan R. Liss Inc,p.111

Lotan R (1980). Effects of vitamin A and its analogs (retinoids) on normal and neoplastic cells. Biochem Biophy Acta 605:33.

MacCabe JA, Parker BW (1976). Polarizing activity in the developing limb of the syrian hamster. J Exp Zool 195:311.

Maden M (1982). Vitamin A and pattern formation in the regenerating limb. Nature Lond 295: 672.

Pitts JD, Burk RR, Murphy JP (1981) Retinoic acid blocks junctional communication between animal cells. Cell Biol Int Rep Suppl A 5:45.

Saunders JW, Gasseling MT (1968). Ectodermal-mesodermal interactions in the origin of limb symmetry. In Fleischmajer R, Billingham RE (eds): "Epithelial-Mesenchymal Interactions," Baltimore: Williams and Wilkins, p.78.

Shellswell GB, Wolpert L (1977). The pattern of muscle and tendon development in the chick wing. In Ede DA, Hinchliffe JR, Balls M (eds): "Vertebrate Limb and Somite Morphogenesis," Cambridge: Cambridge University Press, p.71.

Smith JC (1979). Evidence for a positional memory in the chick limb bud. J Embryol exp Morph 52:105.

Smith JC, Tickle C, Wolpert L (1978). Attenuation of positional signalling in the chick limb by high doses of γ-radiation. Nature Lond 272:612

Smith JC, Wolpert L (1981). Pattern formation along the anteroposterior axis of the chick wing: the increase in width following a polarizing region graft and the effect of X-irradiation. J Embryol exp Morph 63:127

Strickland S, Mahdavi V (1978).   The induction of
    differentiation in teratocarcinoma stem cells by
    retinoic acid.   Cell 15:393
Summerbell D (1974)   Interaction between the proximo-distal
    and antero-posterior co-ordinates of positional value
    during specification of positional information in the
    early development of the chick limb-bud.   J Embryol
    exp Morph 32:227
Tickle C (1980)   The polarizing region and limb development.
    In Johnson MH (ed): "Development in Mammals," Vol 4,
    Amsterdam: North-Holland Biomedical Press, p 101.
Tickle C (1981)   The number of polarizing region cells
    required to specify additional digits in the developing
    chick wing.   Nature Lond 289:295
Tickle C, Alberts B, Wolpert L, Lee J (1982)   Local
    application of retinoic acid to the limb bud mimics the
    action of the polarizing region.   Nature Lond 296:564
Tickle C, Shellswell G, Crawley A, Wolpert L (1976)
    Positional signalling by mouse limb polarizing region
    in the chick wing.   Nature Lond 259:396
Tickle C, Summerbell D, Wolpert L (1975)   Positional
    signalling and specification of digits in chick limb
    morphogenesis.   Nature Lond 254:199
Wolpert L, Hornbruch A (1981)   Positional signalling along
    the antero-posterior axis of the chick wing.   The effect
    of multiple polarizing region grafts.   J Embryol exp
    Morph 63:145

Limb Development and Regeneration
Part A, pages 99–108
© **1983 Alan R. Liss, Inc., 150 Fifth Avenue, New York, NY 10011**

POLARIZING ACTIVITY OF THE AVIAN LIMB EXAMINED ON A CELLULAR
BASIS

Lawrence S. Honig

Laboratory for Developmental Biology, GER 323
University of Southern California, P.O. Box 77912
Los Angeles, CA 90007   U.S.A.

At the posterior margin of the limb bud of the developing amniote embryo, a region called the polarizing region
appears responsible for signalling positional information
along the antero-posterior axis (Tickle, Summerbell, Wolpert
1975). When a small piece of polarizing tissue (ca. 10^4
cells) is grafted anteriorly, duplicated host limbs with digit patterns such as 4 3 2 2 3 4 result (Fig. 1A). The signal
from the polarizing region is not species-specific. It appears to be active during normal development (Summerbell
1979) and to act over a long range (Honig 1981). Recently
Tickle, Alberts, Wolpert and Lee (1982) have found that high
doses of retinoic acid can mimic the effect of the polarizing
region, causing limb reduplications; I have confirmed this
result (e.g. Fig. 1B). Retinoic acid, which also has dramatic effects on pattern formation along a different axis
(proximo-distal) in the regenerating amphibian limb (Maden
1982) seems to have the effect of resetting positional value
at the graft site to a boundary (proximal or posterior)
value. Equally, it is possible retinoic acid is related to
the morphogen postulated in gradient models. The compound
uniquely seems capable of phenocopying the effects of polarizing region grafts. However, a number of experiments suggest that polarizing region cells need nucleus-associated
information and cell integrity to signal positional information.

The polarizing region need not be grafted as an undisturbed piece of tissue. It can be totally dissociated into
single cells, then reaggregated by centrifugation and grafted to an anterior limb site, successfully signalling the

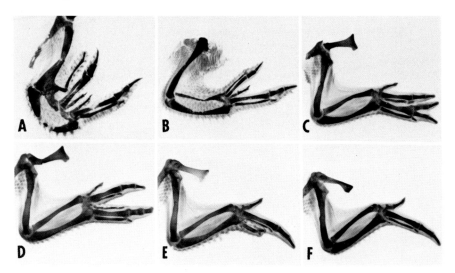

Fig. 1. Duplicated limbs resulting from grafts of reaggregated polarizing region cell pellets (A,C,D,E,F), or a graft
of retinoic acid impregnated filter material (B), opposite
somites 16 (A), 16/17 (B), 17 (C), 17/18 (D), 18 (E), and
19 (F). Mag. ∿4x.

host limb to duplicate its pattern (Saunders 1972; Figs. 1A,
   1C,D,E,F, 2). Intact cell-cell geometric relationships are
unnecessary for signalling: cell cytostructure may be completely disrupted, yet polarizing activity is manifest upon
return of the cells to the limb environment. Polarizing
activity is a cellular not tissue phenomenon. But when I
mixed polarizing region cell homogenates into low gelling
temperature agarose, no positional signalling ensued. Also,
frozen-thawed tissue or cell membrane-type preparations such
as those resulting from Triton X-100 extraction of polarizing
region tissue cells show no activity (Honig, Hornbruch 1982).
Despite the polarizing activity shown by retinoic acid,
attempts have not succeeded to reduce the activity of isolated polarizing region *tissue* to a subcellular level.

SIGNALLING BY POLARIZING REGION CELLS

     Reaggregated cell pellets formed from single cell sus-
pensions of polarizing region mesodermal cells signal as well
as do intact polarizing regions. The full range of dupli-

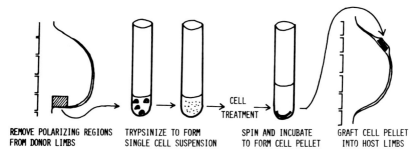

REMOVE POLARIZING REGIONS TRYPSINIZE TO FORM SPIN AND INCUBATE GRAFT CELL PELLET
FROM DONOR LIMBS SINGLE CELL SUSPENSION TO FORM CELL PELLET INTO HOST LIMBS

CELL
TREATMENT

Fig. 2. Reaggregated cell pellet assay of limb polarizing
activity.

cations (Tickle, Summerbell, Wolpert 1975) from 4 3 2 2 3 4,
2 3 4 3 3 4, 2 3 4 4 3 4, to 2 3 4 4 and 2 3 4 is observed
when pieces of cell pellet are grafted at anterior and
increasingly posterior locations on the margin of the host
limb (Fig. 1A,C,D,E,F). Grafts must be performed under the
apical ectodermal ridge to demonstrate full (90-100%) polarizing activity: breaking or removing part of the ridge at the
grafting site causes lower observed activity. Use of reaggregated cell pellets allows grafts of polarizing region to
be spread over a large fraction of the apical limb margin:
an extreme extension of the multiple polarizing region grafts
of Wolpert & Hornbruch (1981) and Summerbell (1981) but without the presence of any intervening responding tissue. Operated limbs might be expected to yield many digits 4 (and 3)
especially if digit pattern formation involved the combination of a periodic antero-posterior signal, and a morphogen
level indicating positional information (e.g. Cooke 1982).
I have grafted strips of reaggregated polarizing region cell
pellet opposite 2-2½ somites antero-posteriorly without
removing host tissue. If placed opposite somites 16 and 17
(16.0-18.0) digit patterns 4 3 3 4 typically result. If the
posterior border of the strip extends into somite 18, digit
patterns 4 3 4 predominate, and if placed still more posteriorly projecting under the ridge opposite somite 19, limbs
of the sort 2 3 4(4) resulted. Thus, grafts with strips
opposite somites 16.0-18.0 or 16.5-18.5 show digit patterns
like those of the digits found posterior to conventional
grafts opposite somite 17/18, although production of any
anterior digits has been suppressed. Strip grafts more posterior (e.g. 17.5-19.5) in which graft polarizing regions are
contiguous with host polarizing regions result in patterns

similar to those found after very posterior polarizing region
cube grafts. Again, the polarizing region cell pellet has
prevented production of extra (posterior) digits. This sup-
pressing activity of the polarizing region cells may be analogous to the inhibition of bud outgrowth observed by Crosby
and Fallon (1975). The strip grafts provided at least two-
thirds of the limb margin with subjacent polarizing activity
(graft plus host) however no excess of posterior digits was
observed. Multiple adjacent digits occurred less often than
with normal, localized (<1 somite), grafts and were limited
to split digits 3 and repeated 4 4. Triplicate digits are
the maximum obtainable from a limb with two polarizing regions;
strip grafts did not produce any of these.

ATTENUATION OF SIGNALLING BY IRRADIATION OF CELLS

    Immersion of polarizing regions in solutions of various
metabolic and biosynthetic inhibitor reduces signalling in a
dose-dependent manner (Honig, Smith, Hornbruch, Wolpert 1981).
The reaggregated cell assay for polarizing activity has been
used to advantage by Honig & Hornbruch (1982) to show that
these drugs act at the cell level. Treatments of cell suspensions followed by cell reaggregation in inhibitor-free media
mostly result in similar inhibition to that observed after
treatment of whole tissue pieces. However, inhibitors of
cytostructural components such as microtubules (colchicine or
vinblastine) or microfilaments (cytochalasin B), adversely
affected polarizing region activity in the conventional assay
but had much milder effects on reaggregated cells. Their
inhibition of polarizing activity seems due to the effects of
drug retention in tissue cubes; cells are more effectively
washed free of agent.

    Attenuation of polarizing activity has been demonstrated
by Smith, Tickle & Wolpert (1978) who showed dose-dependent
loss of the ability to specify extra digits 4,3, and 2 suc-
cessively, following grafting operations in which the donor
eggs had been subject to increasing doses of γ-irradiation.
Although Smith isolated polarizing regions within 15 minutes
of irradiation, it has been mooted that perhaps some systemic
or tissue interactions were involved in maintenance or loss
of polarizing activity. However, the reaggregated cell assay
indicates this is not true. Irradiation of cell suspensions
results in similar inhibition of polarizing activity to irradiation of tissue cubes or whole eggs. For example Smith

found 90%, 90%, 67% and 29% activity at 0, 100, 240, and 640
Gy (1Gy = 100 rad) doses delivered to whole quail eggs, and I
found 95%, 88%, 62% and 0%, activity respectively using cells
from the quail polarizing region, irradiated in suspension.

The effects of γ-radiation are not that specific: a
variety of host cell functions are affected (Honig, Smith,
Hornbruch, Wolpert 1981). Ultraviolet radiation is a more
specific agent , having greatest selectivity for nucleic
acids by at least 100-fold. But the substantial absorption
of ultraviolet by tissue leads to half-attenuation by tissue
of ∿10μm thickness. This precludes effective uniform irradiation of the polarizing region except as a cell suspension.
When this is performed, reaggregated cell pellets of UV-
irradiated suspensions grafted into host chick wings yield a
dose-dependent graded inhibition of polarizing activity similar to that observed with γ-irradiation. The dose required
to reduce activity by a factor of e is ∿18 j/m^2. The extrapolation number on a semi-log plot is ∿3.6, implying that
either the usual graft in these experiments contained an
approximately 4-fold excess of polarizing region cells over
that minimally necessary to cause 4-reduplications or (assuming grafts were about the minimal size) that ∿4 radiation
hits per cell are necessary for loss of polarizing region
activity. Corresponding inhibition of polarizing activity
was produced by 1 j/m^2 ultraviolet for every 32 Gy γ-radiation.
This ratio may be compared to literature studies showing 1
j/m^2 UV causes (a) similar nucleic damage to ∿50 Gy γ-radiation
but (b) comparable cell lethality with 0.3 Gy γ-radiation
(Honig 1982). Thus ultraviolet and γ-radiation eliminate
polarizing region activity at equivalent nucleic acid dosages,
not at equivalent lethal doses. These data, together with that
of biochemical inhibition studies (Honig, Smith, Hornbruch,
Wolpert 1981; Honig, Hornbruch 1982) in which high sensitivity
was observed to transcriptional inhibitors, suggests a nuclear
role in positional signalling.

NUCLEUS AND CYTOPLASM IN SIGNALLING

Many proposed mechanisms of positional signalling involve
cell surface molecules or cytoplasmic functions. But no
homogenized or cell-lysed (e.g. frozen-thawed) preparation
of polarizing region tissue has shown activity. A less disruptive procedure for examining whether positional signalling
can be accomplished without the cell nucleus is to enucleate
cells by centrifugation in warm medium containing cytochalasin

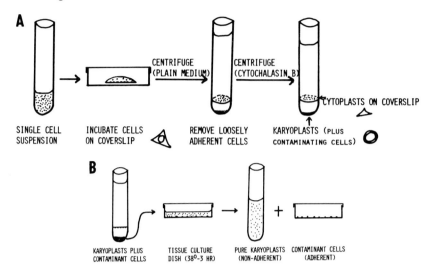

Fig. 3. Enucleation of polarizing region cells (A) followed by purification of karyoplasts (B).

B. Polarizing regions are excised, disaggregated, cells plated onto coverslips, incubated to allow spreading (∿6 hours; Figs. 3, 4A) spun to remove superficially attached cells, and then spun in cytochalasin B-containing medium. Afterwards, the cytoplasts remain on the coverslip (Fig. 4B) and the karyo-plasts (Fig. 4C) are pelleted. Both preparations exclude (>80%) trypan blue, and may be trypsinized, washed, pelleted and grafted into chick host limbs. Grafts of the cytoplasts (enucleated cells) never resulted in any extra digits: occas-ionally single extra nodules of cartilage formed anteriorly, but even if counted as digits 2, the percentage polarizing activity was only ∿5%. Karyoplast preparations, without any purification to remove the few stray cells that contaminated the pellet, yielded polarizing activities of 4-23%. When viable cells were removed from the preparations by incubation on tissue-culture plastic, still ∿10% activity remained. But these purified karyoplast suspensions when cultured at 38°, still produced some viable cells: enucleation of the chick cells creates some regenerable karyoplasts. Other investi-gators have reported (e.g. with L-cells) or denied (e.g. L6, CHO) cell regeneration from karyoplasts. Control cell pre-parations which were either not spun but treated with cyto-chalasin, or spun but in the absence of the drug exhibited 50-85% polarizing activity when otherwise treated in parallel.

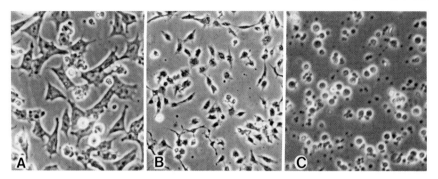

Fig. 4. Polarizing region cells (A) before enucleation, and resulting cytoplasts (B) and purified karyoplasts (C). Mag. 250x.

These experiments suggested that neither cytoplasts nor karyoplasts were capable of signalling on their own. The level of signalling found with karyoplasts was likely due to the contaminating intact cells which were observed, or in the case of the purified preparations, to regenerated cells. The cytoplasts were healthy despite possessing no nucleus: they respread after trypsinization, and they stayed spread for ∿20 hours, a sufficient time for signalling to occur (Smith 1980). To tighten the conclusion that only for lack of their counterpart were cytoplasm and nucleus containing cell fragments unable to signal, it remains to fuse them back together and observe activity. Control cells were fused by means of UV-inactivated Sendai virus or by means of polyethylene glycol solutions and exhibited 16-83% activity. However, fusion of cytoplasts with karyoplasts yielded no polarizing activity. Nor were any viable cells recovered: these cell fusion experiments await perfected technique.

CULTURE OR POLARIZING REGIONS AND THEIR CELLS

The behavior of the polarizing region during *in vitro* culture can show whether signalling ability is a stable cell state outside the embryonic limb environment or whether the capacity is maintained and disappears through tissue or systemic interactions in the embryo. The time course and topographical distribution of polarizing activity in the developing chick wing has been examined by MacCabe, Gasseling & Saunders (1973) and Summerbell and Honig (1982 & *in preparation*). Activity is demonstrable at early limb bud stages,

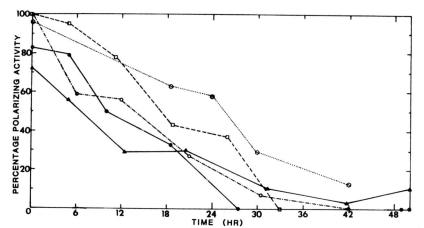

Fig. 5. Decline of polarizing activity during culture at 38°C. Mesodermal polarizing regions from chick (stage 18 ●—·—● , stage 24/25 ▲———▲), quail (stage 18 O—·—··—·—O, stage 22/23 ▫-----▫), and polarizing region cells in monolayer from quail (stage 22/23 ◙···········◙).

and is 100% in the zone of maximal activity for stages 19 through 25. It is present at reduced levels as late as stage 29 (27% max.). If cultured at 4°-24° in serum containing medium (submerged in bacteriological grade plastic culture dishes) polarizing region activity remains at high levels over periods of many days. Wing and leg bud polarizing regions cultured alone, or within limb buds, or as mesodermal cubes of polarizing region devoid of epithelium keep 60-75% activity after 72 hr at 24° (more than half the grafting operations result in 4-duplications). At this temperature, no growth occurs although epithelium shows some spreading over naked mesenchyme.

At 38°, polarizing activity is lost rapidly (Fig. 5). Decay is similar for culture of isolated mesodermal cubes of polarizing region and of monolayers of polarizing region cells. After 30 hr, no 4-duplications and only a rare 3-duplication were observed. The time course of loss of activity was similar whether the polarizing region material originated from stage 18 or stage 24/25 material, through which stages about 30-40 hr elapses. Chick and quail regions also behaved similarly (grafted into chicks). The decay of polarizing activity *in vitro* is much faster than *in ovo* (cf. stage 29, reached 70-80 hr after stage 18). Assuming no improvement

this instability of polarizing region cells provides a timetable for future work at the cellular level. Only short-term
experiments may be conducted at 38°. The lability of the
signalling state suggests that decay is faster than programmed obsolescence since growth *in vitro* is only a fraction of
that *in ovo*. There must be maintaining influences in the
embryo; perhaps the apical ectodermal ridge is involved,
perhaps vasculature or other nutrient supports.

ACKNOWLEDGEMENTS

Supported by Anna Fuller Fund Fellowship #487, the MRC
of Great Britain, and NIH grants DE-02848, DE-07006 and HL-
28325.

REFERENCES

Cooke J (1982). The relation between scale and the complete-
   ness of pattern in vertebrate embryogenesis: models and
   experiments. Amer Zool 22:91.
Crosby GM, Fallon JF (1975). Inhibitory effect on limb
   morphogenesis by cells of the polarizing zone coaggregated
   with pre- or postaxial wing bud mesoderm. Develop Biol 46:
   28.
Honig LS (1981). Positional signal transmission in the
   developing chick limb. Nature (London) 291:72.
Honig LS (1982). Effects of ultraviolet light on the activity
   of an avian limb positional signalling region. J Embryol
   Exp Morphol (in press).
Honig LS, Hornbruch A (1982). Biochemical inhibition of
   positional signalling in the avian limb bud. Differentiation
   21:50.
Honig LS, Smith JC, Hornbruch A, Wolpert L (1981). Effects
   of biochemical inhibitors on positional signalling in the
   chick limb bud. J Embryol Exp Morphol 62:203.
MacCabe AB, Gasseling MT, Saunders JW Jr (1973). Spatiotemporal distribution of mechanisms that control outgrowth and
   anteroposterior polarization of the limb bud in the chick
   embryo. Mech Ageing Develop 2:1.
Maden M (1982). Vitamin A and pattern formation in the
   regenerating limb. Nature (London) 295:672.
Saunders JW Jr (1972). Developmental control of three-dimensional polarity in the avian limb. Ann N Y Acad Sci 193:29.

Smith JC (1980). The time required for positional signalling in the chick wing bud. J Embryol Exp Morphol 60:321.

Smith JC, Tickle C, Wolpert L (1978). Attenuation of positional signalling in the chick limb by high doses of γ-radiation. Nature (London) 272:612.

Summerbell D (1979). The zone of polarizing activity: evidence for a role in normal chick limb morphogenesis. J Embryol Exp Morphol 50:217.

Summerbell D (1981). The control of growth and the development of pattern across the antero-posterior axis of the chick limb bud. J Embryol Exp Morphol 63:161.

Summerbell D, Honig LS (1982). The control of pattern across the antero-posterior axis of chick limb bud by a unique signalling region. Amer Zool 22:105.

Tickle C, Albert B, Wolpert L, Lee J (1982). Local application of retinoic acid to the limb bud mimics the action of the polarizing region. Nature (London) 296:594.

Tickle C, Summerbell D, Wolpert L (1975). Positional signalling and specification of digits in chick limb morphogenesis. Nature (London) 254:199.

Wolpert L, Hornbruch A (1981). Positional signalling along the anteroposterior axis: The effects of multiple polarizing regions grafts. J Embryol Exp Morphol 63:145.

Limb Development and Regeneration
Part A, pages 109–118
© **1983 Alan R. Liss, Inc., 150 Fifth Avenue, New York, NY 10011**

VITAMIN A AND THE CONTROL OF PATTERN IN DEVELOPING LIMBS

Dennis Summerbell and Fiona Harvey

National Institute for Medical Research
The Ridgeway, Mill Hill,
London NW7 1AA U.K.

This paper follows from the discovery of the dramatic effect of Vitamin A on pattern formation (Maden, 1982). When axolotls with limbs amputated through the forearm are kept in a solution of retinol palmitate, instead of regenerating just those parts removed they regenerate an entire new limb. This gives a limb with the original stump elements, plus the regenerated limb, so that there is tandem repetition along the proximo-distal (PD) axis. We set out to obtain the same results in a developing system, the embryonic chick limb. Instead we find that while Vitamin A does affect development, it seems to act primarily on the antero-posterior (AP) axis.

SYSTEMIC EFFECTS

Retinol Palmitate

The simplest approximation to Maden's experiment is to add Vitamin A to the egg. We windowed and staged (15-28, Hamburger & Hamilton, 1951) 254 embryos, then added 0.1 ml of Hank's BSS containing retinol palmitate (Sigma) at concentrations from 10^3 to 10^5 iu/ml. Survival at day 10 was dose dependent. Even at the highest dose there were no obvious general defects and limbs appeared normal. The proportions of individual skeletal elements (Summerbell 1978) lay within normal limits (fig. 1a). There were two exceptions. Fig. 1b shows a reduced wing reminiscent of x-irradiation at st. 13, or of inserting a barrier between responding tissue and the posterior limb organiser (ZPA). The second embryo had supernumerary digits on both wings (fig. 1c). Both these

abnormalities are more typical of an effect on the AP axis
rather than the PD axis.   No results resembled those obtained
in Maden's experiments.

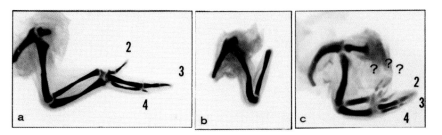

Fig. 1. Dropping on retinol palmitate: (a) no effect (normal
limb), (b) reduced wing, (c) supernumerary digits.

Retinoic acid

     Maden (1982b) has shown that retinoic acid has higher
activity than retinol palmitate.   We added retinoic acid
(Sigma) dissolved in 5µl of DMSO to 0.1ml BSS to give a
concentration between 10µg to 1000µg/ml.   Survival at day
10 was again dose dependent, and there were no defects
except in one case (at the highest survivor/dose level) with
a beak defect;   otherwise the embryos and limbs were normal.

     We also placed small quantities (<1µg/egg) of retinoic
acid in direct contact with the bud (see below).   Despite
the lower dose these normally produced systemic effects,
including the same beak abnormality.   Dropping on may not
be a good method of delivering of retinoic acid.   (But see
Summerbell and Hornbruch, 1981.)

THE PROXIMO-DISTAL AXIS

     The simple analagous experiment failed to reproduce
Maden's results.   We therefore tried a number of variations
aimed at detecting analagous effects on the PD axis.

The Effect of Retinol Palmitate on a Truncated Stump

     When the tip of a wing bud is removed there is no
regulation and it forms a truncated stump.   We removed
the distal tip (st.18-24) at the level of the sub-apical

sinus (approximately 150μm from tip), then added 0.1 ml BSS
with ret.palm. at approximately 10 iu/ml (n=49). 16 survived
and the level of truncation was within the normal range of
control experiments. The single exception had a fairly
normal hand. This could indicate stimulation of regeneration
but as there is only one case it is conceivable that the
excised tip had accidentally reattached. Our results so far
provide no strong evidence for an effect on the PD axis.

Intercalary Regeneration of Defects

The limb bud has a limited ability to regulate intercalary defects along the PD axis (Hornbruch, 1980). As Maden's
experiments could be interpreted as showing a type of 'super-
intercalation' we investigated this phenomenon in the chick
in the presence of retinol palmitate.

We removed the distal tip of the wing bud (200-400μm)
from the stump, then replaced the tip. We chose stages
(20-22) and position of defect (presumptive elbow) to cover
the period during which the bud changes from being regulative
to mosaic. We either added 0.1 ml of BSS containing 5000 iu/
ml after operating (n=27), or soaked the tip for 4-8h in 300
or 600 iu/ml ret.palm. in BSS before pinning back to the
stump (n=22). Control experiments without Vitamin A give a
defect at the elbow involving humerus and/or radius-ulna,
the size of the defect being proportional to the size of the
deletion. We had similar results except for two cases in
which there was no distal tip which had presumably fallen
off or died. There was no evidence for regulation along the
PD axis, but there were two cases with an intercalary defect
and an extra digit 2. This occasionally results from similar
control experiments and is suggestive of modification of
the AP axis.

Soaking Isolated Tips

We also tried removing a distal tip, soaking it in
Vitamin A for several hours, and then returning it to its
original stump at st. 18-24 (n=164). We soaked for 4,8 or
24h and used: ret.palm. in BSS at a concentration of 35 iu/ml
to 1800 iu/ml, and retinoic acid in DMSO:BSS 1:19 at 16-
500μg/ml. The survivors (60%) fell into two main groups:
normal limbs (39%) and loss or reduction of elements (18%).
The latter result was more frequent at high dose or longer
soaking. It is not clear whether the loss of distal parts

is due to disruption of the AP axis or the PD axis. In 5
cases (3%) however there were extra digits (e.g.fig.2a).
None of our experiments on the PD axis gave a consistent
effect analagous to Maden's.

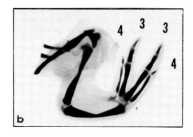 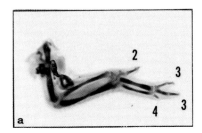

Fig.2. (a) extra digit 3 after soaking tip in ret.palm.
(b) mirror image supernumerary after retinoic acid injection.

THE ANTERO-POSTERIOR AXIS

We will discuss the AP axis in the context of control by
a posterior organising region, the zone of polarising activity
(review Summerbell and Honig, 1982). Much of what we say
would be equally relevant in the context of the Polar Coordinate model (review, Iten, 1982). The results so far did not
suggest any PD effect. The extra digits and the lost digits
reminded us more of experiments interpreted as involving the
ZPA (see above). Our treatments had all affected the tip
uniformly, and when ZPA cells are scattered throughout the
bud (Crosby and Fallon, 1975), they have an inhibitory effect.
We therefore looked for ways of imitating a local ZPA graft.

Injection of Vitamin A

We injected a small bubble of Vitamin A under the apical
ridge at the anterior margin of the limb at st.18-24. The
analogue, solvent, concentration range and results are shown
in table 1. Retinoic acid dissolved in DMSO gave us four
limbs with mirror image supernumeraries (e.g.Fig. 2b) and
five limbs with reduced distal parts that were reminiscent of
an effect on pattern formation across the AP axis. The reduced
limbs were only obtained at the high end of the dose range but
supernumerary limbs were also obtained at the lower end. The
total proportion of affected limbs was very much less than one
would expect from a ZPA graft where the proportion of supernumeraries in the right conditions approaches 100%.

Table 1: Results of injection of Vitamin A analogs

	total	dead	normal	lost radius	lost digits	super-numerary
Retinol palmitate:						
Oil/DMSO 10^6 iu/ml	17	7	8	2		
BSS 10^4 - 10^5 iu/ml	16	5	11			
Retinoic acid:						
Oil 8 - 32 mg/ml	16	4	12			
DMSO 100 mg/ml	21	7	5	3	4	2
DMSO/oil 8-64 mg/ml	25	9	11	2	1	2

Assorted carriers

        We next sought a more reliable method of delivery of
the Vitamin A. We tried (a) small pieces of limb tissue
soaked in 16-250μg/ml retinoic acid in DMSO/BSS for 4-8h;
(b) 300^3 μm agar soaked in 12mg/ml retinoic acid in DMSO;
(c) 300^3 μm or 100x300x300 μm of silastin (Dow Corning) into
which 64mg/ml retinoic acid had been stirred before setting.
In each case the carrier was grafted to the anterior margin
of the host limb at st. 18-21. Both the silastin and the
agar gave consistent results: either the loss of digits
(fig.3a), or the formation of mirror image supernumeraries
(fig.3b, arrow = silastin). Silastin was easier to handle.

Table 2: Application retinoic acid using various carriers

	total	dead	normal	lost radius	lost digits	super-numerary
Ant. limb tissue	50	14	18	18		
DMSO/BSS in agar	20	8	2		7	3
Silastin	38	19	1		9	9

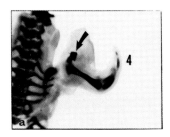 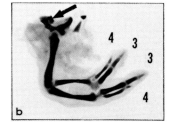

Fig. 3. Results from retinoic acid in silastin carrier:
(a) reduced limb, (b) mirror image supernumerary.

Paper as a carrier

At this time we learned that a group at the Middlesex
Hospital were coincidentally working on Vitamin A using a
paper carrier (subsequently published as Tickle et al. 1982).
To allow better comparison we decided to try paper. We
examined several possible types and finally settled on
newsprint (Richmond and Twickenham Times). We carried out
a series of operations inserting a 500μm square of paper
soaked in retinoic acid (0.25 to 64mg/ml DMSO) into the
anterior margin of the limb bud (st.17-22). The survival
rate (69%) is similar to that for control experiments
without retinoic acid (71%) nor does it vary with stage or
dose.

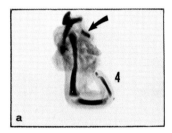 

Fig.4. Results from retinoic acid with paper carrier:
(a) reduced limb, (b) mirror image supernumerary.

There was a wide range of results varying with stage
and dose. Late stage and low dose tend to give normal limbs.
Early stage and high dose often caused total loss of limb
(Fig.4a). In between we observed loss of digits or
supernumerary digits (Fig.4b). The range of supernumeraries
was reminiscent of the dose/response relationship from an
"attenuated" ZPA graft (Smith et al. 1978). The attenuation
data led to the idea of the "strength of activity index"
(Honig et al. 1981). One interpretation of the reduced
limbs is that even higher "enhanced" ZPA activity can raise
the signal above the level that can ever naturally occur so
that even posterior "host" digits are flooded out in an AP
sequence. We have therefore used a modified "strength of
activity index" on a scale from 0-6 with reduced limbs at
the high end. Using this method (among others) we can show
a strong relationship (Fig.5) between the stage and the
result (early stage, high activity) and perhaps between the
dose and the result (high dose, high activity). Analysis
of variance showed a significant effect of stage but not dose.

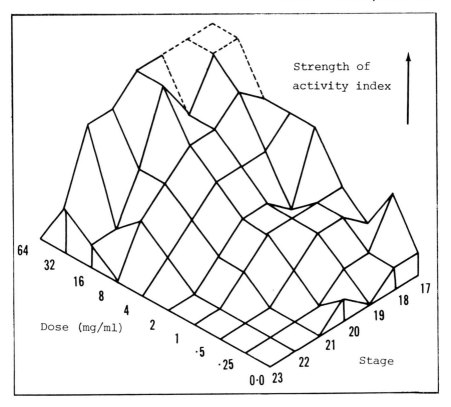

Fig.5. 3-D graph showing strength of activity index (0-6) varying with stage and dose of retinoic acid.

MODE OF ACTION

At least in some ways local application of retinoic acid mimics the action of a graft of the posterior limb organiser (ZPA). The following experiments are based on initial exposure of a st.18-19 limb bud to 500^2 μm paper-soaked in 16 mg/ml retinoic acid in DMSO.

Required Exposure Time

If a ZPA is grafted to the limb bud then later removed it requires 12h minimum exposure before any supernumerary digits are formed, and 24h before a digit 4 is found next to the graft. Retinoic acid gives very similar results.

Table 3: Result of varying periods exposure retinoic acid.

exposure	total	dead	normal	digit 2	digit 3	digit 4	Lost digits
					digit nearest graft		
4 h	6	1	5				
8 h	6	2	3	1			
18 h	6	3			1	1	1
24 h	28	11	3	1	1	4	8

It requires longer exposure to lose host digits than to gain supernumeraries. This is in accord with the notion above that normal ZPA activity gives supernumeraries while "enhanced" activity suppresses host digits. An example of a limb with digit 3 next to the graft after 18 h exposure is shown in Fig. 6a. The paper removed can then be inserted into a second host. So far in 13 cases, at 18h or later, we have always obtained normal limbs. The retinoic acid seems to be available for only a short time.

Direct or Indirect Action

Retinoic acid may act directly on limb tissue inducing supernumerary digits or it may convert adjacent tissue into an effective limb organiser. We tested this idea by removing tissue adjacent to a paper graft after 24h then inserting it into the anterior margin of a st.19 host limb. Preliminary results suggest that retinoic acid may indeed cause host tissue to form a limb organiser. In some cases (Fig.6b) we obtained good mirror image supernumeraries. It is possible that in other cases the extra structures were caused by self differentiation of the graft or by transfer of retinoic acid in the graft.

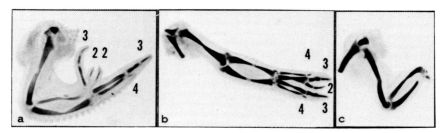

Fig.6. Retinoic acid in paper carrier: (a) result if paper removed after 18h, (b) graft tissue adjacent to paper to host limb after 24h exposure, (c) insert paper in host posterior margin.

ZPA Enhancement

If a second ZPA is grafted to the posterior margin it has no effect on the development of the bud. One interpretation of this is that the ZPA acts as a homeostatic source of morphogen (Summerbell and Honig, 1982). A second is that there is insufficient discontinuity in positional value to cause intercalation (Iten, 1981). However, application of retinoic acid to the same site at dose levels causing supernumeraries elsewhere almost invariably causes loss of digits near the graft (Fig. 6c).

Fig.7. Limb bud 48h after paper with retinoic acid graft: (a) operated side, gave girdle but no wing, (b) control side.

Growth control

We have been examining particularly carefully those cases in which we lose the whole limb. The bud continues to grow apparently normally for many hours after the operation. It can be identical in size and stage to the contralateral control at 24h. At 48h it is still growing, though the stage and size may be reduced compared with the control side (Fig.7). At day 10 the result was the loss of the entire limb with only the scapula remaining.

CONCLUSION

So far it seems that Vitamin A affects amphibian regeneration and chick limb development differently (Maden et al. 1982). In the former it causes tandem reduplication along the PD axis. In the latter there is so far no evidence for an effect on the PD axis. With respect to the AP axis it appears to have two major effects. It mimics the effect of a ZPA graft, perhaps by inducing an organising region next to the source. At high doses it causes original host

digits to be lost. We have tried to explain these paradoxical
results in the text using the notion of 'enhancement'.
However as we have now shown diverse effects in regenerating
axolotl and in developing chick limbs, there seems no reason
to insist on a single mode of action within any one system.

Crosby GM, Fallon JF (1975). Inhibitory effects on limb morphogenesis by cells of the polarizing zone coaggregated with
pre- or post-axial wing bud mesoderm. Develop Biol 46:28.
Hamburger V, Hamilton HL (1951). A series of normal stages
in the development of the chick embryo. J Morph 88:49.
Hornbruch A (1980). Abnormalities along the proximo-distal
axis of the chick wing bud: the effect of surgical intervention. In Merker H, Nau H, Neubert D (eds): "Teratology
of the limbs", Berlin: De Gruyter, p 191.
Honig LS, Smith JC, Hornbruch A, Wolpert L (1981). Effects
of biochemical inhibitors on positional signalling in the
chick limb bud. J. Embryol exp Morph 62:203.
Iten LE (1981). Pattern specification and pattern regulation
in the embryonic chick limb bud. Amer Zool 22:117.
Maden M (1982). Vitamin A and pattern formation in the
regenerating limb. Nature 295:672.
Maden M (1982). Vitamin A and the control of pattern in
regenerating limbs. This Symposium.
Maden, M, Gribbin, MC, Summerbell D (1982). Axial organisation in developing and regenerate vertebrate limbs. In
Goodwin BC, Holder N (eds): "Development and Evolution"
Cambridge. Cambridge University Press, in the press.
Smith, JC, Tickle C, Wolpert L (1978). Attenuation of
positional signalling in the chick limb by high doses of
irradiation. Nature 272:612.
Summerbell D (1978). Normal and experimental variations in
proportions of skeleton of chick embryo wing.Nature 274:472.
Summerbell D , Honig LS (1981). The control of pattern
across the antero-posterior axis of the chick limb bud by
a unique signalling region. Amer Zool 22:105.
Summerbell D, Hornbruch A (1981). The chick embryo: a standard against which to judge in vitro systems. In Neubert D,
Merker H (eds): "In vitro studies of prenatal differentiation and toxicity", Berlin: De Gruyter, p 529.
Tickle C, Alberts B, Wolpert L, Lee J (1982). Local
application of retinoic acid to the limb bond mimics the
action of the polarizing region. Nature 296:564.

Limb Development and Regeneration
Part A, pages 119–130
© 1983 Alan R. Liss, Inc., 150 Fifth Avenue, New York, NY 10011

A FREEZE-FRACTURE AND MORPHOMETRIC ANALYSIS OF GAP JUNCTIONS
OF LIMB BUD CELLS: INITIAL STUDIES ON A POSSIBLE MECHANISM
FOR MORPHOGENETIC SIGNALLING DURING DEVELOPMENT

Robert O. Kelley[1] and John F. Fallon[2]
Departments of Anatomy, University of New Mexico
School of Medicine,[1] Albuquerque, New Mexico 87131,
and University of Wisconsin,[2] Madison, Wisconsin 53706

Fundamental to the problem of spatial patterning within
limb mesoderm is the identification and characterization of
structural mechanisms which mediate intercellular communication between cells of the mesoderm and within the over
lying ectoderm, part of which develops into the apical ridge.
The presence of gap junctions has been correlated with intercellular communication(see Loewenstein, 1981, for review).
The presence of these structures coincides with both lowered
electrical resistance between cells, and with the flow of
small molecules and ions from cell to cell.

Furshpan and Potter (1968) were among the first to
demonstrate that intercellular communication may be an important regulatory requirement in differentiation and morphogenesis. Since then, several investigators have suggested
that gap junctions may be assembled between contacting cell
membranes but may not be permeable to metabolites and ions
(Casper, et al., 1977). Consequently these junctions may
have the potential to serve as regulators in the mediation
of intercellular signalling (Albertini, et al., 1975). Most
recently, Peracchia (1980) has reported that divalent cation
concentration in cells may play a role in effecting
reversible changes in structure of junctions, suggesting
that junctions with closely aggregated particles (approximately 8.5 nm center-to-center spacing) are not permeable
(uncoupled), whereas a slight dispersion of particles (to
approximately 10.5 nm center-to-center spacing) in the
hydrophobic plane of the membrane is a structural indication
of permeability (i.e., cellular coupling).

Using techniques of electron microscopy, we have reported the presence of numerous, large gap junctions in the inductively active apical ectoderm (Kelley, Fallon, 1976; Fallon, Kelley, 1977) which distinguish it from adjacent dorsal and ventral ectoderm. In addition we have shown (Kelley, Fallon, 1978) that mesodermal cells in the subridge, core, anterior and posterior borders of developing chick limbs also have the structural capability for electronic and metabolic coupling (Gilula, et al., 1972). In this paper, we report initial observations on the use of morphometric analyses of gap junctions in replicas of limb ectoderm and mesoderm from stage 22-24 chick embryos following freeze-fracturing and electron microscopy. Based on ultrastructural organization, we conclude that the apical ridge is not coupled with adjacent dorsal and ventral ectoderm, but is free to integrate metabolic and electrotonic signals only within the body of the ridge. In addition, it appears that gap junctions in subridge and core mesoderm may be functionally coupled, whereas the structural appearance of junctions in pre- and postaxial mesoderm suggest that many cells may be functionally uncoupled.

MATERIALS AND METHODS

Fertilized eggs (White Leghorn; 46-53 g per egg; Sunnyside Hatchery, Oregon, Wisconsin) were maintained at 38°C for 3.5 days and opened by the technique described by Zwilling (1959).

For transmission electron microscopy, wing buds (stages 22-24; Hamburger, Hamilton, 1951) were dissected from embryos and fixed for 2 h (4°C) by immersion in a fixative composed of 0.02% trinitrophenol; 2.0% formaldehyde; and 3.0% glutaraldehyde buffered to pH 7.2 with 0.075 M phosphate buffer. Specimens were washed in buffer, postfixed for 2 h (4°C) in 2.0% osmium tetroxide (pH 7.4, 0.1 M S-collidine buffer), stained en bloc with 0.5% aqueous uranyl acetate for 30 min, rapidly dehydrated in ethanol and embedded in Epon 812 prior to thin sectioning and staining.

For freeze-fracturing, buds were placed in Ham's F-10 medium and dissected into four pieces: (a) mesoderm immediately subjacent to the apical ridge; (b) mesoderm of the anterior border; (c) core mesoderm; and (d) mesoderm of the posterior border. In addition, specimens of apical ridge ectoderm were also retained following removal of wing buds from embryos. Tissues were immersed for 30 min in 2.0% glutaraldehyde (pH 7.4 in 0.1 M cacodylate buffer; 2 h; room temperature). Tissues were mounted on gold stubs, rapidly frozen in the liquid phase of partially solidified Freon 22 (monochlorodifluoromethane) and stored in liquid nitrogen. Frozen specimens were fractured at $-115°C$ and replicated without etching in a Balzers apparatus. Replicas were cleaned with methanol and bleach and mounted on uncoated copper grids. All specimens were examined in either an Hitachi HU-11C, AEI 801 or Philips 200 electron microscope.

For morphometric analyses of intramembrane organization, at least seven replicas were examined for the apical ridge and each region of mesoderm. Replicas were photographed at low magnification to permit determination of the number of cells included in each replica. Total numbers of junctions observed per number of cells examined in each region were recorded and the number of particles ("connexons") in junctions in each region were counted directly and averaged. In addition, the average center-to-center spacing between adjacent particles was obtained by calculating the average particle periodicity along each of three rows of particles oriented at approximately $120°$ to each other and by making a final average of the three means. In junctions containing irregularly packed particles, equal numbers of center-to-center spacings were measured in three directions at $120°$ to each other and average spacing was determined. It should be mentioned that because of the variable orientation of junctions in tissues, the fracture planes along which some of the junctions are split may be tilted at variable angles to the axis of the electron beam. This tilt will introduce errors in measurement of particle periodicity. However, since all specimens are subjected to the same experimental artifacts, errors of calculation tend to cancel out.

In this note, the terms P-face and E-face indicate intramembrane particles which remain associated with the inner and outer portions, respectively, of the fractured cell membrane (Branton, et al., 1975). Illustrations of replicas are presented with the platinum shadow direction approximately from the bottom to the top of the plate.

OBSERVATIONS

A transverse section through the wing bud of a stage 23 chick embryo (fig. 1) demonstrates the structure of the apical ridge; a thickened, pseudostratified columnar epithelium composed of closely aggregated cells and overlying periderm. Thin sections (fig. 2) reveal numerous gap junctions, approximately 0.75 to 2.0 μm in sectioned length, at apposed lateral surfaces. Junctions are most numerous between cells at the basal lamina. Fewer and smaller junctions were seen in more distal portions of the apical ridge. No junctions were observed between ridge cells and the periderm. Higher magnification (fig. 3) illustrates the pentalaminar appearance of gap junctions which exhibit a 2 to 4 nm "gap" between outer surfaces of apposed cell membranes and a center-to-center spacing between connexons of approximately 10.5 nm.

Replicas of limb buds freeze-fractured through the apical ridge permit examination of the dimensions and organization of gap junctions viewed en face. Figure 4 illustrates the lateral surface of a ridge cell in contact with a neighboring cell. The fracture plane reveals P-face particles that are approximately 10.0-10.5 nm in diameter and aggregated into clusters which are separated from other groups of particles by "aisles." Several hundred particles comprise each junction located within apical ridge tissue.

In contrast, adjacent dorsal and ventral ectoderm exhibit occasional small gap junctions composed of 35-40 connexons and exhibiting center-to-center spacing of 8.5 to 9.0 nm. These junctions generally measure less than 0.2 μm in diameter.

Table I combines comparative data reflecting the structural variations in the organization of gap junctions between apical ridge and non-ridge ectoderm. It is significant to note the variance in size, particle number and center-to-center spacing of particles in junctions within the apical ridge.

In the mesoderm, we have reported that gap junctions were observed between apposed cell bodies; at points of contact between cell processes and other mesenchymal cell bodies; and between contacting tips of slender cell projections (Kelley, Fallon, 1978). Figure 5 shows mesenchymal cells

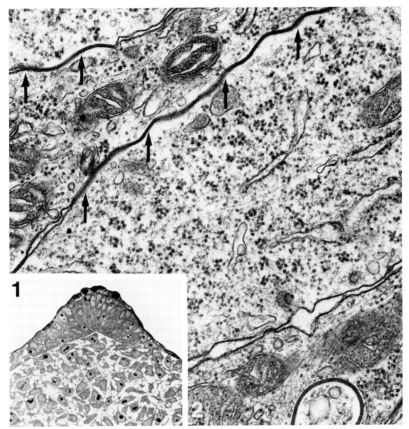

Figure 1. The apical ectodermal ridge in a stage 23
    chick embryo consists of a pseudostratified, columnar
    epithelium capped by a layer of periderm. All basal
    cell surfaces contact a continuous basal lamina. X 400.

Figure 2. All cells within the chick apical ridge are
    coupled by gap junctions (arrows). It is important to
    note that most, gap junctions between ridge cells are
    located along the basal surfaces of cells within the
    pseudostratified tissue. X 35,000

immediately beneath the apical ridge. Cellular processes
extend both towards the epithelial surface and towards
adjacent cells within the limb mesoderm. In addition
processes are known to penetrate deep into the limb mesoderm
forming contacts with other cells considerably removed from

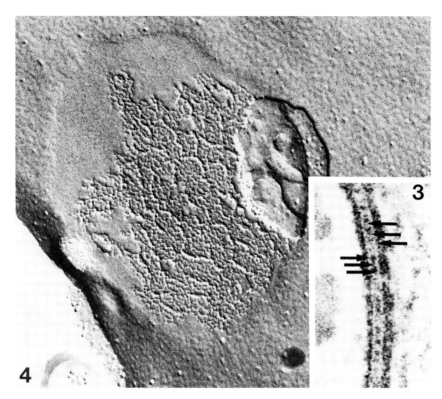

Figure 3. Higher magnification of gap junction between
    ridge cells. Apposed cell membranes are separated by
    a 2–4 nm "gap." Individual connexons (arrows) are
    separated by center-to-center spacing of approximately
    10.0 to 10.5 nm. X180,000.

Figure 4. Replica revealing gap junction particles in the
    P-face of a freeze-fractured membrane of a ridge cell.
    Particles are clustered in aggregates with center-to-
    center spacing of approximately 10.5 nm. Note "aisles"
    separating aggregates. X 100,000.

the immediate subridge zone. Figure 6 depicts the E-face
of the tip of a cellular process, revealing a pattern of
pits where connexon particles have been removed by the
fracturing process.

    Junctions between mesodermal cells in all four regions
examined are considerably smaller and exhibit fewer

Table I

	Chick Limb Ectoderm (St 22-24)		
	Apical Ridge	Dorsal	Ventral
Number of specimens examined	15	13	14
Number of replicas examined	11	9	10
Gap junctions observed/total number of cells examined*	87/165	18/158	16/147
Number of junctions observed/100 cells	52.7	11.3	10.8
Average number of particles/gap junction	<500	51.3±5.6	49.8±6.3
Average center to center spacing	10.4±0.3 nm	9.0±0.1 nm	8.9±0.1 nm

*Data represents the total number of gap junctions observed/total number of cells examined in a minimum of nine replicas of each region of limb ectoderm

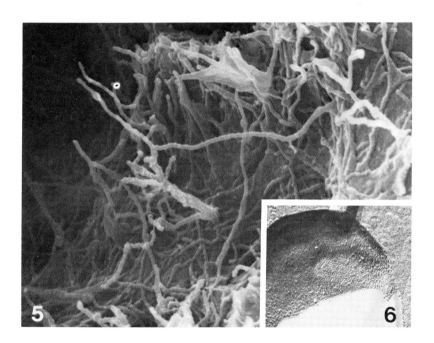

Figure 5.   Scanning electron micrograph of mesenchymal cell surfaces immediately beneath the apical ridge.   Cellular processes extend both towards the epithelial surface and towards cells within the deeper limb mesoderm. X 20,000.

Figure 6. Replica of fracture plane at tip of mesenchymal
cell process. Pattern of pits in the E-face of the cell
membrane denotes a gap junction at the tip of the
projection. X 45,000.

component particles than cells in the apical ridge (see
Table II for comparison). In addition, careful examination
revealed two general classes of organization within each
region of mesoderm. The first, illustrated in figure 7,
consisted of P-face particles clustered in small aggregates,
which were separated from each other by aisles. Center-to-
center spacing in these junctions approximated 10.0 to 10.5
nm. The second class was a group of junctions which did not

Table II

| | Chick Limb Mesoderm (St 22-24) | | | |
	Tip	Core	Preaxial	Postaxial
Number of specimens examined	11	10	11	12
Number of replicas examined	8	9	7	8
Gap junctions observed/total number of cells examined*	31/256	7/82	16/183	11/151
Number of junctions observed/100 cells	12.1	8.5	8.7	7.3
Average number of particles/gap junction	42.4±7.6	38.3±5.2	37.4±6.5	34.5±4.7
Average center to center spacing	10.4±0.3 nm	10.0±0.7 nm	9.1±0.3 nm	8.6±0.1 nm

*Data represent the total number of gap junctions observed/total number of cells examined in a minimum of
seven replicas of each region of limb mesoderm

show separate aggregates with aisles, and which demonstrated
a center-to-center spacing of particles less than 9.0 nm
(fig. 8). The latter group was generally composed of fewer
connexons and was subsequently smaller in diameter than
junctions in the other class. Table II reveals that
mesenchymal cells in the subridge exhibit slightly more
junctions than cells in either the core or pre- and
postaxial regions. In addition, junctions in the subridge
and core mesoderm exhibit greater center-to-center spacing
and, in general, greater degrees of particle dispersion than
do junctions in the pre- and postaxial mesoderm. It should
be noted that cells in the postaxial border exhibit both the
fewest gap junctions, and the highest degree of particle
packing into a tight, hexagonal lattice (fig. 9).

DISCUSSION

The gap junction has now been clearly demonstrated in both ectoderm and mesoderm during limb morphogenesis.

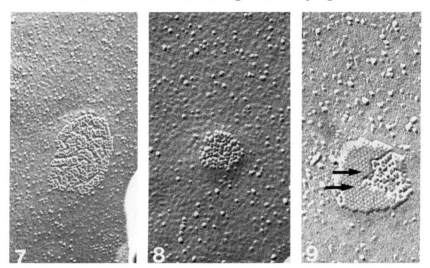

Figure 7. Gap junction in membrane of mesenchymal cell in subridge mesoderm. Particles are clustered into groups separated by particle free aisles. Center-to-center spacing averages approximately 10.5 nm. X 60,000.

Figure 8. Gap junction in membrane of cell located in the postaxial mesoderm. P-face particles are closely aggregated with center-to-center spacing averaging approximately 8.5 to 9.0 nm. Note absence of aisles, suggesting a closed, uncoupled junction. X 65,000.

Figure 9. Gap junction in membrane of cell in preaxial mesoderm. Both P- and E-faces are present from apposing cell membranes and the tight lattice pattern of particle packing is evident in the rows of E-face pits (arrows). X 75,000.

Based on the observations of Peracchia (1980) and Loewenstein (1981), it is becoming equally clear that gap junctions, once assembled, may be capable of opening and closing, thus regulating the flow of metabolic and electrotonic information which is transferred between cells in the course of normal development. The observations of our present study suggest that coupling and uncoupling of gap junctions occurs during limb morphogenesis and that these events may play a central role in intercellular signalling which is required for normal patterning to occur.

In the limb ectoderm, our observations indicate that gap junctions increase both in size and quantity in the apical ectodermal ridge during a developmental period when epithelial-mesenchymal interactions essential to normal limb morphogenesis occur. Compared to adjacent non-ridge ectoderm, the membranes of cells in the apical ridge are different in their response to environments (i.e., signals) created by the subjacent mesoderm. It seems reasonable to point out that the abundance of gap junctions in the apical ridge makes it likely that ridge cells are extensively coupled, both metabolically and electronically. To demonstrate this point, Fallon, Sheridan and Clark (in preparation) injected Lucifer Yellow CH dye into both ridge and non-ridge ectoderm. Dye transfer was extensive between cells of the apical ectodermal ridge but could not be demonstrated to transfer between ridge and adjacent dorsal and ventral ectoderm. These results are in concert with the respective structural organization of gap junctions in ridge and non-ridge ectoderm included in the present report. Junctions within the apical ridge are in a permeable or "coupled" configuration, whereas those in the non-ridge limb ectoderm appear to be in an impermeable or "uncoupled" configuration.

In limb mesoderm, gap junctions are observed at all potential sites of cell-to-cell contact in the four regions examined. In addition, it is important to realize that contact and potential coupling of cells several cell diameters distant from one another exists in the subridge mesoderm. This is less common in other regions of the limb mesoderm where shorter cellular projections extend to neighboring, more closely associated cells. The intriguing observations of the present study suggest that, although junctions may assemble at points of contact between mesenchymal cells, the junction may not be permeable to the transfer of metabolites and ions between cells at all times. Furthermore, it is not unreasonable to suggest that the coupling and uncoupling of intermembrane channels may play an important role in regulating the complex signalling which must occur between mesenchymal cells to define position within the limb and to specify subsequent events of differentiation contributing to individual limb parts. In this regard, it is noted that junctions in the postaxial limb border (which contains the polarizing zone and prospectively necrotic cells) exhibit junctions with uncoupled structural configuration, whereas cells in the subridge region contain the greatest number of junctions in the coupled mode.

Finally, it is clearly premature to speculate on the precise functions of gap junctions within the inductively active epithelium and mesenchyme of early limb development. However, one must consider that the initiation of pattern and, ultimately, differentiation may be related to the assembly, maintenance and regulation of gap junctions which permit constant metabolic or electrotonic signalling between cells. Alternatively, interruption or reduction of signals may also be key components in the mechanisms of development. In this context, we have suggested the hypothesis that gap junctions are regulating as well as communicating structures for intercellular signals in the developing limb and that the apical ridge may ultimately condition the strength of the signal. Coupling of junctions may be the basis for maintaining the embryonic nature of the subridge mesoderm, whereas too rapid uncoupling, as a result of ridge removal or dysfunction, may evoke cell death in subridge mesoderm (Rowe, et al., 1982). In addition, regulation of subtle changes in signals through coupling and uncoupling of gap junctions may be a central mechanism in establishing the cell behavior which leads to differentiation (see the progress zone theory of Summerbell, et al., 1973).

ACKNOWLEDGMENTS

This investigation was supported by NIH grants AG 00191 and HD07402 to Robert O. Kelley, and by NSF grant PCM 8205368 to John F. Fallon.

LITERATURE CITED

Albertini DF, Fawcett DW, Olds PJ (1975). Morphological variation in gap junctions of ovarian granulosa cells. Tissue Cell 7:389-402.
Branton D, Bullivant S, Gilula NB, Karnovsky MJ, Moor H., Muhlethaler K, Northcote DH, Packer L, Satir B, Satir P, Speth V, Staehelin LA, Steere R, Weinstein RS (1978). Freeze-Fracture Nomenclature. Science 190:54-56.

Caspar D., Goodenough DH, Makowski L, Phillips WC (1977). Gap junction structure. I. Correlated electron microscopy and X-ray diffraction. J Cell Biol 74:605-628.

Fallon JF, Kelley R (1977). Ultrastructural analysis of the apical ectodermal ridge during vertebrate limb morphogenesis ii. Gap junctions as distinctive ridge structures common to birds and mammals. J Embryol exp Morph 41:223-232.

Furshpan EJ and Potter D (1968). Low resistance junctions between cells in embryos and tissue culture. Curr Top Devl Biol 3:95-127.

Gilula NB, Reeves OR, Steinbach A (1972). Metabolic coupling, ionic coupling and cell contacts. Nature 235:262-265.

Hamburger V, Hamilton H (1951). A series of normal stages in the development of the chick embryo. J Morph 88:49-92.

Kelley RO, Fallon JF (1976). Ultrastructural analysis of the apical ectodermal ridge during vertebrate limb morphogenesis. I. The human forelimb with special reference to gap junctions. Devel Biol 51:241-256.

Kelley RO, Fallon JF (1978). Identification and distribution of gap junctions in the mesoderm of the developing chick limb bud. J Embryol exp Morph 46:99-110.

Loewenstein WR (1981). Junctional Intercellular Communication: the cell-to-cell membrane channel. Physiol Rev 61:829-913.

Peracchia C, Peracchia LL (1980). Gap junction dynamics: reversible effects of divalent cations. J Cell Biol 87:708-718.

Rowe DA, Cairns JM, Fallon JF (1982). Spatial and temporal patterns of cell death in limb bud mesoderm after apical ectodermal ridge removal. Devel. Biol. 93 (in press).

Summerbell D, Lewis JH, Wolpert L (1973). Positional information in chick limb morphogenesis. Nature 244:492-496.

Zwilling E (1959). A modified chorioallantoic grafting procedure. Transplant Bull 6:238-247.

Limb Development and Regeneration
Part A, pages 131–140
© **1983 Alan R. Liss, Inc., 150 Fifth Avenue, New York, NY 10011**

REGULATION ALONG THE ANTERO-POSTERIOR AXIS OF THE CHICK WING
BUD

B.L. Yallup & J.R. Hinchliffe

Zoology Department, University College of Wales,
Aberystwyth, Wales, U.K. SY23 3DA

Regulation in the chick wing bud is most often assessed
on the basis of skeletal pattern. The development of the wing
skeleton involves successive phases of condensation, chondrogenesis and finally ossification. The skeleton is laid down
within the mesenchyme in a proximal to distal direction as the
wing bud elongates under the inductive influence of the apical
ectodermal ridge (AER). The stage at which the mesenchyme
cells become irreversibly committed to chondrogenesis is a
crucial issue in analysis of chick wing morphogenesis. The
timing of such determination is important in relation to the
models of limb development intended to explain control of
differentiation along the proximo-distal axis (the progress
zone hypothesis, Summerbell et al 1973) and along the antero-
posterior axis (the zone of polarizing activity hypothetically
controlling this axis, Tickle et al 1975).

One indicator of determination is loss of regulative
capacity. Along the proximo-distal (P-D) axis, regulation has
been demonstrated up to stage 22 of development (Summerbell
1977; Hornbruch 1980), and Kieny (1977) has shown gradual loss
of regulative capacity between stages 22 and 24. Stage 22
corresponds with the first signs at zeugopod level of cytodifferentiation, increased sulphate uptake into chondroitin
sulphate (Searls 1965), which precedes the appearance of
mesenchymal condensations at stage 24.

Recent work on the zone of polarizing activity (ZPA)
suggests that the antero-posterior (A-P) axis is controlled by
a concentration gradient of a diffusible morphogen produced by
the ZPA located at the posterior margin of the wing bud.

Distal cells are thought to respond to this gradient and
differentiate according to their position along it (Tickle et
al 1975). When a second ZPA is grafted preaxially, duplication of the digits results. This suggests that the distal
mesenchyme is not determined at stages 18-22 since it responds
to the graft by forming extra digits where none normally
develop. Under different experimental conditions, this
lability during the early stages of development is not seen,
since no regulation is found along the A-P axis. Warren (1934)
inserted impermeable barriers into the wing along this axis
and claimed mosaic development occurred on both sides of the
barrier. Hinchliffe and Gumpel-Pinot (1981) amputated anterior or posterior halves of the wing bud. An anterior half in
isolation developed less than its prospective fate, whilst a
posterior half in isolation formed its normal complement of
skeletal parts without regenerating the missing anterior part.

This study examines this issue of regulation by analysing
the development of wing buds which have either excesses or
deficiencies of tissue along the A-P axis. The contribution
of anterior and posterior parts to the regulation is assessed
by chimeric grafting.

EXPERIMENTAL WORK

Operations were performed on stage 19-25 (Hamburger and
Hamilton 1951) wing buds. Cuts were made parallel to the
proximo-distal axis, the wing buds being divided into approximately 1/3 areas by an anterior cutting level at mid-somite 17,
and a posterior cutting level at intersomite 18/19. Reference
to a fate map of a stage 20 wing bud (Hinchliffe et al 1981)
indicates that the presumptive areas for the major parts of
the humerus and ulna, half the radius and digits 2 and 3 lie
between the two cutting levels (Fig 5a). Most of the presumptive skeleton is thus affected by the operations, and fate
maps of Stark and Searls (1973) indicate that comparable areas
are affected in the other stages used.

A central slice of tissue of $1\frac{1}{2}$ somite widths was removed
or duplicated along the A-P axis. To create a deficiency, the
wing bud was divided by cuts into 1/3rds, the central 1/3
portion being removed then the anterior 1/3 moved posteriorly
and grafted to the posterior portion with platinum wire pins
(Fig 1a). To create an excess, an anterior 2/3 wing bud
portion from a donor embryo was grafted with platinum wire

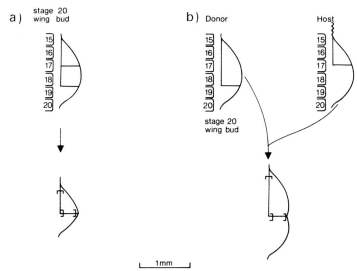

Figure 1.   Diagram of a) deficiency and b) excess operations.

pins to a host posterior 2/3 wing bud.   The flank region anterior to the wing bud of the host was slit prior to operation
to ensure maximal graft-stump contact (Fig 1b).   Control 'repin' operations were also performed using stage 19-22 wing buds.
Either an anterior 2/3 or 1/3 portion was excised and repinned
on to its original stump, to act as 'repin' controls for the
deficiencies and excesses respectively.   On the 10th day of
incubation, both operated (right) and control (left) wing buds
were fixed and stained with Methylene Blue.   The pattern of
skeletal elements was examined and the widths and lengths of
humerus, radius, ulna and digit 3 were measured.

     Pattern regulation was considered perfect where a normal
complement of skeletal elements was obtained and abnormal
where absence, bifurcation or duplication of elements occurred.
The repin operations had a normal skeletal pattern (Fig 2a),
indicating that the cutting levels and pinning procedures had
no effect on pattern.   Good regulation was observed in deficiencies and excesses up to stage 24 (Fig 2).   With deficiencies, at stage 24 there is severe reduction in the size of
radius and ulna which frequently fuse to the distal end of
the humerus and loss of digits occurs, particularly digit 3.
In excesses, bifurcation of the humerus occurs in 50% of cases
at stage 22, reaching 100% at stage 24.   An extra zeugopodial
element of mixed origin and extra digits (digits 2 and/or 3

are repeated) appear at stage 24. Before stage 24, occass-
ionally a small cartilage occurs anterior to digit 2, these
mainly at stage 22 (4 out of 10 results).

Figure 2. Methylene blue preparations of 10 day wing
skeletons. a) a control 'repin' operation, b)-d) deficiencies
and e)-h) excesses. Numbers represent the stage at operation.

An effect on the widths and lengths of elements can be seen before pattern abnormalities occur, so the widths and lengths of elements were examined on a stage by stage basis. The overall widths of the humerus, radius, ulna and metacarpal of digit 3 were calculated by averaging the two epiphyseal widths, then this one measurement with the diaphyseal width. Fig 3 plots the widths of the operated limb elements as a percentage of the controlateral control limb elements with increasing stage at operation. The repin (control points) show there is a negligible effect of the operative procedure on development, the points all lying close to 100%.

Deficient limb elements (Fig 3a) achieve only 80% of the width of the control limb, the humerus and ulna decreasing in widths at stage 23, and digit 3 at stage 24. The radius maintains its width, but is severely shortened at stages 24 and 25. Excess limbs (Fig 3b) show an increased humerus width from stage 21, increasing markedly at stage 24, when 100% of cases show bifurcation. An extra zeugopodial element appears at stage 24, achieving approximately 70% of the width of the ulna which maintains its width whilst the radius decreases slightly. An extra digit 3 also appears at stage 24, attaining 90% of the width of the original digit 3 by stage 25.

The overall length of the skeleton is also affected by the operation. Deficiencies only achieve 80% of the length of the control limb, this length decreasing with increasing stage at operation after stage 22, whilst excesses tend to remain slightly shorter than their control limbs (Fig 2).

In summary, the wing can regulate for deficiency or excess of tissue along the A-P axis. Good regulation of pattern and size is seen at stages 19-22. Size regulation decreases gradually with increasing stage at operation and is poor when pattern regulation virtually ceases at stage 24.

If the graft tissue could be distinguished from that of the host, the relative contribution of each to the regulation observed could be determined. Quail cells are naturally labelled and can be distinguished from chick cells as they possess a distinct nucleolar marker visible with Feulgen staining (Le Douarin 1973). Chimeric chick-quail grafts were performed at stages 19-21, stages when good regulation was achieved in chick-chick grafts. Deficient and excess wing buds were created as described previously, except the anterior tissue was of quail origin. The chimeric wing buds were allowed to

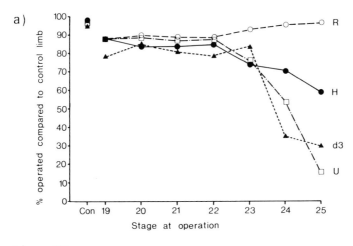

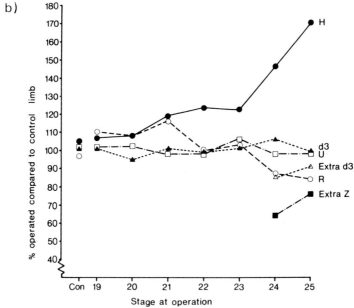

Figure 3. Graphs of overall widths of individual elements
for a) deficiencies and b) excesses. % widths of elements
of operated wings compared to control limbs is plotted against
stage at which the operation was performed. Each point
represents approximately 12 replicates. Con represents
control 'repin' results, H humerus, R radius, U ulna, d3
digit 3 and Z a zeugopodial element of mixed origin.

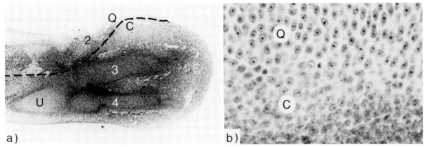

Figure 4. Sections of chimeric limbs resulting from excess
operations. The dotted line represents the boundary (c chick,
Q quail) under low power (a). A sharp boundary is shown at
higher power (b) passing through a digit 2 metacarpal.
(a) x 30, (b) x 320.

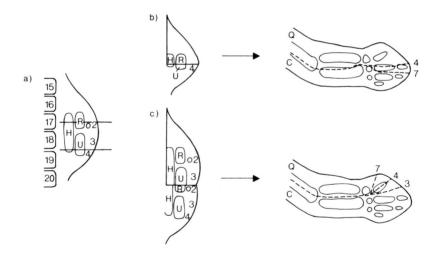

Figure 5. a) Fate map for a stage 20 wing bud (Hinchliffe
et al, 1981), the lines representing anterior and posterior
cutting levels. Diagram of graft compositions and the chimeric results for b) deficiency operations and c) excess
operations. The dashed line represents the chick-quail
boundary, (C chick, Q quail), and the numbers outside the limb
skeleton drawing represent the number of replicates for each
type of result.

develop either for 4 days after operation then fixed in
Carnoys, sectioned and stained by the Feulgen technique, or
for 6-7 days for a whole skeleton preparation.

Chimeric limb buds were found to regulate almost as well
as chick-chick grafts. In 19 Methylene Blue stained chimeric
limbs (10 excesses, 9 deficiencies) the skeleton was essentially normal, apart from a tendency for the humerus to bifur-
cate proximally. Examination of the sections revealed little
mixing of the chick and quail cells, the boundary being distinct (Fig 4). Eleven deficient limbs were examined, and the
boundary line was found to pass through the posterior part of
the humerus, through the anterior part of the ulna and posterior to or through digit 3 (Fig 5). Fourteen excess limbs
were examined and here the boundary passed through the humerus, between the radius and ulna, and either posterior or anterior to digit 2 (Figs 4,5). When these results are compared
with the prospective areas present in the chimeric wing bud
(Fig 5) it is clear that a contribution to the regulation
obtained comes from both graft and host tissues. In excesses,
there is a considerable loss of prospective areas on both
sides of the interface, and with deficiencies there is a
greater contribution from the anterior part than the posterior.

DISCUSSION

These results clearly show that the chick wing bud has the
capacity to regulate for both size and pattern along the A-P
axis at stages 19-23, and are consistent with Kieny (1977)
finding regulation up to stage 24 along the P-D axis. Such
regulation conflicts with the progress zone model (Summerbell
et al 1973) which proposes that while cells within the progress zone are maintained in a labile state as they emerge both
their P-D and A-P levels of differentiation are determined
(Summerbell and Tickle 1977). In our results proximal parts
of the limb such as prospective stylopod and zeugopod which
have left the progress zone well before stage 23 show properties of regulation. This indicates that A-P determination can
still be reversed after the onset of cytodifferentiation, which
can be detected in prospective zeugopod as early as stage 22
(Searls 1965).

The capacity for regulation along the A-P axis in these
experiments is in direct contrast to previous reports of mosaic
development of isolated posterior halves with intact ZPA. It

is possible that for regulation to occur the wing bud needs in addition to the ZPA a second anterior reference point which could be the anterior mesenchyme or AER. Strict AER continuity between host and graft is apparently unnecessary, as normal wings result from operations with small gaps between the two cut ends of the AER. Another possible explanation for regulation may be that the more normal limb configuration in our regulation experiments allows for adjustment and sliding of AER and ectoderm, relative to mesoderm.

According to the theory of ZPA control of the A-P axis, regulation could occur through the evening out of discontinuities in the morphogen concentration profile between host and graft tissues. The chimeric grafts show that contributions to regulation occur on both sides of the host/graft interface, as was found by Kieny (1977) along the P-D axis. To obtain this result the hypothetical morphogen profile would have to adjust in both graft and host tissues.

Recently the ZPA theory has been challenged by Iten and Murphy (1980). They suggest that a polar coordinate system (French et al 1976) operates in the chick limb, in which discontinuities are evened out by intercalation of missing 'positional values' between host and graft tissues through the formation of additional tissue at the interface. With this theory, regulation is epimorphic since it involves increased cell division in a growth zone. The deficiency experiments are not inconsistent with this. However, the polar coordinate theory applied to the excess experiments appears to predict the formation of from 5-7 digits (fig. 5). Since only the 3 normal digits form at stages when regulation occurs this theory is not supported by these experiments.

The terms epimorphic or morphallactic are frequently used to describe alternative forms of regulation. The regulation we observed in deficiencies was not epimorphic as described above, but neither was it truly morphallactic since although positional values in the limb field may be reassigned (as in the ZPA model) morphallaxis should result in a small but perfectly proportioned skeleton. Deficient limb buds in fact attempt to regulate for size and do not achieve perfect proportions. These two types of regulation represent extreme situations, a combination of both may be occurring in the regulation observed.

SERC support for B.L.Y. is gratefully acknowledged.

French V, Bryant PJ, Bryant SV (1976). Pattern regulation in epimorphic fields. Science 193:969.

Hamburger V, Hamilton HL (1951). A series of normal stages in the development of the chick embryo. J Morph 88:49.

Hinchliffe JR, Garcia-Porrero JA, Gumpel-Pinot M (1981). The role of the zone of polarizing activity in controlling the differentiation of the apical mesenchyme of the chick wing-bud. Histochem J 13:643.

Hinchliffe JR, Gumpel-Pinot M (1981). Control of maintenance and anteroposterior skeletal differentiation of the anterior mesenchyme of the chick wing bud by its posterior margin (the ZPA). J Embryol exp Morph 62:63.

Hornbruch A (1980). Abnormalities along the proximo-distal axis of the chick wing bud : the effect of surgical intervention. In Merker HJ, Nau H, Neubert D (eds): "Teratology of the Limbs", Berlin: De Gruyter, p 191.

Iten LE, Murphy DJ (1980). Pattern regulation in the embryonic chick limb : supernumerary limb formation with anterior (non-ZPA) limb bud tissue. Devl Biol 75:373.

Kieny M (1977). Proximo-distal pattern formation in avian limb development. In Ede DA, Hinchliffe JR, Balls M (eds): "Vertebrate Limb and Somite Morphogenesis", Cambridge University Press, p 87.

Le Douarin N (1973). A biological cell labelling technique and its use in experimental embryology. Devl Biol 30:217.

Searls RL (1965). An autoradiographic study of the uptake of S^{35}-sulphate during the differentiation of limb bud cartilage. Devl Biol 11:155.

Stark RJ, Searls RL (1973). A description of chick wing bud development and a model of limb morphogenesis. Devl Biol 33:138.

Summerbell D (1977). Regulation of deficiencies along the proximo-distal axis of the chick wing-bud : a quantitative analysis. J Embryol exp Morph 41:137.

Summerbell D, Lewis JH, Wolpert L (1973). Positional information in chick limb morphogenesis. Nature 244:492.

Summerbell D, Tickle C (1977). Pattern formation along the antero-posterior axis of the chick limb bud. In Ede DA, Hinchliffe JR, Balls M (eds): "Vertebrate Limb and Somite Morphogenesis", Cambridge University Press, p 41.

Tickle C, Summerbell D, Wolpert L (1975). Positional signalling and specification of digits in chick limb morphogenesis. Nature 254:199.

Warren AE (1934). Experimental studies on the development of the wing in the embryo of Gallus domesticus. Amer J Anat 54:449.

SECTION TWO

CELL AND TISSUE INTERACTIONS

Limb Development and Regeneration
Part A, pages 143–154
© 1983 Alan R. Liss, Inc., 150 Fifth Avenue, New York, NY 10011

THE FLUID FLOW DYNAMICS IN THE DEVELOPING CHICK WING

David M. Jargiello and Arnold I. Caplan

Department of Biology
Developmental Biology Center
Case Western Reserve University
Cleveland, Ohio 44106

ABSTRACT
     The results of previous studies on the temporal
sequence of limb vascularization suggest that prospective
morphogenetic areas of the limb are distinguished by a
differential vascularization pattern prior to the overt
expression of distinctive phenotypes (Evans, 1909; Caplan
and Koutroupas, 1972; Feinberg and Saunders, 1982). The
experiments presented here reveal the dynamic aspects of
limb vasculogenesis by detailing how a particulate tracer,
india ink, is dispersed by the complex vascular tree. Data
are presented as a temporal sequence of fluid flow maps
which indicate the direction and specific rate of vascular
flow in the limb. Our observations suggest that the limb is
subcompartmentalized into discrete microenvironments that
are spatially distinct with regard to their capacity for
transporting a particulate tracer. The developmental
significance of these observations may be that limb
mesenchymal cells are granted precise "positional
information" in the form of the specific nutrient and oxygen
levels they encounter during critical phases of limb
morphogenesis.

INTRODUCTION
     Morphological observations have indicated that the
vascular pattern in the developing chick limb unfolds during
a discrete window of time that spans Hamburger and Hamilton
(1951) stages 20-25 (Evans, 1909; Caplan and Koutroupas,
1972; Feinberg and Saunders, 1982). Our observations have
further revealed that this vascularization pattern unfolds
such that prospective myogenic, chondrogenic, and osteogenic

areas are differentially nourished prior to the overt expression of muscle, cartilage and bone phenotypes (Caplan and Koutroupas, 1972). Based on these observations, we have proposed the working hypothesis that the vascular system plays a morphogenetic role in limb development by establishing and maintaining microenvironments which differ in the local availability of nutrients. An implicit assumption in this vascular "pre-pattern" model is that limb mesenchymal cells are sensitive to nutrient thresholds; hence a cell responds to a particular nutrient environment by expressing an appropriate developmental program. Alternatively, one can envision that committed mesodermal cells overtly express their distinctive phenotypes only when the nutrient environment is "permissive" to such an event.

This concept of "positional signalling" as an informative mechanism operative in the unfolding of limb pattern is not novel; indeed, it is a phenomenon predicted by a number of investigators based upon a diversity of experimental data. A "gradient" model, for example, has been put forth by investigators of the Zone of Polarizing Activity (ZPA), a discrete section of posterior mesoderm inherently capable of shifting the axial polarity of wings into which it has been surgically introduced pre-axially (Tickle, et. al.,1975; MacCabe and Parker, 1976; Summerbell and Tickle, 1977). It is supposed, in this case, that the ZPA represents a source of a diffusible morphogen which is distributed as a linear gradient across the limb, ostensibly by means of sequential degradation or use by cells along the way.

In contrast, a model based upon the precisely timed mitoses of developmentally "labile" cells has been proposed from observations of distal wing tip transplants to hosts of different stages of development (Summerbell, et. al., 1973; Summerbell, 1974; Wolpert, 1978, 1979). By this view, a distal "progress zone" exists in which uncommitted mesodermal cells rapidly proliferate and leave behind cells which differentiate according to the time actually spent by their ancestors in this zone.

Our observations reveal that the fluids coursing through the vascular tree of the limb do so at exceedingly high rates; such fluid flows preclude any simple diffusion models involving morphogenetic substances. Indeed, it is clear from a visual observation of the limb vasculature that if a diffusible morphogen is present within the limb, it must be transmitted through cell membranes and not extracellularly. The rationale here is simply that the mesenchyme is so thick with vessel structures through so much of its area that morphogens trickling in linear fashion would undoubtledly be swept away before a stable "gradient" could be established.

The studies reported here attempt to quantify vascular direction and flow in the temporally changing limb bud and to determine if the limb is actually compartmentalized by the vasculature during early development. Specifically, we have attempted to describe and verify in a quantitative manner the information we have obtained previously by analyzing the gross morphology of vascular patterns visualized in whole mounted limbs. Furthermore, we have attempted to determine, if the mesoderm is so differentially vascularized, what the subtleties of this pattern may be. For instance, are there vascular subcompartments within the prospective myogenic and chondrogenic areas? In sum, we have attempted to determine the three dimensional pattern of vascular fluid movement during the time between initial vasculogenesis (stage 19) and overt muscle and cartilage differentiation (stage 26). The data reported here provide a precise description of the vascular pattern;experimentation in progress will attempt to relate the observations presented here with the extravascular fluid movements during development.

MATERIALS AND METHODS
        Fertilized White Leghorn eggs were windowed at three days of development and appropriately staged embryos were kept at 37.5°C by placing them in front of a Sage Instruments air curtain incubator (model 279) to insure a constant rate of heart beat. After exposing the right limb bud by removing the chorionic membrane, finely drawn capillary needles capable of delivering nanoliter quantities of liquid were used to inject sonocated india ink (Pelikan No. 17 Black) directly into the limb mesoderm. The rate of movement of the ink particles from this injection site was monitered by using a stopwatch and measuring the distance from the injection site to some arbitrary distant site with a micrometer. Compilation of the data into fluid flow maps as represented in the figures of this paper were from over one thousand separate injections. Each rate value represents an average of 2-3 injections at exactly the same

point in the limb. The maximum deviation for any pathway was 0.15mm/sec.

All injections carried out were controlled with respect to heart rate of the embryo, the volume of the ink placed in the tissue, the proximo-distal site, and the dorso-ventral depth of the injection, and the method for clocking particle movement. Complete details of these methods will be published elsewhere (Jargiello and Caplan, 1983).

RESULTS
The local introduction of india ink provided a sensitive method for marking the rate and direction of vascular flow in the developing chick limb since clearance fronts from the injection site could easily be followed and clocked. The precision of such measurements is due to the black color of this tracer and from the fact that ink particles cling to the blood vessel walls and thus leave a discernible trail.

Three separate injection localities show different fluid flow characteristics from the earliest stage studied; these are the dorsal, the ventral, and the central aspects of the limb. The resulting fluid flow maps are depicted in Figures 1, 2, and 3 and require detailed study by the reader of this article. Because of space limitations, a detailed description will not be attempted. Some generalizations with regard to individual stages of development can easily be discerned by considering these figures. On the top of Figure 1, the dorsal flow maps can be seen while the bottom panel represents the ventral flow maps. At stage 20, it is already obvious that the dorsal and ventral aspects of the limb show distinct patterns. The ventral pattern is the less complicated with most of the vessels carrying fluid out of the limb. The dorsal pattern is somewhat more complex, however, and this dorso-ventral difference is maintained throughout development, the dorsal pattern consistently being the more complicated of the two. By stage 22, a distinct "null zone" (indicated by the stippled area) can be observed. Injections of india ink into this area result in the maintenance of the volume of ink at that injection site while in all other injection sites in the limb, the india ink is carried away by the vascular tree. The distal part of the limb is also distinguished by the complexity of its vascular pattern. By stage 23, the null zone at the core is more clearly discernible and the distal half of the limb is

FIGURE LEGEND

Figures 1, 2, and 3.
Vascular pathways are depicted as single lines having flow rates measured in mm/sec. Each line represents multiple injections from point of origin of that line with measurement of rate of flow along the path indicated by the arrow in the direction of the arrowhead. Zones into which india ink was injected but was not carried away are represented by the shaded or stippled areas; these are referred to in the text as null zones. Often rates of exit of india ink were too fast to accurately measure, those rates are in excess of 2mm/sec. and are referred to as "fast" on the individual diagrams. In the core area of the limb, single large vessels could be visualized and are depicted as arrows of thicker dimension. Data for the three depths of injection are so indicated by the shaded cross-sections on the left side of the figure; uppermost is the dorsal projection, in the center, the core region and on the bottom of the figure the ventral aspect of the limb.

The figures described above appear on the three succeeding pages.

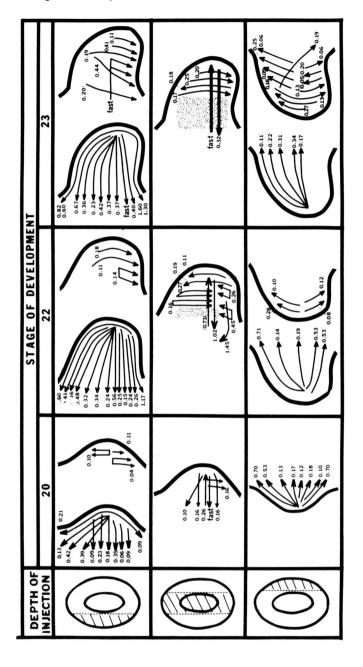

Figure 1

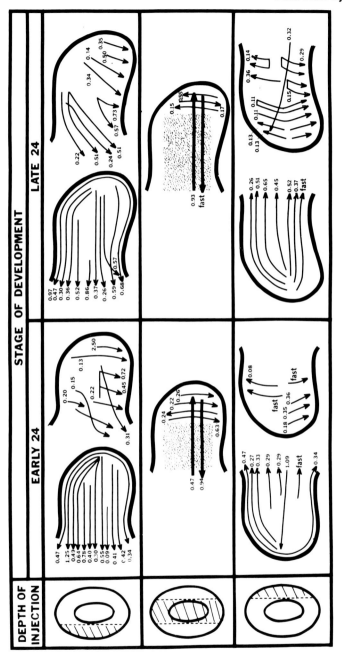

Figure 2

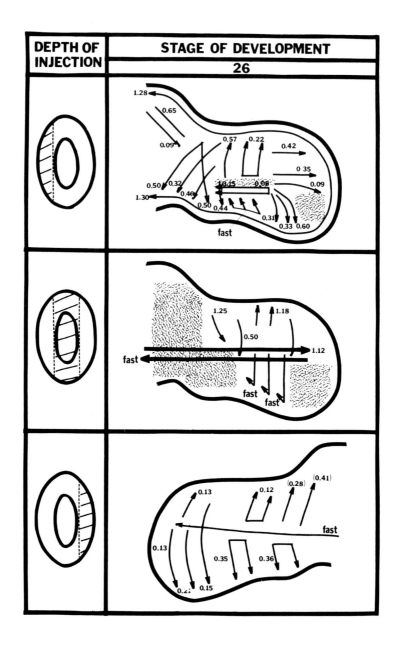

Figure 3

distinguished by a dense vascular flow in a variety of directions. This demarcation may indicate a functional delineation of the "progress zone" postulated by others (Summerbell, et. al., 1973).

By early stage 24, the patterns established in the previous stages become more distinct with the dorsal vascular pattern again more complex than the ventral the central pattern delineated by a large null zone and a distal tip which is heavily vascularized. By late stage 24, this pattern, especially that of core, is even more pronounced.

As seen in Figure 3, a distinct vascular subcompartmentalization, indicating discrete microenvironments, can be seen in the stage 26 limb. Consistent with morphogenesis and the differentiation of distinct phenotypes is the presence of new and isolated null zones in the dorsal sector. This is in contrast to the ventral soft tissues which exhibit a far simpler pattern in that branching anterior-posterior vessels almost exclusively nourish this area. The distal tip of the limb, furthermore, no longer exhibits the complex array of small vessels seen in earlier stages. The core of the limb is distinguished by an apparent "recession" of the null zone, and its "replacement" with a constellation of dense, arching vessels that weave their way to the periphery to feed a large, dorso-medial sinus.

DISCUSSION
The observations presented here indicate that the developing chick wing is progressively segregated into fluid flow microenvironments that are spatially distinct with respect to their capacity for transporting an inert and particulate tracer (india ink). The fluid flow maps presented in Figures 1-3 indicate stage-specific reorganization of the vascular network. Particularly noticeable are differences in pattern complexity in the dorso-ventral sectors and the appearance of a null zone within the core. In addition, the distal tip of the limb is distinguished during pre-stage 25 development by a complex array of small vessels which branch in both an anterior and posterior direction. The temporal-spatial pattern of fluid direction appears to begin with dense, antero-posterior capillary networks in the distal tip of the stage 22 limb and then proceeds toward the dense patterned arrangement of

such small vessels and capillary branches throughout the limb. This pattern seems finalized in the stage 26 limb since none of the proximally orientated flow pathways were observed in this stage.

The differential orientation of flow pathways within the dorso-ventral peripheral zones, in concert with the difference in flow rates, suggest that aspects of these tissue layers are dissimilar in their manner of nutrient supply and dispersal. In considering the central zone of the developing limb, the most striking observation is that the distal tip of the limb remains heavily vascularized and, therefore, an area of high metabolic activity throughout the time span studied. This vascular compartment is consistent with the prediction of a distal "progress zone" at the tip of the limb. Indeed, it has been hypothesized that differentiating mesodermal cells of the limb measure distance by measuring the time they spend in a highly active, forward area known as the "progress zone"; this knowledge of distance, it is believed, constitutes the actual "positional information" by which cells ascertain their discrete locations in the developing wing (Wolpert, 1969; Summerbell, et al. 1973; Summerbell, 1974; Wolpert, 1978, 1979). We interpret our observation of the highly active distal zone in the emerging wing as a means of insuring an appropriate pace and pattern to limb outgrowth. It is conceivable that the distal cells are maintained at a specific rate of division and replication by being emerged in a nutrient environment of specific character as guaranteed by the vasculature; cells leaving this growth zone are thus deposited in areas of different vascular flow qualities which perhaps provide the actual cues for the expression of appropriate phenotypic programs.

We further postulate that complex interactions between the vascular supply (and thus nutrient accessibility) and the differentiating mesoderm could result in the establishment of discrete dorsal,, ventral, central, and distal tip populations of cells. It is our view, in fact, that the vascular pattern serves as the scaffolding within which the precisely timed decisional events of limb morphogenesis occur. By utilizing the intricately formed vascular skeleton as a "blueprint", it is conceivable that the mesenchyme could be organized into discrete pockets of cells that are either directed or permitted along particular phenotypic pathways.

From this view, one might further predict that precise disruptions of the vascular pre-pattern should produce predictable limb malformations. Although histological studies of thallidomide mutants (Jurand, 1966), the development of Brachydactylia in the rabbit (Inman, 1941), and Talpid mutants (Ede and Kelly, 1964), indicate that vascular disruption is in fact the first discernable event in the malformation process, further detailed investigation of this hypothesis is required.

To summarize, the observations presented here detail the vascular aspect of the fluid flow of the developing chick wing. The data indicate that the vasculature emerges such that peripheral mesoderm is subcompartmentalized in an anterior-posterior fashion and the limb core in a proximo-distal manner. The question of whether the compartmentalization of the peripheral mesenchyme relates to patterns of muscle condensation and other phenomena is not clear; however, the progressive enlargement of the central null zones devoid of formalized vessel networks is consistant with the concept of a proximal-distal outgrowth and cytodifferentiation of the cartilagenous aspect of the limb. The observation of the densely vascularized distal tip throughout the period of study provides morphological insight into the possible existence of the previously hypothesized "progress zone".

ACKNOWLEDGEMENT:
We are grateful for the support to N.I.H, the March of Dimes/Birth Defects Foundation, and the Muscular Dystrophy Association.

REFERENCES

Caplan, A.I. and Koutroupas, S. (1973). The control of muscle and cartilage development in the chick limb: The role of differential vascularization. J. Embryol. exp. Morph. 29, 571-583.
Ede, D.A. and Kelly, W.A. (1964). Developmental abnormalities in the trunk and limbs of the talpid mutant of the fowl. J. Embryol. exp. Morph. 12, 339-356.
Evans, H.M. (1909). On the development of the aortae, cardinal and umbilical veins, and the other blood vessels of vertebrate embryos from capillaries. Anat. Rec. 3, 498-518.

Feinberg, R.N. and Saunders, J.W. (1982). Effects of excising the apical ectodermal ridge on the development of the marginal vasculature of the wing bud in the chick embryo. J. Exp. Zool. 219, 345-354.

Hamburger, V. and Hamilton, H.L. (1951). A series of normal stages in the development of the chick embryo. J. Morph. 88, 49-92. Inman, O.R. (1941). Embryology of hereditary brachydactyly in the rabbit. Anat. Rec. 79, 483-505.

Jargiello, D.M. and Caplan, A.I. (1983). The establishment of vascular-derived microenvironments in the developing chick wing. (In preparation).

Jurand, A. (1966). Early changes in limb buds of chick embryos after thallidomide treatment. J. Embryol. exp. Morph. 16, 289-300.

MacCabe, J.A. and Parker, B.W. (1976). Evidence for a gradient of a morphogenetic substance in the developing limb. Develop. Biol. 54, 297-303.

Summerbell, D. (1974). Interaction between the proximo-distal and antero-posterior co-ordinates of positional value during the specification of positional information in the early development of the chick limb-bud. J. Embryol. exp. Morph. 32, 227-237.

Summerbell, D. and Tickle, C. (1977). Pattern formation along the antero-posterior axis of the chick limb bus. In "Vertebrate Limb and Somite Morphogenesis" (ed. D.A. Ede, J.R. Hinchliffe, and M. Balls), pp. 41-54. Cambridge: University Press.

Summerbell, D., Lewis, J.H., and Wolpert, L. (1973). Positional information in chick limb morphogenesis. Nature (London) 244, 492-496.

Tickle, C., Summerbell D., and Wolpert, L. (1975). Positional signalling and the specification of digits in chick limb morphogenesis. Nature (London) 254, 199-202.

Wolpert, L. (1969). Positional information and the spatial pattern of cellular differentiation. J. Theor. Biol. 25, 1-47.

Wolpert, L. (1978). Pattern formation and the development of the chick limb. In "Birth Defects: Original Article Series" (Volume XIV, No. 2), pp. 547-559. The National Foundation.

Wolpert, L. (1979). The development of the pattern of growth. In "Paediatrics and Growth" (5th Unigate Paediatric Workshop), pp. 15-21. Oxford: Blackwells.

Limb Development and Regeneration
Part A, pages 155–163
© **1983 Alan R. Liss, Inc., 150 Fifth Avenue, New York, NY 10011**

REGULATION IN THE CHICK LIMB SKELETON. MORPHOGENETIC RELA-
TIONS BETWEEN LONG BONE RUDIMENTS AND JOINTS

Rodolfo Amprino

Institute of Human Anatomy
University of Bari
70124 - Bari

By means of a variety of recombination experiments of
the proximal and distal halves of the chick limb buds (stages 23 to 25) transected at the presumptive zeugopod mid-
length, a continuity was established between the opposite
moieties of the mesenchymal precursors of heterologous shaft
bones of the zeugopod. In later development, both proximal
and distal components of the resulting unitary skeletal pieces consistently showed marked differences in size and shape
compared with the corresponding parts of control bones. The
changes in the presumptive developmental fate of the opposite parts of these chimaeric bones seemed to show that both
components of the combined rudiments exert reciprocal interactions that affect their growth and morphogenesis (Amprino
1979). On this basis it was assumed that in normal development the distinctive morphology of each part, proximal and
distal, of shaft bones may be the expression not only of an
intrinsic developmental capacity but also of a morphogenetic
control exerted on its cell population by the adjoining complementary portion of the skeletogenous rudiment. Actually,
when the opposite parts of the rudiment of each zeugopodal
bone were separated from one another so that they could not
contact and fuse secondarily, hypoplasic pieces developed
which failed to attain the size and shape typical of the
control segments (Amprino 1980).

Nevertheless, the disjoined moieties of some of the shaft
bone rudiments tested, freed from the influences supposedly
exerted by their complementary portions in normal development,
showed a tendency to regulate the missing part. In the embryos operated at earlier stages (21 to 23), particularly

when more than one half of the entire preskeletal blastema
had been disconnected from its complement, a diminutive but
complete and fairly harmoniously proportioned long bone,
built of a diaphysis and two meta-epiphyses, developed. In
limbs subjected to the same operation at stage 24 or stage
25, regulation, when occurring, appeared much **less marked**,
and the histology showed that reorganization of the differentiation pattern had not involved the whole disjoined segment but was restricted to a region of it close to the plane of division, where from presumptive diaphyseal or metaphyseal material a variously diminutive epiphysis developed.
This newly formed epiphysis, in spite of its underdevelop-
ment, exhibited in some cases a free, fairly typical articular surface, i.e., an area where the epiphyseal small-celled
cartilage was invested by a layer morphologically similar
to the joint cap which covers the articular surfaces in the
normally developing synovial joints. Apparently, in the latter the joint cap takes origin from the interzonal mesenchyme, whereas in the cases under discussion the layers of
flattened cells derived presumably from cells of the perichondrium enveloping the outer aspect of the cut diaphysis,
then sliding over its amputation surface.

Accordingly, it was thought of interest to verify this
assumption. The simplest approach to test whether a joint
may develop without the intervention of the interzonal mesenchyme seemed that of bisecting the blastema of a long bone at its presumptive mid-diaphyseal level, then introducing
a Millipore filter into the slit to prevent secondary fusion
of the two pieces. This device, however, did not prove sufficient to ensure an independent development of the opposed
segments: in fact, the cut surfaces of the diaphysis stuck
on both sides to the **intervening sheet** thus hindering sliding of the surrounding perichondrium on the amputation surfaces, so that apparently the comunication between the opposed segments was not blocked and a practically normal, bone developed.

We resorted then to the following experimental model:
nearly one half, proximal or distal, of the tibia blastema
was proximo-distally and dorso-ventrally inverted in situ by
rotating it, after surgical isolation, around the transverse
axis of the limb bud. Two diminutive tandem long bones and
an intervening variously rudimentary joint developed with some frequency from the bisected tibia rudiment in limbs operated between stages 24 and 26: each piece, built of a diaphy-

sis and two identifiable epiphyseal ends, underwent independent ossification. The heterotopic joint region showed variations from case to case, and local structural differences were also observed in each joint. A continuous cleft between
the opposite articular surfaces was not always present; in
several cases the articular cavity was interrupted at places
by cartilaginous bridges connecting the opposed bones. However, the remaining free surfaces facing the cleft exhibited
a fairly typical articular structure and were covered by variously thick layers of small, densely packed, flattened
cells.

When the proximal piece of the tibia rudiment had been
inverted, the presumptive epiphyseal end of the reoriented
piece together with a various amount of the adjacent interzonal mesenchyme was brought in contact with the cut surface
of the diaphysis of the unturned piece. From this diaphyseal
end an epiphyseal-like formation developed whose surface facing the articular cleft was invested by layers of flattened
cells, continuous all around with the adjacent diaphyseal perichondrium, suggesting that they had taken origin from the
latter. Also the terminal diaphyseal portion of the inverted
proximal piece showed - even in limbs operated upon at stage
$27-27\frac{1}{2}$, i.e., when chondrogenesis of the tibia rudiment was
already under way - a tendency to develop an epiphysis-like
region bearing an articular surface lined by a layer of dense, flattened cells continuous with the surrounding perichondrium. This cut end of the presumptive diaphysis had been
displaced, in general, in proximity to the distal extremity
of the femur so that it lay in a presumptive articular environment. In some cases, however, depending on some tilting
of the reoriented piece, it projected free into the surrounding connective or muscle tissue, and thus had no apparent
relations with the interzonal mesenchyme.

In the cases in which the distal part of the tibia rudiment had been inverted leaving in situ the mesenchyme that
gives rise to the tibiale and the fibulare, a fairly typical
new epiphysis often developed from the cut end of the diaphysis of the unturned, proximal segment of the tibia (Fig. 1).
Its articular surface was facing the originally distal end
of the inverted tibia piece, i.e., presumptive metaphyseal
material that in normal development fuses with the proximal
tarsal plate. Also this metaphyseal end underwent more or
less marked epiphyseal changes in shape and structure. The
opposed surfaces of the two bone pieces facing the interve-

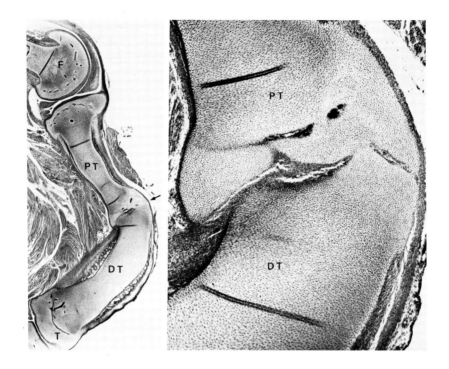

Fig. 1. 180° pd-reorientation of about the distal half of the tibia rudiment, leaving the tarsals in situ (stage 27-). Longitudinal sections (14 i.d.). <u>Left</u>: F, femoral condyle; PT, unturned proximal segment and DT, reoriented distal segment of the tibia; T, proximal tarsal. <u>Arrow</u> points to the site of the heterotopic joint, which is shown in detail (<u>right</u>) in another section 90 μm ca. aside.

ning joint cleft - interrupted at places by cartilaginous bridges - were invested by layers of flattened cells continuous with the surrounding perchondrium. Obviously, in these cases the interzonal mesenchyme could not possibly take part in the formation of the joint and contribute cells to the joint cap.

In a further attempt to exclude the participation of the interzonal mesenchyme in the development of the heterotopic joint, another experiment was carried out in embryos at stages 25-26½. About the proximal fourth of the tibia ru-

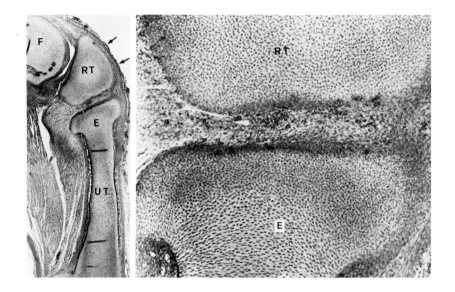

Fig. 2. 90° reorientation of about the proximal one fourth of the tibia rudiment (stage 26½). Longitudinal section (12 i.d.). F, femoral condyle; RT, reoriented piece of the tibia: arrows point to its originally proximal aspect. UT, unturned, distal segment of the tibia: its proximal end is enlarged in the newly formed epiphysis, E. At the right: detail of the heterotopic joint developing between the reoriented piece, RT, and the newly formed epiphysis, E.

diment was isolated and turned 90° around the antero-posterior axis of the zeugopod so as to shift to a ventral position its presumptive proximal, articular surface, to a distal position, i.e., opposite to the articular portion of the distal femoral end, the originally dorsal surface, and bring the ventral surface of the reoriented meta-epiphysis to face the proximal end of the cut diaphysis of the unturned distal segment of the tibia rudiment (Fig. 2). From this diaphyseal end an epiphysis developed: its proximal aspect, opposite to the originally ventral surface of the reoriented meta-epiphysis was, much as the latter, invested by perichondrium. Cartilaginous bridges did not form, in general, between these two opposed surfaces which, in specimens fixed between the 12th and the 15th day, appeared lined by flattened cells; be-

tween these two layers looser connective tissue containing
degenerating cells, blood vessels and variously extensive
clefts was present. Also the originally dorsal aspect of the
reoriented piece, covered by perichondrium, which was brought
opposite to the articular end of the femur, showed a joint
cap, which again was apparently originated from the perichondrium. Thus an heterotopic, variously atypical or rudimentary joint - or, respectively, an unmatched, free articular
end - may develop from parts of the bone rudiment which in
normal development are not articular without intervention of
interzonal mesenchyme, i.e., of presumably joint-determined
cells.

Experiments have been carried out also to clarify other
questions prompted by the study of the development of the tibia and its joint regions. One question, for instance, is
raised by consideration of the fusion of the proximal tarsal
with the distal end of the tibia rudiment; after fusion the
tarsal becomes in fact the distal epiphysis of the tibia and,
in turn, it articulates with the distal tarsal. The original
distal end of the tibia behaves, therefore, as a presumptive
metaphyseal region rather than an epiphysis - although in later development (Stocum, Davis, Leger, Conrad 1979) a part
of the distal tibia end seems to contribute to the proximal
portion of the distal epiphysis. Does fusion of the two mentioned skeletal regions require a mutual affinity, or is it
merely the consequence of the intimate contact and reciprocal pressures of two structures that in normal development
lie opposite to one another?

In embryos between stage 23 and stage 27 attempts have
been made to invert the whole blastema of the tibia by 180°,
leaving however in situ the presumptive tarsal material. The
operation was fully successful only in some cases, whereas
in other specimens a variously large part of the proximal
tarsal mesenchyme was included in the turned slice. When the
tarsal mesenchyme was left in situ and the entire tibia blastema turned 180° around its transverse axis, the originally
proximal end, i.e., the presumptive proximal epiphysis, was
brought adjacent to the condensations of the tibiale and fibulare: fusion of the latter masses to form the tarsal plate
and sealing of this plate with the proximal epiphysis of the
reoriented tibia took place. In 12-14 day embryos, the originally proximal tibial epiphysis had undergone marked structural changes. Instead of its usual small-celled structure,
presence of vascular channels and of the joint cap, this zone

had acquired metaphyseal features becoming avascular and containing flattened chondrocytes with their long axes perpendicular to the long axis of the bone rudiment. On the plane of
fusion with the tarsal plate a few layers of small, flattened
cells, presumably remnants of the joint cap of the displaced
epiphysis, persisted for some time and were gradually incorporated into the growing metaphyseal cartilage. Vascular
channels invaded the deep layers of the tarsal plate whose
structure was also in other respects similar to that of an
epiphysis. Thus a presumptive epiphysis which in normal development is typically articular may fuse with the rudiment of
the proximal tarsal and undergo morphological changes into a
metaphyseal zone. Conversely, the originally distal end –
presumptive metaphysis – of the inverted tibia blastema, that
in some specimens articulated with the femoral condyles, underwent clearcut epiphyseal changes. In a few cases, probably as a consequence of a defective alignment of the inverted tibia rudiment, the originally proximal epiphysis did
not contact the tarsal plate; the latter developed as an independent piece bearing two articular surfaces invested by
joint caps, i.e., a proximal one that developed as a typical
articular zone facing the opposed tibia end, and a distal
surface more or less regularly articulated with the distal
tarsal.

In conclusion, the results of our experiments can be
viewed through a common interpretative framework since they
all seem to be the outcome of the regulative ability of the
skeletogenous mesenchyme. It has long been known that the regulation capacity of the skeletal blastemata decreases markedly during early stages of limb development; however, as
the present findings show, it is not completely exhausted
even at the onset of their chondrification. In these later
stages (stages 25-27) the reorganization of the original differentiative pattern does not involve, as it does in earlier
stages, the whole preserved part of the rudiment but only a
region adjacent to the plane of amputation, wherefrom a new
epiphysis bearing a free articular surface arises from presumptive non-epiphyseal material. Sliding of the surrounding
perichondrium over the amputation surface seems to be a prerequisite for the regulation process, though the role actually played by this layer and the extent of the involvement of
its cells in the formation of the new epiphysis has not been
ascertained. Expression of regulative ability also appear, some of the results of the various series of 180° pd-reorientation of the whole tibia rudiment leaving or not in situ the

precursors of the proximal tarsal plate.

The more or less extensive modifications of the original developmental pattern of variously large portions of the skeletogenous mesenchyme observed in the experiments reported here seemed to take place as a direct consequence of positional changes undergone by some part of the developmental system. Such positional changes should be viewed under various aspects. They may involve: 1) alterations of the presumable morphogenetic influences between different parts of a skeletal blastema, that apparently vary according to the location of each part with respect to the center or the ends of the rudiment; 2) modifications of the mechanical conditions related to the pressures intrinsic and extrinsic to the growing blastema; and 3) affect the metabolic conditions of the displaced portion within rudiments which - except the epiphyseal ends in relatively late stages - are avascular and thus depend on the materials diffusing from neighboring blood vessel networks. All these conditions, in turn, are likely to exert more or less marked influence depending on the susceptibility of the various parts of the rudiment in relation to the degree of differentiation attained at operation time.

Clearly, our data do not bring any contribution to the question of whether the differentiative pathway followed by the cells of the interzonal mesenchyme in normal joint formation is governed by instructions given to them during the early stages of limb development, as recently maintained (Holder 1977); they show, however, that under given experimental conditions which result in the formation of heterotopic, somehow rudimentary but recognizable articulations, perichondrial cells - that cannot be looked upon as a specific cell type receiving information at early stages for its participation in joint development - may apparently play the role normally exerted by the interzonal mesenchyme.

Some of the present findings seem compatible with the conclusion drawn from classic in vitro experiments (Fell, Canti 1935) showing that an articulation may form in the absence of the presumptive joint region provided sufficient undifferentiated material remains at the ends of the bone rudiments (Fell 1954). However, in their cases the undifferentiated material was a displaced part of the interzonal mesenchyme, whereas in most of our experiments it was of perichondrial origin. According to the above authors, the articular surfaces are automatically produced as a consequence of the gradual incorporation of the soft tissue of the undifferentiated joint region into one or the other of the opposed chondrifying epiphyses until no free cells remain. In line with this view on the mechanism of formation of normal articular surfaces it seems reasonable to interpret some of our findings where epiphyses bearing a free articular surface were seen to develop from non-epiphyseal regions of the bone rudiments and to grow by incorporation of undifferentiated cells which, moreover, did not belong to the interzonal tissue but to the perichondrium.

Amprino R (1979). Developmental interactions between the adjacent parts of combined heterologous skeletogenous territories. Anat Embryol 155:135.

Amprino R (1980). The early rudiment of the shaft bone.A unitary developmental system. Anat Embryol 160:53.

Fell HB (1956). Skeletal development in tissue culture. In Bourne GH (ed): "The Biochemistry and Physiology of Bone", London:Academic Press, p 401.

Fell HB, Canti RG (1935). Experiments on the development in vitro of the avian knee joint. Proc Roy Soc Lond B 116:316.

Holder N (1977). An experimental investigation into the early development of the chick elbow joint. J Embryol exp Morph 39:115.

Stocum DL, Davis RM, Leger M, Conrad HE (1979). Development of the tibiotarsus in the chick embryo:biosynthetic activities of histologically distinct regions. J Embryol exp Morph 54:155.

Limb Development and Regeneration
Part A, pages 165–174
© **1983 Alan R. Liss, Inc., 150 Fifth Avenue, New York, NY 10011**

SHOULDER FORMATION, ROTATION OF THE WING, AND POLARITY OF
THE WING MESODERM AND ECTODERM

Robert L. Searls, Ph.D.

Biology Department
Temple University
Philadelphia, PA   19122

Experiments done in conjunction with Dr. Gail Yander
have indicated that two cell movements in the lateral body
wall of the embryonic chick have a significant influence on
normal wing development (Yander and Searls, 1980a; 1980b).
In this paper, I shall attempt to summarize these experiments and then describe some further experiments that have
been done concerning the significance of the polarity of
the ectoderm and mesoderm. The investigation concerning
the development of the shoulder involved use of the
scanning electron microscope, photography of live embryos
in ovo, marking of specific regions of the embryo with
carbon particles (Fig. 1), or with cells obtained from the
same region of a chick labeled with tritiated thymidine,
and examination of serial sections of the marked embryos.

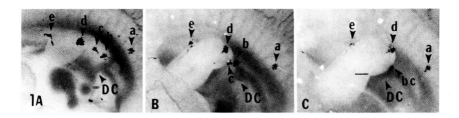

Figure 1. A stage 22 embryo was marked in 5 places with
carbon particles (marks a to e), and then photographed (A).
The same embryo was photographed 24 hours (B) and 48 hours
(C) later. DC is the Duct of Cuvier.

CELL MOVEMENTS CRANIAL TO THE WING.

The wing is ventral to the 16th to 19th somites.
Until stage 22, the cranial end of the coelom is ventral to
the 10th somite. The posterior cardinal vein is dorsal to
the coelom and curves around the cranial end of the coelom
to join the anterior cardinal vein, forming the Duct of
Cuvier. The location of the Duct of Cuvier marks the
cranial end of the coelom in living embryos (Fig 1).

During stage 22, the Duct of Cuvier and the cranial
end of the coelom begin to move in a caudal direction
relative to the somites (Fig. 1). By stage 27, the Duct of
Cuvier is ventral to the 16th somite but is concealed under
the mass of the shoulder. Marks that were placed between
the wing and the Duct of Cuvier during stage 22 (marks b
and c in Figure 1) were observed to move into the shoulder.
If the mark was placed close to the cranial margin of the
wing (mark c), the mark moved into the shoulder only a few
hours later and the portion of the mark near the ectoderm
was far out on the shoulder at stage 27. If the mark was
placed close to the Duct of Cuvier (mark b), the mark moved
into the shoulder many hours later and the portion of the
mark near the ectoderm was on the shoulder close to the
neck at stage 27. Marks between the wing and the Duct of
Cuvier (marks b and c) moved in a caudal direction relative
to a mark on the somites (mark a). When the location of
the mark was examined using histological sections, it was
found that the mark remained in contact with the coelom
about the same distance from the cranial end of the coelom
as at the time of marking. The marks extended under the
shoulder girdle cartilages from the coelom to the cranial
surface of the shoulder with the earliest cells to enter
lateral to the cells that entered later.

By stage 27, all of the cells that had been in the
lateral body wall over the coelom and cranial to the wing
had moved into the base of the wing. The addition of cells
from the lateral body wall to the base of the wing caused
the base of the wing to become thicker in a lateral direction and wider in a dorsal-ventral direction. This increase
in size of the base of the wing at its cranial margin is
the process of shoulder formation. The migration of the
cells of the lateral body wall into the base of the wing
increases the size of the shoulder but does not change the
caudal wing. The dorsal surface of the wing rotates so that
it slopes from the thickening of the shoulder (mark d in

Fig. 1B) to the caudal margin of the wing (mark e in Fig. 1B). These cell movements are diagramed in Figures 2 and 3.

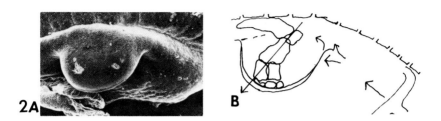

Figure 2A. Scanning electron micrograph of a stage 22 wing bud photographed by Dr. Gail Yander. Figure 2B. Fate map of a stage 22 wing bud taken from Stark and Searls (1973). Arrows indicate cell movements.

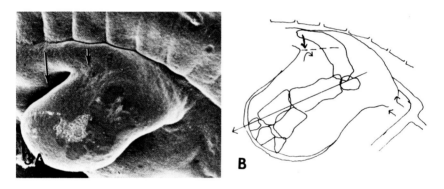

Figure 3A. Scanning electron micrograph of a stage 24 wing bud photographed by Dr. Gail Yander. Figure 3B. Fate map of a stage 24 wing bud taken from Stark and Searls (1973). Arrows indicate cell movements.

We do not know as yet what causes the caudal movement of the Duct of Cuvier. Marks placed cranial to the Duct of Cuvier do not move relative to marks placed on the somites. We interpret this to mean that the "pharynx" cranial to the 10th somite does not grow in length, displacing the "esophagus" beneath the 10th to 16th somites to more caudal positions. The movement of the Duct of Cuvier can have nothing to do with the hyoid arch because the movement of the Duct of Cuvier is almost complete before the hyoid arch has completely overgrown the cervical sinus. Clearly it is

not due to the progressive attachment of the lateral body
wall to the splanchnopleure surrounding the gut because the
cells of the lateral body wall are displaced into the
shoulder. Both neural crest cells and somite cells migrate
into this area but these migrations are said to occur long
before stage 22 (Hazelton, 1970; Noden, 1975; Tosney,
1978). This question is under investigation.

CELL MOVEMENTS CAUDAL TO THE WING.

At stage 22, the caudal margin of the wing is an
extension of the ventral surface of the lateral body wall
while the cranial margin of the wing is an extension of the
dorsal surface of the lateral body wall (Figs. 2A and 3A).
A groove is present ventral to the 19th somite where the
dorsal surface of the wing meets the lateral surface of the
lateral body wall.

At stage 24, the presumptive sternal cells lie in the
lateral body wall ventral to the 18th to 23rd somites
(Fell, 1939). During stage 25, the presumptive sternum begins to move in a ventral and cranial direction (arrows on
Fig. 3). The cells of the sternum begin to differentiate
as cartilage during stage 30 and the paired halves of the
sternum fuse at the ventral midline during stage 34 to 35.

At stage 25, the groove at the junction of the dorsal
surface of the wing and the lateral body wall ventral to
the 19th somite begins to become deeper. The groove ventral to the 19th somite becomes so deep and so narrow that
the wing seems to separate from the lateral body wall ventral to the 19th somite (Fig. 1C). However, deep within the
groove the caudal margin of the wing remains attached to
the lateral body wall ventral to the 19th somite at stage
27 (Fig. 4), and the caudal margin of the wing is still attached to the lateral body wall at the level of the 2nd rib
as late as stage 33. The groove ventral to the 19th somite
separates the distal tip of the humerus from the scapula.

When the lateral body wall proximal to the caudal margin of the wing was marked during stage 22 with carbon
particles, the marks were not observed to move until stage
25 (mark e in figure 1). After stage 25, the marks moved
into the groove and by stage 27 were deep in the groove
under the elbow. By stage 31, these marks were deep under
the wing at the ventral end of the 2nd vertebral rib.

Marks placed at the base of the wing in the groove became part of both the elbow and the lateral body wall (Fig. 1).

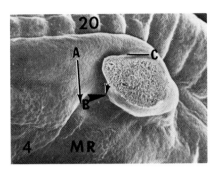

Figure 4. Stage 27 wing bud with the distal wing removed photographed by Dr. Gail Yander. The presumptive sternum has just begun to move in a ventral direction as indicated by arrow A. The caudal margin of the wing is attached to the lateral body wall ventral to the 19th somite (arrow B). The base of the wing ventral to the 18th and 19th somites is rotating down into the groove (arrow C). MR is the membrana reunions.

Movement of the lateral body wall ventral to the 18th to 23rd somites in a ventral and cranial direction has been rather thoroughly investigated by Fell (1939). It was found that the lateral body wall was not pushed in a ventral direction by cells moving in from the somites because the movement took place normally in vitro after the somites had been removed. It was found that the lateral body wall was not pulled in a ventral direction by cell death or some other process occurring in the membrana reunions ventral to the lateral body wall because the movement occurred normally in vitro after the membrana reunions had been removed. Further investigation of the ventral and cranial movement of the lateral body wall will not be possible until experiments more sophisticated than those of Dr. Fell (1939) have been devised.

DEVELOPMENT OF MYOGENIC TISSUE GRAFTED IN NORMAL ORIENTATION.

The previous sections have suggested that the dorsal surface of the wing rotates, first due to the formation of

the shoulder and second due to the ventral movement of the
lateral body wall ventral to the 18th and 19th somites. To
investigate further the rotation of the dorsal surface of
the wing, dorsal right wing myogenic tissue from a stage 22
to 23 embryo that had been labeled with tritiated thymidine
was grafted into the dorsal myogenic region of an unlabeled
host stage 22 to 23 embryo in normal orientation. A camera
lucida drawing was made at the time of the operation so
that the location and shape of the graft on the dorsal
surface of the host wing was recorded. After 1 to 5 days
of further development, the host wing was fixed, sectioned,
prepared as autoradiographs and the host wing was
reconstructed from the serial sections to discover the
location of the labeled cells.

I have suggested that the dorsal ectoderm may have
some influence on normal development of the wing (Stark and
Searls, 1974; Searls, 1976; Searls and Cioffi, 1977).
Grafts made with or without their original ectoderm
developed identically when the grafts were in normal
orientation. Therefore, grafts with and without ectoderm
will be described together. When the ectoderm had been
removed, the host ectoderm grew over the graft from
proximal to distal so that the grafts were completely
covered by host ectoderm by 3 days in 6 out of 10 cases.
When the ectoderm was included in the graft, grafted
ectoderm moved distally with respect to grafted mesoderm
but only about 0.3 mm in 3 days.

When embryos were fixed 1 day after the operation the
graft had neither grown nor contracted. Comparison of the
axis of outgrowth of the wing with carbon marks placed at
the cranial and caudal margins of the wing, or with the
shape of the graft, indicated that the wing grew out on the
axis of the humerus as described in the fate maps in
Figures 2 and 3 (long arrows in Figures 2B and 3B).

When embryos were fixed 3 days after grafting, the
graft had grown on the axis of the humerus. When the graft
extended proximal to the base of the wing ventral to the
18th somite, the grafted cells participated in forming the
groove. If the host wing was sectioned on its long axis,
grafted cells that were on the proximal border of the
groove appeared to be on the lateral body wall caudal to
the wing because the bottom of the groove was moving
ventrally almost at right angles to the long axis of the
wing (Fig. 3). If the host wing was sectioned

perpendicular to the long axis of the wing, grafted cells
that were on the distal border of the groove appeared
posterior to the humerus.

By 3 days after the operation, grafted cells that had
been placed proximal to the base of the wing and ventral to
the 17th somite were found to have penetrated very deeply
into the wing between the humerus and the scapula. The
grafted mesoderm over the radius and ulna, over the humerus
distal to the groove, and on the proximal and distal borders of the groove had not doubled in thickness since the
time of grafting. Grafted cells cranial to the groove
penetrated into the wing until they were as deep as the
ventral edge of the humerus and scapula forming a very
thick layer of grafted cells. Deep penetration of grafted
cells into the wing was not observed 2 days after grafting,
at stage 27. The deep penetration of cells in the wing
cranial to the groove suggests that cells of the lateral
body wall cranial to the groove also move in a ventral
direction.

By 4 and 5 days after the operation grafted cells
covered the posterior border of the humerus, passed over
the humerus just short of the elbow, and extended down the
dorsal surface of the radius and ulna a distance determined
by the exact location of the graft. The grafts grew on the
axis of the humerus with a doubling time of 2.4 to 2.8 days
for grafts with or without ectoderm, respectively. The
grafts grew on the axis perpendicular to the axis of the
humerus with a doubling time of 5.6 to 9.1 days,
respectively. This is less than one doubling in width
during the maximum of 5 days that grafts were kept before
fixing, the grafts may not have grown at all on the axis
perpendicular to the humerus.

DEVELOPMENT OF MYOGENIC TISSUE GRAFTED IN REVERSED
ORIENTATION.

These experiments were done in exactly the same way as
those experiments in which grafts were placed in normal
orientation. In this case, the presence or absence of the
ectoderm that had originally been with the mesoderm had an
effect on the result and the two sets of grafts will be
discussed separately.

One day after grafting mesoderm with ectoderm in

reversed orientation, a protuberance was seen on the dorsal
surface of the host wing ventral to somites 18 and 19.
When embryos were fixed 1 day after grafting, it was found
that the graft had increased in thickness to produce a
rather symmetrical protuberance on the dorsal surface of
the wing. The graft had neither increased nor decreased in
length or width and had the same shape as at the time of
the operation.

By 3 days after the operation, the protuberance had
become much more pronounced. The graft protruded from the
dorsal surface of the wing at its distal, caudal and
proximal margins but not at its cranial margin. The graft
did not participate in the formation of the normal groove
ventral to the 19th somite. When the graft extended
proximal to the base of the wing, the groove ventral to the
19th somite came to an abrupt end where it met the caudal
margin of the graft. However, a groove was always present
caudal and proximal to the protuberance. Grafts made in
reversed orientation grew symmetrically so that after 3
days the graft had become larger but the graft had not
changed shape. Growth that occurred perpendicular to the
axis of the humerus was predominantly into the proximal and
caudal corner of the wing. The lateral body wall moved
normally in the ventral direction and cells of the graft
penetrated deep into the wing between the humerus and the
scapula even though the normal groove was absent.

The development of the host wing 4 and 5 days after
the operation depended on the exact location of the graft.
If the graft extended even a little proximal to the base of
the wing at the time of the operation, the groove came to
an abrupt end early in development and the tip of the
humerus did not separate from the scapula. A webbing of
grafted cells extended from the tip of the humerus to the
lateral body wall. The humerus was short. By 5 days, this
webbing sometimes extended from the tip of the humerus to
the lateral body wall at the level of the 3rd to 4th rib
although during normal development marks placed proximal to
the caudal margin of the wing end at the level of the 2nd
rib. If the graft had been placed distal to the base of
the wing, the groove came to an end much later in development and the wing had a flap of grafted mesoderm posterior
to the humerus but the humerus was of normal length.

Almost all limbs carrying grafts in reversed
orientation developed ectopic cartilage. The ectopic

cartilage differentiated from grafted myogenic cells that had grown into the proximal and caudal corner of the wing. The grafted cells differentiated cartilage in about the same relationship to the ectoderm as the relationship of the humerus to the ectoderm during normal differentiation. Five days after the operation, the grafts were still growing symmetrically so that the grafts had increased in size but had not changed shape. The grafts increased in length on the axis of the humerus with a doubling time of about 3.8 days and increased in width on the axis perpendicular to the axis of the humerus with a doubling time of about 3.8 days. The grafts had also increased in thickness, but this was not studied further because the thickness of the graft varied with location on the wing.

Grafts without ectoderm in reversed orientation did not protrude from the dorsal surface of the wing. The grafts grew on the axis of the humerus at about the same rate as grafts in normal orientation ($T_{1/2}$ = about 2.1 days). If the graft without ectoderm extended proximal to the base of the wing, it blocked the formation of the groove. Unlike grafts with ectoderm, grafts without ectoderm did not have a groove proximal to the edge of the graft either, so the dorsal surface of the wing did not have a groove at all. Grafts without ectoderm grew on the axis perpendicular the humerus at about the same rate ($T_{1/2}$ = 3.7 days) as grafts with ectoderm in reversed orientation ($T_{1/2}$ = 3.8 days).

Grafts with ectoderm in reversed orientation showed two pronounced differences from grafts with or without ectoderm in normal orientation. First, grafts with ectoderm in reversed orientation did not grow only on the axis of the humerus; they grew in all directions. They grew so as to protrude from the dorsal surface of the wing, they grew on the axis of the humerus and they grew on the axis perpendicular to the humerus. Second, grafts with ectoderm in reversed orientation did not participate in the formation of the groove ventral to the 19th somite. If they blocked the formation of that groove early enough, the tip of the humerus did not separate from the scapula and the host wing developed with a short humerus. Grafts made without ectoderm did not protrude and grew on the axis of the humerus at the normal rate. It may be suggested that protruding and slow growth on the axis of the humerus may be due to the reversed ectoderm. Grafts without ectoderm grew on the axis perpendicular to the axis of the humerus

and did not participate in the formation of the groove
ventral to the 19th somite. Reversed mesoderm could not be
completely "entrained" by the host wing, presumably because
of properties inherent in the mesoderm cells. The precise
nature of these properties is unknown.

Fell, HB (1939). The origin and developmental mechanics of
     the avian sternum. Phil Trans R Soc Lond, Ser B,
     229:407-463.
Hazelton, RD (1970). A radiographic analysis of the
     migration and fate of cells derived from the occipital
     somites in the chick embryo with specific reference to
     the development of the hypoglossal musculature. J
     Embryol Exp Morph 24:455-466.
Noden, DM (1975). An analysis of the migratory behavior of
     avian cephalic neural crest cells. Dev Biol
     42:106-164.
Searls, RL (1976) Effect of dorsal and ventral limb
     ectoderm on the development of the limb of the
     embryonic chick. J Embryol Exp Morph 35:369-381.
Searls RL, Cioffi M (1977). Production of skeletal
     abnormalities in the proximal part of the limb bud of
     the embryonic chick. In Ede DA, Hinchliff JR, Balls M
     (eds): "Vertebrate Limb and Somite Morphogenesis,"
     Cambridge University Press, pp 105-122.
Stark RJ, Searls RL (1973). A description of chick wing
     development and a model of limb morphogenesis. Dev
     Biol 33:317-333.
Stark RJ, Searls RL (1974). The establishment of the
     cartilage pattern in the embryonic chick wing, and
     evidence for a role of the dorsal and ventral ectoderm
     in normal wing development. Dev Biol 38:51-63.
Tosney KW, (1978). The early migration of neural crest
     cells in the trunk region of the avian embryo: an
     electron microscope study. Dev Biol 62:317-333.
Yander G, Searls RL (1980a). A scanning electron microscope
     study of the development of the shoulder, visceral
     arches and the region ventral to the cervical somites
     of the chick embryo. Am J Anat 157:27-39.
Yander G, Searls RL (1980b). An experimental analysis of
     shoulder and lateral body wall development in the chick
     embryo. J Exp Zool 214:79-92.

Limb Development and Regeneration
Part A, pages 175–182
© 1983 Alan R. Liss, Inc., 150 Fifth Avenue, New York, NY 10011

ROLE OF CELL DEATH IN THE MORPHOGENESIS OF THE AMNIOTE
LIMBS

Suresh C. Goel

Department of Zoology
University of Poona
Poona 411 007   India

During development an embryo tries its utmost to
increase the cell number by very fast cell divisions
(Balinsky 1970). Cell death at this time means frittering
away the limited resources of energy and nutrition, and
perhaps information, so assiduously accumulated during
ovogenesis (Davidson 1976). Cell death during embryonic
development is thus a paradox. Yet cell death is not un-
common during the morphogenesis and histogenesis of organs
and tissues (Glucksmann 1951; Bowen and Lockshin 1981).
Development of the limb in amniotes is accompanied by large
scale cell death (Fell and Canti 1934; Saunders et al.
1962; Ballard and Holt 1968; Goel and Mathur 1978;
Hinchliffe 1982). This paper is concerned principally with
the role of cell death in amniote limb morphogenesis but
will generally review also the incidence, mechanism and
control of cell death.

The development of the amniote limb begins with the
proliferation of the lateral plate mesoderm in the flank
region. This causes the formation of a limb bud which has
a core of mesoderm covered with a thin two-cell-layered
ectodermal epithelium. The basal layer of the ectoderm
soon gets thickened at the margin of the bud to form an
apical ectodermal ridge (AER). The AER has an inductive
role towards the mesoderm and is vital also for the further
growth of the limb mesoderm (Saunders 1948; Hinchliffe and
Johnson 1980). The limb bud grows to form the limb paddle
and becomes regionalised in stylopodium, zeugopodium and
autopodium. In the autopodium the digits are formed partly
by sculpturing of existing tissue and largely by outgrowth

at the digital tips. The skeleton of the limb is formed entirely in the mesoderm in a proximodistal sequence. It is generally believed that the mesodermal cells receive positional information on at least two coordinates for their determination of a particular skeletal element (Wolpert 1978). The proximodistal coordinate is specified by the amount of time a cell spends (measured in terms of number of mitotic divisions) in the sub-AER region, the progress zone. The anteroposterior coordinate is specified by the distance of the cells from the zone of polarising activity (ZPA), an area of mesoderm at the posterior junction of the limb bud with the adjoining body wall.

The areas of cell death have been studied by vital staining of the embryos with neutral red or Nile blue sulphate and by histology. During vital staining the dye accumulates in the lysosomes of cells and phagosomes of macrophages. The macrophages which engulf the dead and dying cells are thus easily located (Pexieder 1975). In histological sections dead cells or cell fragments yet to be engulfed by the macrophages can be located in not only superficial but also deep mesoderm. Histochemical preparations for acid phosphatase indicate the presence of dying cells (Dawd and Hinchliffe 1971).

The areas of massive cell death during limb development include the anterior necrotic zone (ANZ), posterior necrotic zone (PNZ), interdigital necrotic zone (INZ) and opaque patch (OP). In the lizard Calotes versicolor ANZ first appears at the limb bud stage at the anterior proximoventral margin of the bud in the superficial mesoderm. It perhaps passes as a wave on the anterior margin of the developing limb and comes to occupy a location first at the base of the autopodium, and later on the anterior margin of the first digit where it is observed till the digits are well formed. In Calotes the cell death in the fore limb spreads also in the ectoderm (Mathur 1974). The extent and duration of cell death varies not only from species to species but also between the fore and hind limbs of the embryos of the same species; further, in the literature there is a fair amount of confusion on the precise amount and duration of cell death in the same species (for chick compare Saunders et al. 1962, Hinchliffe and Ede 1967, and Hinchliffe 1974). Anyway, the ANZ is reported to be present in all the amniote species studied so far.

The PNZ is seen first in the posterior proximal mesoderm and like ANZ it passes as a wave on the posterior margin of the bud. Like ANZ the extent and duration of the PNZ is also species specific. It may be noted, however, that the PNZ in the proximal mesoderm is either much less pronounced or missing in the hind limbs of both Calotes and the chick (Hinchliffe 1974; Goel and Mathur 1978).

The INZ is the site of most prominent cell death. It appears between all the digits in all the amniotes studied so far (Hinchliffe 1982). The duration and extent of INZ is also species specific, being less in species with webbed feet (Hinchliffe and Johnson 1980). Furthermore, in a number of mutants the reduction in INZ is associated with presence of the webbed feet. The OP is reported in the deep mesoderm of chick limbs between the zeugopod skeletal elements (Fell and Canti 1934). But it is absent in Calotes, though present in the rat and mouse (Hinchliffe 1982).

The minor areas of cell death occur in the AER, on the digital tips and along the middorsal line of digits, in the superficial tissues. In the deep mesoderm necrotic cells are observed around the blastemal condensation of the stylopodium in Calotes (Goel and Mathur 1978), and in those carpal and tarsal condensations of the lizard and the chick which undergo degeneration (Goel and Mathur 1978; Hinchliffe 1982).

The mechanism of cell death in the limbs as also in the heart (Pexieder 1975) has been extensively studied. In all cases of cell death, the prospective necrotic cells undergo degenerative changes in their cytoplasm as well as nucleus, and the autophagic vacuoles of these cells become active (Dawd and Hinchliffe 1971). The autolytic fragmentation is followed by phagocytic digestion of the fragments by the neighbouring viable mesodermal cells which are transformed into large macrophages. The dying cells may be engulfed by the macrophages even before the fragmentation.

The cause of cell death has been experimentally investigated only in a few cases. In the classical experiments of Saunders and Fallon (1966) it has been shown that the cells of chick wing PNZ are genetically programmed to die at a specific time (stage 24) though a 'death-clock' is set much in advance (stage 17) of the actual time of death. It was also shown that the explantation of the

prospective PNZ cells in suitable locations on the embryo
or under suitable set up in vitro can prevent cell death,
that is, 'death-clock' can be stopped until a certain
developmental stage. For this reason Saunders and Fallon
(1966) compared the events during cell death to those during
cell differentiation since both involved determination. In
comparable experiments on heart bulbar cushions in the chick,
Pexieder (1975) on the other hand, found that the manifestation of cell death in vitro is considerably reduced in the
absence of haemodynamics; in other words, the flow of blood
has a role in causing the cell death. These results do not
agree with the 'death-clock' model of Saunders and Fallon
(1966).

The role of non-genetic factors in cell death as shown
by Pexieder (1975) is widely known. For example, Janus
green inhibits the INZ cell death and causes soft tissue
syndactyly in the chick (Menkes and Deleanu 1964). Excision
of the posterior half of the wing bud in the chick causes
extensive cell death in the ANZ area; no increased cell
death occurs in posterior half of the wing bud following
amputation of the anterior half (Hinchliffe and Gumpel-
Pinot 1981). Hinchliffe and Gumpel-Pinot (1981) suggest
that in order to survive and differentiate, the anterior
part of the wing bud needs a factor from the posterior part
of the bud. In this connection, however, the role of AER
in controlling the cell death should not be overlooked. In
the wing bud experiments the extention of ANZ occurs concurrently with the regression of the AER, and it is already
known that AER degeneration always preceeds the mesodermal
cell death in the limb buds of apodous reptiles (Anguis
fragilis, Python) and the INZ cell death in all the species
studied so far (Raynaud 1977; Milaire 1965).

The role of genetic factors in the cell death, however,
is beyond any doubt. The most cogent argument for this is
the presence of cell death in precisely the same location
at the same developmental stage during the normal development of all the embryos of a species. It is also important
to note that only some cells in a given area die while the
adjoining cells not only live but also proliferate. Further,
the areas where cells die have neither the scarcity of
nutrition (cells die close to blood vessels) nor space.

Before we consider the role of cell death it is worthwhile to compare this type of cell death with the large

scale cell death observed at the time of metamorphosis. The
cell death at metamorphosis always is under hormonal control,
depends on gene activity (can be prevented by actionmycin D
treatment), and serves to recycle the differentiated tissues
which were earlier essential but have now become redundant
(Berrill and Karp 1976). The normal cell death under consideration, on the other hand, does not seem to be under
hormonal influence, nor it apparently requires the immediate expression of genes, and does not involve differentiated
cells which have already served an essential function but
affects apparently undifferentiated cells.

Two different types of roles have so far been suggested
for this type of cell death. First, the cell death removes
the superfluous cells, which in turn may lead to shaping of
an organ (role for INZ or OP), or restrict the supply of
cells to form an organ (role for ANZ; Hinchliffe 1982), or
integrate various parts of an organ (Pexieder 1975), or
remove the unused cells, as in the case of neurons (Knyihar
et al. 1978). Second, the cell death is an ancestral legacy
(Goel and Mathur 1975), that is, it occurs since it had a
role in ancestors.

An important aspect of normal embryonic cell death
remains unexplained by the roles suggested above. It is
enigmatic as to why the embryos should continue to fritter
away their scarce resources of energy and nutrition. The
embryos could have easily evolved alternate strategies to
deal with the excess of cells at a specific location. The
neural crest cells, the germ line cells and the chick
epiblast cells do not die at the site of their origin but
migrate considerable distances (Balinsky 1970). Further,
the morphogenesis of a pentadactyl limb can be accomplished
in the absence of cell death, as in the case of amphibians
(Cameron and Fallon 1977). So the presence of cell death
in similar locations at comparable periods of embryonic
development in species that are widely separated on the
evolutionary tree for millions of years must have some
significance.

The pressure of natural selection operates as much on
the embryos as on the adults, and cell death cannot be beyond the control of selection forces. An example from the
limb system may be given to underscore this point. The
skeletal pattern of the pentadactyl limb, that is, the
number of digits and carpal and tarsal elements, differs

very considerably from species to species (Hinchliffe and
Johnson 1980). Recent work on the lizard (Mathur and Goel
1976), the chick and others (Hinchliffe and Griffiths 1982)
has clearly shown that the embryos have a cartilage pattern
conforming to the adult pattern of bony elements; there are
no significant degenerations or fusions of cartilage
elements as reported by earlier workers (Holmgren 1933;
Montagna 1945) under the influence of biogenetic law. This
means that natural selection apparently acting in the adult
on its bony pattern has suitably changed the cartilage
pattern of the embryo. This can be explained on assuming
that the cartilage pattern is also subject to natural
selection. In the embryo the transformation of the cartilage pattern into the bone pattern may be achieved either
by each cartilage element receiving a separate message for
ossification or by all cartilage elements receiving a
uniform message. The latter, of course, is more probable :
uniform messages in such situations are sent, for example,
from the AER, the ZPA or the thyroid (hormones at the time
of metamorphosis). If one assumes a uniform message for
ossification hypothesis then the embryo whose cartilage
pattern conforms to the desired bone pattern of the adult
(perhaps, due to mutations) will survive better. The
natural selection can thus operate to change the cartilage
pattern. If the natural selection has been successful in
changing the cartilage pattern of the limb, then it is
rather odd that the cell death patterns should remain nearly
constant in all amniote embryos, particularly when no such
cell death occurs in the ancestral amphibians, unless the
cell death has a functional significance.

The hypothesis is put forward that only sometimes the
cell death is responsible for the removal of the superfluous
cells, but more often the dying cells release signals for
the neighbouring cells to follow a particular developmental
pathway. The dying cells in that case are undifferentiated
only in their morphology, and not in their biochemical make
up. The hypothesis receives support from the roles of PNZ
and ANZ. The PNZ is coextensive with the ZPA, which is
generally considered a source of diffusible morphogen. That
the ANZ may act as a second signal point for the determination of anteroposterior axis of limb skeleton is also
becoming clear from the recent experiments (Hinchliffe
personal communication) involving excision of the limbs in
the chick.

Cell death apparently is a multifaceted phenomenon and at present it is difficult to propose a unified hypothesis regarding its role(s) in embryonic development, metamorphosis, adult organism, and situations like overproduction of sperms. It is quite probable that different types of normal cell death have roles that are not comparable.

REFERENCES

Balinsky BI (1970). "An Introduction to Embryology." Philadelphia : Saunders.

Ballard KJ, Holt SJ (1968). Cytological and cytochemical studies on cell death and digestion in foetal rat; the role of macrophages and hydrolytic enzymes. J Cell Sci 3: 245.

Berrill NJ, Karp G (1976). "Development." New York: McGraw-Hill.

Bowen ID, Lockshin RA (1981). "Cell Death in Biology and Pathology." London : Chapman and Hall.

Cameron JA, Fallon JF (1977). The absence of cell death during development of free digits in amphibians. Devl Biol 55: 331.

Davidson EH (1976). "Gene Activity in Early Development." New York: Academic Press, P 322.

Dawd DS, Hinchliffe JR (1971). Cell death in the 'opaque patch' in the central mesenchyme of the developing chick limb: a cytological, cytochemical and electron microscopic analysis. J Embryol exp Morph 26: 401.

Fell HB, Canti RG (1934). Experiments on the development in vitro of the avian knee joint. Proc R Soc Lond B 116:316.

Glucksmann A (1951). Cell death in normal vertebrate ontogeny. Biol Rev 26: 59.

Goel SC, Mathur JK (1975). Cell death in reptilian limb morphogenesis. Proc 62nd Ind Sc Cong Part III: 187.

Goel SC, Mathur JK (1978). Cell death in reptilian limb morphogenesis. Ind J exp Biol 16: 653.

Hinchliffe JR (1974). The patterns of cell death in chick limb morphogenesis. Libyan J Sc 4A: 23.

Hinchliffe JR (1982). Cell death in vertebrate limb morphogenesis. In Harrison RJ, Navaratnam V (eds):"Progress in Anatomy vol 2," Cambridge : Cambridge University Press,p 1.

Hinchliffe JR, Ede DA (1967). Limb development in the polydactylous talpid mutant of the fowl. J Embryol exp Morph 17: 385.

Hinchliffe JR, Griffiths PJ (1982). The prechondrogenic patterns in tetrapod limb development and their phylogenetic significance. Brit Soc Devl Biol Symp on Devl and Evol, p 24.

Hinchliffe JR, Gumpel-Pinot M (1981). Control of maintenance and anteroposterior skeletal differentiation of the anterior mesenchyme of the chick wing bud by its posterior margin (the ZPA). J Embryol exp Morph 62: 63.

Hinchliffe JR, Johnson DR (1980). "The Development of the Vertebrate Limb." Oxford : Clarendon Press.

Holmgren N (1933). On the origin of the tetrapod limb. Acta Zoologica 14 : 187.

Knyihar E, Csillik B, Rakic P (1978). Transient synapses in the embryonic primate spinal cord. Science 202 :1206.

Mathur JK (1974). "Studies on the Development of the Limbs in Calotes versicolor." Ph D Thesis, Poona University.

Mathur JK, Goel SC (1976). Patterns of chondrogenesis and calcification in the developing limb of the lizard, Calotes versicolor. J Morph 149 : 401.

Menkes B, Deleanu M (1964). Leg differentiation and experimental syndactyly in chick embryo. Rev Roumaine Embryol Cytol 1 : 69.

Milaire J (1965). Aspects of limb morphogenesis in mammals. In DeHaan RL, Ursprung H (eds): "Organogenesis," New York: Holt, Rhinehart and Winston, p 283.

Montagna W (1945). A re-investigation of the development of the wing of the fowl. J Morph 76 : 87.

Pexieder T (1975). "Cell Death in the Morphogenesis and Teratogenesis of the Heart." Berlin : Springer-Verlag.

Raynaud A (1977). Somites and early morphogenesis of reptile limbs. In Ede DA, Hinchliffe JR, Balls M (eds): "Vertebrate Limb and Somite Morphogenesis," Cambridge : Cambridge University Press, p 373.

Saunders JW Jr (1948). The proximo-distal sequence of origin of the parts of the chick wing and the role of the ectoderm. J exp Zool 108 : 363.

Saunders JW Jr, Fallon JF (1966). Cell death in morphogenesis. In Locke M (ed) : "Major Problems in Developmental Biology," New York : Academic Press, p 289.

Saunders JW Jr. Gasseling MT, Saunders LC (1962). Cellular death in morphogenesis of the avian wing. Devl Biol 5:147.

Wolpert L (1978). Pattern formation in biological development. Scient Am 239 : 124.

Limb Development and Regeneration
Part A, pages 183-193
© 1983 Alan R. Liss, Inc., 150 Fifth Avenue, New York, NY 10011

DEVELOPMENT OF MOTOR INNERVATION OF CHICK LIMBS

Margaret Hollyday
Department of Pharmacological and Physiological
    Sciences
The University of Chicago
Chicago, Illinois   60637

The developmental processes responsible for the formation of specific patterns of connections between motoneurons
and the muscles of the limb have received considerable attention in recent years (for reviews see: Hamburger 1977;
Hollyday 1980a; Landmesser 1980; Hollyday and Grobstein 1981;
Landmesser 1981; Summerbell and Stirling 1982).   In this
paper, I describe recent studies from my laboratory on the
normal development of motor projections to the chick limb.
I begin by describing studies on the early development of
spinal motoneurons in relation to the differentiation of the
limb tissue.   These provide a picture of the behavior of axons
at the earliest stages of outgrowth as well as of the conditions in the limb bud at the time connections first begin to
form.   They also provide a description of the morphogenesis
of peripheral nerve branching patterns in the limb which is
important for an analysis of the precision with which axons
initially innervate the limb.

Limb muscles are innervated by groups of motoneurons,
"motor pools," located in stereotyped positions within the
lateral motor columns (Romanes 1964; Landmesser 1978a; Hollyday 1980b).   One possible model for explaining the highly
specific and stereotyped patterns of connections between
motoneurons and limb muscles seen in adult animals is that
motor axons initially project into the limb bud in a diffuse
pattern which is subsequently refined.   It is known that a
substantial number of motoneurons die during normal development at a time when the individual muscles are separating
from the precursor tissues, the embryonic pre-muscle masses
(Hughes 1968; Hamburger 1975).   The possibility exists that

this cell death serves to remove aberrently projecting motor
axons. Hence, it is of interest to know the extent to which
the initial motor projections resemble those of the adult.
If the motor projection pattern observed before the period
of motoneuron loss differs substantially from that in the
mature animal, then cell death could serve to transform an
initially diffuse projection pattern into the precise adult
pattern. In the second section of the paper I describe
studies in which we assessed the accuracy of the early projection patterns relative to those found in the mature animal.

We have studied the development of limb innervation
using horseradish peroxidase (HRP) as an orthograde tracer
of axonal projections. Chick embryos at various stages of
development (staged according to Hamburger and Hamilton 1951)
were removed from the shell and placed into a dish containing
oxygenated Tyrode's solution (Landmesser 1978a). The embryos
were eviscerated and the notochord removed to expose the
ventral surface of the spinal cord and the various nerves
projecting to the limbs. Injections of concentrated solutions
of HRP were made with a glass micropipette directly into the
spinal cord or ventral roots of identified segments. Neurons
and axonal processes in the vicinity of the injected HRP take
up the injected enzyme. Because neurons transport material
both towards and away from their cell bodies, the entire
neuron including its soma and growing axon can be labeled in
this way. After a suitable survival period to allow transport of the enzyme, the embryo is fixed, sectioned on a
vibratome and reacted with cobalt and diaminobenzidine according to the procedure of Adams (1977). Neurons labeled via
this technique are densely filled and resemble Golgi-stained
preparations. Unlike the Golgi technique, or other silver
staining methods, it is possible by appropriate placement of
injections to select particular neuronal populations to be
visualized.

DEVELOPMENT OF MOTOR PROJECTIONS TO LIMBS

A summary of the major developmental events of limb
innervation and their relation to the morphogenesis of the
limb muscles is given in figure 1.

The youngest embryos we have studied were at stage 17.
This is only a few hours after the earliest-born motoneurons
have completed their final round of DNA synthesis (Hollyday

DEVELOPMENTAL TIMETABLE

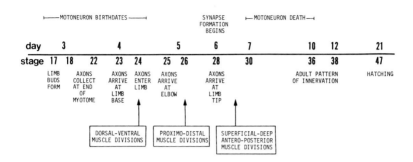

Fig. 1. Summary of major events of chick limb innervation. (Modified from Whitelaw 1981).

and Hamburger 1977) and is during the time when motoneurons are migrating into the lateral motor column from the ventricular zone. HRP injections into the neural tube revealed motor axons which were directed laterally into the somites and had growth cones at their tips (see fig. 2). Axons from several adjacent motoneurons could be seen bundled together as they emerged from the neural tube. Some of the bundled axons separated from each other within the somite and had large, membranous growth cones and filopodial processes extending in all directions. A few of the axons had grown through the somite, as far as the medial surface of the myotome. The appearance of these axons with very prominent, active-looking growth cones suggests to us that they were exploring their environment as they emerged from the neural tube.

In stage 18 embryos, both active-looking growth cones and also axons ending in club-shaped structures, or having a beaded appearance, were present at the medial border of the myotome. It seems likely that the latter at least represent processes which are either degenerating or being withdrawn in favor of the more correctly (ventrally) directed axons seen for the first time in stage 18 embryos. The trajectories of the limb directed axons were highly variable, suggesting that their outgrowth was not guided by channels or some other oriented substrate within the somites.

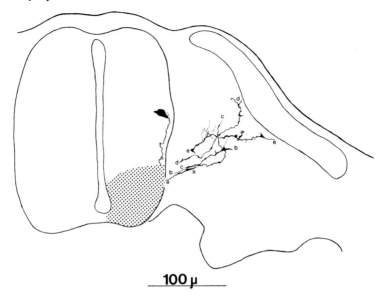

100 µ

Fig. 2. Motor axons with growth cones extending from neural tube into the somite and reaching as far as the medial surface of myotome. Stage 17 chick embryo. HRP injection site shown by heavy stippling. Individual axons marked with letters a-e.

A progressive accumulation of motor axons at the ventral tip of the myotome was observed between stages 18 and 22 (between approximately 72 and 88 hours of incubation). As more axons reached the ends of the myotomes, they grouped together forming the spinal nerves. The axons from late-born motoneurons probably grow along the earlier arriving axons. After collecting at the ventral tip of the myotome as separate nerves, the axons converged at the base of the limb bud at stage 23. Motor axons were first observed growing into both the wing bud and the hindlimb bud at stage 24. This is when the central condensation of pre-cartilage first becomes distinguishable from the cells forming the pre-muscle masses. There is thus a substantial "waiting period" (from stages 18 to 23) in which axons collect first at the base of the myotome and then come together at the base of the limb before entering as a group. The significance of this waiting period is not known. It is tempting to speculate that the early differentiation of the limb bud coincident with the separation of the limb mesenchyme into pre-cartilage and pre-muscle masses provides permissive conditions for axonal ingrowth.

After entering the limb bud, the groups of axons first project as two sheets of fibers, one directed towards the dorsal pre-muscle mass and the other towards the ventral pre-muscle mass. The clear time difference between motoneuron birthdates for dorsally- and ventrally-destined motoneurons (Hollyday and Hamburger 1977) is not reflected in a difference in the order of axonal ingrowth toward these two muscle masses suggesting that motoneuron birth order is not critical for patterning motor projections. The hindlimb is innervated by two limb plexuses while the wing is supplied by only one. Despite this difference a similar pattern of early motor projections to the dorsal and ventral pre-muscle masses is readily apparent. In the wing, the separation of axons from the brachial plexus into a dorsally directed and a ventrally directed sheet of nerve fibers is seen at the proximal head of the humerus. In the hindlimb, the proximal head of the femur marks the point of divergence; axons forming the crural plexus diverge around the anterior head of the femur while those from the sciatic plexus diverge around the posterior head of the femur.

The characteristic pattern of peripheral nerve branches derived from the common pattern of a dorsally projecting sheet and a ventrally projecting sheet of axons becomes apparent between stages 26 and 28. This corresponds to the time when the dorsal and ventral pre-muscle masses are beginning to separate to form muscles. At stage 26, short axons with growth cones at their tips can be seen emerging from the two sheets of nerve fibers in places appropriate for the proximal peripheral nerves. The youngest embryos in which a characteristic brachial plexus is recognizable have been at stages 26+ to 27+. By stage 28, the formation of individual muscle nerves is more distinct, especially in the shoulder and upper arm region, and the hip. The major divisions of the nerves to the forearm and the calf are also apparent, but few individual muscle nerves have yet formed in these regions. The first morphological signs of synapse formation are seen at stage 28 which is in agreement with the electrophysiological studies by Landmesser and Morris (1975) and Landmesser (1978b) on the early innervation of hindlimb muscles.

Other experiments involving injections of the primary sensory neurons in the dorsal root ganglia have shown that the early time course of sensory innervation is very similar to that of motor innervation. The extended period of ganglion cell proliferation (Carr and Simpson 1978) relative to that of the motoneurons suggest that sensory axons continue to innervate the limb after motor innervation is complete.

SPECIFICITY OF INITIAL MOTOR PROJECTIONS

The specificity of the early motor projections to the wing has been studied by injecting HRP into single spinal segments of the brachial cord of embryos at various stages. An example of the results obtained in a single, stage 28 embryo is illustrated in the reconstruction of the brachial plexus shown in figure 3. Comparable results were obtained

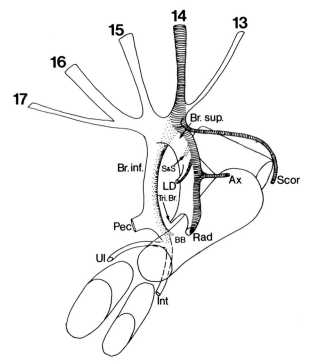

Fig. 3. Reconstruction of right brachial plexus of stage 28 chick embryo showing distribution of labeled axons emerging from the 14th spinal segment. Note that labeled axons were found in only some of the peripheral nerves formed by brachial segments 13-17. Abbreviations: Ax-axillaris; BB-biceps brachii; Br.inf.-brachialis longus inferior; Br.sup.-brachialis longus superior; Int-interosseus; LD-latissimus dorsi; Pec-pectoralis; Rad-radialis; S&S-subscapularis and scapulohumeralis; Scor-supracoracoideus; Ul-ulnaris.

when other segmental nerves were labeled. From the time when axons first grow into the wing bud, those emerging from a single spinal segment were distributed to a subset of the total number of peripheral nerves formed by the brachial plexus. At no stage were axons from a single spinal segment observed to project widely in the limb. Indeed, even before the nerve branches typical of the mature innervation pattern emerged from the two sheets of nerve fibers, axons labeled from injections of a single spinal segment grouped together in a restricted part of the sheet. This suggests a fairly high degree of order in the early projection pattern.

Comparisons of the early projection patterns with the post-hatching innervation pattern were made to assess the accuracy of the initial motor projection pattern. The set of muscle nerves expected to be supplied by motoneurons in each spinal segment was inferred from motor pool mapping experiments done on hatched chicks (Hollyday and Jacobson, unpublished results). In the vast majority of experiments done at stages 26+ to 27+, and at stages 28 to 29 (see fig. 4), labeled axons were found in the expected nerves exclusively. However, in a few cases, labeled axons were found in unexpected spinal nerves. More instances of labeled axons in unexpected nerves were found in the younger group of embryos than in the older group. The numbers of labeled axons in unexpected nerves were few, sometimes only a single axon was seen, but their presence was unequivocal. By stages 28 and 29, the number of such observations was reduced to 6 in a total of 40 embryos studied.

Our interpretation of these findings is that in general axon outgrowth is not random, but rather is quite specific. The occasional projection errors that were seen involved axons going to neighboring regions of the dividing muscle masses. The number of early projection errors detected was small at stage 26+ to 27+, and was reduced even further by stages 28 and 29. The majority of neurons that die during normal limb innervation die after this time, between stages 29 and 34 (Hamburger 1975; Oppenheim and Majors-Willard 1978). Hence cell death does not play a major role in patterning motoneuron projections. However, since some dying motoneurons have been reported as early as day 5 of incubation (Chu-Wang and Oppenheim 1978) it is possible that the small numbers of growth errors we observed are indeed removed by cell death. An alternative explanation is that these erroneously project-ing axons are simply retracted.

STAGE 28 to 29	Supra - Coracoideus	Biceps	Pectoralis	Brachialis inferior (Ulnaris & Interosseus)	Axillaris (Deltoid & Cbr. ant.)	Subscap. Scapulohum	Latissimus dorsi	Triceps	Brachialis superior (Radialis)
12 N=7*									
13 N=5	5/5				4/5	4/5			
14 N=9	9/9	9/9	3/9	1/9	9/9	9/9	3/9		4/9
15 N=6		2/6	6/6	6/6	4/6	4/6	6/6	6/6	6/6
16 N=8			8/8	8/8			4/8	7/8	8/8
17 N=5				5/5					

* 1 case joined 13 and entered plexus

Fig. 4. Summary of experiments in which HRP was injected into individual spinal segments 12 thru 17 in chick embryos at stages 28 to 29. The nerves examined are given across the top of the figure. Cross-hatched squares indicate nerves expected to contain labeled axons based on results from experiments in hatched chicks in which the positions of motor pools for wing muscles was determined by retrograde transport of HRP. The number of cases in which labeled axons were found in given wing nerves is shown. Note that the pattern of labeled nerves is very similar to that of the mature animal.

Our conclusion that initial axon outgrowth to the wing is quite specific is similar to that of Lance-Jones and Land-messer (1981) who used orthograde tracing techniques similar to ours to study the chick hindlimb, and with the conclusions from other studies on the hindlimb using both electrophysio-logical and retrograde labeling techniques (Landmesser and Morris 1975; Landmesser 1978b). Our conclusions differ from those of an earlier study on the chick wing (Pettigrew, Lindeman and Bennett 1979) which concluded that some muscles in the chick wing are initially supplied by motoneurons in abnormal spinal segments. The reasons for this discrepancy are not clear.

CONCLUSIONS

Our analysis of early projection patterns indicates that from very early stages in the development of limb innervation the relation between motoneurons and muscles displays a specificity not greatly different from that which characterizes the adult pattern. The explanation for the adult specificity thus cannot be found by analysis of relatively late developmental events such as motoneuron death but rather must be sought by analysis of events during or even prior to initial innervation of the developing limb bud. Our characterization of the behavior of outgrowing motoneurons represents the beginnings of such an analysis. Several aspects of it are worth particular note in this regard. First, the initial segregation into a dorsally directed and a ventrally directed sheet may relate to the finding that in some experimentally perturbed situations motoneurons display a preference for muscles derived from the same (dorsal or ventral) primary embryonic muscle mass as that from which their normal target muscle is derived (Hollyday 1981). Second, prior to this segregation there is a "waiting period" when motoneuron axons cluster at the base of the limb bud. How important this is for normal innervation and whether it represents a time of some sorting of axons or of some signalling between the developing limb and spinal cord remains to be determined. Third, motoneurons have axons in the periphery at very early stages, both relative to their own development and to that of the limb. Whether important interactions take place at these very early times also remains to be determined.

ACKNOWLEDGEMENTS

Some of the experiments described in this paper were done with Heather Perry and Bradley Lau. The research in my lab is supported by PHS grant #NS-14066 and by the Spencer Foundation. I thank Dr. Paul Grobstein for his assistance with this manuscript.

REFERENCES

Adams JC (1977). Technical considerations on the use of horseradish peroxidase as a neuronal marker. Neuroscience 2:141.
Carr VM, Simpson SB (1978). Proliferative and degenerative events in the early development of chick dorsal root ganglia.

J Comp Neurol 182:727.
Chu-Wang I-W, Oppenheim RW (1978). Cell death of motoneurons in the chick embryo spinal cord. I. A light and electron microscopic study of naturally occurring and induced cell loss during development. J Comp Neurol 177:33.
Hamburger V (1975). Cell death in the development of the lateral motor column of the chick embryo. J Comp Neurol 160:535.
Hamburger V (1977). The developmental history of the motor neuron. Neurosci Res Prog Bull Vol 15 Suppl:1.
Hamburger V, Hamilton H (1951). A series of normal stages in the development of the chick embryo. J Morph 88:49.
Hollyday M (1980a). Motoneuron histogenesis and the development of limb innervation. In Hunt RK (ed) "Current Topics in Developmental Biology Vol 15," New York: Academic Press, p181.
Hollyday M (1980b). Organization of motor pools in the chick lumbar lateral motor column. J Comp Neurol 194:143.
Hollyday M (1981). Rules of motor innervation in chick embryos with supernumerary limbs. J Comp Neurol 202:439.
Hollyday M, Grobstein P (1981). Of eyes and limbs and neuronal connectivity. In Cowan WM (ed) "Studies in Developmental Neurobiology: Essays in Honor of Viktor Hamburger," New York: Oxford Univ Press, p 188.
Hollyday M, Hamburger V (1977). An autoradiographic study of the formation of the lateral motor column in the chick embryo. Brain Res 132:197.
Hughes AFW (1968). "Aspects of Neural Ontogeny." London: Logos.
Lance-Jones C, Landmesser L (1981). Pathway selection by chick lumbosacral motoneurons during normal development. Proc Roy Soc Lond B 214:1.
Landmesser L (1978a). The distribution of motoneurons supplying chick hind limb muscles. J Physiol Lond 284:371.
Landmesser L (1978b). The development of motor projection patterns in the chick hind limb. J Physiol Lond 284:391.
Landmesser L (1980). The generation of neuromuscular specificity. Ann Rev Neurosci 3:279.
Landmesser L (1981). Pathway selection by embryonic neurons. In Cowan WM (ed) "Studies in Developmental Neurobiology: Essays in Honor of Viktor Hamburger," New York: Oxford Univ Press, p 53.
Landmesser L, Morris DG (1975). The development of functional innervation in the hind limb of the chick embryo. J Physiol Lond 249:301.
Oppenheim RW, Majors-Willard C (1978). Neuronal cell death in the brachial spinal cord of the chick is unrelated to the loss of polyneuronal innervation in wing muscle. Brain Res

154:148.

Pettigrew AG, Lindeman R, Bennett MR (1979). Development of segmental innervation of the chick forelimb. J Embryol exp Morph 49:115.

Romanes GJ (1964). The motor pools of the spinal cord. In Eccles JC, Schade JP (eds): Progress in Brain Research Vol. 11," Amsterdam: Elsevier, p 93.

Summerbell D, Stirling RV (1982). Development of the pattern of innervation of the chick limb. Am Zool 22:173.

Whitelaw V (1981). Developmental mechanisms involved in chick limb innervation. PhD Thesis. University of Chicago.

Limb Development and Regeneration
Part A, pages 195–205
© **1983 Alan R. Liss, Inc., 150 Fifth Avenue, New York, NY 10011**

THE CONTROL OF AXON OUTGROWTH IN THE DEVELOPING CHICK WING

Julian Lewis, Loulwah Al-Ghaith, Gavin Swanson
and Akbar Khan

Department of Anatomy,
King's College University of London,
London WC2R 2LS, and
Department of Biology as Applied to Medicine,
Middlesex Hospital Medical School,
London W1P 6DB.

The vertebrate limb is assembled from several distinct and originally separate cell lineages – epidermis from the ectoderm, connective tissues from the lateral plate meso-derm, muscle from the somites, and nerve fibres, Schwann cells and melanocytes from the neural tube and neural crest. Each of these components has its own precise pattern, whose development is accurately coordinated in space and time with the development of the others to give the final complex yet orderly structure. How is this coordination achieved? Can any one class of cells be singled out as the master com-ponent, controlling the pattern of the rest? So far as the non-neural tissues are concerned, the non-muscle mesenchyme derived from the lateral plate appears to play this key role: its cells seem to be the repository of positional information in the limb (Wolpert, 1978). This paper will consider to what extent the same mesodermal component can be said also to determine the pattern and timetable of in-nervation.

Before turning to the nervous system, however, it is helpful to summarise the control relationships between the non-neural components of the limb. At early stages, the limb mesenchyme depends on signals from the ectoderm, which, though crucial, are thought to be of a relatively simple and non-specific nature. The detailed structure of the skeleton represents a pattern of positional values within the mesenchyme set up largely through complex responses of the mesenchymal cells to these simple cues operating in conjunction with interactions of the mesenchymal cells with one another (Rubin and Saunders, 1972; Summerbell, Lewis and Wolpert, 1973; Kieny, 1977; Tickle, Summerbell and Wolpert, 1975). The pattern of positional values having been assigned within the mesenchyme, this tissue defines the pattern of epidermal appendages formed subsequently from the ectoderm (Sengel, 1976).

The respective roles of the non-muscle mesenchyme cells and the somite-derived myoblasts can be analysed by deletion and transposition experiments. If the somites are destroyed by X-irradiation so as to deprive the limb bud of myoblasts, the non-muscle mesenchyme cells nevertheless form a normal pattern of connective tissues, including skeleton, dermis and even tendons (Chevallier, Kieny and Mauger 1978; Kieny and Chevallier, 1979). If the somitic mesoderm at the wing level is replaced by somitic mesoderm from elsewhere, the displaced myoblasts enter the wing and there form a normal wing pattern of muscles (Chevallier, 1979). In an analogous way, grafted melanoblasts will form the pattern of pigmentation appropriate to the limb (Rawles, 1948). All these experiments indicate that the non-muscle mesenchyme orchestrates the patterning of the other commponents of the limb, and in developing its own pattern does not depend on them.

If the neural tube and neural crest at the brachial level are destroyed in the early embryo – as can be done, for example, using a focussed beam of ultraviolet light (Lewis, 1980) – the adjacent pair of limbs will develop a normal pattern of bones, muscles and feather papillae even though devoid of innervation. Nerves, therefore, do not control the patterning of the rest of the limb. Given this simplifying fact, one can proceed to ask how the rest of the limb controls the patterning of the nerves.

The nerves in a limb form a pattern whose precise and predictable nature is demonstrated by a comparison of the right and left limbs of a normal embryo: there is near-perfect symmetry with respect to the routes of the main nerve trunks and the positions of the branches that diverge from them to innervate individual muscles and patches of

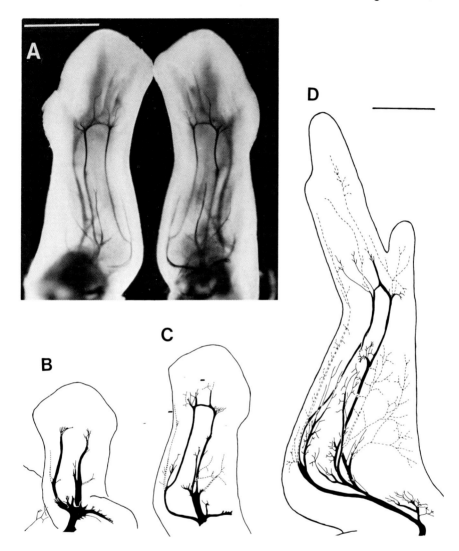

Fig. 1. (A) The right and left wings of a stage 31 embryo, silver-stained as a whole mount and viewed from the ventral side with dark-field illumination. The dorsal system of nerves is out of focus. (B), (C) and (D) Camera lucida drawings of the ventral system of nerves in wings at stages 29, 30/31 and 35, respectively; cutaneous nerve branches are shown dotted. Scale bars = 1 mm. Staining method as in Lewis et al. (1981); (B) – (D) from Swanson and Lewis (1982).

skin (Fig. 1). Asymmetries become apparent only within those
target muscles and skin regions, where the pattern of arbor-
isation is more or less random. The first crucial point to
be made about the control of the regular nerve pattern is
that it is determined (to a good approximation) by the limb
tissues rather than by the specific characters of the
nerves. That is, when axons that would normally innervate
some other part of the body are caused experimentally to
innervate a limb, they follow practically the same set of
routes that would be followed by normal limb nerves (see,
for example, Hamburger (1939) and Straznicky (1963)). Thus
the limb may be said to provide "public highways" for nerve
outgrowth - public in the sense that they are open to axons
from any source, highways in the sense that they constitute
a restricted set of permitted routes through the terrain of
the limb.

The highway system does not, however, define fully the
trajectory of a given axon. The highways branch, and a
growth cone entering the limb faces a series of discrete
choices at the branch points. Landmesser and Lance-Jones
(1981a) have shown that these choices are normally made
according to strict rules, such that motor axons from specific motor neuron pools in the spinal cord are regularly
directed to specific target muscles. The selectivity appears
to be governed by differences of intrinsic character between
the different motor neuron pools on the one hand and the
different branches of the highway system on the other.
Experimentally misplaced portions of spinal cord still send
their axons to the appropriate muscles, provided that the
axons are funnelled into the appropriate limb plexus. With
more extreme disturbances, such that the axons are funnelled
into the wrong plexus, selectivity usually seems to be lost:
the axons are still generally confined to the standard
public highways, but within that system they make their
choices more or less randomly, showing no regular preference
for specific branches, as though unable to recognise the
signposts on unfamiliar roads (Landmesser and Lance-Jones,
1981b). The specificity of nerve connections is not our main
theme in this paper, however. We have digressed to discuss
it in order to emphasize that two levels of axon guidance
can be distinguished: one - our main concern here - defining
the public highway system, the other, more delicate, controlling the specific choices that growth cones make between
the different branches of the public highway system.

Some insight into the way in which the limb tissues define the routes of nerves is provided by a study of the early stages of limb innervation (Al-Ghaith and Lewis, 1982). The first parts of the nerve pattern to develop are the main mixed nerve trunks, such as the radialis profundus on the dorsal side of the wing and the interosseous and median nerves on the ventral side. In the period from stage 26 to stage 28, the pioneer growth cones of these three nerves advance from the neighbourhood of the elbow to the neighbourhood of the wrist. Close examination of the tips of the nerves shows that the pioneer growth cones follow routes which are indeed precisely defined, but not quite so precisely defined as the paths of the eventual mature nerve fascicles. The pioneer growth cones behave in a somewhat exploratory fashion, giving the tip of the developing nerve a "frayed" appearance (Fig. 2(A)). The extent of the fraying varies as the growth cones proceed through the different regions of the wing: it is barely perceptible in the mid-forearm, for example, but very marked in the neighbourhood of the wrist. Thus at stage 28 the pioneer fibres approaching the wrist diverge from one another to the extent of about 60 um, though the fascicle formed by the bundling-together of the axons just proximal to the frayed end is only about 10 um in diameter. It is noteworthy also that the proximal nerve trunks at very early stages are broader and less compact than they will be a little later. These observations suggest that lateral forces of cohesion bind the axons of the developing nerve together into a fascicle, whose course is thereby more narrowly and precisely constrained than the path of a single isolated nerve fibre would be: the course of the fascicle represents an average of the courses taken by the individual pioneer growth cones.

But how are these pioneer growth cones themselves guided through the virgin mesenchyme? What are they crawling over? We have examined the growing end of the interosseous nerve by transmission electron microscopy in transverse sections cut serially at closely spaced intervals from the proximal tightly fasciculated region out to the furthest part of the frayed end. The pioneer growth cones travel singly or in small bundles (Fig. 2(B)), and it is striking that they make close (10 - 20 nm) contact with other cells over practically the whole of their surface; any part that is not in contact with other growth cones is closely enfolded by mesenchymal cells. This suggests that the routes for axon outgrowth are marked out by the local surface

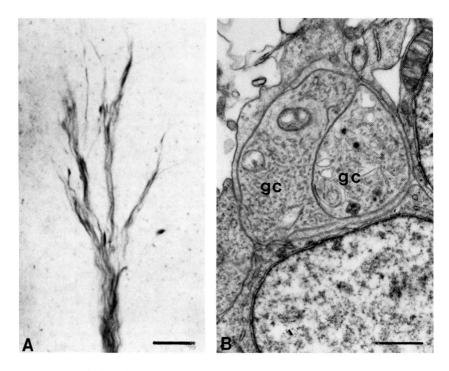

Fig. 2. (A) The growing end of the interosseous nerve at stage 28, as the pioneer growth cones are approaching the wrist. Silver-stained whole mount; scale bar = 20 um. (B) Electron micrograph of a pair of pioneer growth cones closely enfolded by mesenchyme cells. Scale bar = 1 um. From Al-Ghaith and Lewis (1982).

properties of the mesenchymal cells.

The close wrapping of mesenchymal cells around the pioneer growth cones is continued in the more proximal parts of the fascicle, forming a primitive perineurial sheath. Growth cones travelling out a little later than the pioneers can be seen inside this sheath. They show a strong tendency to be located in the outermost part of the fascicle, so as to have part of their surface in contact with axons and part in contact with the perineurial sheath cells. Inside the fascicle, the individual axons lie in close contact with one another – myelination does not begin until about 12 days of incubation.

The intimate contact between the nerve fibres and the perineurial cells raises the question of whether the latter might be immature Schwann cells. Though we do not have a direct answer to this question, we do at least have evidence suggesting that Schwann cells do not play any essential part in defining the routes of nerve outgrowth. This has come from extirpation experiments using a focussed beam of ultraviolet light from a low-pressure mercury lamp to destroy the neural crest in the two-day embryo. The lamp produces radiation chiefly in the far UV (at 254 nm), which penetrates and kills only the most superficial layers of cells. After irradiating the neural crest over an extent of at least 12 somite widths centred on the brachial levels, we leave the embryos to develop for another 6 – 7 days and examine the pattern of innervation by silver staining. Though the results are rather variable, we achieve success, or at least the appearance of success, in about 10% of cases (out of a series of just over 300 embryos). In these, the cutaneous nerve branches are largely eliminated, while the pattern of main nerve trunks and muscle nerve branches is largely normal. The absence of cutaneous nerve branches implies that the neural crest (from which sensory neurons derive) has been destroyed, and this is confirmed by examination of serial sections of the trunk region: in most cases, the dorsal root ganglia are absent or reduced to less than 10% of their normal size. Two deductions can be drawn. First, it is evident that the motor axons do not depend on sensory axons to find the way through the limb for them. Second, it seems unlikely that Schwann cells or any other class of cells derived from the neural crest are required in order to guide the motor axons along the standard routes; though clearly further studies are required before this conclusion can be stated firmly.

Given that the limb mesenchyme defines the spatial pattern of innervation, what controls the timing of nerve outgrowth – the capacities of the neurons or the opportunities offered by the mesenchyme? To find out, we have surgically exchanged right wing buds between young (typically stage 18) and somewhat older (typically stage 22) embryos, and have assessed the maturity of the pattern of innervation some three days later by silver staining (Swanson and Lewis, 1982). The contralateral undisturbed wings serve as con-

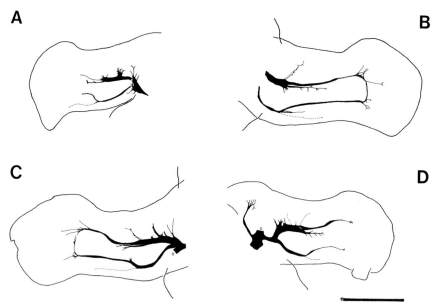

Fig. 3. Camera lucida drawings of the ventral pattern of innervation in wing buds exchanged between embryos of different ages. The embryos were at stages 17 and 22 respectively at the time of the operation, and at stages 28 and 30 respectively at the time of fixation. (A) The young wing grafted onto the old host; (B) the old control wing; (C) the old wing grafted onto the young host; (D) the young control wing. Scale bar = 1 mm. From Swanson and Lewis (1982).

trols. We find that the extent of the innervation corresponds to the age of the wing bud, not to the age of the host embryo that has supplied its nerves: compared with the host control limbs, the innervation of the young bud grafted onto the old host is retarded while that of the old bud on the young host is precociously developed (Fig. 3). This implies that the timing of nerve development, as well as its spatial pattern, is governed by the opening-up of highways for axon outgrowth in the mesenchyme. Both the mixed nerve trunks and the side branches that go to individual muscles develop according to the mesenchymal timetable.

The causal analysis can be taken a step further since the limb mesenchyme is itself a composite tissue. It turns out that the muscle cells and the connective tissue cells are both important in controlling axon outgrowth, but in different ways. The muscle cells, as might be expected, are essential for the development of the muscle nerve branches; the non-muscle mesenchyme cells define the routes of the mixed nerve trunks and cutaneous nerve branches. The evidence comes from experiments in which the brachial somitic mesoderm on one side of the embryo is destroyed by X-irradiation at two days of incubation, so as to deprive the wing bud of myoblasts (Lewis, Chevallier, Kieny and Wolpert, 1981). In this way a muscleless wing is obtained, with a practically normal pattern of connective tissues. When this limb is fixed at a relatively late stage and silver-stained, it is found that the main nerve trunks and cutaneous nerve branches are present and essentially normal, while the usual muscle nerve branches are absent.

The meaning of this observation becomes clearer when one examines specimens at earlier stages. In a normal wing, the various muscle nerve branches first become visible a little later than the main nerve trunks, between stages 27 and 32, at about the same time as the corresponding myotubes are beginning to form, and a little before the individual muscles split off from the dorsal and ventral muscle masses. Most of the muscle rudiments at first lie very close to the main nerve trunks, and the muscle nerve branches originate as short, relatively unfasciculated tufts of nerve fibres

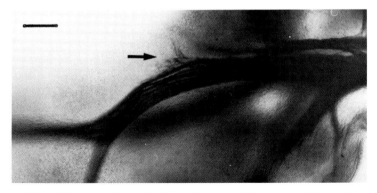

Fig. 4. The early origins of a muscle nerve branch as a short tuft of nerve fibres side-tracked from an established nerve trunk into closely adjacent differentiating muscle. The arrow points to the rudiment of the nerve branch to the muscle extensor metacarpi radialis, as seen at stage 28 in a silver-stained whole mount. Scale bar = 100 um.

diverging only slightly from the main nerve trunks to enter the target tissue (Fig. 4). Subsequently, as the limb grows and the muscles shift relative to the nerve trunks, these rudimentary muscle nerve branches become drawn out into well-formed fascicles which often stretch for a considerable distance between the main nerve trunk and the muscle. In wings devoid of muscle, we find that the main nerve trunks form more or less on schedule, but the muscle nerve branches do not even begin to develop.

Clearly, the absence of muscle nerve branches in the muscleless limbs does not imply that the muscle cells must necessarily exert a long-range guiding influence on the growth cones of the motor axons. The simplest hypothesis suggested by the data is rather that the muscle cells provide something in the nature of a "stop" signal: the non-muscle mesenchyme cells define the highways for axon outgrowth, and the muscle cells beside the highway simply act as sites where axon terminals, having travelled along the highway, form lasting attachments. Once such attachments have been formed, other axons coming later in development can be guided to the same target muscle by following the nerve fascicle, even if the muscle itself shifts.

What answer then can be given to the question posed at the beginning of this paper? Evidently, the non-muscle mesenchyme cells play the master role in the patterning of limb innervation, as they do in the patterning of the other features of the limb. Some aspects of their control of nerve development are indeed exerted indirectly, via control of the disposition of muscle cells; but it seems that the functions of these intermediaries in nerve guidance may be only of a very simple sort.

We thank the MRC, the Government of Kuwait, and King's College London for financial support.

REFERENCES

Al-Ghaith LK, Lewis JH (1982). Pioneer growth cones in virgin mesenchyme: an electron-micrscope study in the developing chick wing. J Embryol Exp Morphol 68:149.
Chevallier A (1979). Role of the somitic mesoderm in the development of the thorax in bird embryos. II. Origin of thoracic and appendicular musculature. J Embryol Exp Morphol 49:73.

Chevallier A, Kieny M, Mauger A (1978). Limb-somite relationship: effect of removal of the somitic mesoderm on the wing musculature. J Embryol Exp Morphol 49:73.

Hamburger V (1939). The development and innervation of transplanted limb primordia of the chick embryo. J Exp Zool 80:347.

Kieny M (1977). Proximo-distal pattern formation in avian limb development. In Ede DA, Hinchliffe JR, Ball M (eds): "Vertebrate Limb and Somite Morphogenesis," Cambridge: Cambridge University Press, p. 87.

Kieny M, Chevallier A (1979). Autonomy of tendon development in the embryonic chick wing. J Embryol Exp Morphol 49:153.

Lance-Jones C, Landmesser L (1981a). Pathway selection by chick lumbosacral motoneurons during normal development. Proc Roy Soc Lond B 214:1.

Lance-Jones C, Landmesser L (1981b). Pathway selection by embryonic chick motoneurons in an experimentally altered environment. Proc Roy Soc Lond B 214:19.

Lewis J (1980). Defective innervation and defective limbs: causes and effects in the developing chick wing. In Merker H-J, Nau H, Neubert D (eds): "Teratology of the Limbs," Berlin: de Gruyter, p. 235.

Lewis J, Chevallier A, Kieny M, Wolpert L (1981). Muscle nerve branches do not develop in chick wings devoid of muscle. J Embryol Exp Morphol 64:211.

Rubin L, Saunders JW (1972). Ectodermal-mesodermal interactions in the growth of limb buds in the chick embryo: constancy and temporal limits of the ectodermal induction. Dev Biol 28:94.

Sengel P (1976). Morphogenesis of Skin. Cambridge: Cambridge University Press.

Straznicky K (1963). Function of heterotopic spinal cord segments investigated in the chick. Acta Biol Hung 14:145.

Summerbell D, Lewis JH, Wolpert L (1973). Positional information in chick limb morphogenesis. Nature 244:492.

Swanson GJ, Lewis J (1982). The timetable of innervation and its control in the chick wing bud. J Embryol Exp Morphol, in press.

Tickle C, Summerbell D, Wolpert L (1975). Positional signalling and specification of digits in chick limb morphogenesis. Nature 254:199.

Wolpert L (1978). Pattern formation in biological development. Sci Am 239(4):124.

Limb Development and Regeneration
Part A, pages 207–216
© **1983 Alan R. Liss, Inc., 150 Fifth Avenue, New York, NY 10011**

THE RESPONSE OF AVIAN HINDLIMB MOTOR AND SENSORY
NEURONS TO AN ALTERED PERIPHERY

L.T. Landmesser, M.J. O'Donovan and M. Honig

Biology Department
Yale University
New Haven, CT 06511

Motoneuron Projection Patterns Following Displacement
from Peripheral Targets.

In an earlier set of experiments, the response of
embryonic chick motoneurons was explored, following
surgical manipulations that displaced hindlimb
motoneurons from their original targets (Lance-Jones and
Landmesser 1980, 1981b). It was found that after
moderate lengths of St 15-16 neural tube were reversed
about the a-p axis (this is before lumbosacral
motoneurons are born and therefore prior to axon
outgrowth), motoneuron axons invariably reached their
original muscle.

This is shown in the serial reconstructions of
hindlimb nerve patterns depicted in Figure 1. In A, the
extent of the tube reversal is indicated by the large
arrow, and the numbers refer to the lumbosacral spinal
(LS) nerves. LS3 motoneurons, placed into the position
of LS1, altered their axon trajectories in the crural
plexus to reach their original muscle, the femorotibialis
(femoro). They bypassed the sartorius (sart) nerve which
would normally receive axons from motoneurons in the
position of LS1. LS3 motoneurons had been labeled with
horseradish peroxidase (HRP) (indicated by stippling) and
their axons could therefore be traced to termination
sites within the limbs. Similar results can be seen in B
where displaced (LS1) sartorius axons cross anteriorly
within the plexus to project out their original nerve.
We concluded that motoneurons must be specified or have

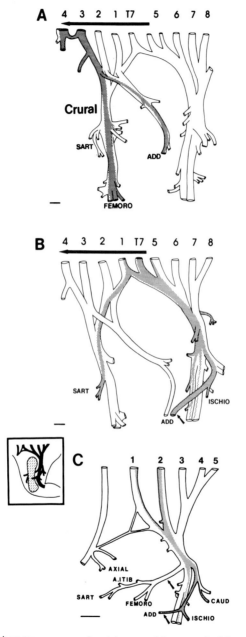

Fig. 1. Motoneuron projection patterns following A. small cord reversal B. large cord reversal C. limb shift.

an identity prior to axon outgrowth, and further that they could make use of this identity and presumably chemical cues within the developing limb bud to compensate for moderate displacements.

We found however, that when motoneurons were displaced greater distances and their axons entered the wrong plexus, they exhibited two rather different sorts of behavior. Some axons were still able to reach their original muscle, and did this by taking totally novel pathways within the limb (note the adductor axons in 1B indicated by a small arrow). However, the majority of axons when confronted with an extremely foreign limb region, projected to a variety of wrong muscles, (Lance-Jones and Landmesser, 1981b). These observations are consistent with a number of studies in which motoneurons formed wrong connections following limb displacements (Morris, 1978, Hollyday, 1981, Stirling and Summerbell, 1979, 1981).

We found similar results when the limb was displaced (Lance-Jones and Landmesser, 1981b). For example, when the limb is shifted forward, anterior spinal nerves (LS1, 2 and 3) that would normally form the crural plexus, are displaced posterior to the femur where they contribute to the sciatic plexus (see inset in Figure 1C). Again, some axons take novel pathways to reach their original muscle (small arrows in Figure 1C) but a significant proportion project to wrong muscles.

Inappropriately projecting motoneurons form functional synapses with wrong muscles which persist through the normal motoneuron cell death period (Lance-Jones and Landmesser 1981b). Since cell death has been implicated in the process of error removal during embryonic development (McGrath and Bennett, 1979; Lamb, 1976, 1977, 1979, 1981) we wondered why such errors in peripheral connectivity were not removed in this system either by motoneuron cell death or by some alternative mechanism. Lamb (1976, 1977) has shown that wrongly projecting amphibian motoneurons are removed by cell death, and that this process occurs without any competition from correct motoneurons, suggesting a direct incompatibility between a motoneuron and an incorrect muscle (Lamb, 1979).

Several possible reasons why wrongly projecting motoneurons were not removed in our system come to mind. One possibility is that following such extreme displacements of limb or neural tube, the spinal cord is no longer activated properly. Oppenheim and his co-workers have shown that blocking neuromuscular activity with paralytic agents, prevents normal motoneuron cell death (Pittman and Oppenheim, 1979; Oppenheim, 1981). Thus failure of motoneurons to be activated in our experimental animals, could result in the survival of wrongly projecting motoneurons. Another rather different explanation, is that the central nervous system may be sufficiently plastic to compensate for such peripheral derangements. For example, if extensor motoneurons innervating a flexor muscle, developed central connections appropriate for flexor motorneurons, the muscle would be activated in a behaviorally appropriate manner. Since the peripheral connectivity errors would be without behavioral consequence their removal would be unnecessary. To further test these possibilities and to determine whether the periphery affects motoneuron connectivity, we assayed the central connections of motoneurons projecting to wrong muscles.

Activation Patterns of Wrongly Projecting Motoneurons

Although wrong connections were formed following both cord and limb manipulations, we chose to use the latter because it avoided surgery to the cord that might also affect motoneuron activation, and it allowed us to use the normally innervated contralateral limb as a control. We studied physiologically the behavior of 35 inappropriately innervated muscles in 23 embryos (St 34-36) which had had one limb shifted anteriorly, or rotated about the a-p or d-v axis at St 16-18.

Although we did not attempt to quantify the activity, visual observations through windows in the shell prior to sacrifice, indicated that the experimental limb appeared to be moved vigorously during the spontaneous embryonic movements that occur at frequent intervals. The activation of such limbs was studied in more detail in isolated spinal cord-hindlimb preparations (O'Donovan, et al, 1981), by making electromyographic (EMG) recordings from selected muscles.

Such isolated cord preparations exhibit periodic
sequences in which the hind limbs are moved a number of
times. This can occur spontaneously, or be elicited by a
single snock to the cord. During such movements,
motoneuron pools are activated in bursts of 1-2 seconds,
eacn burst corresponding to a kick. Their activation
pattern is qualitatively similar to that during normal
walking following hatching (Jacobson and Hollyday, 1982)
For example, in Figure 2A, the control sartorius (a
flexor) gives a characteristic series of eight bursts.
Normally, the posterior iliotibialis, an extensor (not
shown in Figure 2A) would be activated out of phase or
during the silent periods between sartorius bursts.

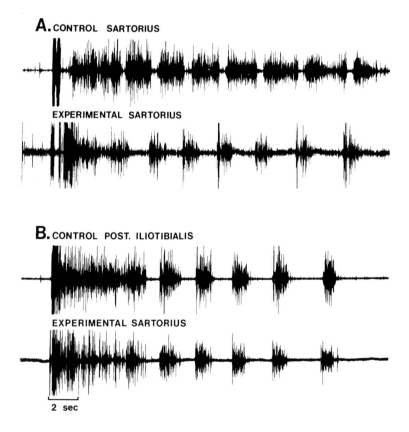

Fig. 2. EMG recordings or inappropriately innervated
sartorius compared to A. control sartorius and B.
control. P. iliotibialis.

Our main finding is that motoneurons are activated in a manner consistent with their original identity, irrespective of what muscle they synapse with. In the majority of cases (28) however the activation pattern of the muscle was at least generally appropriate (i.e. flexor vs extensor) for the muscle innervated even though it was innervated by inappropriate motoneurons. This may result in part because the rules for setting up peripheral projection patterns in the limb (Lance-Jones and Landmesser, 1981b; Hollyday, 1981) may restrict a motoneuron's choice of peripheral target. For instance, the main thigh muscles derived from the ventral muscle mass are all extensors. Medial motoneurons that normally innervate muscles derived from this muscle mass, will preferentially innervate other ventrally derived muscles, following a variety of manipulations (Lance-Jones and Lanndmesser, 1980, 1981b; Hollyday, 1981; Ferguson, 1981 but see Stirling and Summerbell, 1981). This will in most cases insure at least grossly appropriate behavior (i.e., extensor).

However in seven cases we obtained clear examples of inappropriate behavior, one being shown in Fig. 2. The experimental, right sartorius muscle (originally a flexor) is activated (Fig. 2A) as an extensor. In isolated cord preparations the two hind limbs are moved in tight synchrony so that the right and left homologous muscles are activated exactly in phase (O'Donovan, et al., 1981; M. Cooper, in preparation). In this figure it is clear that the experimental sartorius muscle is activated out of phase with its contralateral homolog. The behavior of the experimental sartorius is instead quite like that of the control posterior iliotibialis (shown in 2B). In fact when the motoneurons innervating this experimental sartorius were retrogradely labeled with HRP, they were found to be posterior iliotibialis motoneurons. Due to their characteristic location (Landmesser, 1978) it is possible to determine the original identity of a motoneuron based on the location of its cell body in the cord. We found similar results in six additional cases. It is therefore clear that synapsing with a muscle of opposite function does not alter the basic central connectivity of a motoneuron pool.

We conclude that the activation pattern of motoneurons is not influenced by the periphery. The central connections which are responsible for the basic pattern of muscle activation during walking and during spontaneous embryonic motility are interneuronal. It is possible that afferent projections from the limb, or descending input from higher centers might be altered, and these possibilities will have to be independently assessed. However, these results allow us to exclude several explanations of why peripheral projection errors are not removed. Since motoneurons are activated normally following these manipulations, lack of activation which would prevent cell death (Oppenheim, 1981) and possibly error removal by this mechanism can be excluded. Preliminary estimates of motoneuron numbers also indicate that cell death has not been prevented in this situation (Landmesser, unpublished observation). Since central connections are not altered by the muscle synapsed with, we can also exclude the possibility that central rearrangement functionally compensates for errors in peripheral connectivity. A similar lack of central plasticity has been shown for proprioceptive afferents in the chick hindlimb (Eide, et al., 1982).

In light of these observations, the mechanisms that normally ensure that peripheral projections develop with only minor errors assume added importance.

Sensory Projection Patterns Following Motoneuron Removal

Honig (1982) showed that hindlimb cutaneous and muscle afferents grew into the limb at approximately the same time as motoneurons, that they projected precisely from early times, and that they, like motorneurons, were able to project to displaced muscles in some cases. The last result suggests that the neural crest cells making up the segmental ganglia may also be specified (at least on a segmental basis) prior to axon outgrowth. Alternatively, muscle afferents may simply associate with motoneurons from the same segment and thereby be guided to their correct target.

We tested this possibility by selectively removing motoneuron containing regions of the neural tube at St 16-20. We determined the effect of this procedure on 16 limbs and found in general that muscle

nerves were either lacking (compare Sart, A. itib & Fem
nerves in removal and control side in Fig. 3) or reduced
(obt nerve in same fig.). (Motoneuron removal was often
incomplete with some remaining in at least one segment.)
Since sensory axons make up 14-40% of the axons

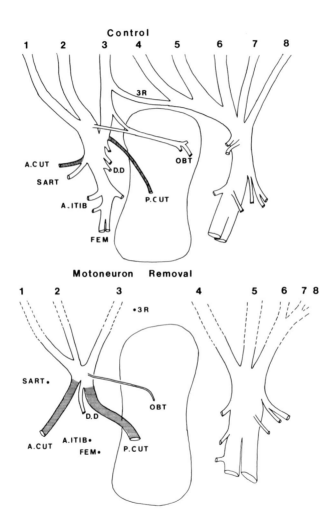

Fig. 3. Hindlimb nerve pattern in St30 embryo on control
side (top) and side where motoneurons had been removed
from first three segments.

in a muscle nerve normally, and since the sensory ganglia
formea were often of normal size in these preparations,
it would appear that sensory axons are unable to project
to muscle when motoneurons are lacking. In addition,
pure cutaneous nerves (indicated by wavy lines in Fig. 3)
were in some cases considerably enlarged, suggesting that
sensory axons that would originally have projected out
muscle nerves, are now projecting out cutaneous nerves.

Unfortunately, these results do not allow us to
resolve whether sensory afferents are specified prior to
axon outgrowth. They do however, indicate that sensory
axons have an unexpected dependency on motoneurons in
forming their peripheral projection. The details of this
interaction between sensory and motor axons remain to be
explored. (Supported by NIH Grant NS10666)

Eide A, Jansen, J, Ribchester, R (1982). The effect of
lesion in the neural crest on the formation of synaptic
connections in the embryonic chick spinal cord. J
Physiol 324:453.
Ferguson B (1981). Development of motor innervation of
the chick following dorsal-ventral limb bud rotations.
Dissertation: Yale University.
Hollyday M (1981). Rules of motor innervation in chick
embryos with supernumerary limbs. J Comp Neurol
202:439.
Honig M (1982). The development of sensory projection
patterns in carpryonic chick hindlimb. J Physiol (in
press).
Jacobson RD, Hollyday M (1982). A behavioral and
electromyographic study of walking in chick. J
Neurophysiol (in press).
Lamb AH (1976). The projection patterns of the ventral
horn to the hind limb during development. Dev Biol
54:82.
Lamb AH (1977). Neuronal death in the development of the
somatotopic projections of the ventral horn in xenopus.
Brain Res 134:145.
Lamb AH (1979). Evidence that some developing limb
motoneurons die for reasons other than peripheral
competition. Dev Biol 71:8.
Lamb AH (1981). Target dependency of developing
motoneurons in xenopus laevis. J Comp Neurol 203:157.
Lance-Jones C, Landmesser L (1980). Motoneuron
projection patterns in the chick hindlimb following
early partial spinal cord reversals. J Physiol, London
302:581.

Lance-Jones C, Landmesser L (1981). Pathway selection by embryonic chick motoneurons in an experimentally altered environment. Proc R Soc Lond B 214:19.

Landmesser L (1978). The distribution of motoneurons supplying chick hind limb muscles. J Physiol, London 284:371.

McGrath PA, Bennett MR (1979). The development of synaptic connections between different segmental motoneurons and striated muscles in the axolotl limb. Dev Biol 69:133.

Morris DG (1978). Development of functional motor innervation in supernumerary hind limbs of the chick embryo. J Neurophysiol 41:1450.

O'Donovan MJ, Cooper MW, Landmesser L (1981). Electromyographic activity patterns of embryonic chick hindlimb muscles. Soc for Neurosci 7:688.

Oppenheim RW (1981). Cell death of motoneurons in the chick embryo spinal cord V. evidence on the role of cell death and neuromuscular function in the formation of specific connections. J Neurosci 1:141.

Pittman RH, Oppenheim RW (1979). Cell death of motoneurons in the chick embryo spinal cord IV. evidence that a functional neuromuscular interaction is involved in the regulation of naturally occurring cell death and the stabilization of synapses. J Comp Neurol 187:425.

Stirling RV, Summerbell D (1979). The segmentation of axons from the segmental nerve roots to the chick wing. Nature, London 278:640.

Stirling RV, Summerbell D (1981). The innervation of dorsoventrally reversed chick wings: evidence that motor axons do not actively seek out their appropriate targets. J Embryol Exp Morphol 61:233.

Limb Development and Regeneration
Part A, pages 217–226
© **1983 Alan R. Liss, Inc., 150 Fifth Avenue, New York, NY 10011**

FAMILIARITY BREEDS CONTEMPT: THE BEHAVIOUR OF AXONS IN
FOREIGN AND FAMILIAR ENVIRONMENTS

R. Victoria Stirling
Dennis Summerbell
National Institute for Medical Research
The Ridgeway, Mill Hill,
London NW7 1AA.

INTRODUCTION

The pattern of nerves in the vertebrate limb is
remarkably constant between individuals, as is the pattern
and distribution of motor neurones (pools) supplying
particular muscles in the limb. How do these patterns
arise? The peripheral nerve pattern is controlled by the
local environment through which the nerves are growing; thus
a wing bud transplanted (before axons have reached it) to
lumbar region or even head region will develop normally with
its characteristic pattern of wing nerves.

The pattern in the spinal cord is determined by the
behaviour of growing axons taking particular branches in
the periphery (for reviews see Landmesser, 1980, and
Summerbell and Stirling, 1982).

How do axons behave in a foreign environment? If we
perturb the system we can see if axons have selective
preferences for particular targets. Following dorsoventral reversal of the wing bud, before axons have reached
it, we found that motor pools to dorsal muscles now
occupied positions characteristic of ventral muscles and
vice versa. The axons in these cases showed no evidence
for selective choice, growing in an orderly fashion in the
foreign territory, innervating alien targets by passive
deployment (Summerbell and Stirling, 1982).

Experiments by Lance Jones and Landmesser (1981) in
which the lumbar spinal cord was reversed about the antero-

posterior axis in general supported this view.   However,
when they reversed only a small length of spinal cord they
found that frequently some groups of axons had reached their
appropriate targets.   The difference between achieving
appropriate or inappropriate innervation in these
experiments we assumed to be related to the distance axons
had been displaced from their normal environments.

In this paper we look at the motor pools and axon
trajectories in wings reversed about the antero-posterior
axis (AP reversals) and compare them with results from
dorsoventral wing reversals (DV reversals) and AP cord
reversals.   The results suggest that both passive
deployment and selective growth, and possibly cell death
play a part in defining the motor innervation of the
manipulated chick wing.

METHODS

Operations were performed on windowed embryos at
stages 18 to 20 (Hamburger and Hamilton, 1951), buds from
opposite sides of two embryos were exchanged to produce
reversals of the antero-posterior axis (AP rev.) or dorsoventral axis (DP rev.); buds were attached to host with
fine pins, as described in Summerbell and Stirling (1981).
Six or seven days after operation (between stages 34 and 35),
the embryos were bled, eviscerated and the ventral brachial
spinal cord exposed.   Selected wing muscles were injected
with horseradish peroxidase (HRP) as described previously.
Transverse 60 µm frozen sections of spinal cord were stained
using the benzidene dihydrochloride blue reaction of
Mesulam (1976), while the limbs were sectioned at 100 µm
with a vibroslice and the sections stained with a modification of the Hanker-Yates reaction (Perry, 1982).   The areas
of the motor horns were measured from camera lucida drawings
made at the level of the 14th, 15th and 16th segmental roots
using a graphics tablet, in all cases the unoperated
contralateral side was used to determine the normal
innervation pattern and horn size.

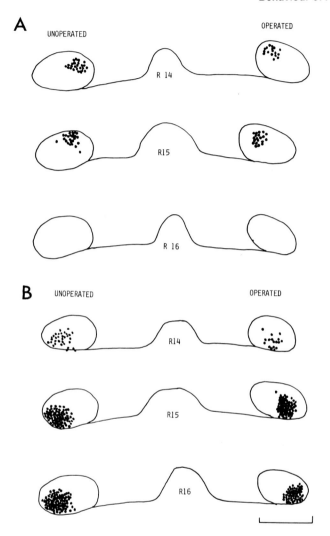

Fig.1. The distribution of heavily labelled cells in the
ventral horn following bilateral injection of (A) biceps
(6 animals) and (B) triceps (14 animals) on unoperated and
operated sides after AP limb reversal. The position of
labelled cells from camera lucida drawings of single 60um
sections at the level of the 14th, 15th and 16th segmental
roots have been superimposed. The distribution of labelled
cells is remarkably symmetrical in these operated
animals. Bar 200um

RESULTS

Antero-posterior axis reversal

.Out of 45 embryos with good reversed wings, 30 had
symmetrical motor pools on operated and unoperated sides
following bilateral injection of a wing muscle.   Only 8
had abnormal pool positions on the operated side, shifted
anteriorly (3), shifted mediolaterally (2) or shifted in
both axes (3).   In seven animals the labelling was not
good enough to tell.

The similarity of pool positions on operated and
unoperated sides of the 30 cases was striking (Fig. 1).
We compared reversals made at the shoulder with those made
closer to the somites at the flank.   Significantly,all
18 flank rotations yielding mappable cords showed normal
pool distributions on the operated side, whereas of the 26
shoulder rotations 12 gave normal and 8 abnormal
distribution.

When the nerve endings in the muscle are damaged at
injection the axons become filled with HRP and can be
followed in sectioned limbs.   In figure 2, the labelled
fibres from roots 15 and 16 on the operated side appear to
split into two bundles at the plexus. One of these clearly
changes its position so that it occupies a more anterior
position to reach the reversed triceps muscle.   Similar
changes in trajectory have been observed in other cases
where axons had reached the appropriate targets, but the
exact nature of the change in position is often hard to
assess.

Those animals in which the pool position was altered
on the operated side tend to show severe disruption of the
nerve supply to the grafted wing, often with either the
dorsal or ventral nerve missing in the graft.

Dorsoventral reversals

The majority of motor pools supplying muscles on
the operated side in these animals were in inappropriate
positions (see Summerbell and Stirling, 1982).   Injection
of triceps always labelled medial instead of lateral
motoneurones; injection of biceps muscle on the operated

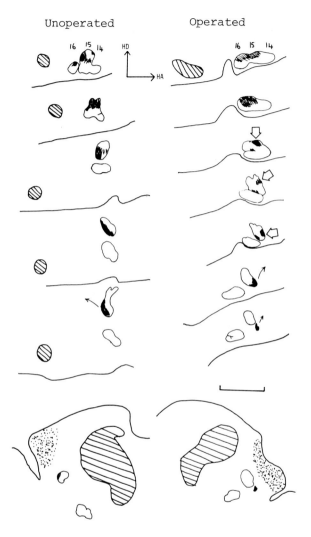

Fig.2. Camera lucida drawings of transverse sections of unoperated and operated limbs after AP reversal showing position of axons labelled by triceps injection. The drawings are arranged so that host axes for both sides are the same, HD-host dorsal and HA-host anterior. There are clearly two groups of filled axons on the operated side, the more dorsal group (open arrow) clearly moves from a posterior to a more anterior position to reach reversed muscle, small arrow shows distal change in nerve position. Bar 500μm.

side labelled lateral rather than medial motoneurones in
threequarters of the animals, while the remaining quarter
were appropriately located, being symmetrical on operated
and unoperated sides.

   In most animals for upper arm muscles the mediolateral
change in pool position on the operated side was also
accompanied by a shift in the rostrocaudal axis.
The similarity of pool position for injected grafted biceps
and normal triceps on the one hand and between grafted
triceps and normal biceps is striking (Fig. 3).

   Figure 4 shows the trajectories of axons labelled
from biceps injection on operated and unoperated sides after
dorso ventral limb reversal.   There is no change in
the position of labelled axons on the operated side, dorsal
caudal axons stay dorsal and caudal to reach the biceps in
the reversed limb.   Such absence of change in trajectory is
usually seen in these animals.

The size of the motor horn supplying dorsoventral and
antero-posterior reversed wings.

Table 1. The area of the motor horn on the operated compared
to the unoperated (control) side taken from single
transverse sections at the level of the 14th,15th and 16th
segmental roots after various unilateral limb reversals.

Operation	pool posn	no of cases	Area of operated / Area of unoperated X 100		
			R14(s.d.)	R15(s.d.)	R16(s.d.)
sham	app.	12	103(13)	100(14)	115(13)
DV rev	inapp.	17	96(17)	97(19)	92(14)
	app.	7	80(13)	79(16)	71(13)
AP rev shoulder	inapp	8	85(22)	83(18)	74(18)
	app	11	83(15)	75(11)	76(14)
AP rev flank	inapp	18	75(17)	84(16)	73(14)

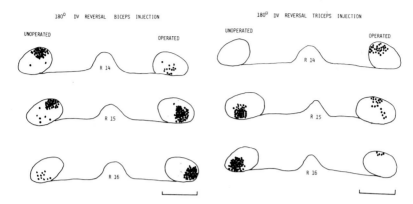

Fig.3. The distribution of heavily labelled cells in the ventral horn following bilateral injection of biceps (8animals) and triceps (9animals) after DV limb reversal. Bar 200um.

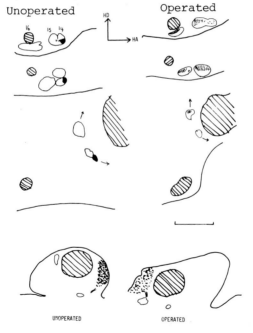

Fig.4. Camera lucida drawings of transverse sections of unoperated and operated limbs after DV reversal showing position of axons labelled from biceps injection, the drawings arranged as in Fig.2. Labelled axons run from dorsal caudal position to innervate the reversed biceps, showing no change in position relative to the host axes. Bar 500um.

Table 1 shows measurements of the difference in area of operated compared to unoperated sides at level of 14th, 15th and 16th segmental root in various operated and control animals.

We have two classes of results; those where labelled pool position following bilateral muscle injection is symmetrical and those where it is asymmetrical. The table clearly shows that the size of the motor horn is certainly no smaller than the operated side in the latter group, but may be in the former!

DISCUSSION

In both our experiments (where the periphery is manipulated) and those of Lance-Jones and Landmesser (1981) (where the cord is manipulated), the axons behave in two distinct ways. Following AP limb reversal and 'short' AP cord reversal the axons find their way to their appropriate targets (specificity). However, following DV limb reversal there is no evidence for growth selectivity, axons enter foreign territory and innervate alien targets by passive deployment.

Any hypothesis explaining limb innervation must take into account these different modes of behaviour. We will consider here how nerves might behave in territory that they would normally expect to encounter ('familiar'), and in alien ('unfamiliar') territory. Our discussion is based on the premise that the peripheral pattern of nerves is based on loosely defined tracts running through the tissue, "a system of environmental cues that map out the position of future nerves" (Summerbell and Stirling, 1982).

Familiarity Breeds Contempt

Following AP limb reversal and 'short' cord reversal, the axons still find themselves in familiar nerve tracts as they enter the limb. They can use expected cues or signals at branch points to reach their appropriate targets. The situation is quite different following DV limb reversal or 'long' cord reversals, where axons find themselves in alien tracts. In the former group, axons from the dorsal

tract are channeled into the ventral limb, while in
the latter axons from the lumbar plexus are channelled into
the sacral plexus. These axons are therefore in unfamiliar
territory and cannot recognise guidance cues for particular
branches, but file down the nearest branch by passive
deployment.

We do not know what defines the tract or what the
guidance cues are (they may be identical). They may be
cell bound, short, or long range. There is no reason to
suppose that such signals identify muscles by name, or even
that motoneurones are prespecified to innervate particular
targets. A short range signal would explain economically
much of the data. Alternatively, it is interesting that
while there is abundant evidence for the presence of graded
positional values across the AP limb axis, there is no
good evidence for a similar system across the DV axis.

Cell Death

One powerful hypothesis has been the idea that every
motoneurone has a specific target, those that make
inappropriate connections die, resulting in discrete
motor pools in the cord (Pettigrew, Lindeman and Bennett,
1979 Lamb, 1977). This led Alan Lamb to predict that
we should find substantial cell loss in our DV limb
reversals compared to the AP reversals. The absence of
significant cell loss (as shown by the similarity of motor
horn size on the two sides) on the operated side of the DV
reversed animals argues against this. Indeed, the actual
values suggest it is more likely that there is increased
cell death in the appropriately innervated AP reversals.

Cell death may, however, have a different role to play
in the formation of motor pools. In normal development,
axons from a particular position in the cord innervate
a particular target muscle in the periphery. Similarly,
after DV reversal passive deployment also results in
a relatively localised cluster of cells innervating
a common target. However, in the AP reversals, where
a proportion of axons show evidence of selective growth,
there may be a far greater scatter of cells supplying
a common target. If cell death ensures that muscles are
innervated by groups of neighbouring cells, one would expect
an increase in cell death in these cases. It is also
possible that this cell loss might reflect the disruption

of the axon tracts into the limb after AP reversal, whereas
in the DV reversals the continuity of tracts is much better
preserved.

It is not clear how important the observed difference
in axon behaviour in foreign and familiar environments is
for the normal development of limb innervation since in
the normal situation axons are ipso facto in familiar
environments. However, we believe that detailed study of
the behaviour of axons entering manipulated limbs (varying
degrees of foreignness) does give some idea of the nature of
the factors which affect the way axons reach their targets
in normal development.

We thank Alan Lamb for his crate of beer which has
at last stimulated us to start looking at the size of
the motor pools.

REFERENCES

Hamburger V, Hamilton H (1951).   A series of normal stages
in the development of the chick embryo.   J Morph 88:49.
Lamb AH (1977).   Neuronal death in the development of
the somatotopic projection of the ventral horn in Xenopus.
Brain Res 134: 145.
Lance-Jones C, Landmesser L (1981).   Pathway selection by
embryonic chick motoneurones in an experimentally altered
environment.   Proc R Soc Lond B 214:19.
Landmesser L (1980).   The generation of neuromuscular
specificity. Ann Rev Neurosci 3:279.
Mesulam M-M (1976).   The blue reaction product in horseradish
neurohistochemistry: incubation parameters and visibility.
J Histochem Cytochem 24:1273.
Perry VH, Linden R (1982).   Evidence for dendritic competi-
tion in developing retina.   Nature (in press).
Pettigrew AG, Lindeman R, Bennett MR (1979).   Development
of the segmental innervation of the chick forelimb.
J Embryol exp Morph 49:115.
Summerbell D, Stirling RV (1980).   The innervation of
dorsoventrally reversed chick wings: evidence that motor
axons do not actively seek out their appropriate targets.
J Embryol exp Morph 61:233.
Summerbell D, Stirling RV (1982).   Development of
the pattern of innervation of the chick limb.   Amer Zool
22:173.

Limb Development and Regeneration
Part A, pages 227–236
© **1983 Alan R. Liss, Inc., 150 Fifth Avenue, New York, NY 10011**

NERVE MUSCLE SPECIFICITY IN THE DEVELOPING LIMB

Alan H. Lamb

Department of Pathology
University of Western Australia
Perth, Australia.

In adult animals, motor nerves can be induced to
innervate foreign muscles. Generally speaking there seems
to be no impediment to functional connections and the
resulting muscle contractions are normal. Thus it seems, on
this criterion, that motoneurons do not have a particular
requirement for any given muscle type. The same is not true
for developing motoneurons. Several studies in the 1960s on
tadpoles and chick embryos demonstrated that thoracic
motoneurons which normally innervate trunk muscles would not
successfully innervate limb muscles (Hughes 1964; Straznicky
1967; Morris 1978). Though there was evidence of some
functional connections being formed these were never long
lasting and the motoneurons eventually died. Nor it seems
can developing limb motoneurons permanently innervate trunk
muscle. Though the motoneurons form temporary functional
connections with trunk muscle, they fail to mature (Letinsky
1974) and subsequently die (P. Harrison and R.F. Mark,
personal communication). Thus developing motoneurons have
a specific requirement for certain kinds of muscle and in
this sense they can be said to have specificity.

The questions I wish to address in this paper are
whether developing limb motoneurons are specified more
precisely, for particular regions of the limb musculature,
and whether precise specificity may be the origin of
motoneuron death in normal embryos.

In adult animals the limb muscles are each innervated
by a particular region of the lateral motor column which is
constantly sited, not only within individuals of a species

but also among different species and even classes, e.g. frog
(Cruce 1974) and chick (Landmesser 1978a). Does this reflect
that motoneurons are specified for particular limb regions?
Using the same rationale by which motoneurons were shown to
be specified for trunk or limb muscle, it is possible to
observe what happens when limb motoneurons are induced to
project to foreign regions. Several workers have carried
out this experiment in chick embryos and Xenopus tadpoles.
Unfortunately the results are not consistent between the two
species.

Most of the experiments have been conducted on the chick
embryo. Manipulations have included rotation of the spinal
cord (Lance Jones and Landmesser 1981), rotation of the limb
bud (Summerbell and Stirling 1981), displacement of the limb
bud (Lance Jones and Landmesser 1981) and addition of
supernumerary limb buds (Hollyday et al., 1977; Morris 1978;
Lance Jones and Landmesser 1981). All manipulations were
carried out before motor axons reach the limb bud, and in
most cases before the motoneurons are generated. The results
are conclusive; following such manipulations many motoneurons
make erroneous projections and survive beyond the normal
period of motoneuron death.

In Xenopus, a different result was obtained. Using a
manipulation not directly comparable with the work on chick,
erroneous projections were induced by removing proximal or
distal segments of the limb bud just as the motor axons were
beginning to reach it and during the period of motoneuron
generation. In all cases, motoneurons making erroneous
projections died and motoneuron numbers were diminished
(Lamb 1981a).

The different results in frog and chick seem to imply
a species difference. However, there are some points which
need further consideration. An important question is whether
in the chick, the motoneurons making the inappropriate
projections were in fact doomed though still alive at the
time of examination. In none of the studies were embryos
raised past Stage 42 and most were raised only to Stage 36
or less, only shortly after the normal period of motoneuron
death. It is worth drawing an analogy from the behaviour
of thoracic motoneurons induced to innervate limb muscle.
After an initial period of functional innervation, function
is lost, the musculature atrophies and the motoneurons die
(Hughes 1964; Straznicky 1967; Morris 1978). The specificity

of the thoracic motoneurons is thus expressed not in an inability to make connections but an inability to sustain them. Should the same hold for limb motoneurons projecting to inappropriate regions, then it may not be possible to judge whether motoneurons express regional specificity without rearing animals to sufficient maturity to ensure that the motoneurons have had time to die. I am not aware of anyone having succeeded in doing this though Landmesser (personal communication) has observed that motoneuron numbers per transverse section appear normal in hatchlings after early spinal cord rotations. This suggests that the proportion of motoneurons dying is not increased as might be expected if erroneously projecting motoneurons were dying.

Another explanation for the different results may relate to the different procedures used. In the study on Xenopus, the partial limbs received both correct and erroneous projections whereas in all the experiments on chick, at least some limb regions would have received only incorrect projections. Error may be detected by some form of interneuronal referencing similar to that postulated for optic neurons (e.g. Prestige and Willshaw 1975; Overton and Arbib 1982). Without correct projections, the reference frame may be so distorted that the system is fooled into accepting incorrect projections as correct. However, another study in Xenopus throws some doubt on this explanation. When the correct projection for the future knee flexors was prevented by removing part of the spinal cord, a group of caudal motoneurons which normally project incorrectly was still lost (Lamb 1979) suggesting that direct nerve-muscle incompatability is the cause of death.

ROLE OF MOTONEURON SPECIFICITY IN NORMAL DEVELOPMENT

The deaths of erroneously projecting motoneurons in Xenopus indicate that in this species at least, motoneurons express regional specificity by their inability to survive in foreign limb regions. Does this observation have relevance to normal embryos? Hughes (1965) postulated that motoneuron death was a mechanism to eliminate errors of projection by outgrowing motor axons. He envisaged that the initial invasion of the limb bud may be somewhat disorganised and that only motoneurons projecting to correct regions would be allowed to survive. Such a mechanism would require some form of specific nerve muscle recognition. However, studies

on both Xenopus and chick have shown that motor axons are
actively guided to their correct regions (Lamb 1976, 1981b;
Landmesser 1978b; Lance-Jones and Landmesser 1981) The
accuracy of the guidance is however not perfect and a minority
of axons make apparently erroneous projections. The sorts of
errors that have been detected are exemplified by the early
("primitive") projections to Xenopus hindlimb (Fig. 1). The
errors are gross in that certain regions of the developing
thigh receive segmentally inappropriate projections. This
appears to happen because of disparities in the timing of the
development of the lateral motor column and of the distal
parts of the limb which make it physically impossible for
some axons to reach their correct regions (Lamb 1976). The
motoneurons making the errors subsequently die (Lamb 1977).
The similarity between the death of these motoneurons and
the death of motoneurons induced to make segmental errors
after partial deletion of the limb bud is obvious. But it
does not necessarily follow that nerve muscle specificity is
the basis of the loss of primitive projections in normal
development. For example motoneurons making primitive
projections may serve a temporary function which does not
require their projections to be somatotopic. Then, when they
are no longer needed, they die.

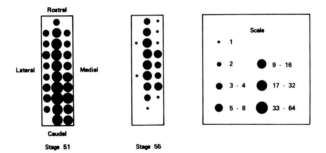

Fig. 1. Projections of developing motoneurons to the
presumptive knee flexors before (Stage 51) and after
(Stage 55) the period of motoneuron death. Rectangle
represents the lateral motor column. At Stage 51 many
motoneurons in the most caudal part of the lateral motor
column project to the knee flexors and constitute a "primitive"
projection. Between Stages 53 and 55 the primitive
projection disappears as the motoneurons die. Surviving
motoneurons lie approximately within the region of the
lateral motor column that projects to the knee flexors in
the adult (Redrawn from Lamb 1977).

Recently, we have been investigating whether the loss
of the primitive projections is indeed due to nerve muscle
specificity. The aim has been to rescue the caudal
motoneurons that project to the thigh, by providing them
with mesenchyme of foot paddles. Three kinds of operation
have been tried including insertion of foot paddle mesenchyme
into the host thigh, addition of a supernumerary foot paddle
to the host thigh, and substitution of a foot paddle in place
of the whole limb bud. To identify the caudal motoneurons
to be rescued they were first labelled at Stage 51 with
horseradish peroxidase injected into the presumptive knee

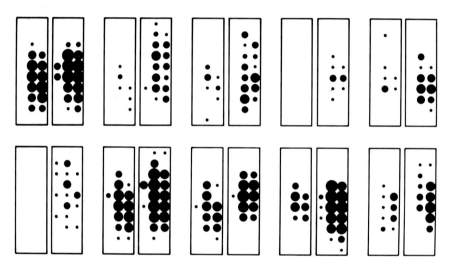

Fig. 2. Distributions of HRP labelled motoneurones surviving
to Stage 55 in tadpoles with the left limb bud replaced by
a distal segment of limb bud at Stage 51. Each pair of
rectangles represents the lateral motor columns of the
operated (left) and unoperated sides of one animal.
Motoneurones were labelled by HRP injected into the
presumptive knee flexors the day before operating. In the
first 6 animals (top row and bottom left), the limb bud was
replaced by its own distal segment, in the 7th animal by a
distal segment from a Stage 55 donor, and in the last 3
animals by distal segments from Stage 53 donors. No caudal
motoneurons survived on either side in any of the animals.
The paucity of labelled motoneurons more rostrally on the
left in many animals reflects the massive loss of motoneurons
rostrally that results from the absence of proximal limb
segments (Lamb 1981a).

flexors. The operations were done the next day. The results were consistent throughout; no caudal labelled motoneurons were rescued (Fig. 2). However, rostral labelled motoneurons survived which shows that the axotomy made necessary by the need to label the motoneurons before the operation was not the cause of death (see also Lamb 1981c).

Does this result mean that the caudal motoneurons are programmed to die after a certain maturity and that their death is not an effect of nerve muscle specificity? The first part of the question can be answered. In some tadpoles the foot paddle used to replace the limb bud did not adhere, and a new limb bud regenerated instead. Such regenerated limb buds lag behind in development and usually consist only of mesenchymal cells when the normal limb has already developed muscle fibres and movements. In these cases, caudal labelled motoneurons were found still alive on the side with the regeneration though they had been lost on the unoperated control side (Fig. 3). Though it is not clear how the regenerating limb bud can maintain the caudal motoneurons, the fact that they are maintained shows they are not programmed to die after a certain period.

The failure to rescue caudal motoneurons with foot paddle mesenchyme should not be construed as evidence that nerve muscle specificity is not involved in the death of the cells. Although a positive result could have proved the error/death hypothesis the negative result does not refute it. There are reasons why the motoneurons may have died despite the rescue attempts, for example, failure to find appropriate sites. This could occur if the axon guidance

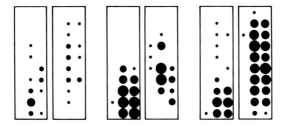

Fig. 3. HRP labelled motoneurons surviving to Stage 55 in tadpoles with a regenerating limb bud on the left following amputation at Stage 51. Caudal motoneurons are still alive on the regenerating side.

mechanism were disturbed by the operation. The fact that the labelled motoneurons had already invaded the limb bud before the operation could also be relevant. Contact with inappropriate tissue may begin an irreversible chain of events that takes some time to culminate in motoneuron death. Unfortunately there seems to be no easy way of avoiding these problems and so while it is reasonable that the loss of primitive projections is a consequence of error, the case remains unproved.

A criticism often levelled at the specificity hypothesis is that even if we accept that errors of projection lead to motoneuron death, there are not enough detectable errors to account for more than a small minority of the 50% or greater motoneuron loss in normal embryos. However, until recently the methods used for determining motor projections have been crude, and capable of detecting only the grossest errors. Recent work by Bennett and his colleagues suggests that there may be numerous errors of projection at a much finer scale than had hitherto been examined (Bennett and Lavidis 1982). By making intracellular recordings from the glutaeus muscle of the Australian frog *Lymnodynastes peronii* it was shown that there is a distinct ordering of the motor projection across the muscle in the adult which is not present when the muscle first becomes innervated. The period when the ordering emerges coincides with that of motoneuron death. There is therefore justification for suspecting that the fine ordering comes about through motoneuron death just as it does in the proximo distal axis of the limb.

SPECIFICITY IN THE FORMATION OF MOTOR UNITS

There is another kind of error which could occur at the small scale, not of the spatial variety but in terms of the muscle fibre type with which the motoneuron connects. Muscles are composed of motor units of several functional types which can be classified by several criteria (e.g. fast and slow). In any given motor unit all muscle fibres are of the same type (Burke 1980). Muscles vary in composition and distribution of muscle fibre types (hence motor unit types) though the patterns are remarkably similar from one animal to the next (Fig. 4); so similar that we have been considering how the pattern is imposed. The dogma handed down from adult regeneration studies is that muscle fibre

type is determined by the motoneuron contacting it
(Jacobson 1978, p.322). The implication for development is
that muscle fibres are initially unspecified thus presenting
the axons with a featureless target field. If that were true
it would mean that the motoneurons, each with a certain type
inherent within it to stamp on the muscle fibres, must
organise their projections with extraordinary precision in
order to get the same pattern every time. It seemed to us
more rational to propose that muscle fibre type is
intrinsically determined. N. Laing and I have therefore
begun experiments to test this notion in chick embryos. In
one experimnt the wing bud was transplanted into the lumbar
region in place of or alongside the leg bud. The reasoning
is that lumbar axons growing into a wing would be unlikely
to be able to distribute themselves in the correct manner to
yield a normal distribution of fast and slow fibres without
some sort of intrinsic muscle fibre specification to guide
them. With very minor exceptions we found that the fast/slow
distributions were normal which does suggest that muscle
fibre type is intrinsically specified (Fig. 4). Other
experiments in progress including examination of aneural limbs
should give a more definite answer.

A major attraction of intrinsic muscle fibre
specification is that it is easy to postulate a role for
motoneuron death during the formation of motor units. We can
begin by remembering that all muscle fibres belonging to a
motoneuron must eventually be of the same type. The simplest
way to achieve that would be to give the motoneuron and the
muscle fibre an intrinsic specificity and allow them to match
in the limb. But because muscle fibre composition varies
widely between muscles it is likely that the proportions of
motoneuron types invading any given muscle are not going to
match the proportions of muscle fibre types. Thus motoneuron
types in excess would find too few fibres and die.

We may also consider the case where motoneurons are not
intrinsically specified, but induced by muscle fibres to take
on matching properties. This leads to an interesting
analysis for it means that initial contacts would be with a
random mixture of muscle fibre types. Out of that situation
the motoneuron would then have to decide which motor unit
type it was going to become and break contact with all fibre
types not conforming. The simplest way would be to have a
voting system with the majority muscle fibre type taking
control. Inevitably some motoneurons will by chance

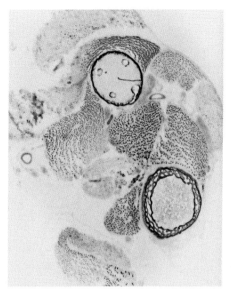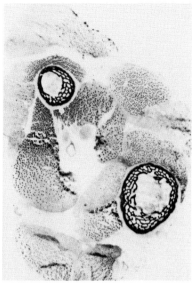

Fig. 4. Cross sections of 17d incubated chick embryo wings
at the forearm level stained for ATPase activity after
preincubation at pH 4.35 to distinguish the slow muscle
fibres (black). The left frame is from a normal wing, the
right from a wing innervated by lumbar motoneurons.
Allowing for the fact that the sections are not from
precisely the same level, there is a remarkable similarity
in distributions of fast and slow fibres.

initially acquire a clear majority of one fibre type over all
others. The decision will be quick, and the motoneuron, now
committed, will be in a position to acquire any muscle fibres
of the same type in its vicinity (bearing in mind that muscle
fibres are polyneuronally innervated at the time). However,
in the case of the motoneuron that finds itself with an even
distribution of different fibre types, voting will be
indecisive. During the indecision fibres will be lost to
adjacent motoneurons that have been determined quickly.
Eventually it will have too few fibres to sustain it and it
will die.

   Further progress in these ideas will depend on the

development of new experimental approaches. We are
optimistic that studies now in progress may provide some of
the answers.

REFERENCES

Bennett MR and Lavidis N (1982) Developmental Brain
    Research, in press.
Burke RE (1980) Trends Neurosci 3:255.
Cruce WLR (1974) J Comp Neurol 153:59.
Hollyday M, Hamburger V, Farris JMG (1977) Proc Natl
    Acad Sci USA, 74:3582.
Hughes AFW (1964) J Embryol exp Morph 12:27.
Hughes AFW (1965) J Embryol exp Morph 13:9.
Jacobson M (1978) Developmental Neurobiology, Plenum NY.
Lamb AH (1976) Dev Biol 54:82.
Lamb AH (1977) Brain Research 134:145.
Lamb AH (1979) Dev Biol 71:8 .
Lamb AH (1981a) J Comp Neurol 203:157.
Lamb AH (1981b) J Embryol exp Morph 65:149.
Lamb AH (1981c) Brain Research 204:315.
Lance-Jones C and Landmesser L (1981). Proc. R. Soc. Lond.
    B. 214:19.
Landmesser L (1978a) J Physiol 284:371.
Landmesser L (1978b) J Physiol 284:391.
Letinsky MS (1974) Dev Biol 40:129.
Morris DG (1978) J Neurophysiol 41:1450.
Overton KJ and Arbib MA (1982) Proc Roy Soc, Lond. B
    in press.
Prestige MC and Willshaw DG (1975) Proc Roy Soc Lond B.
    190:77.
Straznicky C (1967) Acta Biol Acad Sci, Hung. 18:437.
Summerbell D and Stirling RV (1981) J Embryol exp Morph
    61:233.

Limb Development and Regeneration
Part A, pages 237–244
© **1983 Alan R. Liss, Inc., 150 Fifth Avenue, New York, NY 10011**

THE CONTROL OF AXIAL POLARITY: A. A LOW MOLECULAR WEIGHT
MORPHOGEN AFFECTING THE ECTODERMAL RIDGE B. ECTODERMAL
CONTROL OF THE DORSOVENTRAL AXIS

J. A. MacCabe, K. W. Leal and C. W. Leal

Department of Zoology
University of Tennessee
Knoxville, TN   37996

This report summarizes current research in this laboratory on the control of anteroposterior and dorsoventral
spatial patterns in the developing chick limb. The studies
reported here are ongoing and thus the information is incomplete. It represents, therefore, a current report of
our progress.

Considerable attention has been paid to the control of
limbs' anteroposterior (a-p) pattern. Results obtained
long ago by Zwilling (1956) indicated a-p polarity is controlled within the mesoderm, though the plane of the axis
is determined by the ectodermal ridge. The finding by
Saunders and Gasseling, (1968) that posterior mesoderm cells
are capable of inducing polarized, supernumerary limb
structures when transplanted anteriorly further implicate
the mesoderm in polarity control. Considerable evidence
now exists demonstrating anterior limb bud tissue will
respond to posterior tissue by developing polarized limb
structures in a variety of experimental situations (c.f.
Saunders and Gasseling 1968; Summerbell and Tickle, 1977;
Tickle et. al. 1975; MacCabe, et. al 1979; Iten and Murphy,
1980; Javois and Iten, 1981). This posterior polarizing
tissue apparently acts on the adjacent anterior mesoderm,
making it capable of maintaining a thickened apical ridge
and of outgrowth (MacCabe and Parker, 1979). This close
relationship between ridge maintenance and the polarizing
activity has made it difficult to distinguish them experimentally. We developed a culture system wherein anterior
ectodermal ridge is maintained by posterior tissue or by
homogenates of posterior tissue (MacCabe and Parker, 1975;

Calandra and MacCabe, 1978). Using intact polarizing tissue _in vitro_ the response includes an increase in mesodermal mitosis and an increase in ridge thickness (MacCabe and Taylor, in preparation). Cell free preparations, on the other hand, have been able to maintain, but not thicken, ectodermal ridge _in vitro_. Thus _in vitro_ it has also proven difficult to distinguish between polarizing activity and ridge maintenance activity. Both are apparently involved in a-p pattern control.

Cell homoginates from posterior, but not anterior, limb halves maintain the ectodermal ridge _in vitro_. The active fraction is of high molecular weigh (HMW), binds concanavalin A and is sensitive to proteolysis (MacCabe and Richardson, 1982).

Ridge maintenance activity can also be obtained by conditioning culture medium with posterior (but, again, not anterior) limb tissue (Calandra and MacCabe, 1978). The posterior cells are thus exporting the morphogenetically active molecule to the medium. This molecule is also trypsin sensitive, but of rather low molecular weight (LMW). It will pass through dialysis tubing with a molecular weight cutoff of 12,000 but not of 3,500. This LMW ridge maintenance factor can be changed to a HMW one by a brief incubation with a high molecular weight fraction from anterior limb halves. Since the anterior homogenate has no ridge maintenance activity _in vitro_, it is likely that the LMW active molecule is binding to a large inactive molecule from anterior homogenates. The size of this large molecule is about the same as the HMW ridge maintenance factor found in homogenates of posterior limb halves. Both are found in the same fraction from a Sepharose 6B chromatography column (Table 1, Figure 1). This binding is presumably why the morphogenetic activity is found in the HMW form from posterior homogenates. Whether the factor is present within the cell in HMW form, or binds as cells are homogenized, is not known. It also is not clear whether the binding has biological significance or is the result of destroying compartmentalization when these cells are disrupted. Our results suggest that while the LMW factor is produced only by posterior cells, the HMW inactive factor to which it binds is distributed throughout the limb. We are now doing experiments to determine whether the inactive HMW factor is on the cell surface or is inside mesodermal cells.

	No. Cultures	Ave.No. Macro- phages	Ridge Thick	Ridge Thin or Absent
a. Medium	10	53.7	0	10
b. Conditioned medium	12	2.3	12	0
c. Conditioned medium (dialyzed)	10	48.2	0	10
d. Anterior super- natant	12	54.6	0	12
e. Conditioned medium & anterior supernatant	12	0	12	0
f. Conditioned medium & anterior super- natant (dialyzed)	12	0.4	12	0
g. Condition medium & fraction #9 of anterior super- natant	12	4.8	10	2
h. Conditioned medium & anterior super- natant minus fraction #9	12	47.5	0	12

Table 1. Morphogenetic activity indicated by an inhibition
of cell depth (low number of macrophages) and
thick ectodermal ridge in the responding anterior
limb tissue. A lack of activity (medium, a.) is
indicated by approximately 50 macrophages and a
very thin or no ectodermal ridge after 24 hrs in
culture. Medium conditioned with posterior limb
tissue (b.) has morphogenetic activity that is
lost by dialysis (c.). A 100,000 xg supernatant
from sonicated anterior limb halves is without
activity (d.) unless first incubated 90 minutes
with conditioned medium (e). This activity is not
lost upon dialysis (f.). The binding of the LMW
morphogen in conditioned medium is found only in
fraction #9 from a Sepharose 6B column (g. and h.
this table and figure 1.).

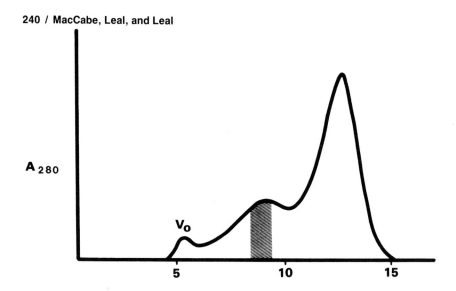

Figure 1. Absorbance profile showing the fraction (cross-
           hatched) having ridge maintenance activity in
           posterior limb halves and LMW morphogen-binding
           activity in anterior limb halves.

     The control of dorsoventral (d-v) pattern specification
has been relatively neglected over the past few years. From
experiments reversing the d-v polarity of the ectoderm
relative to the mesoderm, a confusing story emerges. When
done during the limb bud stages, the limb tip develops d-v
polarity corresponding to the ectoderm, whereas more
proximal structures develop the polarity of the mesoderm
(Patau and Kieny, 1973; MacCabe, et.al., 1974). The
reversal of d-v polarity by reversing the ectoderm, includes
only the distal-most structures at later stages (stgs 24 &
25) but more proximal structure when done at earlier stages
(stgs 19-21). This might suggest that at very early, pre-
bud stages, d-v polarity would more completely conform to
the ectoderm. Such experiments with the prospective wing
however, show a mesodermal control of d-v polarity (Saunders
and Reuss, 1974). However the tissue recombination technique used for these early stages (stgs 12-17) differed from
that used for limb buds. For the early stages, prospective
limb mesoderm is separated from its' ectoderm after incubation in trypsin or EDTA, then transplanted to a host embryo

such that host ectoderm could heal over the graft, thereby
effecting the recombination. Tissue recombinations with
limb buds involve isolating the ectoderm with trypsin, the
mesoderm with EDTA, then combining the two mechanically.
In addition to the technical differences, the experiments
with prospective limb tissue were done with wings, while
the ones with limb-buds were done with legs. Are the
differing results due to the different techniques used, to
the stage differences or are there basic differences in the
way wings and legs develop? We are currently using the
recombination method developed by Zwilling (1955) to re-
orient the d-v polarity of the ectoderm of the prospective
leg. Ectoderms were isolated after a 1-1½ hr. incubation
in 1% trypsin at 1-3°C and the mesoderms after 20 minutes
in EDTA at 37.5°C. The ectoderms were then placed onto
mesoderms with their d-v polarities opposed. The recombinants were then transplanted to the somites of stage 21 host
embryos and allowed to develop for 10-14 days. Thirty nine
hosts lived to the time they were sacrificed. Twenty five
of these were distally complete and of these 25, d-v characteristics could be diagnosed in 17. Twelve of these had
d-v polarity conforming to the ectoderm, including ventral
rotation of the halux and the directions of flexure of the
tibiotarsal joint. The knee retained the flexure characteristic of mesodermal polarity (Figure 2). The remaining

Figure 2. A profile of a leg that developed after combining a stage 15 right leg mesoderm with stage 15
left ectoderm so their d-v polarities were
opposed. The toes and tibiotarsal joint display
d-v polarity corresponding to the ectoderm but
the knee (arrow) displays mesodermal polarity.

five recombinants had tibiotarsal flexures conforming to the ectoderm but had biventral toes. Since there is some shrinkage of the ectoderm upon trypsinization, slightly larger tissues were excised as ectoderm donors, including more ventral, (but not more dorsal) tissue. Thus the biventral toes may result when the prospective apical mesoderm is covered with ventral ectoderm. This presupposes that, unlike prospective wing, the prospective leg mesoderm can induce ectodermal ridge formation in "non-healing" ectoderm. To test both the above assumptions, prospective leg mesoderm was combined with ventral flank ectoderm and with back ectoderm. Those combined with back ectoderm developed bidorsal toes whereas those combined with ventral ectoderm developed biventral toes. In addition it is clear that prospective leg mesoderm is fully capable of inducing ectodermal ridge in intact (i.e. non-healing) non-limb ectoderm.

Our results with the prospective leg differ from those of Saunders and Reuss (1974) using the prospective wing. During development of the leg the ectoderm exercised considerable control over d-v polarity, even as early as stage 14. While this study is incomplete and still in progress, it is clear that ectodermal reversal affects more than just ectodermal derivatives, including the rotation of the halux to a ventral position and the flexure of the leg at the tibiotarsal joint. Apparently, the earlier the ectodermal reversal is done, the more proximal its effects extend. Whether complete d-v reversal can be obtained with earlier stages remains to be tested.

In further preliminary experiments prospective leg mesoderm was transplanted in reversed d-v orientation to stage 15-16 hosts thus allowing host ectoderm to heal over the transplant. The legs developed d-v polarity corresponding to the mesoderm along their entire length. Prospective wings whose ectoderms were reversed mechanically developed dorsoventrally reversed tips.

Our preliminary results thus suggest the early control of d-v polarity resides with the ectoderm, but the dorsoventral information may be lost when the ectoderm proliferates over limb mesoderm.

This work was supported by Grant HD07282 from the NIH.

References

Calandra AJ, MacCabe JA (1978). The in vitro maintenance of the limb-bud apical ridge by cell-free preparations. Develop Biol 62:258.

Iten LE, Murphy DJ (1980). Pattern regulation in the embryonic chick limb: Supernumerary limb formation with anterior (non-ZPA) limb bud tissue. Develop Biol 75:373.

Javois LC, Iten LE (1981). Position of origin of donor posterior chick wing bud tissue transplanted to an anterior host site determines the extra structures formed. Develop Biol 82:329.

MacCabe JA, Errick J, Saudners JW Jr (1974). Ectodermal control of the dorsoventral axis in the leg bud of the chick embryo. Develop Biol 39:69.

MacCabe JA, Lyle PS, Lence JA (1979). The control of polarity along the anteroposterior axis in experimental chick limbs. J Exptl Zool 207:113.

MacCabe JA, Parker BW (1975). The in vitro maintenance of the apical ectodermal ridge of the chick embryo wing bud: An assay for polarizing activity. Develop Biol 45:349.

MacCabe JA, Parker BW (1979). The target tissue of limb-bud polarizing activity in the induction of supernumerary structures. J Embryol Exp Morph 53:67.

MacCabe JA, Richardson KEY (1982). Partial characterization of a morphogenetic factor in the developing chick limb. J Embryol Exp Morph 67:1.

Patau M-P, Kieny M (1973). Interaction ectomesodermique dans l'establissement de la polarite dorsoventrale du pied de l'embryo de ponlet. C R Acad Sci Ser D 227:1225.

Saunders, JW Jr, Gasseling MT (1968). Ectodermal-mesenchymal interactions in the origin of limb symmetry. In Fleischmajer R. Billingham RE (eds): "Epithelial-Mesenchymal Interactions," Baltimore: Williams and Wilkins Co, p 78.

Saunders JW Jr, Reuss C (1974). Inductive and axial properties of prospective wing-bud mesoderm in the chick embryo. Develop Biol 38:41.

Summerbell D, Tickle C (1977). Pattern formation along the antero-posterior axis of the chick limb bud. In Ede DA, Hinchliffe JR, Balls M (eds): "Vertebrate Limb and Somite Morphogenesis", Cambridge: Cambridge University Press, p 41.

Tickle C, Summerbell D, Wolpert L (1975). Positional signalling and specification of digits in chick limb morphogenesis. Nature, London 254:199.

Zwilling E (1955). Ectoderm-mesoderm relationship in the development of the chick embryo limb bud. J Exptl Zool 128:423.

Zwilling E (1956). Interaction between limb bud ectoderm and mesoderm in the chick embryo I. Axis establishment. J Exptl Zool 132:157.

Limb Development and Regeneration
Part A, pages 245–249
© 1983 Alan R. Liss, Inc., 150 Fifth Avenue, New York, NY 10011

CELLULAR AND COLLAGEN FIBRILLAR POLARITY IN DEVELOPING
CHICK LIMB TENDON

Robert L. Trelstad, M.D.
David E. Birk, Ph.D.
Frederick H. Silver, Ph.D.

Department of Pathology
UMDNJ-Rutgers Medical School
Piscataway, New Jersey 08854

The conditions which determine the orderly deposition
of matrix components during morphogenesis involve both
physicochemical interactions among the matrix components
and regulation by the cells. There is little reason to
believe that 'self assembly' of matrix components is the
sole determinant of matrix architecture even though
isolated matrix components have the capacity to aggregate
under a variety of conditions in vitro into structures
similar to those found in vivo. It is more probable, given
our present understanding of the process, to conclude that
the physicochemical forces serve to form intermediate
aggregates which are instrumental in the stepwise assembly
of the matrix and these forces also serve to stablilize the
various aggregates once they have formed (Trelstad, Silver
1981; Trelstad, 1982). But it is the cells which dictate
when, where and in what spatial pattern matrix components
are discharged and assembled in the extracellular space.

Studies of chick limb tendon formation have shown that
the process of collagen fibril assembly occurs in deep
cellular recesses and that the process is under close
cellular regulation (Trelstad, Hayashi 1979). The manner in
which these recesses interface with the forming collagen
fibrils suggests that the fibrils are 'spun' from the
fibroblast surface and that the forces associated with limb
growth generate tensions which are sensed by the fibroblast
at these sites of fibril anchorage and formation. This
mechanism of fibril formation provides a means by which
cells might transduce the tension associated with limb

growth into an appropriate length of tendon.

In associated studies of collagen fibrillogenesis in cells and from solutions (Trelstad, Hayashi, Gross 1976; Williams, Gelman, Piez 1978; Bruns, Thierey, Hulmes, Gross 1979; Silver, Langley, Trelstad 1979; Silver, 1981) it has been shown that collagen fibril assembly is a multistep process. The number of steps in the complete process has not been determined. A similar conclusion can be drawn from in vivo observations in which cells package secretory products into various intermediate aggregates. This is effected by cells by appropriate partitioning during the process of secretory packaging. The cell is able to produce a number of different matrix components simultaneously and to package them in various secretory vacuoles. The stochiometry as well as qualitative character of such mixtures is clearly under cellular control.

We know that the mesenchyme cells in the early limb are arranged in orderly patterns (Holmes, Trelstad 1977; Ede, Flint, Wilby, Colquhoun 1977; Holmes, Trelstad 1980). By using a marker for one pole of the cell, such as the Golgi apparatus, and another pole of the cell, such as the nucleus, an arbitrary axis and vector can be assigned to each cell. Using light or electron microscopy it is then possible to make maps of the spatial orientation of the cells during early morphogenetic stages (Trelstad, 1977).

The relationship between the patterns of cellular orientation which can be detected in this manner and the three dimensional architecture of the matrix has yet to be established. From what we know of the process of matrix secretion, however, we suspect that the patterns of the cells influence the architecture of the matrix just as the matrix influences and stabilizes the spatial positions of the cells.

In order to pursue this relationship we have examined the orientation of mesenchyme cells and collagen fibrils in the embryonic chick limb tendon. The orientation of the tendon mesenchyme in 14 to 17 day chick embryo foot flexor tendons has been studied using the Golgi impregnation technique described elsewhere (Trelstad 1977). The axis of the cell which passed through the nucleus and Golgi was assessed in tissue sections as well as the 'vector' of the Golgi apparatus. Tendons of similar aged embryos were studied by transmission electron microscopy. The orientation of the individual collagen fibrils in sections was determined by direct inspection of micrographs printed at a magnification of 50,000. Computer assisted analyses of such preparations is possible (Bender, Silver, Hayashi, Trelstad 1982).

In approximately 70% of the tendon mesenchyme cells, the Golgi pole was directed toward the distal end of the limb. Nearly all of the cells in the developing tendon were aligned with the long axis of the cell in a proximo-distal orientation. The thin sections of the tendon which were analyzed for the polarity of the tendon collagen fibrils indicated a slight predominance (53.5 %) for fibrils whose amino termini were directed toward the distal portion of the limb. This value is similar to that reported for a comparable number of fibrils studied (400) by Parry and Craig (1977) who concluded that the pattern was random. Although the value we have obtained is close to that expected for a random distribution, there is reason to believe that the small difference from random is significant.

One reason to suspect that the structural elements of the tendon, the collagen molecules, are not in a random pattern is based on the pyroelectrical and piezoelectrical properties of such tissues (Athenstadt, 1974). These physical properties require that the tissue possess a crystal-like structure with some anisotropy. The extent to which structures such as tendon need to be anisotropic in order to be piezoelectric has not been determined. Our observation that 53.5% of the collagen molecules are oriented in one direction is not a great difference, but perhaps sufficient for the effect.

The piezoelectrical property of tendons may be dependent on summation of the intra and intermolecular dipoles, dipoles generated by charged pairs of amino acid side chains which are important for the formation and stabilization of the collagen fibril (Silver, 1982).

These data indicate the coexistence of a structural aniostropy of the tendon with a non-random pattern of cellular alignment. At the moment, the causal relationship between the two patterns is not apparent. The quantitative difference in extent of the two polarities appears to be

248 / Trelstad, Birk, and Silver

substantial. However, until the causal relationship, if
any, between these two patterns is established we can only
note and wonder about their interrelatedness.

Athenstadt HH (1974). Pyroelectric and piezoelectric
  properties of vertebrates. Ann NY Acad Sci 238:68.
Bender E, Silver FH, Hayashi K, Trelstad RL (1982). Type I
  Collagen SLS Banding Patterns: Evidence that the α2 chain
  is in the reference or 'A' position. J Biol Chem In
  Press.
Bruns RR, Hulmes DJS, Therrien SF, Gross, J (1979).
  Procollagen segment-long-spacing crystallites: Their role
  in collagen fibrillogenesis. Proc Natl Acad Sci USA
  76:313.
Ede DA, Flint OP, Wilby OK, Colquhoun P (1977). The
  development of pre-cartilage condensations in limb bud
  mesenchyme in vivo and in vitro. In: Ede DA, Hinchliffe
  JR, Balls M, (eds). "Vertebrate Limb and Somite
  Morphogenesis". Cambridge Univ. Press, pp. 161-179.
Holmes LB, Trelstad RL (1977). Patterns of cell polarity in
  the developing mouse limb. Develop Biol 59:164.
Holmes LB, Trelstad RL (1980). Cell polarity in precartilage mouse limb mesenchyme cells. Develop Biol
  78:511.
Parry DAD, Craig AS (1977). Quantitative electron
  microscope observations of the collagen fibrils in
  rat-tail tendons. Biopolymers 16:1015.
Silver FH, Langley KH, Trelstad RL (1979). Type I collagen
  fibrillogenesis: Initiation via a reversible linear
  growth step. Biopolymers 18:2523.
Silver FH (1981). Type I collagen fibrillogenesis in vitro.
  Addition evidence for the assembly mechanism. J Biol Chem
  256:4973.
Silver FH (1982). A molecular model for linear and lateral
  growth of type I collagen fibrils. Coll Res 3:219.
Trelstad RL (1977). Mesenchymal cell polarity and
  morphogenesis of chick cartilage. Develop Biol 59:153.
Trelstad RL (1982). Multistep assembly of Type I collagen
  fibrils. Cell 28:197.
Trelstad RL, Hayashi K (1979). Tendon fibrillogenesis:
  Intracellular collagen subassemblies and cell surface
  changes associated with fibril growth. Develop Biol
  71:228.
Trelstad RL, Hayashi K, Gross J (1976). Collagen
  fibrillogenesis: Intermediate aggregates and
  suprafibrillar order. Proc Natl Acad Sci USA 73:4027.

Trelstad RL, Silver, FH (1981). Matrix Assembly. In: Hay ED, (ed): "Cell Biology of the Extracellular Matrix" Plenum Press, New York pp. 179-215.

Williams BR, Gelman RA, Poppke DC, Piez, KA (1978). Collagen fibril formation, optimal in vitro conditions and preliminary kinetic results. J Biol Chem 253:6578.

Limb Development and Regeneration
Part A, pages 251–266
© 1983 Alan R. Liss, Inc., 150 Fifth Avenue, New York, NY 10011

AGAINST PROGRAMS:
LIMB DEVELOPMENT WITHOUT DEVELOPMENTAL INFORMATION

Stuart A. Newman and Claire M. Leonard

Department of Anatomy
New York Medical College
Valhalla, New York 10595

INTRODUCTION: PROGRAMMATIC VS. NONPROGRAMMATIC VIEWS
OF DEVELOPMENT

The formation of limbs and other organs during ontogeny
or regeneration depends on the emergence of various types
of cells which, in turn, interact to influence one another's
fates. Within this commonly accepted framework two general
approaches characterize most of the recent work in organogenesis. One point of view makes use of the analogy of a
digital computer, and seeks to identify repositories of
"information" whereby the organism formally codifies the form
and pattern of developing structures in a hierarchy of
abstract languages. The most extreme "molecular" version of
this paradigm holds that the "developmental program" is
literally written in the sequence of bases of the DNA of the
zygote or the cells of the regeneration blastema. The cells
in question are supposed to execute their genetic instructions very much like a computer processes its data tape.
Another popular version of the computer paradigm acknowledges
that biological information can be encoded in forms other
than DNA base sequence, and hypothesizes that the developing
or regenerating organism sets up fields of "positional information", (Wolpert, 1969; 1981) each item of which calls
up a specific phenotype from the cells under its influence,
much as a command in a computer program can call up a specific
subroutine.

In contrast to the computer paradigm is the theoretical
approach which treats development as a reproducible but nonprogrammatic phenomenon. The return of Halley's Comet to

the earth's vicinity every 75 years follows definite laws, but not a script. While we can codify the comet's trajectory in a language interpretable by a computer and thereby predict its return, the comet itself makes use of no such codification to guide its path. Similarly, a developing or regenerating organ need not follow a program, but only physical and chemical laws (constrained by evolved relationships such as the genetic code) to yield a structure which is similar each time it arises (see also discussion by Stent, cited in Lewin 1981).

These theoretical distinctions have practical consequences for the experimental study of organogenesis in general and limb formation in particular. As one example of the ways in which the different approaches can influence the questions addressed, consider Figure 1, a cross section of a Hamburger-Hamilton (Hamburger, Hamilton 1951) stage 27-28 chick wing bud in the process of establishing the pattern of cartilage elements that provide the embryonic model of the limb skeleton. At the distal end of the wing certain of the mesenchyme cells are participating in the formation of the cartilages of the digits, while other mesenchyme cells are preparing to die off to free up the interdigital spaces.

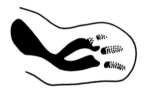

Fig. 1. Drawing of cross section of Hamburger-Hamilton stage 27-28 chick wing bud, indicating regions of definitive cartilage (black) and early cartilage and mesenchymal condensations (stippled). Adapted from Newman and Frisch (1979).

The programmatic approach sees this developmental episode as a question of the deployment of information: for instance, how can the organism establish an informational "map" such that a group of cells a certain distance from the anterior edge of the limb bud forms digit 2, while a group of cells just posterior to this position dies off, the next posterior group forms digit 3, and so on? The fact that the digits are all established as cartilage is incidental to the

pattern forming process in this interpretation. In fact under
one version of this model the cartilage cells of the various
digits are considered "non-equivalent", that is, distinct
subroutines of the genome (Lewis, Wolpert, 1976). The
spatially distributed morphogen postulated by this model only
marks locations. It is not per se a specific inducer of a
cellular behavior, such as movement, aggregation, or death.

In contrast, the nonprogrammatic approach to limb pattern
formation sees the phenomena described as a problem in
spatially-dependent cartilage differentiation. Specifically,
how can those factors and processes stimulatory to the early
stages of chondrogenesis arise in a reproducible, spatially
heterogeneous fashion during the course of limb development?
We have suggested that cell-cell interaction is obligatory
to limb chondrogenesis (Newman 1977), that concentrations of
fibronectin could encourage such interactions (Newman, Frisch
1979; Newman et al, 1981b; Tomasek et al, 1982) and that reproducible patterns of fibronectin concentration could be
achieved by a reaction-diffusion mechanism (Newman, Frisch
1979). This is one nonprogrammatic model; others are possible.
What such models have in common is 1) pattern is not codified
as information in any chronological or material variable
prior to its emergence (although features of a developmentally later pattern may be based on aspects of an earlier one);
2) precursor cells with a very limited array of potentials
are inclined along one or another developmental pathway by
stereotypical interactions with extracellular factors; and
3) regulation of pattern is tied to regulation of cytodifferentiation; that is to say, factors and processes encouraging
the emergence of a specific cellular phenotype in competent
cells are relevant variables in the control of pattern. It
must be recognized that all of these features are inconsistent
with most programmatic models, and are explicitly excluded
from the positional information framework (Wolpert 1969;
1981).

The postulation, or lack of it, of "prepatterns", that
is, direct correspondence between the distribution of morphogenetic factors and the final pattern of differentiated
cells, is not a decisive point of distinction between programmatic and nonprogrammatic models of development. There
is no reason, for example, why a monotonic gradient of a
substance could not lead to periodic cellular patterns under
appropriate conditions of cellular response, in a nonprogrammatic framework (For example, see Cooke 1981). What is

really at issue, as argued in another context (Newman 1974), is whether position per se is encoded in a developmental field as a material or chronological variable, or whether complex patterns arise instead from interactions between precursor cells of limited repertoire and heterogeneously distributed factors capable of eliciting one or another standard behavior.

In pursuing the consequences of the second viewpoint, our research group, along with our collaborators, has attempted to define the in vivo conditions governing spatially-dependent cartilage differentiation in the embryonic chick limb. Our studies have encompassed four aspects of this problem:

1) What are the options of the precursor cell population that participates in limb skeletal pattern formation?

2) What types of cell-cell and cell-matrix interactions encourage local chondrogenesis in competent tissue, and how might these conditions become nonuniformly distributed?

3) What molecular changes at the nuclear level characterize the switch from the precartilage to the cartilage cellular phenotypes?

4) How do pattern-forming cellular interactions trigger the nuclear changes leading to cartilage differentiation?

Results of our work in each of these areas are summarized in what follows.

SUBPOPULATIONS OF PROGENITOR CELLS IN THE EMBRYONIC LIMB

Although precartilage and premuscle cells of the embryonic chick limb appear similar to one another both under light and electron microscopy, the hypothesis that they constitute a homogeneous population of precursor cells up to the point of overt cytodifferentiation (Zwilling 1968) has not been sustained by recent studies. Our tissue culture experiments (Newman 1977) indicated that the stage 25 chick wing bud contains predifferentiated cells with a chondrogenic, but not a myogenic option. Furthermore, work in laboratories using surgically-produced chick-quail chimeric embryos

showed that the cells of the limb muscle and cartilage lineages have different points of origin in the embryo and do
not switch fates during normal development (Chevallier et
al,1977; Christ et al,1977). The tissue culture and chimera
work both strongly suggested that the cellular subpopulation
within which skeletal pattern formation occurs throughout
limb development is the somatopleure-derived, nonmyogenic
precartilage mesenchyme. Our collaborative work with the
group of Dr. M. Kieny (Newman et al, 1981 b) confirmed that
the wing tip at stage 25 was not yet populated with somite-
derived (i.e. myogenic) cells, thus establishing the tissue
of this region as appropriate for cellular and molecular
studies of cartilage differentiation without the complications inherent in heterogeneous cell populations.

A fibroblastic option for the somatopleure-derived
mesenchyme of the limb was also indicated by the chick-
quail chimera studies (Chevallier et al, 1977; Christ et al,
1977). Our studies on the cells of the stage 25 wing tip
showed that this option is expressed in culture under dispersed
conditions of growth (Newman 1980). (In contrast, most of
these cells differentiate into cartilage when grown as
aggregates (Newman 1977)). But whether the fibroblastic and
chondrogenic progenitor cells are disjoint populations, or
the same cell type with alternative fates (including death),
depending on growth conditions, remains an open question.

THE NECESSITY OF CELLULAR INTERACTIONS AND THE POSSIBLE
INVOLVEMENT OF FIBRONECTIN IN SKELETAL PATTERN FORMATION

Because precartilage cells undergo a transient phase of
close proximity in regions of the limb mesenchyme fated to
form cartilage (Fell and Canti 1934; Thorogood, Hinchliffe
1975), and because precartilage cells only differentiate
into cartilage under high density conditions (Newman 1977),
we suggested that cellular contact or close interaction is
an obligatory step in limb chondrogenesis (Newman 1977).
Studies of chondrogenesis in mixtures of chick and quail limb
cells isolated from limbs of different stages also point to
a need for homotypic cellular interactions in limb cartilage
histogenesis (Solursh, Reiter 1980). The nature of the
obligatory interactions are puzzling however, since the
mesenchymal cells appear to have extensive gap junctional
contacts with one another prior to any overt condensation
(Kelley and Fallon, 1978). Possibly the broad contacts that
form between precartilage cell surfaces during the condensa-

tion phase permit an interaction between receptors that are
not located at the pre-existing focal contacts. It has been
suggested that cyclic AMP is transiently elevated in precartilage cells during the condensation phase and might play
a role in establishing the cartilage phenotype (Solursh et
al, 1979; Kosher et al, 1979; Newman, Leonard 1982). This
metabolic event could well be potentiated by the "interacting
mode" of precartilage mesenchyme cells established during
the condensation phase of skeletal pattern formation in the
limb (Newman 1977).

It therefore seems important to understand how loci of
mesenchymal cell condensation are established in the
appropriate positions during limb development. In contrast
to a programmatic model which would characterize condensation
as just one early event in chondrogenesis executed by precartilage cells once their fates have been set, our nonprogrammatic view treats condensation as a causal link in chondrogenesis; it therefore seeks a biochemical signal that
stimulates precartilage cells to condense, rather than one
that directly instructs the cells to express a specific
genetic subroutine.

Fibronectin is a widely distributed extracellular
glycoprotein that mediates cellular attachment to substrates
and possibly to other cells (Ruoslahti et al, 1981). We and
others have found that fibronectin is concentrated between
precartilage cells undergoing condensation during limb
development (Tomasek et al, 1982; Dessau et al, 1980; Silver
et al, 1981; Melnick et al, 1981). Because the glycoprotein
is uniformly distributed in mesenchyme prior to the condensation phase (Tomasek et al, 1982) it is possible that its
redistribution during development plays a role in establishing
the pattern of condensations, and therefore of the cartilage
elements (Newman, Frisch 1979). It is significant that the
hyaluronate of the mesenchymal extracellular matrix is
progressively hydrolyzed as skeletal patterning proceeds
(Toole 1972). This may cause bound fibronectin to be
released, allowing it to diffuse and take up new configurations
according to the reaction-diffusion law (Turing 1952;
Newman, Frisch 1979). Indeed, under reasonable mathematical
assumptions and empirically-based estimates of limb bud
dimensions and macromolecular diffusion rates, a model based
on the redistribution of fibronectin and a hypothesized
cellular aggregation response to this glycoprotein predicts
an arithmetically increasing number of skeletal elements, in

proximodistal sequence (Saunders 1948) for the chick limb
bud (Newman, Frisch 1979). Decisive experimental tests of
these ideas are possible because candidates for the signal
molecule, the mechanism of its patterned distribution, and
the initial cellular response to its concentrated presence,
are all specified in the model.

NUCLEAR EVENTS IN CHONDROGENESIS

The availability of a virtually pure population of
cartilage progenitor cells (viz the mesenchyme of the stage
25 chick wing tip (Newman 1977; Newman et al, 1981a) has
allowed us to analyze molecular changes at the level of the
cell nucleus during chondrogenesis with the confidence that
cell selection is not an important factor in the observed
transitions (Newman et al, 1976; Perle, Newman 1980; Perle
et al, 1982; Newman, Leonard 1982). One finding that
intrigued us was that the two most abundant nonhistone
proteins of precartilage cell chromatin (each present in up
to 10^6 copies per nucleus) were progressively lost during
chondrogenesis in vitro and in vivo (Newman et al, 1976;
Perle, Newman 1980). Because the molecular weights of these
proteins are 35,500 and 125,000, we designated them PCP
(precartilage chromatin protein) 35.5 and PCP 125, respectively.

Apart from the striking change in their amounts during
development, another set of findings with regard to PCP 35.5
and PCP 125 suggested a role for them in the actual regulation of chondrogenesis. The chicken mutant talpid[2] exhibits
a highly perturbed pattern of cartilage differentiation in
the axial and appendicular skeletons when the corresponding
gene is present in the homozygous state (Abbott et al, 1960).
Recombination experiments with normal or mutant limb mesoderms
packed into normal or mutant limb ectoderms established that
the mesodermal tissue was the target of the talpid[2] gene in
the developing limb (Goetinck, Abbott 1964). We therefore
compared the chromatin proteins of talpid[2] limb precartilage
mesenchyme with those of normal tissue (Perle, Newman 1980).
Among the numerous peptides characterizing these preparations,
only PCP 35.5 and PCP 125 exhibited any detectable differences:
the mutant nuclei contained much less of the former, and an
altered version of the latter (approximate molecular weight
120,000). We do not know how far removed the synthesis and
processing of these two proteins are from the presumed
aberrant gene product coded by the talpid[2] gene, but since

these were the only evident differences between the mutant
and normal tissues (cartilage chromatin proteins being
identical in both strains) further analysis of the roles of
these molecules seemed warranted.

By exposing precartilage chromatin to high ionic strength
buffers containing the protein denaturant urea, we determined
that PCP 35.5 was among the small number of proteins remaining
bound to DNA (Perle et al, 1982). This placed it in the
previously identified class of tight DNA-binding proteins
to which possible regulatory significance has been attributed
(Pederson, Bhorjee 1975; Bekhor, Mirell 1979). Using the
deoxyribonuclease DNAase II, we were able to isolate fragments
of precartilage chromatin comprising about 12% of the total
nuclear DNA, and about half the cells' complement of both
PCP 35.5 and PCP 125 (Perle et al, 1982). The DNA of these
excised domains is highly enriched in sequences sensitive
to another deoxyribonuclease, DNAase I. Such sensitivity has
been associated with gene activation in virtually all cell
types in which it has been assayed with appropriate molecular
probes (Weintraub, Groudine 1976; Garel, Axel 1976; Groudine
et al, 1978; Storb et al, 1980). Thus PCP 35.5, which because
of its DNA binding properties is unlikely to redistribute from
its native chromatin sites during DNAase II fractionation,
and possibly PCP 125, are concentrated in chromatin domains
of probable developmental significance. Because the two precartilage-specific chromatin proteins are not released from
chromatin treated only with DNAase I, it seems likely that
these molecules are located nearby, but not within, the DNAase
I-sensitive domains (Perle et al, 1982). They are thus
candidates for components of the "domain attachment points"
postulated to play a key role in the regulation of gene
expression in eukaryotic cells (Igo-Kemenes, Zachau 1977;
see also Benyajati, Worcel 1976 and Paulson, Laemmli 1977).
A schematic diagram representing a working model for the
organization of precartilage chromatin is shown in Figure 2.

The PCP-associated active domains represented in Figure
2 might contain genes expressed in the precartilage cell (such
as fibronectin (Linder et al, 1975; Lewis et al, 1978; Tomasek
et al, 1982) and type I collagen (von der Mark, von der Mark
1974)),or genes that are "preactivated" (Stalder et al, 1980)
for expression in the cartilage cell (such as chondroitin
sulfate core protein (Vertel, Dorfman 1978) and type II
collagen (Miller, Matukas 1979; von der Mark, von der Mark
1974)). The significance of the developmentally-regulated

loss of the putative regulatory proteins PCP 35.5 and PCP 125
would depend strongly on which (if either) of these possibilities proves to be the case. This question is currently
under investigation using molecular probes for several genes
regulated during chondrogenesis (Fagan et al, 1980; Fuller,
Boedtker 1981; Vuorio et al, 1982).

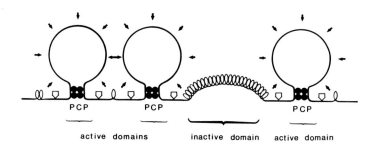

Fig. 2. Hypothesized domain organization of precartilage
cell chromatin. Regions sensitive to DNAase I (indicated by
black arrows) are presumed to be activated or preactivated
domains. DNAase II excises some of these domains along with
nearby DNA sequences bound to PCP 35.5 and PCP 125. (Presumed
sites of DNAase II attack are indicated by white arrows).
PCPs may therefore be targets for domain regulation.

MEDIATION OF EXTRACELLULAR AND NUCLEAR EVENTS BY CYCLIC
AMP: PRELIMINARY RESULTS

Whether the presence of PCP 35.5 and PCP 125 in specific
regions of chromatin plays a role in the expression of
precartilage-specific genes, or whether their absence from
these regions facilitates cartilage-specific gene expression
(two possibilities suggested by the previous results), it
seems likely that the removal of these proteins constitutes
a central event in chondrogenesis. We suspected that the
precipitating cause of this removal might be developmentally
regulated covalent modifications of precartilage chromatin
proteins, and tested this possibility in culture and in
isolated nuclei (Newman, Leonard 1982).

Phosphorylation of nonhistone proteins during chondrogenesis was monitored by culturing stage 25 wing tip

mesenchyme in the presence of $[^{32}P]$orthophosphate. Nuclei
were prepared from mesenchyme cultured for 48 hrs, a period
of time sufficient to initiate but not complete chondrogenesis
in the tissue mass (Newman 1977). It can be seen from this
gel that PCP 35.5 is heavily phosphorylated in the mixture
of precartilage, incipient cartilage, and cartilage nuclei
present in this preparation. Similar gels prepared from
cells at earlier and later phases of differentiation indicate
that PCP 35.5 is increasingly phosphorylated concomitant
with its developmentally-regulated removal from specific
chromatin domains (Newman, Leonard 1982, and unpublished
results).

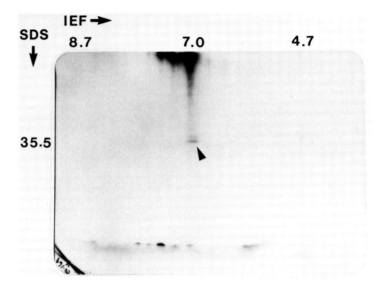

Fig. 3. Autoradiogram of two-dimensional gel electrophoresis
(O'Farrell 1975) of chromatin proteins prepared from precartilage mesenchyme cultured for 48 hr in the presence of
$[^{32}P]$orthophosphate. Note that PCP 35.5 (arrowhead) is
phosphorylated under these conditions. Horizontal axis is
in pH units, vertical axis molecular weight $X\ 10^{-3}$.

Evidence has accumulated from several laboratories that
a transient elevation in intracellular cyclic AMP levels of
precartilage cells may be a link between cell-cell interaction
and the transition to the cartilage cellular phenotype (Solursh
et al, 1979; Kosher et al, 1979; Kosher, Savage 1980; Solursh

et al, 1981). It was of interest, therefore, to determine whether any of the nuclear proteins that become phosphorylated during chondrogenesis are targets for a cyclic AMP-dependent protein kinase (Greengard 1978). When isolated precartilage nuclei are incubated with micromolar amounts of cyclic AMP along with the radioactive phosphate donor $[^{32}P]$ ATP, PCP 35.5 is uniquely and intensely phosphorylated (Figure 4).

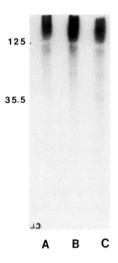

Fig. 4. Autoradiogram of one dimensional gel electrophoresis (Laemmli 1970) of precartilage cell nuclei incubated in the presence of $[^{32}P]$ ATP and (A) O μM (B) 2 μM and (C) 5 μM cyclic AMP for 10 min at 31°C. Numbers are molecular weight X 10^{-3}.

(PCP 125 becomes very heavily phosphorylated in the presence or absence of cyclic AMP under these conditions; the band corresponding to this protein had to be cut out of the three lanes of the gel pictured in Figure 4 to prevent overexposure). Figure 4 also indicates that the phosphorylation of PCP 35.5 occurs at 5 μM, but not 2 μM cyclic AMP. This apparent threshhold has been a relatively consistent feature of replicate experiments of this design (unpublished results).

Under the assumption that the cyclic AMP-dependent protein kinase targeting PCP 35.5 in isolated nuclei is also operative in the intact cell, these results suggest a mechanism by which cellular interactions, through the intermediary of elevated intracellular cyclic AMP concentrations, could cause modifications in the structure of precartilage chromatin i.e., via the phosphorylation of PCP 35.5, which may lower its affinity for DNA) leading to the expression of the cartilage cellular phenotype in a position-dependent fashion.

Clearly further studies are needed to establish the validity of these intriguing possibilities.

CONCLUSIONS

The view of limb development put forward in the previous sections attempts to take serious account of the findings of modern cell biology, biochemistry, and molecular biology in formulating an interpretation of position-dependent carti- lage differentiation and pattern formation. Our work has centered on interactions of cells with their globally organized microenvironments (corresponding to the field phenomena of classical embryology), as well as on cellular competences and response mechanisms (corresponding to the classical questions of differentiation and induction). It is, of course, conceivable that subtle differences in the forms of the radius and ulna, or the various digits of a vertebrate forelimb, might be due to a grid of positional distinctions that is somehow superimposed on the cells of the limb during the patterning process. However, we think it is untenable to maintain that these differences are more important than the similarities between these structures, and that the various skeletal elements are not established by similar cellular mechanisms during ontogeny or regenera- tion. In view of this, our interpretation conceptualizes the various cartilage elements as essentially equivalent from their inception. Asymmetries along the dorsoventral, antero- posterior, and proximodistal axes are presumed to arise as second-order modifications of a basically symmetrical, biochemically-based process that apportions cartilaginous structures from competent mesenchyme (Newman, 1982).

It is our impression that programmatic models of develop- mental processes such as limb formation make the classic error of confusing "information" available for use by a physical or biological system with "information" we require to characterize that system (Newman 1970). As pointed out by Goodwin (1970), in no sense does a cell compute its own state, look at the DNA program for further instructions, and then change its state accordingly. We would take this even further and state that is unlikely that any abstract represen- tation of a biological structure or activity is available prior to the realization of that structure or activity in a living system. We hope that the direction represented by the studies outlined above indicates that artifical, machine-based analogies are not required for the framing of testable explanatory hypotheses for the pheonomena of limb development.

REFERENCES

Abbott UK, Taylor LW, Abplanalp H (1960). Studies with talpid[2], an embryonic lethal of the fowl. J Hered 51:195.

Bekhor I, Mirell CJ (1979). Simple isolation of DNA hydro-phobically complexed with presumed gene regulatory proteins (M[3]). Biochemistry 18:609.

Benyajati C, Worcel A (1976). Isolation, characterization, and structure of the folded interphase genome of Drosophilia melanogaster. Cell 9:393.

Chevallier A, Kieny M, Mauger A (1977). Limb-somite relationship:origin of the limb musculature. J Embryol Exp Morph 41:245.

Christ B, Jacob HJ, Jacob M (1977). Experimental analysis of the origin of the wing musculature in avian embryos. Anat Embryol 150:171.

Cooke J (1981). The problem of periodic patterns in embryos. Phil Trans Roy Soc Lond B 295:509.

Dessau W, von der Mark H, von der Mark K, Fischer S (1980). Changes in the patterns of collagens and fibronectin during limb bud chondrogenesis. J Embryol Exp Morph 57:51.

Fagan JB, Pastan I, de Crombrugghe B (1980). Sequence rearrangement and duplication of double stranded fibronectin cDNA probably occurring during cDNA synthesis by AMV reverse transcriptase and Escherichia coli DNA polymerase I. Nucleic Acids Res 8:3055.

Fell HB, Canti RG (1934). Experiments on the development in vitro of the avian knee-joint. Proc Roy Soc Lond B 116:316.

Fuller F, Boedtker H (1981). Sequence determination and analysis of the 3' region of chicken pro-α1(I) and pro-α2(I) collagen messenger ribonucleic acids including the carboxy-terminal propeptide sequences. Biochemistry 20:996.

Garel A, Axel R (1976). Selective digestion of transcription-ally active ovalbumin genes from oviduct nuclei. Proc Nat Acad Sci USA 73:3966.

Goetinck PF, Abbott UK (1964). Studies on limb morphogenesis. I. Experiments with the polydactylous mutant, talpid[2]. J Exp Zool 155:161.

Goodwin BC (1970) Biological Stability. In Waddington CH (ed): "Towards a Theoretical Biology 3. Drafts, " Chicago:Aldine, p.1.

Groudine M, Das S, Neiman P, Weintraub H (1978). Regulation of expression and chromosomal subunit conformation of avian retrovirus genomes. Cell 14:865.

Hamburger V, Hamilton HL (1951). A series of normal stages in the development of the chick embryo. J Morph 88:49.

Igo-Kemenes T, Zachau HG (1977). Domains in chromatin structure. Cold Spring Harbor Symp Quant Biol 42:109.

Kelley RO, Fallon JF (1978). Indentification and distribution of gap junctions in the mesoderm of the developing chick limb bud. J Embryol Exp Morph 46:99.

Kosher RA, Savage MP (1980). Studies on the possible role of cyclic AMP in limb morphogenesis and differentiation. J Embryol Exp Morph 56:91.

Kosher RA, Savage MP, Chan SC (1979). Cyclic AMP derivatives stimulate the chondrogenic differentiation of the mesoderm subjacent to the apical ectodermal ridge of the chick limb bud. J Exp Zool 209:221.

Laemmli UK (1970). Cleavage of structural proteins during the assembly of the head of bacteriophage T4. Nature 227:680.

Lewin R (1981). Seeds of change in embryonic development. Science 214:42.

Lewis CA, Pratt RM, Pennypacker JP, Hassell JR (1978). Inhibition of limb chondrogenesis in vitro by vitamin A: Alterations in cell surface characteristics. Develop Biol 64:31.

Lewis JH, Wolpert L (1976). The principle of non-equivalence in development. J Theoret Biol 62:479.

Linder E, Vaheri A, Ruoslahti E, Wartiovaara J (1975). Distribution of fibroblast surface antigen in the developing chick embryo. J Exp Med 142:41.

Melnick M, Jaskoll T, Brownell AG, MacDougall M, Bessem C, Slavkin HC (1981). Spatiotemporal patterns of fibronectin during embryonic development I. Chick limbs. J Embryol Exp Morph 63:193.

Miller EJ, Matukas VJ (1969). Chick cartilage collagen: a new type of α1 chain not present in bone or skin of the species. Proc Nat Acad Sci USA 64:1264.

Newman SA (1970). Note on complex systems. J Theoret Biol 28:411.

Newman SA (1974). The interaction of the organizing regions in hydra and its possible relation to the role of the cut end in regeneration. J Embryol Exp Morph 31:541.

Newman SA (1977). Lineage and pattern in the developing wing bud. In Ede DA, Hinchliffe JR, Balls M (eds): "Vertebrate Limb and Somite Morphogenesis", Cambridge:Cambridge Univ Press, p 181.

Newman SA (1980). Fibroblast progenitor cells of the embryonic chick limb. J Embryol Exp Morph 56:191.

Newman SA (1982). Embryonic development as a mode of activity of the living state: The case of the vertebrate limb. In Mishra RK (ed) "Proceedings of the International Seminar on the Living State" in press.

Newman SA, Birnbaum J, Yeoh GCT (1976). Loss of a non-histone chromatin protein parallels in vitro differentiation of cartilage. Nature 259:417.

Newman SA, Frisch HL (1979). Dynamics of skeletal pattern formation in developing chick limb. Science 205:662.

Newman SA, Frisch HL, Perle MA, Tomasek JJ (1981a). Limb Development: Aspects of differentiation, pattern formation, and morphogenesis. In Connelly TG, Brinkley LL, Carlson BM (eds): "Morphogenesis and Pattern Formation", New York: Raven Press, p 163.

Newman SA, Leonard CM (1982). Cyclic AMP-dependent phosphorylation of a developmentally regulated DNA binding protein. Submitted for publication.

Newman SA, Pautou M-P, Kieny M (1981b). The distal boundary of myogenic primordia in chimeric avian limb buds and its relation to an accessible population of cartilage progenitor cells. Develop Biol 84:440.

O'Farrell PH (1975). High resolution two-dimensional electrophoresis of proteins. J Biol Chem 250:4007.

Paulson JR, Laemmli UK (1977). The structure of histone-depleted metaphase chromosomes. Cell 12:817.

Pederson T, Bhorjee JS (1975). A special class of nonhistone protein tightly complexed with template-inactive DNA in chromatin. Biochemistry 14:3238.

Perle MA, Leonard CM, Newman SA (1982). Developmentally regulated nonhistone proteins: Evidence for deoxyribonucleic acid binding role and localization near deoxyribonuclease I sensitive domains of precartilage cell chromatin. Biochemistry 21:2379.

Perle MA, Newman SA (1980). Talpid[2] mutant of the chicken with perturbed cartilage development has an altered precartilage-specific chromatin protein. Proc Nat Acad Sci USA 77:4828.

Ruoslahti E, Engvall E, Hayman EG (1981). Fibronectin: Current concepts of its structure and functions. Collagen Rel Res 1:95.

Saunders JW Jr (1948). The proximo-distal sequence of origin of the parts of the chick wing and the role of the ectoderm. J Exp Zool 108:363.

Silver MH, Foidart JM, Pratt RM (1981). Distribution of fibronectin and collagen during mouse limb and palate development.

Differentiation 18:141.

Solursh M, Reiter RS (1980). Evidence for histogenic interactions during in vitro limb chondrogenesis. Develop
Biol 78:141.

Solursh M, Reiter RS, Ahrens PB, Pratt RM (1979). Increase
in levels of cyclic AMP during avian limb chondrogenesis
in vitro. Differentiation 15:183.

Solursh M, Reiter RS, Ahrens PB, Vertel BM (1981). Stage-
and position-related changes in chondrogenic response of
chick embryonic wing mesenchyme to treatment with dibutyryl
cyclic AMP. Develop Biol 83:9.

Stalder J, Groudine M, Dodgson JB, Engel JD, Weintraub H
(1980). Hb switching in chickens. Cell 19:973.

Storb U, Wilson R, Selsing E, Walfield A (1981). Rearranged
and germline immunoglobulin genes: Different states of
DNAase I sensitivity of constant κ genes in immunocompetent
and nonimmune cells. Biochemistry 20:990.

Thorogood PV, Hinchliffe JR (1975). An analysis of the condensation process during chondrogenesis in the embryonic
chick hind limb. J Embryol Exp Morph 33:581.

Tomasek JJ, Mazurkiewicz JE, Newman SA (1982). Nonuniform
distribution of fibronectin during avian limb development.
Develop Biol 90:118.

Toole BP (1972). Hyaluronate turnover during chondrogenesis
in the developing chick limb and axial skeleton. Develop
Biol 29:321.

Turing AM (1952). The chemical basis of morphogenesis.
Phil Trans Roy Soc Lond B 237:37.

Vuorio E, Sandell L, Kravis D, Sheffield VC, Vuorio T,
Dorfman A, Upholt WB (1982). Construction and partial
characterization of two recombinant cDNA clones for
procollagen from chicken cartilage. Nucleic Acids Res
10:1175.

Wolpert L (1969).Positional information and the spatial
pattern of cellular differentiation. J Theoret Biol 25:1.

Wolpert L (1981). Positional information, pattern formation
and morphogenesis. In Connelly TG, Brinkley LL (eds):
"Morphogenesis and Pattern Formation", New York: Raven
Press, p 5.

Zwilling E (1968). Morphogenetic phases in development.
Develop Biol Suppl 2:184.

Limb Development and Regeneration
Part A, pages 267–278
© **1983 Alan R. Liss, Inc., 150 Fifth Avenue, New York, NY 10011**

THE EARLY GROWTH AND MORPHOGENESIS OF LIMB CARTILAGE

C.W. Archer, P. Rooney, and L. Wolpert

Department of Biology as Applied to Medicine
The Middlesex Hospital Medical School
London W1P 6DB, UK

The form of the vertebrate endoskeleton is a result
of two intimately related processes: cartilage
morphogenesis and growth. For example, just prior to
osteogenesis the chondrogenic rudiments of the developing
limb can be viewed, essentially, as models of the future
ossified elements. As such, the morphogenesis of the
overall shape of the rudiments is virtually complete with
secondary remodelling occurring through ligament and tendon
attachment, and bone remodelling once osteogenesis has
begun. In addition, the continued elongation of the
ossified elements can be attributed to cartilage growth
associated with the epiphyseal plate. Viewed in these
terms, it becomes important to understand fully, events
which specify and control early chondrogenesis in order
to answer basic questions – what makes long bones long?
Why do wrist elements hardly grow at all?

Pattern Formation and Programmed Growth

The importance of pattern formation and the concept of
positional information (Wolpert 1969) in the development of
the separate skeletal rudiments has been reviewed by
Wolpert (1982). In this view, the growth programme of the
individual elements is specified at the same time as the
pattern is laid down. Thereafter, growth and early
morphogenesis is largely autonomous.

This autonomy was classically shown by Fell and
Canti (1934) who removed transverse slices of $4\frac{1}{2}$–5 day

embryonic chick hind limb buds (prechondrogenic) and grew them in culture. They found that recognisable cartilage elements developed from the mesenchymal anlagen which corresponded to different parts of the femur, tibia and fibula inclusive of the knee joint primordium.

Clearly, the acquisition of the growth programme is a crucial step in the morphogenesis of the cartilaginous rudiments of the limb.

Condensation and Chondrogenesis

Initially, the mesenchyme within a developing limb bud is evenly distributed within the ectodermal jacket. However, prior to overt chondrogenesis, the position of the future skeletal elements is preceded by mesenchymal condensations. In the chick limb, the process of condensation has been studied by a number of workers (Searls 1972: Gould et al.1972: Thorogood, Hinchliffe 1975). Whilst controversy surrounds the extent of cell packing in comparison with the surrounding peripheral mesenchyme, and whether or not there is an increase or decrease in cell/cell contact within the condensation, all concur that a general increase in cell density occurs up to the expression of the chondrogenic phenotype.

The precise role of the prechondrogenic condensation is unclear. It has long been recognised that limb mesenchyme will only become chondrogenic in culture if the cells are plated at high density and above confluence. In such cultures, aggregates form in which the central cells are rounded or polygonal, and these represent the sites of initial cartilage matrix secretion. In contrast, mesenchyme cells plated at low density spread and become fibroblastic and secrete collagen types I and III (but not cartilage specific type II collagen). Consequently, it has been proposed that histogenic interaction is required for cartilage differentiation (Solursh, Reiter 1980). We, by contrast, have thought that cell shape may be equally as important as cell contact.

By using different concentrations of poly-hydroxyethyl-methacrylate (poly (HEMA)) (Folkman, Moscona 1978) as cell substrata, we have been able to control cell shape (by nature of the variable adhesivity of

the differing poly (HEMA) concentrations) and have shown
that a rounded cell configuration is conducive to the
synthesis of a sulphated matrix (Fig.1) even at low density
(Archer et al.1982).   We would argue, therefore, that high
cell density brings about the desired cell shape which is
conducive to the expression of the chondrogenic phenotype.
However, we have yet to demonstrate that the initiation of
type II collagen synthesis is dependent in any way on cell
shape.   Moreover, von der Mark et al.(1977) using
fluorescent antibodies have concluded that there was no
clear cut relationship between cell configuration and the
type of collagen synthesised by mature chondrocytes
in vitro.   The maintenance of a rounded cell shape in
culture also markedly depressed proliferation (Folkman,
Moscona 1978: Archer et al.1982), and it has been shown
that exposure to agents which suppress DNA replication
such as dibutryl cAMP have led to the precocious
appearance of cartilage in chick limb buds maintained
in vitro (Kosher, Savage 1980).   It seems likely,
therefore, that there is an integral relationship between
cell shape, proliferation and phenotypic expression.

     Our earlier studies on the cellular basis of the
formation of the perichondrium suggest it results from
the pressure exerted by the secretion of the cartilage
matrix flattening the outer cells of the chondrogenic
condensation (Gould et al.1974).   The present results also
support the idea that the perichondrium is mechanically
specified and may explain why the elongated cells of the
perichondrium (not dissimilar in appearance to fibroblasts
in confluent culture) do not secrete cartilage, but collagen
types I and III (von der Mark et al. 1976) even though they
may be in intimate contact with a chondrogenic matrix.

     There is also evidence that a threshold number of cells
is required for a condensation to become chondrogenic
(Wolpert et al. 1979).   For example, if a limb bud is
exposed to agents which kill cells such as X-rays, certain
elements are affected or lost before others.   In the hind
limb of the chick, the splinter-like fibula is more
sensitive than its neighbouring and larger tibia (Wolff,
Kieny 1962).   In order to account for such phenomena,
Wolff (1958),proposed a 'principle of competition'.   The
essence of the idea is that developing blastemas may compete
for a definitive number of mesenchymal cells.   Thus the
anlagen which exerts a greater 'field of influence' will

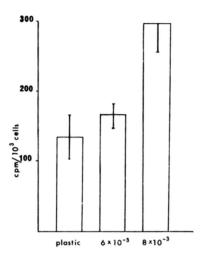

Figure 1. The incorporation of $^{35}SO_4{}^{--}$ into hyaluronidase sensitive material deposited by limb mesenchyme cells after 64h <u>in vitro</u>. Cells plated on dilutions of 6 x 10^{-3} and 8 x 10^{-3} poly (HEMA) showed a 18% and 112% increase in labelled material deposited over cells grown on plastic, but only the cells grown on 8 x 10^{-3} poly (HEMA) showed a significant difference (p$<$0.02)

recruit mesenchyme cells at the expense of a neighbour with a smaller 'field of influence'. He therefore argued that the tibial blastema acquires cells which might originally have been designated to the fibula.

In support of this interpretation, Hampé (1959,1960) showed that the insertion of a mica plate or additional mesenchyme into the developing bud sometimes resulted in an elongated fibula. Consequently, it was argued that the mica plate 'blocked' the acquisition of fibula mesenchyme by the tibia. In the second experiment, Hampé argued that the extra mesenchyme compensated for the loss of fibula cells to the tibia. We have repeated the former experiment of Hampé and have verified his result (Fig.2). However, we have correlated our findings with a detailed histological analysis of the development of the fibula.

We conclude that the increase in length observed in the
fibula is not a result of 'competition' but due to the
distal epiphysis being present whereas, normally, it is
absent.   Therefore a quite different explanation is
possible.   We have found that in normal development an
asymmetric protrusion develops from the posterior aspect of
the tibial distal epiphysis and forcibly detaches the
adjacent distal epiphysis of the fibula which subsequently
fuses to the tibia as the fibulare.   A group of flattened
cells which secrete little matrix, and are situated between
the epiphysis and diaphysis of the fibula, seems to
facilitate the detachment process, probably by acting as a
'weak' point.   Thus, there is no distal epiphysis of the
fibula and its growth is truncated.   Insertion of a barrier
merely inhibits normal morphogenesis and prevents detachment of the distal epiphysis.   Further evidence for such
an interpretation comes from the reduplication experiments
of Tickle (personal communication) who found that grafting
an additional polarizing region to the anterior margin of
the hind limb resulted in the specification of two fibulae
(rather than a tibia and fibula), both of which were
elongated and possessed distal epiphyses.   The classic
experiment of Hampé thus has a very simple explanation.

Growth and Morphogenesis

     The form of cartilage in a developing embryo is very
varied, and can be present as simple sheets, rods or complex
three-dimensional structures such as vertebrae.   What are
the mechanisms that control such diversity?

     Cartilage morphogenesis occurs through growth,which
itself can be viewed as comprising three distinct processes
- cell division, extracellular matrix secretion and cell
hypertrophy.   The extent to which any one of these
processes will contribute to morphogenesis will depend on
the type of skeletal element.   Thus,Biggers & Gwatkin(1964)
showed that blocking cell division in 7-day embryonic chick
tibiae with X-irradiation retarded their in vitro elongation
by only 20%.   Therefore, 80% of the longitudinal growth in
this particular long bone rudiment can be attributed to
matrix secretion and cell hypertrophy.   In contrast,
during normal development, the wrist elements of the wing
hardly elongate at all, probably because the cells do not
hypertrophy.

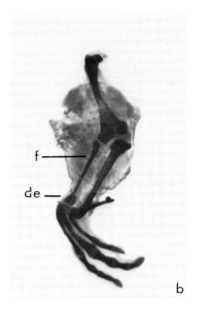

Figure 2.   a. Whole mount of normal embryonic chick leg after 10 days of incubation.   Note the truncated nature of the fibula (f) lacking a distal epiphysis.   b. Whole mount of embryonic chick leg at 10 days of incubation after barrier insertion at stage 21.   Note the increased length of the fibula (f) and the presence of a distal epiphysis (d e).

Perhaps it is important to note that the prechondrogenic condensations of future long bones are themselves elongated, whilst those of the wrist elements are more rounded and less distinct.   Once chondrogenesis is initiated, growth of the future long bones is rapid.   Within a day, the rudiments can be divided into three zones (Fell 1925). The central diaphyseal zone comprising of hypertrophic chondrocytes which largely lose the ability to divide. The epiphyseal zone containing small rounded cells which seem to divide fairly frequently, and an intermediate zone between the epiphysis and diaphysis whose cells are flattened in a direction perpendicular to the long axis of the rudiment.   These cells divide infrequently.   The cells of the wrist on the other hand, remain rounded and behave

as cells in the epiphyseal zone of long bone rudiments
(Holder 1977).

Two important questions arise from these basic
morphological observations. What confers directionality to
the growing rudiment? To what extent is cartilage form
determined by intrinsic properties or cell-to-cell
relationships? We would like to present evidence which
suggests that the spatial pattern of long bone rudiments
and their morphogenesis is specified and controlled by
mechanical influences. Chondrogenesis in a long bone
anlagen is initiated in the central region which subsequently
spreads towards either end (Fell 1925). The cells which
secrete matrix first, rapidly become flattened at right
angles to the long axis of the limb. We are, at present,
uncertain of the mechanism which brings about this initial
cell flattening. At about the same time, the cells around
the cylinder of central matrix secreting cells subsequently
flatten to form a rudimentary perichondrium (Gould et al.
1974).

Meanwhile, chondrogenesis proceeds towards the
'epiphyseal' ends of the condensation. At about the same
time, the central flattened diaphyseal cells begin to
hypertrophy. Our ultrastructural observations suggest
that hypertrophy consolidates theperichondrium, and presses
the cells together forming numerous tight junctions. In
contrast, there is no evidence of a perichondrial structure
towards the epiphyseal ends of the young rudiment. Thus,
the perichondrium becomes a constraining sheath, thus
favouring longitudinal expansion. Another factor which
promotes longitudinal growth derives from the initial
flattening of the chondroblasts perpendicular to the long
axis of the developing rudiment. Consequently, when
these cells hypertrophy (become large and round) expansion
will occur mainly along the longitudinal axis(Skalaslek,
personal communication). We have termed the combined
effects of these two processes 'directed dilation'. As
development progresses, the diaphyseal perichondrium is
strengthened by the deposition of collagen, whilst a
distinct but less robust structure forms around the epiphysis.
The essence is that the diaphyseal perichondrium is 'tight'
and prevents circumferential expansion, whilst the
epiphyseal perichondrium is'weak' thus expansion can occur
there. Moreover, this expansion is facilitated by
immature cartilage being visco-elastic. Pressure is

generated by the constrained expansion and this can be
visualised by making a nick in the perichondrium of the
epiphyseal region.   Over a period of 12 hours, a protrusion
develops which contains typical cartilage cells.   If the
nick is made in the diaphysis however, no protrusion
occurs because hypertrophic cartilage is far more solid.
The constraining role of the perichondrium can also be
demonstrated by biochemically weakening it.   If rudiments
are enzymatically treated (e.g., by collagenase), on
subsequent culture there is a distinct circumferential
expansion of the elements, especially in the epiphyseal
regions.   In addition, the mechanical removal of the
perichondrium from stage 29 chick tibia promotes the
elongation of these rudiments in vitro.   Thus it appears
that the perichondrium constrains both circumferentially
and longitudinally (Rooney, Archer & Wolpert, unpublished).

     In these terms the form of the cartilage is largely
determined by mechanical interactions.   Other reports
suggest that cartilage form is related to an intrinsic
property of the constituent cells.

     It has been suggested by many workers that cell
polarity may play a prominent role in morphogenesis.
Holmes and Trelstad (1980) measured cell orientation (in
terms of nucleus-Golgi complex polarity) in the
prechondrogenic mesenchyme of the mouse limb bud.   They
found that prior to matrix secretion the cells are
orientated towards the central longitudinal axis of each
condensation.   However, their initial measurements after
matrix secretion had begun, showed that the cell
orientation is less obvious.   Clearly, the secretion of
matrix in a polarized fashion could contribute significantly
to the overall morphogenesis and spatial pattern of a
skeletal element, but, it is difficult to see how this
particular orientation facilitates the longitudinal expansion
of the rudiments.

     In a similar vein, Weiss and Moscona (1958) argued that
prechondrogenic mesenchyme from different sites within an
embryo showed type-specific tissue morphogenesis.   In
support of this view, they demonstrated that chick limb bud
mesenchyme when dissociated, and plated at high density
in vitro, formed cartilage in whorls in a similar manner to
early chondrogenesis in the limb when viewed in
cross-section.   But when periocular mesenchyme (which forms

a thin cartilage sheet or capsule, the sclera, around the
eye) was treated in a similar fashion, flat sheets of
cartilage resulted.   It is important to note that such an
interpretation would require cells either to re-aggregate
in a polarized fashion or to re-polarize after settling.
We have repeated the experiments of Weiss and Moscona and
have verified their results.   However, we have also varied
the inoculation density and mixed non-chondrogenic cell
types (e.g., chick heart fibroblast) with the prechondrogenic
mesenchyme.   We found that when plated at low density or
mixed with heart fibroblasts, scleral mesenchyme could form
cartilage whorls.   Limb mesenchyme, on the other hand,
never formed flat sheets even at high density.

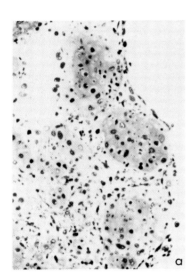 

Figure 3.   a. Low density culture of epiphyseal chondrocytes
from 8 day old chick embryo tibiae.   Note the appearance
of cartilage whorls surrounded by "fibroblastic' cells.
Cells were cultured under Micromass conditions but with an
initial inoculum of 5×10^4 cells/10 µl drop.   b. High
density culture of cells as above.   Note the homogenous
appearance of cartilage.   Cells were cultured under
Micromass conditions with an initial inoculum of 2×10^5
cells/10 µl drop.   Both cultures were sectioned
perpendicularly to the culture dish and stained with
toluidine blue.

Our results suggested to us that cartilage formation in vitro was not tissue specific, but was dependent on two factors, cell type and cell density. In this view, limb mesenchyme would not form a flat sheet because of the contamination of peripheral non-chondrogenic mesoblasts which intersperse the chondrogenic foci. Scleral mesenchyme by contrast, can be obtained as a pure cell preparation which will always form sheets providing the cells are not allowed to flatten on the plastic and become fibroblastic. At high density therefore, all the cells remain rounded and secrete matrix which appears as a homogenous sheet. At lower densities, however, some cells will spread, and from our results using poly (HEMA), fail to form cartilage: but in other areas, aggregates form which maintain the cell morphology and subsequently form nodules of whorled cartilage. Mixing non-chondrogenic cell types with the scleral mesenchyme effectively mimics the situation in limb mesenchyme and produces a similar result.

If this interpretation is correct, then one might predict that by varying the inoculation density of mature chondrocytes, then both whorled and sheet morphologies could be achieved. This is in fact what we obtained using proliferative epiphyseal chondrocytes from 9 day chick embryos (Fig.3).

In conclusion we would like to argue that the differential structure of the perichondrium plays an early and crucial role in the morphogenesis of long bone rudiment. Our initial observations on the development of the rounded wrist elements suggests the pattern of events differs significantly from those described above. The pre-chondrogenic condensation is rounded. Chondrogenesis begins in the centre and progresses outwards evenly (as in a long bone rudiment viewed in transverse section). Consequently, the perichondrium which forms is structurally (and functionally?) similar around the whole element. In addition, during embryonic life at least, cell hypertrophy does not occur (Holder, 1977).

Therefore, we wish to stress the importance of mechanical influences during early growth, especially in relation to the constraining role of the perichondrium, particularly in long bone rudiments, which brings about morphogenesis via differences in its structural integrity.

This work is supported by the Medical Research Council, U.K. We thank Dr. C. Tickle for reading the manuscript.

Archer CW, Rooney P, Wolpert L (1982). Cell shape and cartilage differentiation of early chick limb bud cells in culture. Cell Diff (in press)

Biggers JD, Gwatkin RBL (1964). Effect of X-rays on the morphogenesis of the embryonic chick tibiotarsus. Nature 202:152

Fell HB (1925). The histogenesis of cartilage and bone in the long bones of the embryonic fowl. J Morph 40:417

Fell HB, Canti RB (1934). Experiments on the development in vitro of the avian knee joint. Proc Roy Soc London Ser B 116:316

Folkman J, Moscona A (1978). Role of cell shape in growth control. Nature 273:345

Gould RP, Day A, Wolpert L (1972). Mesenchymal condensation and cell contact in early morphogenesis of the chick limb. Exp Cell Res 72:325

Gould RP, Selwood L, Day A, Wolpert L (1974). The mechanism of cellular orientation during early cartilage formation in the chick limb and regenerating amphibian limb. Exp Cell Res 83:287

Hampé A (1959). Contribution a l'étude du développement et de la régulation des deficiences et des excédents dans la patte de l'embryon de Poulet. Arch Anat micr Morph exp 48:345

Hampé A (1960). La competition entre les elements osseux du zeugopode de Poulet. J Embryol exp Morph 8:241

Holder N (1977). 'The Control of Cytodifferentiation and Growth in the Developing Chick Limb' Ph.D. thesis, University of London

Holmes LB, Trelstad RL (1980). Cell polarity in precartilage mouse limb mesenchyme cells. Dev Biol 78:511

Kosher RA, Savage MP (1980). Studies on the role of cAMP in limb morphogenesis and differentiation. J Embryol exp Morph 56:91

Searls RL (1972). Cellular segregation: a 'late' differentiative characteristic of chick limb bud cartilage cells. Exp Cell Res 73:57

Solursh M, Reiter RS (1980). Evidence for histogenic interactions during in vitro limb chondrogenesis. Dev Biol 78:141

Thorogood PV, Hinchliffe JR (1975). An analysis of the condensation process during chondrogenesis in the embryonic chick hind limb. J Embryol exp Morph 33:581

von der Mark H, von der Mark K, Gay S (1976). Study of differential collagen synthesis during the development of the chick embryo by immunofluorescence. II. Localisation of Type I and Type II collagen during long bone development. Dev Biol 53:153

von der Mark K, Gauss V, von der Mark H, Muller P (1977). Relation between cell shape and type of collagen synthesised as chondrocytes lose their cartilage phenotype in culture. Nature 267:531

Weiss P, Moscona A (1958). Type-specific morphogenesis of cartilages developed from dissociated limb and scleral mesenchyme in vitro. J Embryol exp Morph 6:238

Wolff, E (1958). Le principe de compétition. Bulletin de la Societe Zoologique de France 83:13

Wolff E, Kieny M (1962). Mise en évidence par l'irradiation aux rayons X d'un phénomène de compétition entre les ébauches du tibia et du péroné chez l'embryon de poulet. Dev Biol 4:197

Wolpert,L (1969). Positional information and the spatial pattern of differentiation. J theor Biol 25:1

Wolpert,L (1982). Cartilage morphogenesis in the limb. In, Bellairs R, Curtis A, Dunn G (eds): 'Cell Behaviour' Cambridge University Press, p 359

Wolpert L, Tickle C, Sampford M, Lewis J (1979). The effect of cell killing by X-irradiation on pattern formation in the chick limb. J Embryol exp Morph 50:175

Limb Development and Regeneration
Part A, pages 279–288
© 1983 Alan R. Liss, Inc., 150 Fifth Avenue, New York, NY 10011

RELATIONSHIP BETWEEN THE AER, EXTRACELLULAR MATRIX, AND
CYCLIC AMP IN LIMB DEVELOPMENT

Robert A. Kosher, Ph. D.

Department of Anatomy
University of Connecticut Health Center
Farmington, Connecticut 06032

A variety of studies in vivo (Stark, Searls 1973;
Summerbell et al 1973), in organ culture (Kosher et al
1979a), and in monolayer culture (Globus, Vethamany-Globus
1976; Solursh et al 1981b) have demonstrated that one of the
major functions of the AER is to maintain limb mesenchymal
cells directly subjacent to it in an actively outgrowing,
labile, undifferentiated condition. Throughout early limb
development in vivo, mesodermal cells extending 0.4 mm or
so from the AER retain the characteristics of unspecialized
mesenchymal cells, whereas cells located farther than 0.4
mm or so from the AER are engaged in cartilage differentiation (Stark, Searls 1973). Similarly, when the non-differentiating subridge mesoderm is subjected to organ culture
in the presence of the AER, the cells undergo quite normal
morphogenesis characterized by progressive polarized proximal-distal outgrowth and do not initiate chondrogenic differentiation until they have become located greater than 0.4 mm
or so from the AER (Kosher et al 1979a). When the subridge
mesoderm is cultured in the absence of the AER, the cells
fail to undergo morphogenesis and rapidly and precociously
initiate chondrogenic differentiation (Kosher et al 1979a).
Furthermore, when the AER is placed upon limb mesenchymal
cells that have been subjected to monolayer culture, the
cells underlying the AER accumulate under it in an organized
fashion, appear to initiate outgrowth, but fail to differentiate into cartilage (Globus, Vethamany-Globus 1976;
Solursh et al 1981b). In contrast, cells in regions of the
same monolayer that are not in contact with the AER differentiate into cartilage nodules (Globus, Vethamany-Globus
1976; Solursh et al 1981b).

The above studies indicate the AER exerts a negative
effect on the differentiation of mesenchymal cells directly
subjacent to it, and that when cells are removed from the
AER's influence either artificially or as a result of polarized proximal-distal limb outgrowth, they are freed to commence cytodifferentiation. When the cells in the proximal
central core of the limb have left the AER's influence,
they undergo a cellular condensation process and intimate,
homotypic cell-cell interaction that triggers cartilage
differentiation (see below). The differentiation of non-
chondrogenic tissues in the proximal peripheral regions of
the limb may be influenced at least in part by the non-AER
containing dorsal-ventral ectoderm of the limb. When the
subridge mesoderm is subjected to organ culture in the absence of the AER and dorsal/ventral limb ectoderm, virtually
all of the cells of the explant rapidly differentiate into
cartilage, and non-chondrogenic tissues are not discernible
(Kosher et al 1979a). However, when the subridge mesoderm
is cultured in the presence of the dorsal/ventral limb ectoderm, non-chondrogenic tissues form along the periphery of
the explant subjacent to the ectoderm, and precocious cartilage differentiation only occurs in the central core of the
explant (Kosher et al 1979a). Similarly, limb mesenchymal
cells in high density monolayer culture differentiate into
non-chondrogenic cell types when dorsal/ventral limb ectoderm is placed upon them (Solursh et al 1981b). These
studies suggest that the dorsal/ventral limb ectoderm promotes the formation of non-chondrogenic tissues in the limb
periphery, once the cells have been removed from the negative differentiative influence of the AER.

MECHANISM OF AER ACTION

Toole (1973) has observed that the relative amount of
the glycosaminoglycan (GAG), hyaluronate (HA) synthesized
is higher in whole stage 23 chick limb buds than in older
limb buds in which overt cartilage formation has begun.
Furthermore, hyaluronidase activity becomes detectable at
about the time a metachromatic cartilage matrix is detectable in the proximal regions of the limb (Toole 1973).
Thus, Toole (1973) has suggested that the synthesis of HA
by limb mesenchymal cells is associated with inhibition of
their differentiation, and that the removal of HA may be
necessary for organized chondrogenic differentiation. These
observations prompted us to investigate the possibility that

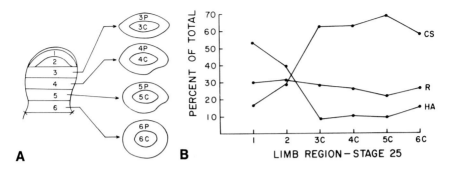

Fig. 1. A. The various segments into which stage 25 limb
buds have been dissected. Note that the proximal segments
were separated into central core (3C,4C,5C,6C) and peripheral (3P,4P,5P,6P) regions (from Kosher et al 1981).
B. Relative amount of hyaluronate (HA), chondroitin sulfate (CS), and other GAG (R) accumulated by various limb
regions during a 90 min labeling period with ^{3}H-glucosamine.
The relative and total (cpm/μg DNA) amount of HA accumulated
in the non-differentiating distal subridge region (segment
1) is about 2-fold greater than in the proximal subridge
region (2) and 4-5-fold greater than in the central core of
segment 3 (3C) and more proximal central core regions (4C-
6C) in which condensation and cartilage matrix deposition
are proceeding (see Kosher et al 1982).

HA may play a role in the negative effect the AER exerts on
the differentiation of the mesenchymal cells directly subjacent to it. Since there is a gradation of differentiation
along the proximal-distal axis of the limb bud, we have
dissected stage 24 and 25 limb buds into various well-defined regions along the proximal-distal axis in which the
cells are in different phases of differentiation (see Fig.
1 and Kosher et al 1981), and have biochemically studied
the accumulation of HA and other GAG in each region. We
have found that there is a gradation of HA accumulation
along the proximal-distal axis of the limb which correlates
with the distance of cells from the AER and the state of
differentiation of the cells. HA is by far the predominant
GAG synthesized by the non-differentiating mesenchymal cells
directly subjacent to the AER and there is a progressive
decrease in HA accumulation by more proximal cells in the
central core of limb which correlates with the initiation
of the critical condensation phase of chondrogenesis (Fig. 1

and Kosher et al 1981). The striking decrease in HA accumulation in the proximal central core regions is accompanied by a striking increase in the accumulation of chondroitin sulfate, one of the major constituents of cartilage matrix (Fig. 1). In contrast to the central core regions of the limb, HA accumulation by the peripheral non-chondrogenic regions remains relatively high (Kosher et al 1981). On the basis of these observations, we have suggested that the AER may exert its negative effect on cartilage differentiation at least in part by causing the immediately subjacent mesenchymal cells to synthesize and secrete a considerable quantity of HA, which, by accumulating extracellularly, may maintain the cells physically separated from one another thus preventing the cellular condensation and resultant cell-cell interaction that is required to trigger differentiation. HA may also feed back upon the cells inhibiting their synthesis of cartilage matrix components (Wiebkin, Muir 1973). In turn, the striking decline in HA accumulation that occurs as cells leave the AER's influence may facilitate the critical cellular condensation process (Kosher et al 1981; see also Shambaugh, Elmer 1980).

We have recently utilized indirect immunofluorescence to study the temporal and spatial distribution of the high molecular weight, adhesive glycoprotein, fibronectin during the course of embryonic chick limb morphogenesis in vivo (Kosher et al 1982). We have found that at all stages of development from 19 through 25, fibronectin (FN) is uniformly and diffusely distributed throughout the non-differentiating mesenchymal tissue directly subjacent to the AER (Kosher et al 1982). Tomasek et al (1982) have also reported the mesenchyme at the distal tip of the embryonic chick and duck wing bud is very rich in FN. The presence of FN throughout the prechondrogenic mesenchyme directly subjacent to the AER is of considerable interest in view of the high rate of HA accumulation by cells in this region (Kosher et al 1981). We have suggested this extracellular HA may be preventing a cell-cell interaction necessary to trigger differentiation and/or may be feeding back upon the cells inhibiting their production of cartilage matrix components. It has recently been demonstrated that FN has a specific HA binding site (Yamada et al 1980). Therefore, an intriguing possibility is that cell surface FN in the distal subridge mesenchyme may serve as a HA receptor and/or that an interaction between HA and FN may in some other manner be involved in maintaining the cells undifferentiated.

It is likely that the effects of the AER are mediated
by molecules synthesized and secreted by it. We have analyzed the pattern of GAG synthesis by the AER, and compared
it to the pattern of GAG synthesis by non-AER containing
dorsal/ventral limb ectoderm and non-limb ectoderm, since
the latter tissues do not support the outgrowth and formation of distal limb structures by limb mesenchymal cells
(Kosher, Savage 1981). The major difference in the pattern
of GAG accumulation between the AER and the other ectodermal
tissues is that there is a selective, quantitative increase
in the amount of HA produced by the AER. The amount of HA
accumulated by the AER is about 3 times greater than the
amount accumulated by the dorsal/ventral limb ectoderm or
non-limb (flank) ectoderm during a 90 min labeling period
with ^{3}H-glucosamine. In contrast, there is little, if any,
difference in the amount of chondroitin sulfate, heparan
sulfate, or glycoprotein accumulated by the tissues (Kosher,
Savage 1981). Thus we have suggested that the excess HA
produced by the AER may not only be involved in the unique
outgrowth promoting effect of the AER, but may also be involved in inhibiting differentiation by acting on the immediately subjacent mesenchymal cells causing them to maintain their own high rate of HA synthesis, or perhaps by
simply contributing to the HA-rich matrix surrounding the
subjacent mesenchymal cells.

REGULATION OF LIMB CARTILAGE DIFFERENTIATION

When limb mesenchymal cells leave the influence of the
AER, they undergo a cellular condensation or aggregation
process prior to depositing a cartilage matrix, i.e., the
cells become closely packed and apposed to one another and
extensive contact between adjacent cell surfaces occurs
(see, for example, Thorogood, Hinchliffe 1975). A variety
of studies indicate that during this condensation process
an intimate, homotypic cell-cell interaction occurs which
is necessary to trigger chondrogenic differentiation (see
Kosher 1982 for review). For example, the chondrogenesis
of dissociated limb mesenchymal cells in monolayer culture
is dependent upon the formation of cellular aggregates
during the initial period of culture (see, for example,
Caplan 1970).

As described above, we have suggested that the critical condensation process is initiated, at least in part, by

a progressive decrease in accumulation of extracellular HA (Kosher et al 1981). An accumulation of FN along the surfaces of closely apposed cells during the condensation phase of chondrogenesis in vivo (Dessau et al 1980; Silver et al 1981; Kosher et al 1982; Tomasek et al 1982) and in vitro (Lewis et al 1978) has been reported, and it has been suggested FN may be involved in promoting cell adhesions during condensation. Relatively high amounts of type I collagen as well as FN have been found in the condensing central core of the limb suggesting an interaction between type I collagen and FN which has a high affinity binding site for type I collagen, may be involved in the condensation phase of chondrogenesis (Dessau et al 1980; Silver et al 1981). In addition, our preliminary studies suggest an interaction between cell surface galactosyltransferases and acceptors on adjacent cell surfaces may occur during the condensation process (Shur et al 1982).

A variety of studies indicate cyclic AMP plays a key role in the critical cell-cell interaction occurring during condensation. Agents that elevate cyclic AMP levels elicit a dose dependent stimulation of the already precocious chondrogenic differentiation subridge mesoderm explants undergo in organ culture in the absence of the AER (Kosher et al 1979b). These agents also promote the chondrogenic differentiation of limb mesenchymal cells in high density monolayer culture (Ahrens et al 1977), and stimulate the chondrogenesis of cells from the non-chondrogenic, peripheral regions of the limb (Solursh et al 1981a). Furthermore, in the presence of cyclic AMP derivatives, subridge mesoderm explants cultured in the presence of the AER fail to undergo the striking proximal to distal outgrowth and contour changes characteristic of control explants, and the cessation of AER-directed morphogenesis in the presence of these agents is accompanied by precocious cartilage differentiation (Kosher, Savage 1980). Thus, agents that elevate cyclic AMP levels enable subridge mesenchymal cells to overcome the negative influences on cartilage differentiation and the positive influences on morphogenesis being imposed upon them by the AER (Kosher, Savage 1980). It has, therefore, been suggested that when limb mesenchymal cells leave the AER's influence, their cyclic AMP content increases triggering chondrogenic differentiation (Kosher, Savage 1980). The increase in cyclic AMP content is thought to result from the cellular condensation and resultant intimate cell-cell interaction that precedes overt cartilage formation

Table 1. Effect of PGE_2 and theophylline on the accumulation of sulfated GAG by subridge mesoderm explants

Treatment	1 day		3 days	
	cpm/µg DNA	treated/ control	cpm/µg DNA	treated/ control
none	1,408	--	6,474	--
10^{-5} M PGE_2	1,634	1.16	6,740	1.04
10^{-3} M Theophylline	1,784	1.27	7,881	1.22
10^{-5} M PGE_2 and 10^{-3} M Theophylline	2,637	1.87	10,363	1.60

(Kosher, Savage 1980; see also Solursh et al 1978).

In summary, the above studies indicate that when mesen-chymal cells leave the AER's influence, they undergo an intimate cell-cell interaction that triggers chondrogenic differentiation by elevating cyclic AMP levels. It is, therefore, of considerable interest that when limb mesen-chymal cells are subjected to high density monolayer culture, they synthesize several prostaglandins including PGE_2 and 6-keto $PGF_{1\alpha}$ (Chepenik et al 1980). Prostaglandins (PGs) are oxygenated fatty acids that have been demonstrated to be important local regulators of a variety of cellular pro-cesses in a number of biological systems, and the regulatory effects of PGs particularly of the E-type are mediated by cyclic AMP (Kuehl 1974; Samuelsson et al 1978). That is, PGs interact with specific cell surface receptors, activate adenylate cyclase, and thus stimulate cyclic AMP accumula-tion. These observations prompted us to consider the pos-sibility that endogenously produced PGs might be involved in regulating limb cartilage differentiation by acting as local modulators of cyclic AMP levels.

In our initial experiments to investigate this possi-bility, we have examined the effect of exogenous PGE_2 on the in vitro chondrogenic differentiation of the subridge mesoderm (Kosher, Walker 1982). The combined effects of PGE_2 and theophylline were also examined, since if PGE_2 is acting via cyclic AMP its effect might be potentiated by a phosphodiesterase inhibitor. As shown in Table 1, we have found that exogenous PGE_2 or theophylline alone elicit only a rather mild stimulation of sulfated GAG accumulation by subridge mesoderm explants (Kosher, Walker 1982). However, when theophylline is added along with PGE_2, a striking stimulation of chondrogenesis is observed which is reflected in a close to 2-fold increase in sulfated GAG accumulation (Table 1 and Kosher, Walker 1982). The effects of PGE_2 and theophylline are not additive, but rather are synergistic. The fact that the stimulatory effect of PGE_2 is greatly potentiated by theophylline strongly suggests its regulatory effect is mediated by its ability to increase cyclic AMP levels. The stimulatory effect of PGE_2 is dose-dependent and can be detected at a concentration at least as low as 10^{-8} M (Kosher, Walker 1982). PGE_1, which is also a potent stimulator of adenylate cyclase, is just as effective as PGE_2 in stimulating in vitro chondrogenesis, while PGA_1 and $PGF_{1\alpha}$ elicit, at best, only a mild effect and thromboxane B_2 has no effect (Kosher, Walker 1982). These initial experiments thus provide strong incentive for considering and further investigating the role of PGs in the regulation of limb cartilage differentiation. For example, it is possible that PGs produced during, or as a consequence of, the critical cell-cell interaction occurring during condensation may regulate chondrogenic differentiation by locally acting on the mesenchymal cells, and, thereby, elevating their cyclic AMP levels.

The work reported here has been supported by grant PCM-7925907 from the National Science Foundation.

REFERENCES

Ahrens PB, Solursh M, Reiter RS (1977). Stage-related capacity for limb chondrogenesis in cell culture. Develop Biol 60:69.
Caplan AI (1970). Effects of the nicotinamide-sensitive teratogen 3-acetylpyridine on chick limb cells in culture. Exp Cell Res 62:341.
Chepenik KP, Waite BM, Parker CL (1980). Prostaglandin synthesis during chondrogenesis in vitro. J Cell Biol 87:21a.
Dessau W, von der Mark H, von der Mark K, Fischer S (1980). Changes in the patterns of collagens and fibronectin during limb-bud chondrogenesis. J Embryol Exp Morph 57:51.
Globus M, Vethamany-Globus S (1976). An in vitro analogue of early chick limb bud outgrowth. Differentiation 6:91.

Kosher RA (1982). The chondroblast and the chondrocyte. In Hall BK (ed): "Cartilage", San Diego: Academic Press (in press).

Kosher RA, Savage MP (1980). Studies on the possible role of cyclic AMP in limb morphogenesis and differentiation. J Embryol Exp Morph 56:91.

Kosher RA, Savage MP (1981). Glycosaminoglycan synthesis by the apical ectodermal ridge of chick limb bud. Nature 291:231.

Kosher RA, Savage MP, Chan S-C (1979a). In vitro studies on the morphogenesis and differentiation of the mesoderm subjacent to the apical ectodermal ridge of the embryonic chick limb bud. J Embryol Exp Morph 50:75.

Kosher RA, Savage MP, Chan S-C (1979b). Cyclic AMP derivatives stimulate the chondrogenic differentiation of the mesoderm subjacent to the apical ectodermal ridge of the chick limb bud. J Exp Zool 209:221.

Kosher RA, Savage MP, Walker KH (1981). A gradation of hyaluronate accumulation along the proximodistal axis of the embryonic chick limb bud. J Embryol Exp Morph 63:85.

Kosher RA, Walker KH (1982). The effect of prostaglandins on in vitro limb cartilage differentiation (submitted).

Kosher RA, Walker KH, Ledger PW (1982). Temporal and spatial distribution of fibronectin during development of the embryonic chick limb bud. Cell Differentiation (in press).

Kuehl FA (1974). Prostaglandins, cyclic nucleotides and cell function. Prostaglandins 5:325.

Lewis CA, Pratt RM, Pennypacker JP, Hassell JR (1978). Inhibition of limb chondrogenesis in vitro by vitamin A: alterations in cell surface characteristics. Develop Biol 64:31.

Samuelsson B, Goldyne M, Granstrom E, Hamberg M, Hammarstrom S, Malmsten C (1978). Prostaglandins and thromboxanes. Ann Rev Biochem 47:997.

Shambaugh J, Elmer WA (1980). Analysis of glycosaminoglycans during chondrogenesis of normal and brachypod mouse limb mesenchyme. J Embryol Exp Morph 56:225.

Shur BD, Vogler M, Kosher RA (1982). Changes in endogenous cell surface galactosyltransferase activity during in vitro limb bud chondrogenesis. Exp Cell Res 137:229.

Silver MH, Foidart J-M, Pratt RM (1981). Distribution of fibronectin and collagen during mouse limb and palate development. Differentiation 18:141.

Solursh M, Ahrens PB, Reiter RS (1978). A tissue culture analysis of the steps in limb chondrogenesis. In vitro 14:51.

Solursh M, Reiter RS, Ahrens PB, Vertel BM (1981a). Stage- and position-related changes in chondrogenic response of chick embryonic wing mesenchyme to treatment with dibutyryl cyclic AMP. Develop Biol 83:9.

Solursh M, Singley CT, Reiter RS (1981b). The influence of epithelia on cartilage and loose connective tissue formation by limb mesenchyme cultures. Develop Biol 86:471.

Stark RJ, Searls RL (1973). A description of chick wing development and a model of limb morphogenesis. Develop Biol 33:138.

Summerbell D, Lewis JH, Wolpert L (1973). Positional information in chick limb morphogenesis. Nature New Biol 244:492.

Thorogood PV, Hinchliffe JR (1975). An analysis of the condensation process during chondrogenesis in the embryonic chick hind limb. J Embryol Exp Morph 33:581.

Tomasek JJ, Mazurkiewicz JE, Newman SA (1982). Nonuniform distribution of fibronectin during avian limb development. Dev Biol 90:118.

Toole BP (1973). Hyaluronate and hyaluronidase in morphogenesis and differentiation. Am Zool 13:1061.

Wiebkin OW, Muir H (1973). The inhibition of sulphate incorporation in isolated adult chondrocytes by hyaluronic acid. FEBS Letters 37:42.

Yamada KM, Kennedy DW, Kimata K, Pratt RM (1980). Characterization of fibronectin interactions with glycosaminoglycans and identification of active proteolytic fragments. J Biol Chem 255:6055.

Limb Development and Regeneration
Part A, pages 289–297
© 1983 Alan R. Liss, Inc., 150 Fifth Avenue, New York, NY 10011

MORPHOGENESIS OF THE DERMAL SKELETON IN THE TELEOSTS FINS

Jacqueline Géraudie

Equipe de recherche "Formations squelettiques"
Laboratoire d'Anatomie comparée, Paris VII,
2,Place Jussieu, 75251 Paris Cedex 05, France.

   In spite of the variety of problems related to the tetrapod limb bud morphogenesis as explained in other sections of
the present book, any researcher in teleost fin morphogenesis
faces original questions due to the structural particularities
of this type of appendage in the adult.

   The prominent cutaneous fin lobe is sustained by bony fin
rays or lepidotrichia (Goodrich 1904) the distal ends of which
are internally overlapped by a tuft of unmineralized actinotrichia (Ryder 1886) made of elastoidin (Krukenberg 1886;
review in Chandross et Bear 1979) (Fig 1).

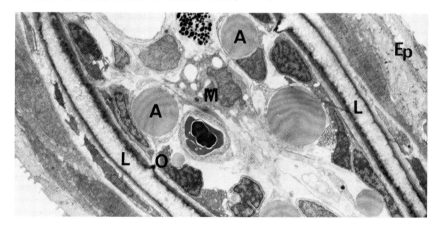

Fig 1. Cross section of the fin apex in *Salmo*. Actinotrichia A
are located between the two hemirays L of a lepidotrichium.
Ep: fin epidermis; M: mesenchymal cell; O: osteoblast? (X2560)

Both skeleton: the actinotrichia and the lepidotrichia are dermal in origin and such a feature will never be found in any tetrapod appendage. Besides, unlike any type of tetrapod limb, the prominent part of the fin lacks of striated muscles. These are divided in two major groups (Grenholm 1923) confined in the trunk for the pelvic fins, on both sides of the endoskeletal girdle and its radials. Consequently, it is obvious that the morphogenesis of the fin skeleton presents original features which shall be reported here.

Like any tetrapod limb bud, the early fin bud is constituted by a core of undifferentiated cells of both, somatopleural and somitic origin. It is covered by a thickened epidermis with differentiated cells such as ionocytes and mucous cells in the trout pelvic fins buds. Apically, this epidermis presents a fold, the so-called "pseudoapical ridge" in *Salmo* (Géraudie et François 1973) by comparison with the tetrapod apical ectodermal ridge found in most of the tetrapod limb buds. A complex acellular interface separates both components.

MORPHOGENESIS OF THE ACTINOTRICHIA

Actinotrichia are long transparent tapered rods lying parallel to each other and mostly orientated in the longitudinal axis of the fin. Ultrastructural studies of the elastoidin actinotrichia reveals a $\simeq 60$ nm periodic cross-striation through the whole giant fiber. Between two electron-dense bands, five intermediate transversal subbands show (Fig 2).

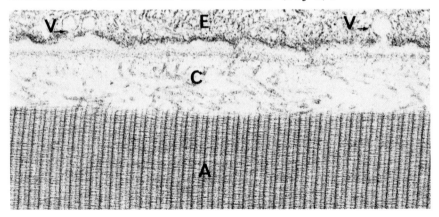

Fig 2. Longitudinal section of a subepidermal actinotrichia A. C: collagenous basal lamella; E: epidermis; V: vesicle (X52000).

First identified rudiments of elastoidin actinotrichia
are found in the subepidermal space of the elongated apical
epidermal fold in *Salmo* (pectoral bud,Bouvet 1974a; pelvic bud,
Géraudie, 1977) as well as in an other teleost studied
Hemichromis bimaculatus (Fig. 3).

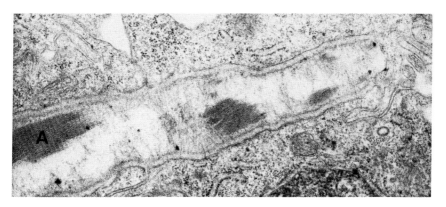

Fig 3. Actinotrichia rudiments appear as two subepidermal
rows under the thin continuous basal lamina (X25600).

That space, devoid of mesodermal cells or their filopods
is filled with a dense filamentous meshwork. Some microfilaments are perpendicular to the basal lamina and connect the
epidermal cell plasmalemma to the already differentiated
actinotrichia rudiments. Such a topographical location suggests
that the pseudoapical epidermal cells could be involved in the
earliest elastoidin deposits. The typical cross-striation
appears only after a delay, in more proximal regions of the
space, and suggests biochemical modifications within the
developing actinotrichium.

There are ultrastructural indications that epidermal
cells could be involved in the proteinaceous elastoidin
synthesis (Géraudie 1977,1978). Besides, in the trout pelvic
fin bud, studies of the incorporation of tritiated thymidine
and treatment with the antimitotic vinblastine sulfate suggested that the pseudoapical epidermal cells could be temporarily
synchronized. At the time of the appearance of the first actinotrichia, they would not incorporate DNA as well as they would
not divide. Those physiological characteristics could be
related to a metabolism orientated more toward protein

synthesis than DNA synthesis. Further studies should test this hypothesis. Nevertheless, one should keep in mind that elastoidin actinotrichia appears in the subepidermal space, so a mesodermal compartment the influence of which has to be evaluated in the elastoidin genesis (Géraudie 1980).

MORPHOGENESIS OF THE LEPIDOTRICHIA

Each bony lepidotrichium is made of two parallel hemi-rays (Fig 1), the development and growth of which are synchronous on each side of the fin paddle.

In *Hemichromis* (unpublished) as well as in *Salmo* fin buds each hemiray appears at the level of the acellular collagenous lamella of the epidermal-dermal interface (Goodrich 1904; Géraudie and Landis 1982). It consists of two regions determined on electron density after contrast of thin sections with heavy metals. In the outer margins, collagen fibrils are easily distinguished in longitudinal and transverse profiles. The other central region is dense and consists of both collagen fibrils and a fine granular ground substance (Fig 4). A solid phase of mineral (needle- or plate-like particles) is more prominent in the central region than in the peripheral ones. Consequently, the early lepidotrichial bone is acellular although in the adult trout, it will be a typical cellular bone except in the ever growing apical tip.

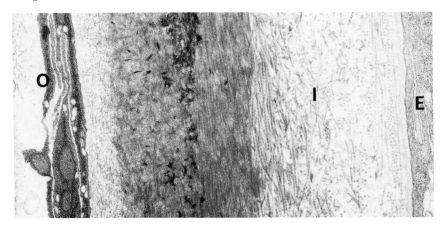

Fig 4. Longitudinal section of a trout lepidotrichium. At the fin apex, it remains in the collagenous layer of the epidermal-dermal interface I.Note its partial decalcification (X16000).

Growth of any lepidotrichium proceeds proximo-distally. While the growing apex always remains inserted within the collagenous lamella of the interface, its proximal part separates from its subepidermal location. A single layer of flattened mesodermal cells infiltrates the interface at the level of the collagen fibrils located close to the basal lamina. A thin epidermal-dermal interface is reconstructed but is thinner than in interradials regions.

At the distal end of the bone, numerous electron-lucent extracellular vesicles are present between the collagen fibrils and in the vicinity of the mesodermal cells. Further studies are necessary to assert their identity: are they matrix vesicles or an other type of vesicles related to calcification?

The origin of the osteoblasts associated with lepidotrichia morphogenesis is not fully elucidated. In paired fins of *Salmo* (Bouvet 1968; Géraudie et François 1973) the first mesenchyme to occupy the subepidermal region is issued from the somatopleura (outer somatic layer). Consequently, osteoblasts could arise from a somatopleural cell population. On the other hand, in median fins (François 1958) the subepidermal cells come from the dorsal edge of the somites which proliferates. So, according to the fins studied, either paired or median, osteoblasts could have a different embryological origin. Actinotrichia also occupy a subepidermal location at the free edge of the fin. Mesenchymal cells are involved in their growth in lenght and width (Géraudie 1982). Within this subepidermal cell population, two subpopulations could exist. One could be involved in bone collagen synthesis and the other one in elastoidin synthesis. An alternative to such a segregation implies the biochemical influence of the extracellular matrix in which mesodermal cells are immersed. The fate of the given single type of collagen would depend upon the environment. Further studies are necessary to uncover the metabolism of these cells.

A growing lepidotrichium provides a useful tool in studying mineral maturation in calcification. Preliminary studies (Landis and coll. 1981) using fixation and further handling of fin buds in anhydrous conditions showed the existence of a mineral phase the quality of which varies from the base to the apex and probably from the periphery to the central core. Microanalysis techniques revealed this evolution. Proximally, poorly cristalline hydroxyapatite is present while apically the solid phase mass did not generate a coherent electron

diffraction pattern. Electron probe microanalysis showed a
progressive increase in Ca/P ratios. Distal regions generated
a molar ratio of 1.1 $\pm$ 0.1 (n=15) and proximal regions a ratio
of 1.3-1.4 $\pm$ 0.1 (21< n <38) (Landis and coll. 1981). Earliest
calcium phosphate solid phase deposists are different from
hydroxyapatite.

RELATIONSHIPS BETWEEN ACTINOTRICHIA AND LEPIDOTRICHIA

Actinotrichia develop in advance of the lepidotrichia
(Harrison 1893). During the regeneration of the amputated
fin, they also appear before the lepidotrichia (Kemp and
Park 1970; Géraudie, unpublished results). During experimental
and morphological studies on trout pelvic fin buds, some
relationships appeared between both dermal skeletal elements.

Fusion between Actinotrichia and Lepidotrichia

Actinotrichia can be observed within the lepidotrichia.
The elastoidin fiber is disorganized especially at the periphery (Fig 5). A mineral phase is present adjacent to the
elastoidin but not within the fiber itself. Such a fusion
already suggested by Goodrich (1904) suggests that actinotrichia can be involved in the lepidotrichial growth.

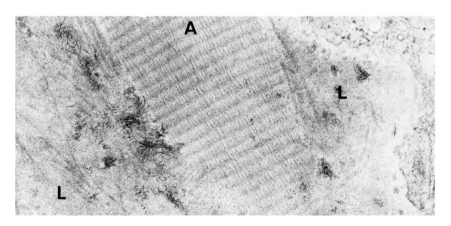

Fig 5. Longitudinal section of an actinotrichium partially
engulfed in the lepidotrichial bone. The elastoidin is
apparently altered and mineral deposits are present at the
interface between both dermal skeleton (X52800).

Developmental Interactions

The treatment of trout embryos during the pelvic fins development with various concentrations of the cytotoxic drug nitrogen mustard (1.25µg/ml to 20µg/ml) altered the skeletal morphogenesis in proportion of the concentration used and the stage of development (Géraudie 1981). This study showed that:-(1)- actinotrichia develop in the deficient fin bud even if the lepidotrichia and the endoskeleton fail to form. -(2)- lepidotrichia develop only if the endoskeleton is present. Consequently, perturbations of the fin development suggest that the morphogenesis of the actinotrichia is independant while morphogenesis of the lepidotrichia is related in some fashion to the development of the endoskeleton. The existence of such developmental interactions was suggested earlier by microsurgery in the trout paired fins (pectoral bud: Bouvet 1970, 1974b; pelvic bud: Géraudie 1975).

The development of the dermal skeleton of the fins is original. Its subepidermal location suggests that the epidermis could be involved and further studies should precise relationships between epidermis and mesoderm in that system.

REFERENCES

Bouvet J (1968). Histogenèse précoce et morphogenèse du squelette cartilagineux des ceintures primaires et des nageoires paires chez la Truite (*Salmo trutta fario L.*). Arch Anat micros Morph exp 57:35.
Bouvet J (1970). Etablissement de la carte des territoires présomptifs du bourgeon de la nageoire pectorale chez la Truite indigène (*Salmo trutta fario L.*) à l'aide d'excisions et de marques colorées. Ann Embryol Morph 3:315.
Bouvet J (1974a). Différenciation et ultrastructure du squelette distal de la nageoire pectorale chez la Truite indigène (*Salmo trutta fario L.*) I Différenciation et ultrastructure des actinotriches. Arch Anat micros Morph exp 63:79.
Bouvet J (1974b). Différenciation et ultrastructure du squelette distal de la nageoire pectorale chez la Truite indigène (*Salmo trutta fario L.*) II Différenciation et ultrastructure des lépidotriches. Arch Anat micros Morph exp 63:323.
Chandross RJ, Bear RS (1979). Comparison of Mammalian collagen and Elasmobranch elastoidin fiber structures, based on electron density profiles. J Mol Biol 130:215.

François Y (1958). Recherches sur l'anatomie et le développement de la nageoire dorsale des Téléostéens. Arch Zool exp gen 97:1.

Géraudie J (1975). Les premiers stades de la formation de l'ébauche de la nageoire pelvienne de la Truite (*Salmo fario* et *Salmo gairdneri*) III Capacités de régulation. J Embryol exp Morph 34:407.

Géraudie J (1977). Initiation of the actinotrichial development in the early fin bud of the fish, *Salmo*. J Morph 151:353.

Géraudie J (1978). The fine structure of the early pelvic fin bud of the trouts *Salmo gairdneri* and *S. trutta fario*. Acta Zool 59:85.

Géraudie J (1980). Mitotic activity in the pseudoapical ridge of the trout pelvic fin bud. J Exp Zool 214:311.

Géraudie J (1891). Consequences of cell death after nitrogen mustard treatment on skeletal pelvic fin morphogenesis in the trout *Salmo gairdneri* (Pisces,Teleostei). J Morph 170:181.

Géraudie J (1982). Contrôles morphogénétiques du développement et de la régénération des nageoires paires des Téléostéens. Thèse de doctorat d'Etat. Paris VII.

Géraudie J, François,Y (1973). Les premiers stades de la formation de l'ébauche de la nageoire pelvienne de la Truite (*Salmo fario* et *Salmo gairdneri*) I Etude anatomique. J Embryol exp Morph 29:221.

Géraudie J, Landis WJ (1982). The fine structure of the developing pelvic fin dermal skeleton in the trout *Salmo gairdneri*. Amer J Anat 163:141.

Goodrich ES (1904). On the dermal fin-rays of fishes-living and extinct. Quart J Micros Sci 47:465.

Grenholm A (1923). Studien über die Flossenmuskulatur der Teleostier. Uppsala Univ Ars 1.

Harrison RJ (1893). Ueber die Entwicklung der nicht Knorpelig vorgebildeten Skelettheile in den Flossen der Teleostier. Arch Microsc Anat 42:248.

Kemp NE, Park JH (1970). Regeneration of lepidotrichia and actinotrichia in the tailfin of the teleost *Tilapia mossambica*. Devel Biol 22:321.

Krukenberg CF (1886). Ueber die chemische Beschaffenheit der sog. Hornfäden von *Mustelus* and über die Zusammensetzung der keratinösen Hüllen um die Eiern *Scyllium stellare*. Mittheil Zool Sta Neapel 6:286.

Landis WJ, Géraudie J, Paine MC, Neuringer JR, Glimcher MJ (1981). An electron optical and analytical study of mineral deposition in the developing fin of the trout,*Salmo gairdneri*. 27th Annual ORS, Las Vegas, Nevada, Feb 24-26.

Ryder JA (1886) On the origin of heterocercy and the evolution of the fins and the fin-rays of fishes. US Comm Fish Fisheries. Rept Fish Comm for 1884 12:981.

SECTION THREE
LIMB MALFORMATIONS

Limb Development and Regeneration
Part A, pages 301–310
© **1983 Alan R. Liss, Inc., 150 Fifth Avenue, New York, NY 10011**

PATHOGENESIS OF LIMB MALFORMATIONS IN MICE: AN ELECTRON
MICROSCOPIC STUDY

Mineo Yasuda, M.D., Dr.Med.Sci. and
Harukazu Nakamura, B.S., Dr.Med.Sci.

Department of Anatomy, Hiroshima University
School of Medicine, Hiroshima 734, Japan

The pathogenesis of genetic and teratogen induced limb
malformations in experimental animals has been extensively
studied by electron microscopy. Many of the transmission
electron microscopic studies dealt with intracellular
changes related to mesenchymal cell death (for example,
Nakamura et al 1974). Scanning electron microscopic studies
mainly described changes in gross morphology of the limb bud
and the apical ectodermal ridge (AER)(for example, Imagawa
1980).

There may be various cellular changes other than cell
death in the pathogenesis of limb malformations. Ede et al
(1974) reported changes of cell surface morphology in the
pathogenesis of *talpid*[3] malformations in chick embryos.

The present study was undertaken to describe submicro-
scopic changes in cellular morphology in the pathogenesis of
limb malformations in mice induced with a mutant gene,
meromelia, and two teratogens, 5-fluorouracil (5FU) and
vitamin A (VA).

MATERIALS AND METHODS

C57BL/6 *meromelia* mice maintained in our laboratory and
JCL-ICR mice obtained from CLEA Japan were used. *Meromelia*
is an autosomal recessive trait found in Japan, and the
homozygotes express various types of limb deformities ranging from polydactyly to severe reduction deformities (Yasuda
1977). Females were placed with males overnight, and 0:00

a.m. of the day of detection of a vaginal plug was denoted
as the start of day 0 of gestation.

Homozygous *meromelia* embryos were obtained by matings
of homozygous or heterozygous females with homozygous males.
Limb malformations were induced in JCL-ICR embryos by maternal intraperitoneal administration of a single dose of 20
mg/kg of 5FU (Nakarai Chemicals) or 600,000 iu/kg of VA
palmitate (Chocola A, Eisai) at day 10.5 of gestation.
Wild type C57BL/6 mice and untreated JCL-ICR mice served as
controls.

At various intervals from day 10.5, the pregnant animals were killed by cervical dislocation, and embryos were
harvested for observations by scanning electron microscopy
(SEM), light microscopy (LM), and transmission electron
microscopy (TEM). Specimens for SEM and TEM were processed
as described previously (Nakamura, Yasuda 1979). For determination of the AER development, the specific features of
periderm cells covering the AER observed by SEM (Nakamura,
Yasuda 1979) were utilized. For SEM observation of the limb
mesenchyme, the epidermis was peeled off with adhesive tape
after drying. For LM, semithin sections were cut from
specimens embedded in Epon with a Sorval MT-1 ultramicrotome,
and stained with toluidine blue.

Several dams per group were sacrificed at day 15.5 or
18.5 of gestation. Their fetuses were observed under a
dissection microscope for description of gross malformations.
Some of the fetuses were fixed in 95% ethanol and skeletal
specimens doubly stained for cartilage and bone were prepared by the method of Inouye (1976).

RESULTS

Pathogenesis of Limb Malformations in *Meromelia*

Fig. 1 shows some examples of limb malformations observed in homozygous *meromelia* fetuses near term. Both
excessive and reduction deformities were observed.

The most conspicuous feature in the pathogenesis of
meromelia malformations was abnormalities of the AER, such
as general hypoplasia, local hypoplasia, local hyperplasia,

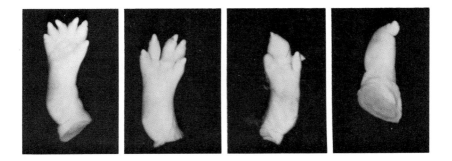

Fig. 1. Various types of limb malformation in near term *meromelia* fetuses (×5).

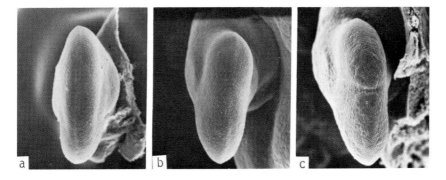

Fig. 2. Apical views of a control forelimb at day 11.5 (a), a *meromelia* forelimb at day 11.5 with local hypoplasia of the AER (b), and a *meromelia* forelimb at day 12.0 with a preaxial protrusion covered with a hyperplastic AER (c)(×40).

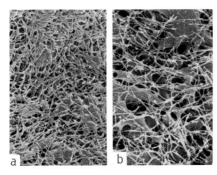

Fig. 3. The subridge CPM of a *meromelia* forelimb at day 11.5, underneath the hypoplastic AER (a) and the well developed AER (b)(×2000).

and delayed involution (Fig. 2). Removal of the epidermis by application of adhesive tape revealed the surface of the limb mesenchyme, which was covered with a meshwork of fine processes of mesenchymal cells. In control limb buds, the mesenchymal cell process meshwork (CPM) underneath the AER was looser than that underneath the dorsal or ventral non-ridge epidermis. When a *meromelia* limb bud with local hypoplasia of the AER was examined, an interesting relationship between the AER development and the CPM was noted. The CPM underneath the hypoplastic AER was tighter than that underneath the well developed AER (Fig. 3). LM and TEM observations showed no indication of abnormal cell death related to the *meromelia* teratogenesis.

Pathogenesis of 5FU Induced Limb Malformations

Maternal treatment with 5FU induced preaxial and axial polydactyly in the hindlimb of the fetuses in a high incidence (Fig. 4). The initial pathological change in the embryonic limb after the treatment was mesenchymal cell death. Pyknotic cells were observed throughout the limb mesenchyme at 12 hours after the maternal injection. The distribution of the necrotic cells was not uniform. The necrotic cells were more frequently found in the periphery of the limb than in the core.

Fig. 4. Polydactyly induced by maternal ip injection with 20 mg/kg of 5FU at day 10.5 of gestation. A cartilage stained hindlimb at day 15.5 (×10).

While the size of the treated limb bud became smaller than controls, the AER became hyperplastic (Fig. 5). The involution of the AER, which normally occurs by day 12.5, was delayed. The density of the mesenchymal cells and the

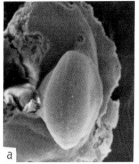 

Fig. 5. Apical views of a control hindlimb at day 11.5 (a) and a 5FU treated hindlimb with a hyperplastic AER at day 11.5 (b) (×40).

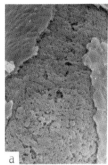 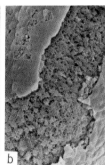

Fig. 6. The subridge CPM at day 11.5 in a control (a) and a 5FU treated (b) hindlimb (×200).

CPM underneath the hyperplastic AER was looser than that in controls (Fig. 6). Observations of the mesenchymal side of the peeled epidermis verified that the observed difference in the density of the CPM was not due to the difference in the amount of fine cell processes attached to the basement membrane of the epidermis.

Pathogenesis of VA Induced Limb Malformations

Maternal injection with VA induced reduction deformities in the hindlimb of the fetuses (Fig. 7). The incidence was 100%. The proximal structures such as the femur, tibia and fibula were more severely deformed than the distal structures, although the number of digits was reduced.

Mesenchymal cell death was the most conspicuous initial change in the VA induced limb teratogenesis, too. However,

Fig. 7. A reduction deformity
induced by maternal ip injection with 600,000 iu/kg of VA
at day 10.5 of gestation.
A cartilage stained hindlimb
at day 15.5 (×10)

Fig. 8. An apical view
of a VA treated hind-
limb at day 11.5 (a)
(×40) and its subridge
CPM (b)(×200).

the distribution of initial cell death was notably different
from that induced by 5FU. At 12 hours after the injection,
the necrotic cells were more abundant in the core mesenchyme
of the limb than in the periphery.

The AER did not become hypertrophic after the VA treatment. After day 11.5 (24 hours after the injection) the AER
was smaller than that in controls (compare Fig. 8a with Fig.
5a), and its involution occurred earlier than controls.
The density of the mesenchymal cells and the CPM underneath
the AER was not different from that in controls (compare
Fig. 8b with Fig. 6a).

DISCUSSION

The most interesting findings in the present study was the relationship between the CPM density and the AER development. The CPM underneath a well developed AER was loose, while that underneath a hypoplastic AER was tight.

Changes in the CPM development in teratogenesis have been reported. Sulik et al (1979) analyzed pathogenesis of phenytoin-induced cleft lip and palate in mice, and found that the CPM underneath the surface epithelium of the lateral nasal process was undeveloped in the teratogen treated embryos. They suggested that interactions within the CPM and/or involving the CPM and the surface epithelium might play a role in growth stimulation or other aspects of morphogenesis.

Our findings support the hypothesis that the CPM is important in morphogenesis. However, it appears that a well developed CPM does not necessarily mean growth stimulation. In the AER-mesenchymal interaction in the limb development, a dense subridge CPM seems to exert a negative influence on the development and maintenance of the AER.

Kaprio (1977) observed the ectoderm-mesenchymal interspace during the formation of the chick leg bud. He found that there were extracellular fibrils interconnecting the basal lamina and mesenchymal cell processes, and that the highest number of the fibrils was found under the AER. Kaprio's description suggested a possibility that the difference of the CPM density observed by us might be an artefact, resulting from the difference of the density of the extracellular fibrils. This possibility was denied by SEM observations of the mesenchymal side of the peeled epidermis.

There have been conflicting descriptions concerning the morphological relationship between the AER and subridge mesenchyme. In some studies, intimate contact between the epidermal basement membrane and the mesenchymal cell processes, directly or with extracellular fibrils, were described in the vicinity of the AER (Bérczy 1966; Kaprio 1977). Some other studies dealing with mutants with limb malformations suggested that less contact between the mesenchymal cell processes and the epidermal basement membrane might promote the development of the AER, and intimate contact might suppress it (Ede et al 1974; Knudsen, Kochhar

1981; Sawyer 1982). Our findings support the latter view.

It is widely accepted that the AER maintains mesenchymal cells directly subjacent to it in an actively growing, labile, undifferentiated condition (Kosher et al 1981). A well organized sequence of temporal and spacial changes in the epithelial-mesenchymal interaction is necessary in the normal limb morphogenesis. In the *meromelia* mutant, the temporal and spacial control of the CPM development may not be stable as that in the normal limb development, and resulting disturbances in the AER-mesenchymal interaction produce various types of limb malformations. Where the CPM density becomes too tight at the stage of AER development, the AER does not develop well, and the mesenchymal proliferation is suppressed, leading to reduction of a part of the limb. Where the CPM density becomes too loose at the time of AER involution, the AER remains too long, and the mesenchymal cells overgrow, leading to excessive formation of a part of the limb.

Under the present experimental conditions, 5FU induced polydactyly in the hindlimb, whereas VA induced limb reduction deformities. This difference may be explained by a difference in the CPM density, resulting from a difference in the distribution of mesenchymal cell death. In the 5FU treated limbs, initial cell death occurred mainly in the periphery of the limb, whereas initial necrotic cells tended to cluster in the core of the limb in the VA treated limbs. This type of agent specificity in the distribution of mesenchymal cell death has been reported in the pathogenesis of mouse limb malformations induced with cytosine arabinoside and retinoic acid (Kochhar, Agnish 1977). With 5FU, peripheral mesenchymal cells which are undergoing active proliferation are killed, resulting in loosening of the CPM. This stimulates development and maintenance of the AER. The hyperplastic AER stimulates proliferation of the surviving mesenchymal cells, sometimes to an excessive extent, leading to formation of polydactyly. In the present experiment, not only polydactyly but also reduction deformities were induced in the hindlimb. Occurrence of reduction deformities is explained by an insufficient amount of mesenchymal proliferation to recover the initial mesenchymal loss. In the case treated with VA, the cell death occurred in the core mesenchyme does not affect the CPM under the AER. The AER involution occurs according to the normal program, hence no recovery of the initial mesenchymal loss happens. The final amount of the limb mesenchyme becomes deficient and limb reduction deformities result.

For further confirmation of the negative effect of the CPM on the AER development, observations of the pathogenesis of limb malformations with other mutant genes and exogenous teratogens are warranted.

ACKNOWLEDGEMENT

This work was supported by Grants-in-Aid for Scientific Research 348082 and 548076 from the Ministry of Education, Japan. The authors wish to thank Messrs. T. Okimoto, H. Miyoshi, T. Murao, and T. Yamabe, medical students of Hiroshima University, for their assistance.

REFERENCES

Ede DA, Bellairs R, Bancroft M (1974). A scanning electron microscope study of the early limb-bud in normal and *talpid*[3] mutant chick embryos. J Embryol exp Morph 31:761.
Bérczy J (1966). Zur Ultrastruktur der Extremitätenknospe. Z Anat Entwickl Gesch 125:295.
Imagawa S (1980). Symbrachydactyly: Pathogenesis of 5-fluorouracil induced model in mice. Hiroshima J Med Sci 29:169.
Inouye M (1976). Differential staining of cartilage and bone in fetal mouse skeleton by alcian blue and alizarin red S. Cong Anom 16:171.
Kaprio EA (1977). Ectodermal-mesenchymal interspace during the formation of the chick leg bud. A scanning and transmission electron microscopic study. Wilhelm Roux's Arch 182:213.
Kochhar DM, Agnish ND (1977). "Chemical surgery" as an approach to study morphogenetic events in embryonic mouse limb. Develop Biol 61:388.
Kosher RA, Savage MP, Walker KH (1981). A gradation of hyaluronate accumulation along the proximodistal axis of the embryonic chick limb bud. J Embryol exp Morph 63:85.
Knudsen TB, Kochhar DM (1981). The role of morphogenetic cell death during abnormal limb-bud outgrowth in mice heterozygous for the dominant mutation *Hemimelia-extra toe (Hm[X])*. J Embryol exp Morph 65 (Suppl):289.
Nakamura H, Yamawaki H, Fujisawa H, Yasuda M (1974).

Effects of maternal hypervitaminosis A upon developing mouse limb buds. II. Electron microscopic investigation. Cong Anom 14:271.

Nakamura H, Yasuda M (1979). An electron microscopic study of periderm cell development in mouse limb buds. Anat Embryol 157:121.

Sawyer LM (1982). Fine structural analysis of limb development in the wingless mutant chick embryo. J Embryol exp Morph 68:69.

Sulik KK, Johnston MC, Ambrose LJH, Dorgan D (1979). Phenytoin (Dilantin)-induced cleft lip and palate in A/J mice: A scanning and transmission electron microscopic study. Anat Rec 195:243.

Yasuda M (1977). Embryogenesis of human limb malformations compared with those in animals. In Japan Medical Research Foundation (ed): "Gene-Environment Interaction in Common Diseases", Tokyo: University of Tokyo Press, p 155.

Limb Development and Regeneration
Part A, pages 311–316
© **1983 Alan R. Liss, Inc., 150 Fifth Avenue, New York, NY 10011**

USE OF GENE LINKAGE TO DETECT LIMBS DESTINED TO BE MALFORMED
BY THE DOMINANT GENE <u>DOMINANT HEMIMELIA</u>

Lewis B. Holmes

Embryology-Teratology Unit
Massachusetts General Hospital
Boston, MA   02114

If the events in limb development that precede the occurrence of a structural abnormality are to be studied, one
must be able to identify the embryo before it is deformed.
In the study of hereditary limb malformations in mice it has
not been possible to distinguish the mutant, i.e. affected,
embryos from their unaffected littermates before the limb
deformity is visible.

In the mouse the dominant gene <u>Dominant hemimelia (Dh)</u>
causes hypoplasia of the tibia and femur, preaxial oligodac-
tyly or triphalangism of the first toe, absence of the spleen
and a small stomach (Searle 1964). The hindlimb deformity
was reported by Green (1967) to be visible in the heterozygote by day 10.5 of gestation (the limb is first visible
on the tenth day), but that the heterozygote (Dh/+) could
not be distinguished from the homozygote (Dh/Dh) at this age.
Rooze (1977) reported that the mutant (Dh) hindlimb could be
identified in the 11 day embryo, but not one day earlier.

We have used the close linkage on chromosome one between the genes Dh and dipeptidase-1 to enable us to determine which embryos on days 10 and 11 of gestation are mutant
and to distinguish between the heterozygous (Dh/+) and the
homozygous (Dh/Dh) mutant embryos. Roderick (1978) has
found that the two loci are between 2 and 6 map units apart.

METHODS AND MATERIALS

Heterozygous (Dh/+) and wild-type (+/+) adult mice of a

hybrid stain (C57 BL/6J X C3He/Fe^J) were obtained from the Jackson Laboratory, Bar Harbor, Maine. Matings between Dh/+ and +/+ mice were carried out. The newborns were raised to 4 to 6 weeks of age, when a nephrectomy was performed under ether anesthesia.

Kidney tissue was used to determine the enzymatic pheno-type of each mouse for the allelic slow and fast variants of dipeptidase-1, designated dipeptidase-1^a and dipeptidase-1^b (Roderick 1978, Ruddle 1971). The tissue is homogenized in distilled water and centrifuged at 37,000 X g for 60 minutes. The supernatant is applied to cellulose acetate plates (Helena Laboratories) that have been soaked in a tris glycine buffer that consists of 3gms Trisma Base and 14.4 gms glycine per liter. After applying 2.5 mamp current for 30 minutes, the plate is stained in the following manner: 500 mg of gum agar (Sigma No. A-7002) is added to 25 ml tris glycine buffer and heated in a hot water bath for 30 min.; the solution is cooled to 50°C and 10 mg crotalus adamanteus (snake) venom (Sigma No. V-6875), 5 mg peroxidase (Sigma No. L-0501) in 25 ml phosphate HCL buffer (0.2M, pH 7.5) is added; immedi-ately thereafter 0.25 ml manganese chloride solution and 0.5 ml 1% O-dianisidine hydrochloride solution are added in that order. The agar mixture is poured over the cellulose acetate plates. The agar-stain mixture is allowed to harden and is placed in a 37°C incubator until the bands appear. The slow and fast variants can be distinguished easily, with the three possible phenotypes:
dip-1^a/dip-1^a, dip-1^a/dip-1^b and dip-1^b/dip-1^b.

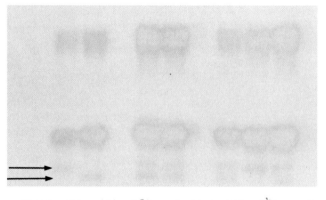

Fig. 1. Shows slow (dip-1^a) and fast (dip-1^b) variants at bottom (positive pole) of cellulose acetate plate after elec-trophoresis. Starting on left, the first three phenotypes are: dip-1^a/dip-1^a, dip-1^b/dip-1^b and dip-1^a/dip-1^b.

To determine whether the dipeptidase phenotype could be identified in day 10 and day 11 embryos, embryos with their placentas were removed from timed pregnancies. The embryos and placentas were homogenized separately and as a combined sample and analyzed for the presence of dipeptidase-1, as described above.

Pedigree analysis of at least three generations was used to determine in the mutant (Dh/+) mouse heterozygous for both the slow and fast variants whether the Dh gene was in coupling with dipeptidase-1^a or dipeptidase-1^b.

To confirm that the two genes are closely linked the frequency of apparent cross overs was determined in informative matings of two types: Dh,dip-1^b/+,dip-1^a X +,dip-1^a/+,dip-1^a and Dh,dip-1^b/+,dip-1^a X +,dip-1^b/+,dip-1^b. Only the findings in mice who survived to ages 4 to 6 weeks will be presented. The reason for this restriction is that non-penetrance of Dh will appear to be a cross over. In this situation in a mouse that has no visible skeletal malformations and is presumed to be normal (+/+) at the Dh locus has offspring with the skeletal malformations characteristic of the Dh/+ phenotype. Most mice who survive to age 4 to 6 weeks will survive long enough to reproduce. The apparent cross overs, such as +,dip-1^b/+,dip-1^a or +,dip-1^b/+,dip-1^b mice from the informative matings listed above, are bred to normal (+/+) mice and their offspring are examined for the skeletal malformations of Dh/+ newborns. Whenever possible, several litters are obtained from mice that are presumed to be examples of crossing over.

RESULTS

Eighteen litters have been obtained from informative matings. Forty mice have survived to 4 to 6 weeks of age. Five of the forty mice have had isozyme phenotypes that could have occurred only as the result of crossing over between the Dh and dipeptidase-1 loci. Three of the five have had offspring of their own which means 38 mice are informative. Each of two female mice, who had no hindlimb malformations, but the same dipeptidase-1 phenotype of their Dh/+ parent, had one litter none of which had any malformations of the hindlimbs; therefore, we must assume that these two mice show crossing over between the genes Dh and dipeptidase-1.

The third presumed cross over, a male, had three litters
by normal (+/+) females with three of the 18 offspring showing severe hindlimb malformations characteristic of Dh/+
mice. We interpret this to mean that this mouse in non-
pentrant for the skeletal effects of the Dh gene. Autopsy
of this animal showed that he had a spleen; his cleared
skeletal showed none of the mild skeletal characteristics
of the Dh/+ phenotype, such as a reduction in the number
of lumbar vertebrae or triphalangism of the first toes
(Searle 1964). Having found that two of 38 mice show crossing
over between the genes Dh and dipeptidase-1 indicates that
the rate of crossing over in these eighteen litters was
2/38 (5.4%), which is consistent with the observation by
Roderick (1978) that the two gene loci are 2 to 6 map units
apart.

Analysis of day 10 and 11 embryos showed that the
dipeptidase-1 isozyme phenotype is clearly visible on electrophoresis. The color reaction is much stronger in the homogenate of the placenta than in that of the embryo (Figure 2).

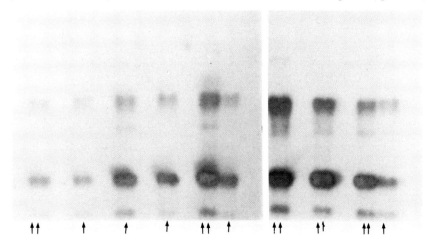

Fig. 2. The cellulose acetate plates after staining of a
litter of day 10 (left side) and day 11 (right side) embryos
and their placentas. The tissue extract of the day 10 embryo
(arrow) was clearly visible, but easier to see if the placenta was added (two arrows). In the day 11 litter both the
embryo alone (asterisk) and the placenta (two asterisks)
were darker than in the d.10 litter. Both the day 10 and day
11 litters were homozygous for dipeptidase-1^b.

DISCUSSION

The process of crossing over is an event that occurs
normally in meiosis during which homologous segments of paired maternal and paternal chromosomes are exchanged. If linkage between two loci is to be useful to identify the developing limbs destined to be malformed, the rate of crossing
over must be low as this rate is in effect the error rate in
this method of distinguishing mutant from normal littermates.
If the frequency of crossing over between dipeptidase-1 and
Dh is 5%, this means our accuracy in identifying Dh/+ and
Dh/Dh embryos is 95%. We present here the feasibility of this
method. We must study the offspring from additional informative matings to determine more precisely the rate of crossing over between these two loci.

The ability to identify mutant embryos with a high degree of accuracy makes it possible to study all events in
early limb development that occur before the deformity is
visible in both Dh/+ and Dh/Dh embryos. These events include
patterns of cell migration from the adjacent somite into the
limb, the appearance of the apical ectodermal ridge on the
preaxial side of the hindlimb, the pattern of mesenchyme
cell necrosis in the "foyer primaire preaxial", and the development of blood vessels and nerves. Rooze (1977) has observed that the pattern of cell necrosis is abnormal in the
hindlimbs of day 11 Dh/+ embryos. Green (1967) postulated
that abnormalities of the anterior splanchnic mesodermal
plate lead to the absence of the spleen and the small stomach.
She noted that the umbilical arteries were smaller in Dh/+
embryos than in +/+ embryos. This technique for identifying
the heterozygote (Dh/+) and the homozygote (Dh/Dh) will allow greater clarification of these phenotypic effects of the
mutant gene.

Absence of the tibia in association with preaxial polydactyly and due to an autosomal dominant gene is a rare human malformation (Clark 1975). It would be of interest to
know whether there was linkage in these families between
the dominant gene causing the limb malformations and pepti-
dase C, the isozyme in humans which is analogous to the
dipeptidase-1 in the mouse (Ruddle 1971). This possibility
is feasible in view of the fact that areas of homology exist
in the sequence of genes in mice and men.

ACKNOWLEDGEMENT

A special thanks to Dr. Margaret C. Green who suggested that we use this technique for identifying mutant embryos and to Mrs. Chris Doane for her technical assistance.

Supported in part by NIH grant No. HD15241.

REFERENCES

Clark MW: Autosomal dominant inheritance of tibial meromelia. J Bone Jt Surg 57-A:262-264, 1975.
Green MC: A defect of the splanchnic mesoderm caused by the mutant gene dominant hemimelia in the mouse. Devel Biol 16:62-89, 1967.
Roderick T: Personal Communication, 1978.
Rooze MA: The effect of the Dh gene on limb morphogenesis in the mouse. Birth Defects: Orig Art Series XIII (No.1): 69-95, 1977.
Ruddle FH, Nichols EA: Starch gel electrophoretic phenotypes of mouse X human somatic cell hybrids and mouse isozyme polymorphisms. In Vitro 7:120-131, 1971.
Searle AG: The genetics and morphology of two "luxoid" mutants in the house mouse. Genet Res, Camb 5:171-197, 1964.

Limb Development and Regeneration
Part A, pages 317–326
© **1983 Alan R. Liss, Inc., 150 Fifth Avenue, New York, NY 10011**

VASCULAR PATTERNS IN THE MALFORMED HINDLIMB OF DH/+ MICE

David J. Zaleske, M.D. and Lewis B. Holmes, M.D.

Pediatric Orthopaedic Unit and
Embryology-Teratology Unit,
Massachusetts General Hospital

The gene Dominant hemimelia (Dh) arose spontaneously in mice (Carter 1954). It causes both skeletal and visceral abnormalities which include hindlimb malformations, a decrease in the number of vertebrae and ribs, absent spleen, small stomach, imperforate vagina, and hydronephrosis (Searle 1964; Green 1967). The hindlimb malformations are a varied pattern of preaxial anomalies, including triphalangism, syndactyly or absence of the hallux, shortening of the tibia and femur, and abnormalities of the pelvis. Homozygotes (Dh/Dh) are more severely affected than heterozygotes (Dh/+) and are not viable. Penetrance of Dh in the heterozygote has been reported to be 100% for the visceral abnormalities and 96% for skeletal malformations (Searle 1964).

Abnormalities of developing blood vessels have been implicated in limb malformations (Chaurasia 1974; Hootnick, Levinsohn, Randall, Packard 1980; Poswillo 1973). Different possible aspects of the vascular contribution to morphogenesis have been emphasized. Chondrogenesis appears to occur in regions with few blood vessels (Caplan, Koutroupas 1973). Absence of a vessel or persistence of an embryonic pattern has been found to be associated with fibular dysplasia in humans (Hootnick, Levinsohn, Randall, Packard 1980). To evaluate these possibilities in Dh/+ and Dh/Dh mice, we have begun a study of mutant mice after birth and during limb development. In this report we will describe the results of the first phase, the study of adult Dh/+ mice.

MATERIALS AND METHODS

Mutant (Dh/+) and normal (+/+) mice have been purchased
from the Jackson Laboratory, Bar Harbor, Maine. TheseB6C3 mice
are a hybrid of two strains C57BL/6J and C3He/Fe^J. Dh/+
mice are identified by the presence of hindlimb malformations. For this study ten Dh/+ mice were selected which
had hindlimb malformations of varying severity. The controls
were littermates who had no visible malformations of their
hindlimbs.

To delineate the vascular pattern of the hindlimbs,
each animal was sacrificed, and the thoracic aorta was can-
nulated with a twenty-two gauge catheter under an operating
microscope. Initially, 0.5 milliliters of a 10 unit/ml.
heparin solution was infused. Then, Batson's Number 17
Anatomic Compound (Polysciences, Inc.; Warrington, PA) or
Microfil (Canton Biomedical; Boulder, CO) was infused until
filling was visible in the blood vessels in the toes. The
carcass was stored for twelve hours at 5° C. The torso was
separated at the level of the lumbar spine. The lower half
of the body was stained with alizarin red S and alcian blue
(McLeod 1980). The soft tissues were then cleared through
graded 1% potassium hydroxide and glycerin mixtures to pure
glycerin. At this stage the vascular pattern could be
easily examined under an operating microscope with trans-
illumination.

RESULTS

To describe the arterial pattern, we have combined the
nomenclature used in several sources (Brookes 1971; Cook
1965; Froud 1959). We found that the vascular pattern of
the ten normal (+/+) mice was constant (Figures 1-4). The
femoral artery branched at the knee. The popliteal supplied
the posterior tibial region and branched into the anterior
and posterior tibial arteries and the peroneal artery
(Figures 1 and 2). There were large collaterals around the
knee with the saphenous artery on the medial aspect originating as one of the branches from the femoral (Figure 1)
and the sural artery on the lateral aspect originating from
the popliteal system (Figure 2). The saphenous artery continued down the medial aspect of the leg dividing into several branches including a deep and superficial medial plantar artery and lateral plantar artery (Figure 3). Nomencla-

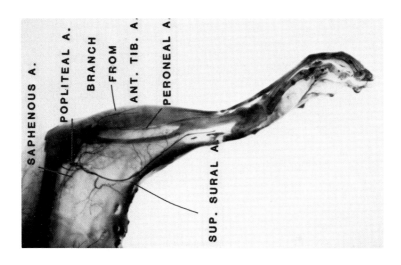

Figure 2. Right (+/+) hindlimb, lateral aspect:

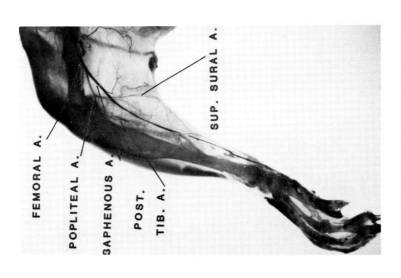

Figure 1. Right (+/+) hindlimb, medial aspect.

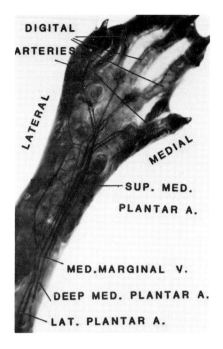
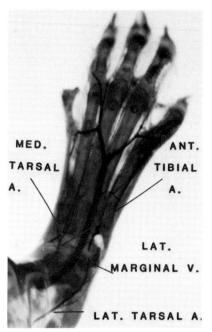

Figure 3. Right (+/+) foot,   Figure 4. Right (+/+) foot,
plantar aspect.   dorsal aspect.

ture for the deep and superficial medial plantar arteries
was not clearly defined elsewhere. We took the more lateral,
slightly deeper, and larger branch as the deep medial plantar
artery. The digital arteries mostly came from this deep
medial plantar artery. The medial digital artery of the
hallux arose from the superficial medial plantar artery. On
the dorsal aspect there was a large marginal vein (Figure 4).
The medial tarsal artery was yet another branch from the
saphenous. The anterior tibial artery came to the dorsal
aspect of the foot. The lateral tarsal artery was a continuation of the superficial sural artery.

The pattern of skeletal malformations in the twenty
hindlimbs of the ten Dh/+ mice ranged from mild to severe.
We have grouped these patterns into five categories: 1) normal limb with malformations of the opposite limb; 2) triphalangism of the hallux with normal tibia; 3) preaxial
polydactyly and syndactyly of the hallux or absence of hallux
with intact tibia; 4) shortened tibia with associated preaxial

foot deformity; and 5) absence of the tibia with associated
preaxial foot deformity. The distribution of these patterns
in the twenty Dh/+ hindlimbs was as follows.

Table 1. Skeletal Patterns of the Studied Dh/+ Limbs

	Group 1	Group 2	Group 3	Group 4	Group 5
Right	1	3	3	2	1
Left	0	3	2	3	2

The extent of the disruption of the vascular pattern
in Dh/+ hindlimbs paralleled the severity of the skeletal
malformations.

Group 1. The phenotypically normal limb (from a Dh/+
mouse with only the contralateral limb affected) had the same
blood supply as described for the normal (+/+) mice.

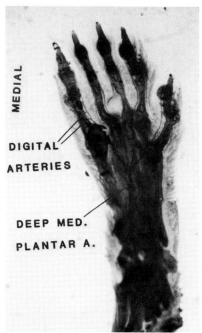

MEDIAL

DIGITAL
ARTERIES

DEEP MED.
PLANTAR A.

Figure 5. Left (Dh/+) foot
with triphalangeal hallux,
plantar aspect

Group 2. In the limbs with
triphalangism, there was no
visible abnormality of the
digital arteries. However,
in these Dh/+ limbs both
digital arteries were derived
from the deep plantar artery
(Figure 5), whereas, in +/+
limbs, one arose from the superficial medial plantar artery and one from the deep
medial plantar artery.

Group 3. In hindlimbs with
preaxial syndactyly, there
was a fusion of the first and
second toes. This fusion
received its blood supply from
a single branch of the deep
medial plantar artery rather
than from two branches
(Figure 6).

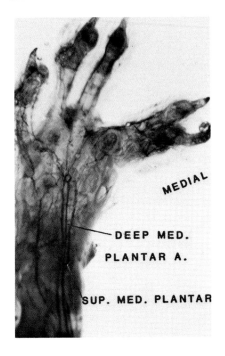

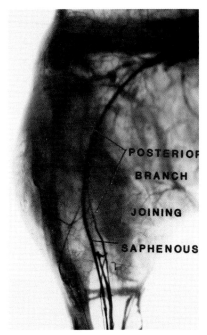

Figure 6. Right (Dh/+) foot with syndactyly of the first and second rays, plantar aspect.

Figure 7. Medial knee and leg region, (Dh/+), proximal to foot in Figure 6.

In some hindlimbs with syndactyly, there was a second, smaller branch of this artery to the first toe. In association with these variations in the plantar arteries, there was a long branch from the posterior tibial region that extended down to join the plantar arterial system (Figure 7). This appeared to be an anomalous artery since in the normal (+/+) hindlimb, the saphenous artery was the only artery that entered the plantar region.

Group 4. In hindlimbs with a shortened tibia, the major arteries ended quite abruptly in a fine network (Figure 8). Below this network an artery arose from a position that was normally occupied by the distal portion of the saphenous artery. Often in association with shortening of the tibia, there was no first toe; there were four branches of the deep medial plantar artery to the four toes that were present.

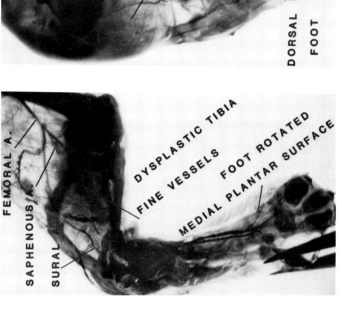

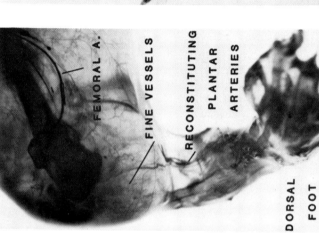

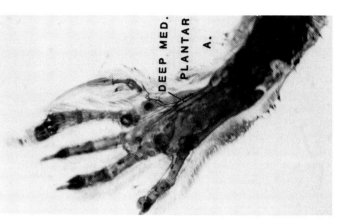

Figure 8. Left (Dh/+) hindlimb, tibial dysplasia, four ray foot.

Figure 9. Right (Dh/+) hindlimb, absent tibia, five ray foot.

Figure 10. Detail of (Dh/+) foot from limb in Figure 9, plantar aspect.

Group 5. When there was no tibia, the vascular abnormalities were visible proximal to the tibia (Figure 9). There was no true saphenous artery and no popliteal arterial system. There was a cluster of fine vessels in the region where the tibia would be located normally and normal size vessels appear to arise in the distal portion of this fine network. In some hindlimbs with no tibia there were four toes with no first toe. When a first toe was present, it was a floating hallux with no first metatarsal (Figure 10). The medial plantar artery did not supply as much of the foot in this group as it does normally.

DISCUSSION

One technical problem in using perfusion to study vessels is the fact that failure for a vessel to fill can be due to an artifact and not represent absence of that vessel. We found that the extent of filling of small vessels varied in the twenty animals studied, a variation that could be important if changes in these vessels were a critical finding. Another problem with perfusion technique is the presence of air bubbles. However, this is easy to recognize and causes no confusion in interpretation.

The consistency with which the pattern of vascular changes correlated with the skeletal malformations was notable. We were surprised to find an extra vessel extending from the posterior tibial region to the plantar system in association with syndactyly of the first and second toes and a normal tibia.

The most striking finding in association with absence and shortening of the tibia was a cluster of small blood vessels. It is tempting to speculate that this represents a persistence of vessels in this region and that this persistence led to the absence of the tibia. This interpretation would be consistent with the hypothesis of Caplan and Koutroupas (1973) who suggested that limb mesenchyme cell aggregation and cartilage formation occurs in regions devoid of vessels. It would be important to know whether this finding in the adult Dh/+ animal reflects events that occurred during limb development. It is possible that significant changes occur in the transition from the embryonic to the adult vascular pattern. In humans, Senior (1918-19; 1919; 1929) noted that variations in arterial pattern were under-

standable in terms of an embryonic blood supply changing
into an adult supply. Knowing embryonic and adult patterns
as well as the sequence of transition may help in understanding the expression of the Dh gene.

We plan to extend this study to the hindlimbs of mutant
(Dh/Dh and Dh/+) mice before the limb deformity is visible.
We will use the technique of gene linkage between Dh and
dipeptidase-1, described elsewhere in this volume to identify
the mutant embryos (Holmes, 1982).

Brookes M (1971). "The Blood Supply of Bone." London:
    Butterworths, pp. 10-11.
Caplan AI, Koutroupas S (1973). The control of muscle and
    cartilage development in the chick limb: The role of
    differential vascularization. J Embryol Exp Morp 29:571.
Carter TC (1954). The genetics of luxate mice.
    IV. Embryology. J Genetics 52:1.
Chaurasia BD (1974). Single umbilical artery with caudal
    defects in human fetuses. Teratology 9:287.
Cook MJ (1965). "The Anatomy of the Laboratory Mouse."
    New York: Academic Press, pp. 113-117.
Froud MD (1959). Studies on the arterial system of three
    inbred strains of mice. J Morph 104:441.
Holmes LB (1982). Use of gene linkage to detect limbs
    destined to be malformed by the dominant gene Dominant
    hemimelia. "Conference on Limb Morphogenesis." New York:
    Alan Liss.
Hootnick DR, Levinsohn EM, Randall PA, Packard DS Jr (1980).
    Vascular dysgenesis associated with skeletal dysplasia of
    the lower limb. J Bone Joint Surg 62A:1123.
Green MC (1967). A defect of the splanchnic mesoderm
    caused by the mutant gene Dominant hemimelia in the mouse.
    Dev Biol 15:62.
McLeod MJ (1980). Differential staining of cartilage and
    bone in whole mouse fetuses by alcian blue and alizarin
    red S. Teratology 22:299.
Poswillo D (1973). The pathogenesis of the first and
    second branchial arch syndrome. Oral Surg 35:302.
Searle AG (1964). The genetics and morphology of two
    "luxoid" mutants in the house mouse. Genet Res Camb 5:
    171.
Senior HD (1918-19). An interpretation of the recorded
    arterial anomalies of the human leg and foot. J Anat 53:
    130.

Senior HD (1919). The development of the arteries of the human lower extremity. Am J Anat 25:55.
Senior HD (1929). Abnormal branching of the human popliteal artery. Am J Anat 44:111.

Limb Development and Regeneration
Part A, pages 327–334
© **1983 Alan R. Liss, Inc., 150 Fifth Avenue, New York, NY 10011**

CONGENITAL TIBIAL APLASIA WITH POLYDACTYLY: IMPLICATIONS OF
ARTERIAL ANATOMY FOR ABNORMAL LIMB MORPHOGENESIS

David R. Hootnick, M.D., David S. Packard, Jr., Ph.D., and
E. Mark Levinsohn, M.D.

Departments of Orthopedic Surgery, Anatomy and Radiology
Upstate Medical Center, Syracuse, N.Y.   13210

The normal development of embryonic bone from mesenchyme is not possible without adequate arterial supply.
Since some human skeletal malformations of the upper (Blauth
and Schmidt, 1969; Van Allen, Hoyme and Jones, 1982) and
lower limb have been associated with abnormal vascular patterns (Ben-Menachem and Butler, 1974) and since seemingly
unrelated conditions such as clubfoot and fibular dysplasia
have been shown to share certain vascular pattern anomalies
(Hootnick et al., 1980; Hootnick et al., 1982), vascular abnormalities may be a common factor in the etiology of some
of these disorders. We recently performed a pre-operative
arteriogram and a post-amputation dissection on a human limb
with complete tibial aplasia (CTA) and diplopodia (duplicated foot) and found vascular anomalies similar to those
associated with the malformations of the lower extremity
mentioned above. Dissection of the arterial system in the
amputated limb confirmed the arteriographic findings. We
propose that the abnormal development of a portion of the
embryonic arterial pattern places the limb at risk for certain bony malformations.

RESULTS

Description of skeletal elements (Figure 1). The tibia
was absent and in its place was a tendinous band with an attached connective tissue mass. This band was attached to
the fibula at the points of normal articulation of the tibia
with fibula. The fibula was thickened and bowed. In the
foot there were six complete metatarsals present.

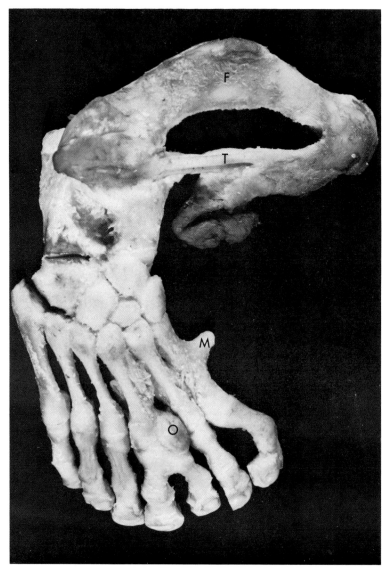

Fig. 1. Photograph of the dissected bony skeleton showing bowed and thickened fibula (F) and absent tibia with tendinous remnant (T). Six complete metatarsals are present as well as an oval metatarsal precursor (O) and a medially projecting bony metatarsal precursor (M). Seven toes are present.

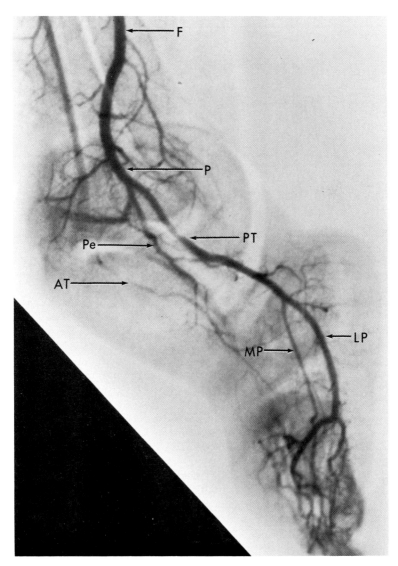

Fig. 2. Arteriogram of the leg showing the anterior tibial and peroneal arteries to be reduced in size terminating near the ankle. The dorsalis pedis artery is absent. AT = anterior tibial artery, F = femoral artery, LP = lateral plantar artery, MP = medial plantar artery, P = popliteal artery, Pe = peroneal artery, PT = posterior tibial artery.

Additionally, a medial bony projection and a central fragment were interpreted as portions of two additional metatarsals. Seven toes were present.

Description of arteries. Arteriography (Figure 2) showed the arterial pattern to be normal with the exception that the anterior tibial artery was reduced in size and terminated near the ankle. The dorsalis pedis artery was absent. These findings were confirmed by dissection. Extra digital branches of the lateral plantar artery and the plantar arch supplied the extra toes.

DISCUSSION

The arterial malformation shown in this case of CTA with polydactyly is characterized by diminution of the anterior tibial artery and absence of the dorsalis pedis artery. This is reminiscent of the pattern associated with skeletal dysplasia of the lower limb with fibular deficiency (Hootnick et al, 1980). Senior (1919) described the normal development of the arterial pattern in the human lower extremity. He found that a distinct embryonic arterial pattern emerges which changes to the adult pattern during the fifth week of embryonic development as a result of the loss of some embryonic arteries and a formation of new arteries. In all limbs with fibular dysplasia which were examined functional embryonic vessels were noted (Hootnick et al., 1980). In patients with clubfoot deformity, deficiency of the anterior tibial and dorsalis pedis arteries was found with the occasional retention of embryonic vessels (Ben-Menachem and Butler, 1974; Hootnick et al., 1982).

Although the arterial pattern deficiencies that we have noted are similar, they are associated with many different bony anomalies. Furthermore, the anterior tibial artery is absent or deficient in 4 to 12% of normal limbs (Huber, 1941; Anderson, 1978). It would then seem unlikely that the arterial abnormalities noted are solely responsible for the bony malformations. Deficiency or absence of the anterior tibial artery would change the pattern of blood flow in the developing limb and reduce the number of potential routes for collateral circulation. Any further loss of vessels could, then, increase the risk of a subsequent malformation. At the same time, the presence of collateral vessels or retained embryonic vessels might serve to ameliorate the

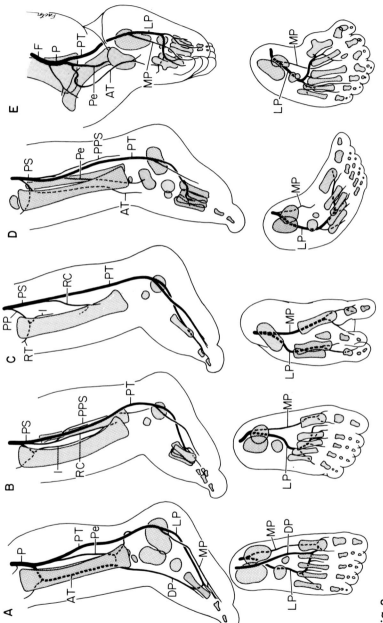

Fig. 3

Fig. 3. Drawing of bony malformations and the associated
arterial patterns.
A - Normal bony and arterial pattern.
B - Skeletal dysplasia with fibular deficiency; fibula is
    present but abnormally short. The anterior tibial and
    peroneal arteries are absent. Functional embryonic
    arteries (I, PPS, RC) are present.
C - Skeletal dysplasia with fibular deficiency; total ab-
    sence of the fibula and midline metatarsal dysplasia
    is noted. The anterior tibial and peroneal arteries
    are absent. Functional embryonic arteries (I, RC) are
    present.
D - Clubfoot; the anterior tibial and peroneal arteries are
    reduced. A functional embryonic artery (PPS) is pres-
    ent.
E - Congenital tibial aplasia with polydactyly; the ante-
    rior tibial artery is absent. No functional embryonic
    arteries are present. The peroneal artery and the
    fibula are present. AT= anterior tibial artery, DP =
    dorsalis pedis artery, F = femoral artery, I = inter-
    osseus artery, LP = lateral plantar artery, MP = medial
    plantar artery, P = popliteal artery, Pe = peroneal ar-
    tery, PPS = peronea posterior superficialis artery,
    PS = poplitea superficialis artery, RC = ramus communi-
    cans artery.

---

vascular loss due to the missing arteries. For example, in
Figure 3, limbs B and C both have deficient fibulae but the
bones of limb B appeared to be more normal than those of
limb C. While both limbs were missing the anterior tibial
and dorsalis pedis arteries, limb B retained several large
embryonic arteries. We suggest that these retained vessels
served as collateral arteries which augmented the blood supply to the developing leg and foot and thus led to a more
normally developed skeleton; the failure to retain larger
embryonic vessels may have predisposed limb C to more profound skeletal deficiency.

    Examination of a sufficient number of limbs with con-
genital bony malformations revealed that there is an approximate correlation between arterial and bony malformations. First, the bony malformations seem to occur at the
level of the arterial malformations. For example, in limb D
the arteries and bones of the leg were normal down to the
ankle unlike those in limbs B and C. The anterior tibial

artery terminated at the ankle and the dorsalis pedis artery was absent. The major bony malformation, an abnormality of the talus, occurred immediately distal to the termination of the anterior tibial artery. Second, given the reduction or absence of the anterior tibial artery, there seems to be a correlation between the absence of additional vessels and the bones adjacent to those vessels. Limbs B and C in Figure 3 were missing both the anterior tibial and peroneal arteries and exhibited fibular deficiencies. Limb E had a peroneal artery and a more normal fibula. On the other hand, given an absent or reduced anterior tibial artery, the presence of an interosseous or ramus communicans artery seems to correlate with the presence of the tibia. In the one case in which the tibia was absent (limb E), these arteries were also absent.

Investigations of congenitally malformed human lower limbs have revealed a significant correlation between defects of the bones and arteries. In almost every case thus far examined, the anterior tibial artery was absent or deficient. Furthermore, there was an approximate anatomical correlation between many of the bony and arterial defects. While these observations are interesting and promising, their meaning must await the results of experimental investigation of the role of the vascular system in abnormal limb morphogenesis.

REFERENCES

Anderson JE, ed. (1978). Grant's Atlas of Anatomy, seventh ed., Williams and Wilkins Company, Baltimore.
Ben-Menachem Y, Butler JE (1974). Arteriography of the foot in congenital deformities. J. Bone Joint Surg. 56A: 1625-1630.
Blauth W, Schmidt H (1969). Uber die bedeutung arteriographischer befunde bei fehlbildungen am radialen randstrahl (radiale dyspladie). Zeit. fuer Orthopaed. und ihre Grenzgebiete 106:102-110.
Hootnick DR, Levinsohn EM, Randall PA, Packard DS (1980). Vascular dysgenesis associated with skeletal dysplasia of the lower limb. J. Bone Joint Surg. 62A:1123-1129.
Hootnick DR, Levinsohn EM, Crider RJ, Packard DS (1982). Congenital arterial malformations associated with clubfoot. Clin. Orth. Rel. Res. 167:160-163.

Huber JF (1941). The arterial network supplying the dorsum
  of the foot. Anat. Rec. 80:373-391.
Senior, HD (1919). The development of the arteries of the
  human lower extremity. Am. J. Anat. 25:55-95.
Van Allen MI, Hoyme HE, Jones KL (1982). Vascular pathogenesis of limb defects: Radial artery anatomy and radial
  aplasia. J. Ped. (in press).

Limb Development and Regeneration
Part A, pages 335–344
© 1983 Alan R. Liss, Inc., 150 Fifth Avenue, New York, NY 10011

DEVELOPMENT OF MOUSE LIMBS IN ORGAN CULTURE: DOSE DEPENDENT
RETINOIC ACID-INDUCED DEFECTS EVALUATED USING IMAGE ANALYSIS

Thomas E. Kwasigroch and Richard G. Skalko

Department of Anatomy, Quillen-Dishner College of
Medicine, East Tennessee State University,
Johnson City, Tennessee 37614

Since the pioneering study of Aydelotte and Kochhar
(1972), it has been shown that mouse limb anlagen, removed
from embryos at appropriate stages of gestation, can be
cultured outside the maternal organism. Under optimal conditions, these explants are capable of cytodifferentiation,
morphogenesis and growth and it is possible to study the influence of embryotoxic agents on these developmental parameters either following in vitro treatment (Aydelotte and
Kochhar, 1972) or by treating the dams directly with subsequent culture of the limb rudiments (Kochhar and Aydelotte,
1974). Kwasigroch et al., (1977) demonstrated that limb
anlagen could be cultured in a submerged medium without
added serum. The use of this method permits an analysis of
limb development in which the influence of a number of experimental variables (gas diffusion gradients, intralitter
variability, dependence on serum proteins) is minized. A
further technological advancement was provided by the work
of Kwasigroch et al. (1981) which employed image analysis
in the evaluation of embryotoxic effects on limbs cultured
following in vivo treatment with either hydroxyurea, bromodeoxyuridine or their combination. This method has the
particular advantage of being quantitative and objective in
nature and is a significant improvement over the somewhat
subjective and semi-quantitative methods that have been
used previously.

Retinoic acid, a potent embryotoxic agent, has been
shown to produce a time-dependent pattern of limb defects
in mice with forelimbs affected earlier in gestation than
are hindlimbs (Kwasigroch and Kochhar, 1980). In this

study, pregnant mice were treated with a range of dosages
of retinoic acid and the fetuses evaluated for malformations
of the limbs and palate using standard methods. In addition, limbs from embryos treated in vivo were cultured 24 h
later and eventually studied for dose-dependent alterations
using image analysis.

Materials and Methods

Nulliparous female mice of the ICR stain were used.
Retinoic acid (Sigma) was suspended in cottonseed oil and
administered, by gavage, at a volume of 0.10 ml/10 g
maternal body weight on E 11. Three dosages were used:
20, 40 and 80 mg/kg. For the in vivo studies, females were
killed on E 17 and their fetuses evaluated for evidence of
embryotoxic effects using standard procedures. For the in
vivo - in vitro studies, females were treated on E 11 as
above. On E 12 (50-53 somites), females were killed, their
embryos dissected free of all extra-embryonic membranes and
then placed into a petri dish containing sterile Hank's
salt solution. Forelimbs and hindlimbs were cut off at a
point immediately adjacent to the body flank (Kwasigroch
and Kochhar, 1975) and then pooled, separately, in 50 ml
Wheaton bottles containing 4 ml of culture medium (Table
1). This medium is supplemented with 0.85 mM ascorbic acid
and 50 µg/ml gentamycin sulfate (Sigma). The bottles were
sealed with a rubber stopper and an aluminum clasp and then
flushed with an appropriate gas mixture (Table 1). The
bottles were then attached to a rotator (speed = 10 rpm)
placed inside an incubator at 37°C. The limbs were cultured for 3 d, with no change in medium. The cultures were
gassed at 24 h intervals according to the protocol shown
in Table 1. Following the culture period, the limbs were
washed in Hank's salt solution and were fixed overnight in
Bouin's fluid. The limbs were then decolorized in ammonia
water, stained in Alcian blue and differentiated in acetic
acid. Dehydration and clearing in xyol and cedarwood oil
then followed. Fore- or hindlimbs from each litter were
then mounted on microscope slides, using DPX (Gurr) as the
mounting medium. Limbs from at least seven litters were
used at each dose range and, in these, at least 7 limbs
(either fore- or hindlimbs) were selected at random for
image analysis. Thus, each dosage was represented by at
least 49 limbs, a sample size previously shown to be sufficient to diminish any effects which might be attributable

to inter- or intra-litter variability (Kwasigroch et al., 1981). The area and form of individual bone anlagen were measured and compared using the Leitz ASM system and the details of the methods used were identical to those described previously (Kwasigroch et al., 1981).

Table 1. Conditions for culturing limbs from E 12 mouse embryos in submerged culture for 3 days.

a.	**Culture medium**				
	BgJ (original formula)		= 75%		
	Salt Solution				
	NaCl	7.0	g/l		
	K Cl	0.42	g/l	= 25%	
	$CaCl_2$	0.67	g/l		
b.	**Gas phase**				
	Forelimbs – each day	– 50% O_2,	5% CO_2,	45% N_2	
	Hindlimbs – day 1	– 20% O_2,	5% CO_2,	75% N_2	
	days 2,3	– 50% O_2,	5% CO_2,	45% N_2	

Results

While retinoic acid is capable of producing a wide range of defects when administered to pregnant rodents, we have restricted this presentation to its influence on two distinct parameters: the occurrence of intrauterine death (resorption rate) and of one specific limb defect, ectrodactyly. As the data show (Table 2), the high dose of retinoic acid (80 mg/kg) produces a significant increase in resorption rate beyond that seen in all other treatment groups. In addition, the data substantiate the observation of Kwasigroch and Kochhar (1980) that, following treatment on E 11, the hindlimb is most severely affected as indicated by the degree of ectrodactyly observed on E 17 at the lower dosages used.

Using the experimental conditions detailed above, both fore- and hindlimbs develop characteristic morphology after 3 days in culture. All bone anlage are represented and, in particular, the morphogenesis of the paw elements is considerably enhanced, a process which, in our laboratory, is particularly sensitive to oxygen tension (Table 1). When representative limbs are observed (Figure 1), there is an indication that, after treatment with retinoic acid, there is a generalized alteration in the degree of morphogenesis

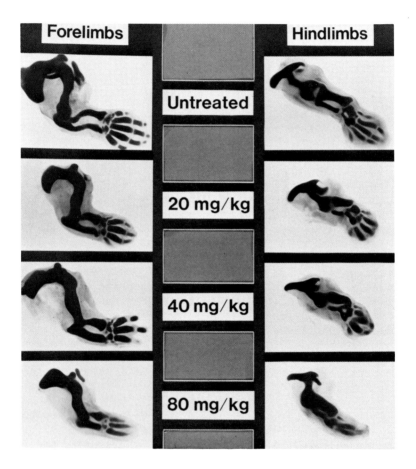

Fig. 1. Representative limbs from all treatment groups. Retinoic acid has its major effect on the morphogenesis of the paw skeleton.

Table 2. Dose-dependent response of the mouse embryo to retinoic acid treatment on E 11.

| | | Viable | % Ectrodactyly | |
Treatment	Resorbed (%)	Fetuses	FL	HL
Control[1]	3.6	135	--	--
Control[1]	4.1	123	--	--
20 mg/kg RA	8.3	121	1.7	4.1
40 mg/kg RA	6.8	124	50.0	86.3
80 mg/kg RA	43.4[2]	77	100.0	100.0

[1] = treated with vehicle at a volume equivalent to 80 mg/kg RA

[2] = $p < 0.05$

attained with the paw elements being most severely affected. As in the in vivo study (Table 2), it appears that the hindlimb is most sensitive to treatment on E 11.

Through the use of the image analysis system, absolute values for total area and from factor for individual components of the limb are obtained (Figure 2). The latter value gives a mathematical indication of the shape of the figure traced with a value of zero indicating a point or a straight line with no enclosed area and a value of 1.0 indicating a perfect circle. Under the conditions of this experiment, each of the anlage analysed has a unique and individual form factor value. In addition to these, the data generated permit the calculation of two additional morphometric parameters, the long bone: paw ratio and the soft tissue: bone ratio.

Retinoic acid, at all dosages, produced a reduction in the total area of the cartilaginous elements of the forelimb (Figure 3) and the hindlimb (Figure 4) with the response being most severe in the hindlimb. When the data were used to calculate the ratio between long bone and paw, this specific value showed a dose-dependent increase in the forelimb (Figure 5). This effect was not as severe in the hindlimb (Figure 6) until the highest dosage (80 mg/kg) was used. Collectively, these results suggest that the primary target for retinoic acid toxicity in the forelimb was the paw skeleton while, in the hindlimb, all cartilag-

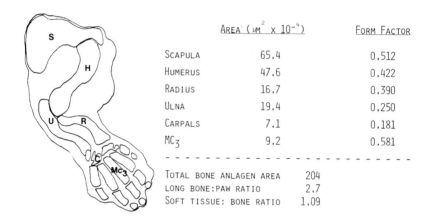

	AREA ($\mu M^2 \times 10^{-4}$)	FORM FACTOR
SCAPULA	65.4	0.512
HUMERUS	47.6	0.422
RADIUS	16.7	0.390
ULNA	19.4	0.250
CARPALS	7.1	0.181
MC$_3$	9.2	0.581

TOTAL BONE ANLAGEN AREA	204
LONG BONE:PAW RATIO	2.7
SOFT TISSUE: BONE RATIO	1.09

Fig. 2. Image analysis of a representative forelimb after 3 d in culture.

inous elements were affected (Figure 1). The statistically significant increase in this ratio following treatment with 80 mg/kg appeared to be the result of an extensive loss of digital elements seen in this specific group.

When the form factor element was calculated, dose-dependent alterations were observed in most of the cartilaginous elements of both the forelimb and hindlimb. These were especially evident in the ulna and carpals and in the

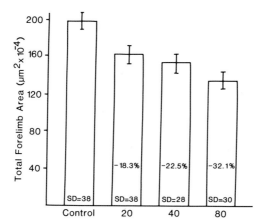

Values = mean ± 95% confidence limits; standard deviation and percentage below control levels are indicated
Retinoic acid treated limbs were significantly different from controls , p<0.05
80mg/kg group was significantly different from 20 and 40 mg/kg groups, p<0.05

Fig. 3.   The effect of retinoic acid on the area of the bone anlagen of the forelimb.   Image analysis values for each component were summed to determine total area per forelimb.

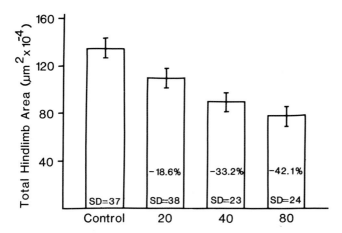

Values = mean ± 95% confidence limits; standard deviation and percentage below control levels are indicated
Retinoic acid treated limbs were significantly different from controls, p<0.05
40 and 80mg/kg groups were significantly different from 20mg/kg, p<0.05

Fig. 4.   The effect of retinoic acid on the area of the bone anlagen of the hindlimb.   Image analysis values for each component were summed to determine total area per hindlimb.

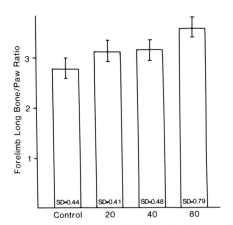

Fig. 5. The effect of retinoic acid on the long bone: paw ratio of the forelimb. The specific ratio was determined by summing the area of the long bone elements and summing the area of all the paw elements and comparing these two values.

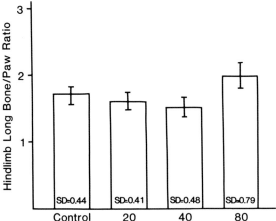

Values = mean ± 95% confidence limits; standard deviation is indicated
80mg/kg group was significantly different from all other groups, p<0.05

Fig. 6. The effect of retinoic acid on the long bone: paw ratio of the hindlimb. The slight decrease which occurred at the low dosages reflects an effect of retinoic acid on long bone morphogenesis.

fibula and tarsals (Table 3). In all groups, there was an indication that retinoic acid tended to produce a "roundedness" (form factor closer to 1.0 in the treated limbs) of the bones and the severity of this response was increased with increasing dosage.

Conclusions

The results presented here suggest that use of the submerged culture system (Kwasigroch et al., 1977) provides a good milieu for the cartilaginous elements of the limb to undergo highly specific and reproducible morphogenesis. The use of image analysis, as described here, is a sensitive and objective tool with which to evaluate the normal development of cultured limbs. It is also useful in the detailed analysis of the response of the many components of the limb

Table 3. Dose-dependent alterations in the form factor for selected cartilaginous elements following treatment with retinoic acid.

a. Forelimb

Treatment	Ulna	Carpals
Control*	$0.33 \pm 0.06_1$	$0.24 \pm 0.05_1$
20 mg/kg RA	$0.39 \pm 0.08_1$	$0.28 \pm 0.05_1$
40 mg/kg RA	$0.41 \pm 0.05_1$	$0.28 \pm 0.05_1$
80 mg/kg RA	$0.45 \pm 0.06_{1,2}$	$0.34 \pm 0.09_{1,2}$

b. Hindlimb

Treatment	Fibula	Tarsals
Control*	$0.46 \pm 0.11_1$	$0.28 \pm 0.07_1$
20 mg/kg RA	$0.57 \pm 0.12_1$	$0.37 \pm 0.01_1$
40 mg/kg RA	$0.68 \pm 0.10_{1,3}$	$0.41 \pm 0.12_1$
80 mg/kg RA	$0.65 \pm 0.07_{1,3}$	$0.50 \pm 0.17_{1,3}$

* = untreated controls
1 = significantly different from untreated controls
2 = significantly different from the 20 and 40 mg/kg groups
3 = significantly different from the 20 mg/kg group

skeleton to the influence of specific embryotoxic agents. In this study, it has been demonstrated that the defects produced in vivo and those observed in cultured limbs following in vivo exposure to retinoic acid were similar. Total area values (Figures 3 and 4) indicate that the hindlimb was most affected by treatment on E 11. The effect of retinoic acid on long bone: paw ratios (Figures 5 and 6) was most pronounced on the paw skeleton in the forelimb while the long bones were most affected in the hindlimb. Collectively, these data provide a quantitative confirmation of the previously described temporal specificity of retinoic acid-induced limb defects, most particularly the fact that these defects mirror the well-known cephalocaudal and proximodistal gradients which occur in the development of the vertebrate limb (Kwasigroch and Kochhar, 1980).

References

Aydelotte MB, Kochhar DM (1972) Development of mouse limb buds in organ culture: chondrogenesis in the presence of a proline analogue, L-azetidine-2-carboxylic acid. Dev. Biol 28: 73–80.

Kochhar DM, Aydelotte MB (1974) Susceptible stages and abnormal morphogenesis in the developing mouse limb, analysed in organ culture after transplacental exposure to vitamin A (retinoic acid). J Embryol exp Morph 14: 233–238.

Kwasigroch TE, Barrach H-J, Tapken S, Rautenberg M, Neubert D (1977) Submerged culture of mammalian limb buds. In: Advances in the Detection of Congenital Malformations (eds EB van Julsingha, JM Tesh, GM Fara), Michael Robbin Printers, Chelmsford, Essex, pp. 257–263.

Kwasigroch, TE, Kochhar DM (1975) Locomotory behavior of limb bud cells. Effect of excess vitamin A in vivo and in vitro. Exp Cell Res 95: 269–278.

Kwasigroch TE, Kochhar DM (1980) Production of congenital limb defects with retinoic acid: phenomenological evidence of progressive differentiation during limb morphogenesis. Anat Embryol 161: 105–113.

Kwasigroch TE, Skalko RG, Church JK (1981) Development of limb buds in organ culture: examination of hydroxyurea enhancement of bromodeoxyuridine toxicity using image analysis. In: Culture Techniques (eds D Neubert and H-J Merker). Walter de Gruyter, Berlin pp. 237–253.

Limb Development and Regeneration
Part A, pages 345–353
© 1983 Alan R. Liss, Inc., 150 Fifth Avenue, New York, NY 10011

RESCUE FROM DISPROPORTIONATE DWARFISM IN MICE BY MEANS OF
CAFFEINE MODULATION OF THE 4-HOUR EARLY EFFECT OF EXCESSIVE
VITAMIN A

Meredith N. Runner

Department of Molecular, Cellular and
Developmental Biology, Campus Box 347
University of Colorado, Boulder, CO   80309

The vertebrate limb skeleton develops out of a covert 3-dimensional pattern. The common event that intervenes between skeletal patterning and overt prechondrogenic differentiation is the sequential, periodic and asymmetric expression of mesodermal prechondrogenic aggregations. At least two components account for the transition from covert pattern to overt cellular and molecular changes. First, cellular interactions by means of extracellular signals impose patterned information on the state of readiness of the central mesodermal cells. Second, the central mesoderm cells attain a level of cascaded readiness that anticipates signalled input and that will initiate molecular syntheses required for subsequent prechondrogenic aggregation. Although the state of readiness had been achieved by prior history of input both from the genome and from extragenomic flow of extracellular signals, the central mesodermal cells permissively await posttranslational modulations.

Retinoic acid at the level of 16 mg/kilo of maternal body weight was administered to mice by gavage at 11.5 days; 43 ± 3 somite stage. Retinoic acid 1) appears to exclusively modify limb development (Kochar, 1973); 2) achieves a modal tissue concentration in less than 4 hours (Newall and Roberts, 1981); 3) has no effect on prenatal mortality or 18.5 day body weight (Runner, 1982a); 4) produces offspring that are viable and express permanent disproportionate dwarfism by virtue of diminutive prechondrogenic aggregations for precursors of the skeletal elements (Runner, 1982a); and 5) produces differential shortening of the long bones of the appendage remarkably coincident with a six-step succession of commitments to prechondrogenic aggregations (Runner, 1982b). The most proximal preskeletal elements are developmentally the most advanced (Forsthoefel, 1963) and are least affected by retinoic acid. The three skeletal elements of the later developing rear limb are more sensitive to retinoic acid than the corresponding skeletal elements of the forelimb. Retinoic acid therefore qualifies as a limb specific and a within-limb specific probe that modulates commitment of mesodermal cells to prechondrogenic aggregations.

We have reported that retinoic acid operationally has an early effect on limb bud development with a modal effective period of less than 4 hours after administering the drug (Runner, 1982b; Newall and Roberts, 1981). We report here that prior administration of caffeine, by in situ modulation of the early effect, can rescue from disproportionate dwarfism induced by retinoic acid. Furthermore, we report that the in situ early effect for retinoic and rescue by caffeine is accomplished by modulation of the proportional distributions among sulfated glycosaminoglycans.

We have investigated interference in situ with prechondrogenic aggregation in an effort to characterize the 4-hour early effect of retinoic acid on central mesodermal cells of the limb bud during their commitment to express prechondrogenic aggregations. The effects of retinoic acid and of caffeine were determined by two methods: 1) linear regressions of body weight against alizarin stained long bone lengths were calculated for 18 day fetuses; 2) limb buds were taken from embryos that had been exposed in utero to retinoic acid for four hours, or to caffeine for five hours, or concurrently to retinoic acid and caffeine. Limb buds were excised at 11.5 days of gestation and incubated with ^{35}S sulfuric acid for 60 minutes. Sulfate incorporation was measured in four subsets of acid precipitable macromolecules; the subsets have been labeled lipid, glycan I, glycan II, and protein hydrolysate (Runner, 1976). The effects of retinoic acid and of caffeine were analyzed as cpm/ug of DNA incorporated into the four fractionated subsets.

The effects of transitory exposure of the 11.5 day embryo to retinoic acid, caffeine and retinoic acid plus caffeine are summarized in Figure 1. Exposure to 16 mg/kilo of retinoic acid resultèd in moving averages for ulnae and tibiae that were selectively and significantly shortened when compared to vehicle treated control fetuses. Prior administration of caffeine resulted in alizarin stained long bones that were significantly longer than in fetuses that had been exposed to retinoic acid alone. Thus, the prospective dwarfism induced by brief exposure to retinoic acid had been prevented by prior administration of caffeine. The treatment level of caffeine did not by itself induce shortening of alizarin stained long bone lengths. In fact, in both fore and rear limbs the long bones for fetuses treated with caffeine were slightly longer than were the corresponding controls.

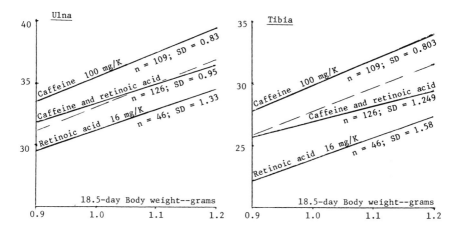

Figure 1: Linear regressions of alizarin stained bone lengths against body weight at 18.5 days of gestation. Retinoic acid dissolved in corn oil was given by gavage and caffeine dissolved in saline was given by intraperitoneal injection. Alizarin stained bone lengths had previously been found to accurately represent total bone lengths (Runner, 1982b). Treatments were given at 11.5 days of gestation where the day of finding the mating plug was day zero. Alizarin stained bone lengths were measured in a dissecting microscope; 12.4 micrometer units equal 1mm. SD = 1 standard deviation from the regression. Nontreated controls are represented by the dashed line (ulna: n = 87, SD = 0.82; tibia: n = 87, SD = 0.69).

The data showing rescue from the early effect of retinoic acid by caffeine, Figure 1, fail to confirm the general potentiation effect of caffeine on "various" teratogens as reported for rats by Ritter et al. (1982). Furthermore, the occurrence of hematomas and blebs, mechanical interference with limb morphogenesis, as reported by Nishimura and Nackai (1960) and Bartel and Gnacikowska (1972) did not appear in our embryos at 5 hours, 24 hours or 7 days after treating with caffeine. The dose level in our experiments was lower than that used by earlier investigators. The interaction of caffeine and retinoic acid, as molecular probes, may be unusually informative about crucial steps in preparation for prechondrogenic aggregations.

Exposure During the Early Effect	Mean cpm/ug DNA Among Four Fractionated Subsets of 11.5 Day Limb Buds; n=16 determinations each			
	Lipid	Glycan I	Glycan II	Protein Hydrolysate
Nontreated	414	1416	4063	490
Retinoic Acid	397	2273	4165	364*
Caffeine	276***	1438**	2267**	361*
Retinoic Acid + Caffeine	231	1180	3529	239

Table 1: The early effect of in utero exposure to retinoic acid and/or caffeine on subsequent incorporation of $H_2{}^{35}SO_4$ by excised 11.5-day (43-somite) mouse limb buds. Retinoic acid, 50 mg/kilo, was given by gavage at 4 hours and caffeine, 100 mg/kilo, was given by intraperitoneal injection at 5 hours before harvest of embryos. Limbs were excised and incubated in glucose phosphate buffered saline containing 125 uC/ml carrier free sulfuric acid. Partition of variance attributable to treatment within each fractionated subset gave the following significances: *P < 0.05; **P < 0.015; ***P < 0.00001.

The results on relative proportions of incorporations of ^{35}S sulfuric acid are given in Table 1. Each limb bud provided four fractionated subsets of cpm/ug of DNA. Partition of variance and significance of differences attributable to treatments were established by two-way analysis of variance within each fractionated subset for each of the four treatment groups. Table 1 shows that four hours of intrauterine exposure to 50 mg/kilo of retinoic acid caused significant reduction in sulfate incorporation into the protein hydrolysate fractions. Five hours of intrauterine exposure to 100 mg/kilo of caffeine on the other hand, caused a general reduction in all of the four fractionated subsets; lipid, glycan I, glycan II and hydrolysate. Because caffeine generally reduced sulfate incorporation into each of the fractionated subsets concomitant with rescue from dwarfism, we conclude that retinoic acid produced dwarfism by its differential effect by unbalanced incorporations. It would appear that the early effect, commitment to prechondrogenic aggregation, may be particularly sensitive to retinoic acid induced imbalance of sulfated glycosaminoglycans.

These results characterize the early 4-hour effect of retinoic acid on sulfation of glycolipids, glycoproteins and proteoglycans. The modulated pattern of sulfation is accompanied by reduced expression of prechondrogenic aggregations, no reduction in labeled thymidine incorporation, abnormally small skeletal elements and permanent disproportionate dwarfism. This early effect of retinoic acid on limb bud mesodermal cells as they acquire capability to express prechondrogenic aggregation, may be modulated in one of two ways. Either the carbohydrate moieties available for sulfation are selectively reduced or the sulfotransferase process for glycolipids and glycoproteins is selectively inhibited by retinoic acid. In either event the results indicate that at least one of the fractionated subsets, glycoproteins, may play a pivotal role for generating the capability for limb bud mesodermal cells to express prechondrogenic aggregations.

The precise cellular mechanisms for induction of dwarfism by retinoic acid is not known. Retinoic acid could act indirectly by causing subliminal patterning signals from the epithelium and/or peripheral mesoderm so that commitment to prechondrogenesis is less than complete. Retinoic acid on the other hand could act directly on the central mesodermal cells to interfere with the commitment process and to thereby reduce the capability to express full participation in prechondrogenic aggregations. Holmes and Trelstad (1980) have demonstrated the significant role played by mesodermal cell polarity and cell contacts during prechondrogenic aggregation. Kosher et al. (1981) and Singley and Solursh (1981) have shown the spatial distribution of extracellular matrix in the early limb bud.

Observations on exposure of limb bud mesodermal cells in vitro to retinoic acid suggest that the compound may have a direct effect on capabilities of prechondrogenic mesoderm to participate in subsequent chondrogenesis. Our observations derived from the in situ effects of retinoic acid indicate that sensitivity to dwarfism occurs at the preparative stage that enables limb bud mesodermal cells to participate in prechondrogenic aggregations.

A number of investigators have determined the in vitro repertoire of prospective capabilities for dissociated and plated limb bud mesodermal cells (Lewis et al., 1978; Pennypacker et al., 1978). Their results indicate that limb bud mesoderm after several days in vitro acquire special metabolic capabilities that enable limb mesodermal cells to express prospective chondrogenic aggregations in vitro. Furthermore, the entrained in vitro sequences can be modulated by continuous exposure to retinoic acid. These "can do" repertoires, in order to be meaningfully applied to developmental events within the limb, must sooner or later be coordinated with in situ "does do" performances; cellular and molecular events that accompany expression of prechondrogenic patterning.

Results so far have shown that retinoic acid induced dwarfism (1) has a modal dwarf inducing period of less than 4 hours (Runner, 1982b), (2) shows a 6-point coincidence with the within-limb stage of prechondrogenic aggregation (Runner, 1982a), and (3) can be rescued by caffeine restoration of balance of glycosaminoglycan syntheses during the early prechondrogenic commitment period. The possibility remains that delayed or late effects of retinoic acid may also contribute to disproportionate dwarfism.

Late effects of exposure to retinoic acid, to caffeine and to combined treatments have been studied in limb buds excised 24 hours after the exposure to retinoic acid and/or caffeine. Experiments parallel to those in Table 1 showed that incorporation of sulfate, 24-hours after exposure to caffeine, was well below control values. The late effects of exposure to retinoic acid on incorporation of thymidine into DNA were also studied, Table 2; (1) the early effect of retinoic acid did not alter ongoing DNA syntheses and (2) maximal delayed inhibition occurred at dose levels of 12.5 mg/kilo and above.

Retinoic Acid — mg/kilo	cpm/ug DNA; mean ± Std. Dev. (n)		% of Controls	"t"
	Nontreated	11-d Exposure		
12.5	647 ± 116 (20)	459 ± 88 (22)	71	4.2
25	579 ± 113 (30)	438 ± 93 (32)	75	5.4
50	621 ± 167 (25)	435 ± 104 (28)	70	4.9

Table 2: Late effect of in utero exposure to retinoic acid on incorporation of tritiated thymidine by excised 12.5 day (50-somite) mouse limb buds. Retinoic acid, 50 mg/kilo, was given by gavage at 11.5 days. Limbs were excised and incubated in glucose phosphate buffered saline containing 5 uC/ml of tritiated thymidine, 6.7 C/mMole. P < 0.01 by student "t" test.

Twenty-four hours after exposure to retinoic acid the limb buds showed a 25-30% decrease in the cpm of thymidine incorporated /ug of DNA, Table 2. These results indicate that although the ongoing DNA syntheses were not affected by exposure to retinoic acid, the capability to initiate new syntheses of DNA after 24 hours had been reduced. This finding that 24 hours after exposure to retinoic acid, a smaller proportion of cells were in DNA synthesis than control cells indicates that new rounds of DNA syntheses were partially inhibited. Furthermore, the dose response test showed that the new syntheses were maximally inhibited by doses of 50, 25 or 12.5 mg/K of maternal body weight. This apparent absence of a dose response effect on DNA synthesis suggests that a finite population of replicating cells was inhibited and that their inhibition had been maximally triggered by an indirect, late effect of retinoic acid.

The late effects of exposure to retinoic acid have shown that caffeine has actually potentiated depression of sulfate incorporation concomitant with rescue from dwarfism. Twenty-four hours after exposure to retinoic acid incorporation of thymidine into DNA was depressed by 25-30%. This depression, however, was not dose dependent as was the induction of disproportionate dwarfism. These 2 observations support the argument that the 24-hour late effects of retinoic acid are irrelevant to and occur after induction of disproportionate dwarfism.

The secondary and delayed responses following the 4-hour early effect of retinoic acid are significant but the early effects, detected 20 hours earlier, are wholly accountable for disrupted prechondrogenic aggregation and for disproportionate dwarfism (Runner, 1982b).

We have investigated in situ interference with prechondrogenic aggregation in an effort to characterize the 4-hour early effect of retinoic acid on central mesodermal cells of the limb bud. We have shown that the induction of and rescue from dwarfism exerts a high degree of stage specificity at the time limb bud mesodermal cells become committed to prechondrogenic aggregations. 1) The 4-hour early effects of retinoic acid are coincident with a 6-point progression toward prechondrogenic aggregation. 2) The early effects of retinoic acid produce an unbalanced reduction of sulfations whereas caffeine, even while rescuing from induced dwarfism, tends to produce a balanced depression of sulfations. 3) The late effects of retinoic acid on sulfate and thymidine incorporation, detectable 24-hours after administration of the compound, seem to be post facto and as such are not relevant to induced dwarfism. Using molecular probes in situ, we have attempted to establish the molecular and cellular state of readiness that accounts for the anticipated prechondrogenic response of mesodermal cells.

The investigation was supported by the National Institutes of Health (HD-13500) and by the National Science Foundation (PCM-15986).

Bartel PM, Griacikowska M (1972). Histologic studies on embryonic development of the limbs in mice. Folia Morph 31:178-184.

Forsthoefel PF (1963). Observations on the sequence of blastomal condensations in the limbs of the mouse embryo. Anat Rec 147:129-137.

Holmes LB, Trelstad RL (1980). Polarity in precartilage limb mesenchyme. Dev Biol 78:511-520.

Kochar DM (1973). Limb development in mouse embryos. I. Analysis of teratogenic effects of retinoic acid. Teratol 7:289-298.

Kosher RA, Savage MP, Walker KH (1981). A gradation of hyaluronate accumulation along the proximo distal axis of the embryonic chick limb bud. J Embryol Exp Morph 63:85-98.

Lewis CA, Pratt RM, Pennypacker JP, Hassell J (1978). Inhibition of limb chondrogenesis in vitro by Vitamin A: Alterations in cell surface characteristics. Dev Biol 64:31–47.

Newall DR, Edwards JRG (1981). The effect of Vitamin A on fusion of mouse palates. I. Retinal palmatate and retinoic acid in vivo. Teratol 23:115–124.

Nishimura H, Nakai K (1960). Congenital malformations in offspring of mice treated with caffeine. Proc Soc Exp Biol Med 104:140–142.

Pennypacker JP, Lewis CA, Hassell JR (1978). Altered proteoglycan metabolism in mouse limb mesenchyme cell culture treated with Vitamin A. Arch Biochem Biophys

Runner MN (1982a). Disproportionate dwarfism in mice caused by excessive Vitamin A. Remarkable correlation between stage of commitment to prechondrogenic aggregation and sensitivity to shortening of long bones. (submitted)

Runner MN (1982b). Disproportionate dwarfism in mice is caused by a window of sensitivity of less than 4 hours. Early effects of retinoic acid. (submitted)

Runner MN (1976). Biosynthesis in the mouse embryo as affected by hydroxyurea. In Neubert, Merker (eds): "New Approaches to the Evaluation of Abnormal Development," Thieme, Stuttgart, p 591–613.

Singley CJ, Solursh M (1981). The spatial distribution of hyaluronic acid and mesenchymal condensation in the embryonic chick wing. Dev Biol 84:102–120.

Ritter DJ, Scott WJ, Wilson JG, Mathinson PR, Randall JL (1982). Potentiative interactions between caffeine and various teratogenic agents. Teratol 25:95–100.

Limb Development and Regeneration
Part A, pages 355–364
© 1983 Alan R. Liss, Inc., 150 Fifth Avenue, New York, NY 10011

CHANGES IN PLASMA MEMBRANE PROTEINS AND GLYCOPROTEINS DURING
NORMAL AND BRACHYPOD MOUSE LIMB DEVELOPMENT

William A. Elmer and J. Thomas Wright
Department of Biology
Emory University
Atlanta, Georgia   30322

Over the past several years morphological and biochemical
studies have demonstrated that distinct changes in cell shape,
cell contacts, and the topography of the cell surface occur
during limb bud chondrogenesis (for review, see Elmer, 1982).
The importance of these changes for the normal differentiation
and development of the limb has come from studies of skeletal
mutations (Ede, 1976). One mutation that has been useful to
study the relationship of cell-cell interactions to the chondrogenic process has been brachypodism (bp^H/bp^H), an autosomal recessive in the mouse. It was initially reported that
in brachypod embryos there is a delay in the formation and a
reduction in the size of the prechondrogenic condensations
(Grüneberg and Lee, 1973). Subsequent studies have demonstrated a correlation between the abnormal process of condensation and aberrations in the surface properties of the mutant
limb cells (Duke and Elmer, 1977, 1978, 1979; Hewitt and Elmer,
1978). The present study is a continuation of our analyses of
the plasma membrane components of normal and brachypod postaxial hindlimb cells during chondrogenesis.

SDS - POLYACRYLAMIDE ELECTROPHORETIC PATTERNS OF LIMB CELL
PLASMA MEMBRANES

Plasma membranes of 11-, 12-, and 13-day stage postaxial
cells from the hindlimbs of normal and brachypod embryos were
isolated using positively charged Cytodex 1 beads (Pharmacia
Fine Chem.). Single cell suspensions were prepared from
ectoderm-free postaxial tissue by dissociation in Ca^{++} and
Mg^{++}-free Tyrode's solution containing the protease inhibitors

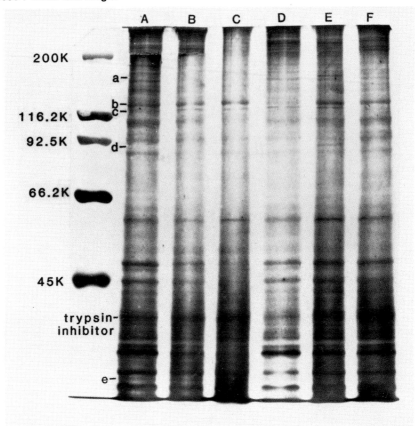

Fig. 1.   SDS - polyacrylamide slab gel stained with silver
stain showing the protein composition of plasma membranes of
11-, 12-, and 13-day postaxial mesenchyme from the hindlimbs
of normal and brachypod embryos.   (A) 11-day normal; (B) 11-
day brachypod; (C) 12-day normal; (D) 12-day brachypod; (E)
13-day normal; (F) 13-day brachypod.   Small letters refer to
components that change: (a) 170 K; (b) 120 K; (c) 117 K;
(d) 90 K; (e) 5 K.

phenylmethyl-sulfonylfluoride (0.1mM) and ovomucoid (20 µg/ml).
Cells were attached to the beads in a sucrose-acetate buffer containing 7 vols 310 mM sucrose, 3 vols 310 mosM sodium acetate, pH 5.2, and the protease inhibitors.   The attached cells were lysed by freeze-thawing in a 0.01 M Tris-
HCl buffer, pH 7.8 (Wright et al., 1982).

Membranes attached to beads were solubilized in SDS sample buffer containing 2% sodium dodecyl sulfate, 10% glycerol, 5% 2 mercaptoethanol, 0.01% bromophenol blue in 0.0625 M Tris-HCl buffer, pH 6.8. Electrophoresis was carried out using the Tris-glycine discontinuous system of Laemmli (1970) on 8.5% acrylamide slab gels (12 cm length, 1.5 mm thick) with a 5% stacking gel. Each sample, containing 100 µg protein, was subjected to electrophoresis in duplicate. Gels were silver stained using the Bio-Rad Silver Stain Kit. Molecular weight markers (Bio-Rad) were: myosin (200 K), β-galactosidase (116.2 K), phosporylase B (92.5K), bovine serum albumin (66.2 K), and ovalbumin (45 K).

The silver staining patterns of the plasma membrane components of the two genotypes over the three day gestation period are shown in Fig. 1. At the 11-day stage there are approximately 43 components which cover a wide molecular weight range. All of the components appear to be present in both the normal and brachypod plasma membranes. By the 12-day stage some differences become apparent. The 120 K component in the brachypod plasma membrane preparation is lighter stained than the normal component suggesting a quantitative difference. In addition, two components with molecular weights of 117 K and 5 K are absent in the mutant. The majority of the differences between the normal and brachypod plasma membrane preparations appear at the 13-day stage. A high molecular weight component (170 K) becomes visible in the normal preparation, but not in the mutant. The 120 K component that stained lighter at 12 days in the mutant appears darker at 13 days. The 117 K and 5 K components are also absent in both genotypes (Fig. 1, lanes E and F). Lastly, a 90 K component that is present in the normal preparation at 13 days is not observable in the mutant.

IODINATION OF CELL SURFACE COMPONENTS

Whole cells were subjected to iodination with diazotized [^{125}I] iodosulfanilic acid (Sp. Act. > 1,000 Ci/m mol, New England Nuclear) to label proteins on the cell surface. Diazotisation was carried out as described by the manufacturer. For each experiment, 100 µCi/reaction mixture was used. The reaction mixture was added to 12 X 10^6 cells suspended in 0.5 ml of 0.1 M phosphate buffer, pH 7.5. Cells were labeled to a final specific activity of 1.0 X 10^5 cpm/100 µg protein. Membranes were prepared as previously described Wright et

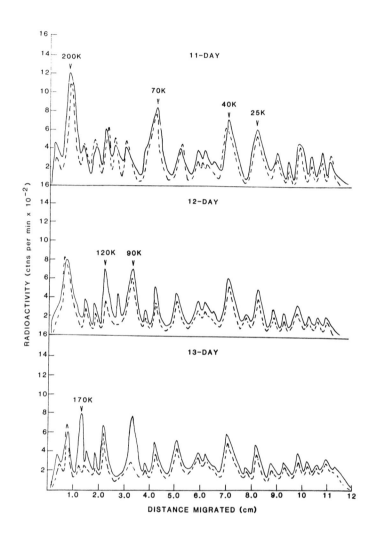

Fig. 2. Polyacrylamide gel electrophoresis of plasma membranes from postaxial hindlimb cells labeled with diazotised [^{125}I] iodosulfanilic acid. Normal (——), brachypod (---).

al., 1982) and subjected to polyacrylamide gel electrophoresis. Each gel was sectioned into 1.2 mm slices and radioactivity was determined using a Packard auto-gamma counter.

The electrophoretic patterns of the iodinated plasma membrane proteins from normal and brachypod 11-day postaxial
hindlimb cells reveal approximately 20 proteins can be
labeled with [^{125}I] iodosulfanilic acid (Fig. 2). Although
there are no apparent differences in the radioactivity at 11
days, both genotypes do exhibit four major components. The
molecular weights of these proteins are 200 K, 70 K, 40 K,
and 25 K, respectively. However, by the 12-day stage several
differences are observable. In the mutant there is a decrease in the amount of radioactivity in the 120 K and 100 K
proteins compared to normal. Both genotypes also exhibit a
significant decrease in the radioactivity of the 200 K and 70
K proteins, whereas the 90 K protein shows a sharp increase
in radioactivity. The major differences in the radioactive
profiles at 13 days occur with the proteins having molecular
weights of 170 K, 120 K, 100 K, and 90 K. The 170 K and 90 K
proteins exhibit a significantly lower amount of radioactivity in the brachypod membranes compared to the controls.
However, no difference in the labeling pattern for the 120 K
protein is observed at 13 days. The absence of the 100 K
protein is also observed at this time period.

BINDING OF PLANT LECTINS TO PLASMA MEMBRANE COMPONENTS

Previous studies showed that limb mesenchymal cells from
normal and brachypod mice differ in their relative adhesive-
ness (Duke and Elmer, 1977) and agglutinability with the
plant lectins, concanavalin A (con A) and wheat germ agglu-
tinin (WGA) (Hewitt and Elmer, 1978). Results obtained from
using FITC lectins revealed that two populations of cells
could be distinguished on the basis of their pattern of
distribution of the lectin binding sites (Hewitt and Elmer,
1978). Cells from 11-day stage limbs exhibited a clustering pattern. This population decreased with a concomitant
increase in a population of cells which maintained an even
distribution of binding sites as development progressed
up to the 13-day stage. In the mutant, a change in the
population of cells from a clustering to an even distribution of their binding sites was delayed until after
the 12-day stage for con A and WGA. In addition trypsinization of the normal and mutant cells restored the

clustering pattern for both plant lectins at the 12-day stage, but failed to significantly affect the redistribution of the WGA binding sites for 13-day normal cells (Hewitt and Elmer, 1978).

From these studies, it was of interest to ascertain whether differences in WGA receptors could be observed during the 3-day gestation period for the two genotypically different cells. The electrophorectically separated components were transferred from the acrylamide gels to nitrocellulose sheets using a Bio-Rad Trans-Blot Cell containing a buffer system composed of 25 mM Tris-HCl (pH 8.3), 192 mM glycine, and 20% methanol. Transfers were accomplished by electrophoresing at 50 V for 4 hours. Detection of the WGA-binding glycoproteins was carried out according to the method of Glass et al. (1981). Nitrocellulose sheets were treated for 1 hour with 3% BSA in phosphate buffered saline to neutralize areas not containing bands. The sheets were then overlaid with 20 μg/ml of peroxidase conjugated WGA (Cal. Med. Chem. Corp.) in BSA-PBS buffer and agitated at R.T. for 1 hour. Controls contained 0.2 M of the inhibitory sugar N-acetyl-glucosamine in all of the solutions. After five washes in PBS, the WGA-peroxidase complexes were localized on the sheets by incubation at R.T. in a solution of 0.05 M Tris-HCl (pH 7.5), 0.3 mg/ml 3,3'-diaminobenzidine, and 0.005% hydrogen peroxide (Glass et al., 1981).

The staining pattern of the peroxidase conjugated WGA shows the presence of several N-acetylglucosamine containing glycoproteins for both genotypes (Fig. 3). The bands below 66K that stained positive should not be considered since they also stained in the controls treated with the inhibitory sugar N-acetylglucosamine. Apparently with this technique nonspecific binding of WGA to proteins can occur (Burridge and Jordan, 1979; Glass et al., 1981). All of the other bands, however, did not appear in the controls. Of these, the three that change the most significantly are those with molecular weight of 135 K, 120 K, and 90 K. The 135 K glycoprotein appears in both genotypes at the 11- and 12-day stages, but is absent in 13-day brachypod membranes. The other two glycoproteins, 120 K and 90 K, are only observed in the 13-day normal membranes (Fig. 3). These data show that differences in N-acetylglucosamine containing glycoproteins occur during limb bud chondrogenesis and that these differences are not expressed in the mutant according to the normal developmental timetable.

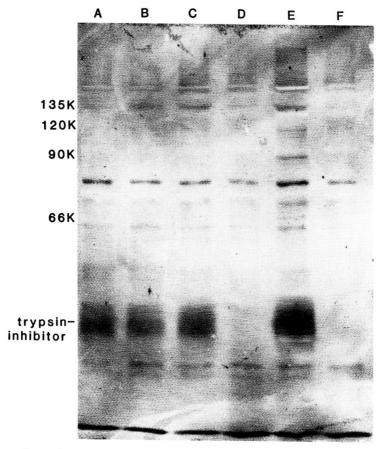

Fig. 3. Nitrocellulose sheet containing plasma membrane glycoproteins with WGA-binding activity. Lanes A-F are the same as in Fig. 1. The N-acetylglucosamine containing glyco-proteins that change during chondrogenesis have approximate molecular weights of 135 K, 120 K, and 90 K.

CONCLUSIONS

The aims of this study were three-fold. First, we wanted to discern whether certain plasma membrane components could be identified as being developmentally regulated during limb bud chondrogenesis. Second, we wanted to ascertain which of these are cell surface components. Third, we wished to determine whether the expression of these components is being

altered in the mouse mutant, brachypodism. For the 11- to
13-day gestation period the expression of several plasma
membrane components varies. For example; most of the components are not developmentally regulated while others
exhibit changes that are comparable for both genotypes. On
the other hand, there seems to be a striking alteration in
the expression of some components in the brachypod membranes compared to normal.

Two components that appear to change to the same extent
in both genotypes are the 200 K and 70 K proteins. Both of
these are cell surface proteins that progressively decreased
in their binding of $[^{125}I]$ between the 11- to 13-day stages.
Although these studies did not identify the nature of these
components, the 200 K one is in the molecular weight range of
fibronectin. The gradual disappearance of fibronectin as
chondrogenesis progresses has been reported (Lewis et al.,
1978; Dessau et al., 1980; Silver et al., 1981; Tomasek et
al., 1982). Other cell surface components that were differentially iodinated in the two genotypes have molecular
weights of 170 K, 120 K, 100 K, and 90 K. The lower radioactivity of the 170 K, 120 K, and 90 K proteins in the mutant
may reflect an effect on their synthesis or degradation since
they did not stain as dark as the normal in the SDS gels.
The decreased iodination of the 100 K protein could be due to
a masking effect in mutant. Finally, the absence of the N-
acetylglucosamine containing glycoproteins at the 13-day
stage in the mutant clearly demonstrates a change occurs in
these components between 12 and 13 days. It has also been
shown during this time period that the brachypod hindlimbs
are deficient in matrix material (Duke and Elmer,1979;
Shambaugh and Elmer, 1980). Furthermore, the 90 K glycoprotein is in the molecular weight range of the 80 K
component that was reported to appear in chondrogenic
cultures, but was absent in those treated with vitamin A
(Lewis et al., 1978).

The data obtained from this study corroborate previous
evidence that differences between the plasma membranes of
normal and brachypod limb cells occur during 11 to 13 days of
gestation (Duke and Elmer, 1977, 1978, 1979; Hewitt and
Elmer, 1978). More importantly, the data were obtained from
analyses of plasma membrane preparations which provides
direct evidence that distinct changes are occurring. It is
not known what regulatory mechanisms are affected in the
brachypod cells which produce these alterations, but it is

apparent that one mechanism may be a defective glycosylation of plasma membrane components.

We thank Mrs. Linda Beard for her excellent typing of the manuscript and Mrs. Nancy Rosenthal for her expert technical assistance. This work was supported by grant HD 10945 from the National Institutes of Health to W.A.E.

REFERENCES

Burridge K, Jordan L (1979). The glycoproteins of Dictyostelium discordeum. Changes during development. Exp Cell Res 124:31.

Dessau W, Von Der Mark H, Von Der Mark K, Fischer S (1980). Changes in the patterns of collagens and fibronectin during limb bud chondrogenesis. J Embryol Exp Morphol 57:51.

Duke J, Elmer W A (1977). Effect of the brachypod mutation on cell adhesion and chondrogenesis in aggregates of mouse limb mesenchyme. J. Embryol Exp Morphol 42:209.

Duke J, Elmer W A (1978). Cell adhesion and chondrogenesis in brachypod mouse limb mesenchyme: fragment fusion studies. J. Embryol Exp Morphol 48:161.

Duke J, Elmer W A (1979). Effect of the brachypod mutation on early stages of chondrogenesis in mouse embryonic hind limbs. An ultrastructural analysis. Teratology 19:367.

Ede D A (1976). Cell interactions in vertebrate limb development. In Poste G, Nicholson, G L (eds): "The Cell Surface in Animal Embryogenesis and Development," Amsterdam: North-Holland, p 495.

Elmer W A (1982). Developmental cues in limb bud chondrogenesis. Coll Res 2:257.

Glass W F, Briggs R C, Hnilica L S (1981). Use of lectins for detection of electrophoretically separated glycoproteins transferred onto nitrocellulose sheets. Anal Biochem 115:219.

Grüneberg H, Lee A J (1973). The anatomy and development of brachypodism in the mouse, J. Embryol Exp Morphol 30:119.

Hewitt A T, Elmer W A (1978). Developmental modulation of lectin-binding sites on the surface membranes of normal and brachypod mouse limb mesenchymal cells. Differentiation 10:31.

Laemmli U K (1970). Cleavage of structural proteins during the assembly of the head of bacteriophage T_4. Nature 227: 680.

Lewis C A, Pratt R M, Pennypacker J P, Hassell J R (1978). Inhibition of limb chondrogenesis in vitro by vitamin A: Alterations in cell surface characteristics. Develop Biol 64:31.

Shambaugh J, Elmer W A (1980). Analysis of glycosaminogly-cans during chondrogensis of normal and brachypod mouse limb mesenchyme. J Embryol Exp Morphol 56:225.

Silver M H, Foidart J M, Pratt R M (1981). Distribution of fibronectin and collagen during mouse limb and palate development. Differentiation 18:141.

Tomasek J J, Mazurkiewicz J E, Newman S A (1982). Nonuniform distribution of fibronectin during avian limb development. Develop Biol 90:118.

Wright J T, Elmer W A, Dunlop A (1982). A simple method for plasma membrane isolation from embryonic tissue using positively charged microcarriers. Anal Biochem (In Press).

Limb Development and Regeneration
Part A, pages 365–375
© 1983 Alan R. Liss, Inc., 150 Fifth Avenue, New York, NY 10011

CORRELATIONS BETWEEN NECROTIC PATTERNS AND LIMB
SKELETAL DEFECTS INDUCED BY ANTIMITOTIC DRUGS
IN THE MOUSE

M. Rooze

Laboratory of Human Anatomy and
Embryology. Free University of Brussels.
2, rue Evers
B. 1000 Brussels.

Three waves of physiological cell death succes-
sively affect definite areas of the undifferentiated limb bud mesoderm of the mouse embryo
(Milaire 1976).During the first wave,the mesodermal "foyer preaxial primaire"(f.p.p.)takes place
in the deep marginal mesoderm facing the preaxial
end of the apical ectodermal ridge(a.e.r.)which
itself exhibits degenerative changes. The second
wave affects a large preaxial and a smaller postaxial areas of the subridge mesoderm. The third
necrotic wave affects the mesoderm of the interdigital areas and the postaxial border of digit V.
Antimitotic drugs have been shown to induce
abnormal cell deaths in the mesoderm of developing
limb buds. According to different teratological
conditions or to the developmental stage attained
by the affected limb buds,the dead cells either
appear diffusely scattered throughout large areas
without apparent relationship with the normal
necrotic pattern, or they may be concentrated in
particular sites where they most frequently
contribute to enlarge normal necrotic sites. The
main purpose of the present study is the search
for comprehensive correlations between the
modified necrotic patterns induced by two
antimitotic drugs in the developing mouse limb
buds and the skeletal defects observed in full-
term fetuses submitted to similar teratological
conditions.
The teratogenic effects of cytosine-arabinoside

(Ara-C)on limb development have been extensively
studied in the rat(Chaube et al.,1968;Scott et
al.,1975),mouse(Kochhar et al.,1978;Rooze 1980),
and chick embryos(Karnovsky & Lacon 1966). Those
of N-formylhydroxyaminoacetic acid or hadacidin
(Had.)were mainly investigated in the rat(Chaube
and Murphy 1963; Milaire 1969; Milaire 1971),but
an extensive survey of the skeletal abnormalities
induced by this drug in the limbs of the mouse
was recently performed in our laboratory(Rooze
1982).

I.Material and methods

   Various doses of Ara-C or of Had. were administered intraperitoneally to pregnant mice(day of
vaginal plug = day 0)according to experimental
conditions which proved the most appropriate for
inducing relatively constant skeletal defects
among the embryos of the same litter(Rooze 1980,
Rooze 1982). Additional control animals were
given cytidine-monophosphate(C.M.P)simultaneously
with Ara-C in accordance with the antiteratogenic
experiments reported by Chaube et al.(1968). The
skeletal abnormalities were studied in bulk preparations of 18-day fetuses submitted to Watson's
staining with alcian blue-alizarin red. The
necrotic patterns were demonstrated macroscopi-
cally in the limb buds of treated embryos
collected 6,24,30 and 54 hours after drug administration and extemporaneously submitted to supra-
vital staining with Nile blue sulphate(1/40000 in
Locke's physiolocal solution).

II.Necrotic patterns demonstrated in the limb
   buds of Ara-C treated embryos.

   Table I summarizes the skeletal defects induced
in both fore and hind limbs by a single dose of
Ara-C(5 or 10mg/K)injected on day 9,10 or 11.
In each experimental condition, abnormal cell
deaths appear a few hours after injection, they
reach a maximal amount between 6 and 24 hours and
thence gradually decrease in number until only
the normal necrotic pattern remains present.
The intensity of abnormal necrosis is dose
dependent. The treated limb buds look morphologically delayed during the first 30 hours after

injection; during the following 24 hours, however, they rapidly recover shape and dimensions in accordance with their age. Their shape appears more or less modified according to the teratogenic schedule. The necrotic patterns observed in the different experimental conditions may be summarized as follows.

1) Injection on day 9.

Although Ara-C does not induce any abnormality in the limb skeleton when injected on day 9 at 5mg/K, a large amount of abnormal necrotic cells have been observed in the limb buds of the treated embryos. Six hours after injection they appear scattered throughout the whole mesodermal field, with a very slight distal predominance. The limb bud shape and dimensions remain normal at all stages, which is consistent with the occurrence of complete and efficient regulative activities. Injected on day 9 at 10mg/K, Ara-C induces skeletal limb defects in 10% of the hind limbs (table I). An increased amount of dead cells was found in the treated limb buds, with the same diffuse distribution; rapid regulation similarly occurs in all 10mg-treated limb buds,so that no distinction could be made between potentially normal and abnormal rudiments.

2) Injection on day 10.

As seen in table I, day 10 is a critical period for the teratogenic influence of Ara-C on limb morphogenesis; the observed abnormal necrotic patterns were, however, found quite similar after the administration of 5 and 10mg/K. In addition to numerous abnormal dead cells scattered throughout the limb mesoderm and ectoderm,a more condensed crescent shaped necrotic site,larger in the 10mg/K treated limb buds,was found in the marginal mesoderm. Another deeper mesodermal necrotic site appears in the distal part of the zeugopod. It provides an explanation for the observed zeugopodal defects. After 30 hours the necrotic area of the autopod has spread out and necrotic-free mesoderm has appeared between it and the a.e.r. Hyperplasia of the a.e.r. was observed mainly on its preaxial side.

After 54 hours, the limbs look deformed with a
shorter zeugopod and a reduced autopod. Some of
them exhibit preaxial overgrowth in the autopod,
which will most probably result in polydactyly
and/or hyperphalangy of digit I.

   3)Injection on day 11.

   Ara-C does not affect the devlopment of the
limb skeleton when injected on day 11 at 5mg/K.
At 10mg/K, however, skeletal defects are induced
in both hands and feet, as well as in the zeugopod of the hind limbs. The abnormal necrotic
patterns observed with either dose are equally
interesting and will therefore be described
separately.
a)Hindlimbs : six hours after injection, a crescent like necrotic site is located in the
postaxial area of the apical mesoderm of the
10mg/K treated limb buds. Transverse sections
through this area reveals a dorsal extension of
the necrotic fields. The normal necrosis affecting
the postaxial portion of the a.e.r. is reduced.
24h. after treatment, the hindlimb buds look
morphologically delayed and abnormally shaped
along their postaxial border. A zone of translu-
cent and unstained tissue has appeared in the
postaxial area of the footplate mesoderm. The
a.e.r. lacks its postaxial necrotic site and its
preaxial end is hyperplastic. After 30h. a sub-
ectodermal bleb has formed in the postaxial clear
area while both mesodermal necrosis and ectodermal
hyperplasia are reduced preaxially. A few hindlimb
buds show a similar bleb on the preaxial side of
the autopod. Between 30 and 54h. after treatment,
the postaxial bleb looks enlarged in some cases
and is no more visible in others. Simultaneously
abnormal necroses decrease in number and both the
zeugopod and autopod segments are progressively
deformed and shortened until the whole limb is
reduced to a scarred stump(fig.1 and 2).
   The 5mg/K treated hindlimb buds show a similar
modified necrotic pattern but the amount of dead
cells present in the marginal mesoderm is lower
and the necrotic field does not extend so far
preaxially and dorsally and the f.p.p. is usually
present. There is no bleb. A local hyperplasia is

sometimes visible in the preaxial portion of the
a.e.r. After 54h. the limb buds recover a normal
appearance.
b)Forelimbs : two mesodermal necrotic sites
located on either side of digit IV were found in
the 5mg/K treated forelimb buds. 6h.after
injection a large one preaxially and a smaller
one postaxially, digit IV itself being selectively preserved. After 24h. the buds appear
slightly delayed and their necrotic areas are
reduced. The normal necrotic pattern is present
in the normally shaped forelimbs 48h. after
injection. The same two necrotic sites described
above were found more important and continuous
accross the IVth digital area in the 10mg/K
treated forelimb buds collected after 6h. after
treatment. After 24h. the amount of abnormal dead
cells is slightly reduced in the marginal mesoderm and the preaxial portion of the a.e.r. may
sometimes be hyperplastic. Pre-and postaxial
blebs appear 30h. after treatment. They are no
longer present after 54h. and the preskeletal
rudiment of digit IV is in most cases the sole
recognizable part of the deformed residual
autopod.

III.Necrotic patterns in Ara-C + CMP treated
limb buds.

A normal skeletal pattern develops after simul-
taneous administration of Ara-C and CMP, provided
the dose of CMP is 10 to 30 times higher than
that of Ara-C. The limb buds of embryos submitted
on day 11 to 10mg/K Ara-C and 100mg/K CMP exhibit
the same abnormal necrotic patterns as those
submitted to Ara-C alone, but the amount of dead
cells is significantly lower after 6 hours and
the necrotic area of the marginal mesoderm does
not extend so far dorsally. Somewhat reduced in
size after 24 hours, the treated limb buds have
recovered normal shape and dimensions after
30 hours and their necrotic pattern is almost
similar to the normal pattern. No hyperplasia
occurs in the a.e.r. and no bleb can be observed.
It is thus evident that CMP prevents Ara-C to
induce abnormal necrosis above the permissive
level.

IV. Necrotic patterns in Had treated limb buds.

Injection of 2gr/K Had. in two successive half
doses on day 10, induces preaxial polydactyly of
the hindlimbs in 92% of the cases and hyperpha-
langia of the first toe in 8% of the cases. The
proximal hind limb skeleton is not affected and
the effects on the fore limbs have not been
studied as yet. Contrary to what was observed
after Ara-C administration, the abnormal necrotic
cells induced by Had are disseminated in the
whole mesoderm of the hindlimbs collected 6h.
after the last injection. 18h. later, the dead
cells remain diffusely scattered in the limb bud
mesoderm though they seem to be concentrated
dorsally. A thin layer of subridge mesoderm looks
selectively healtly. The a.e.r. is hyperplastic
on both its postaxial and preaxial ends but the
hyperplasia is larger and festooned and the
underlying preaxial mesoderm is free of dead cells,
the normal f.p.p. being even absent. 54h after
treatment, an excessive amount of mesoderm
contributes to enlarge the preaxial border of the
footplate. Necrotic cells are still disseminated
in the autopod area. They will disappear after
72h.

V. Discussion.

The present results have confirmed that both
Ara-C and Had exert their deleturious effect upon
limb morphogenesis by inducing abnormal cell
death within the undifferentiated limb bud mesoderm. The fact that abnormal dead cells remain
detectable during a longer period after Had
treatment than after Ara-C exposure suggests that
some additional injury is induced by Had, such as
for example a transient inhibition of the
phagocytic properties in the residual mesoderm.
The overall distribution of abnormal dead cells
observed after Had treatment and after Ara-C
treatment on day 9 indicates that the limb bud
mesoderm does not manifest any particular regional sensitivity to neither drugs. As compared to
the mitotic patterns previously demonstrated in
the chick embryo (Janners and Searls 1970),the
random distribution of abnormal dead cells rather
suggests that the actively dividing cells might

represent the most probable target of both drugs. This idea is strongly supported by the observations made after Ara-C treatment at day 10 and still more so at day 11. The concentration of dead cells in a marginal necrotic site which practically corresponds to the so-called "progress zone" described by Stark and Searls (1973) is indicative of the relative increase in cell proliferation which occurs in this area during the period of active growth of the footplate mesoderm. It must be pointed out that a similar marginal necrotic site can also be obtained with increased doses of Had, the action of which being thus not significantly different from that of Ara-C in this respect.

The changes observed 6,24 and 30 hours after each teratogenic treatments strongly suggest that the final skeletal pattern observed in full term fetuses is the result of a compromise between the number of mesodermal dead cells and the regulatory capacities endowed by the remaining healthy mesoderm. With 5mg/K Ara-C, the latter capacities are sufficient to compensate the meso-dermal damage and the limb buds rapidly recover shape and dimensions compatible with normal skeletogenesis. With the 10mg/K dosage,the amount of dead cells exceeds a threshold value over which regulatory properties are no longer capable to restore normal morphogenesis. The severity of skeletal deficiencies then increases with increasing amount of necrotic mesoderm,and particularly in the marginal area which is the main source of replacement undifferentiated cells. The progressive proximal shift of the marginal necrotic site occuring simultaneously with the appearance of a distal layer of new and healthy subridge mesoderm are very clear indications of the regulatory growth which occurs in this particular area in both teratogenic conditions studied. The antiteratogenic experiments have shown that a sufficient dose of C.M.P. is able to prevent a large number of cells from death and to allow the genesis of a quite normal skeletal pattern.

The significance of two particular changes observed in the treated limb buds remains unclear and certainly require further investigations.

The first one concerns the formation of blebs.
As they usually appear during limited periods of
time in the most severely affected areas of the
limb bud mesoderm, they are most probably involved in a process of rapid elimination of necrotic
tissue. However, similar blebs were never
observed in the normal necrotic sites even in the
heaviest interdigital zones; their formation in the
treated limb buds probably result from some
unknown inhibition of the normal processes involved in elimination of dead cells(macrophage
inhibition,circulatory slackening,...).

The second unexplained change is the hyperplasia
of the a.e.r. which was often observed associated
with a high amount of abnormal dead cells in the
footplate mesoderm in both teratological conditions.
Refering to the experiments performed by Amprino
and Ambrosi (1973) in the chick embryo, the a.e.r.
hyperplasia is most probably the result of
unbalanced growth between the limb bud ectoderm
and mesoderm. Probably less severely affected by
the teratogenic agent than the mesoderm,the
ectoderm continues to grow at a normal rate and
thus provides a normal amount of migratory cells
which contribute to increase the apical
thickening. In normal conditions, the increasing
thickness of the a.e.r. is compensated by the
progressive enlargement of the growing mesoderm.
Inhibited mesodermal growth thus results in
increasing thickness of the a.e.r. which remains
demonstrable until the mesoderm has recovered
normal proliferating activities. The question is
raised,however,whether the hyperplastic ectoderm
does play a role or not in the stimulation of
regulatory growth exhibited by the mesoderm.
Although this problem requires further experimental investigations which presently remain
difficult to apply to mammalian embryos,several
teratological observations support an affirmative
answer to this question(Scott et al.1975;Milaire
1978). Some results obtained after Ara-C treatment
and most of those obtained after Had treatment
support the idea that the hyperplasia of the
a.e.r. might be somehow involved in the excessive
regulatory phenomena responsable for the genesis
of preaxial polydactyly and/or hyperphalangy.

Table I — Skeletal defects induced by AraC in the developing mouse limbs.

	Day of administration		9	9	10	10	11	11
	Dose (mg/K)		5	10	5	10	5	10
FORELIMB	Whole limb	Normal(%)	100	90	31	21	100	-
		Abnormal(%)	-	10	69	79	-	100
	Stylo.	Normal	100	100	66	21	100	98
		Reduced			34	71		2
		Absent				8		
	Zeugo.	Normal	100	100	59	21	100	75
		Reduced			41	75		25
		Absent				4		
	Auto.	Normal	100	90	60	56	100	93
		Reduced		10	40	38		7
		Absent						
		Excess				6		
HINDLIMB	Whole limb	Normal(%)	100	30	14	4	100	-
		Abnormal(%)	-	70	86	96	-	100
	Stylo.	Normal	100	52	59	-	100	51
		Reduced		32	3	28		46
		Absent		16	38	72		3
	Zeugo.	Normal	100	40	55	-	100	36
		Reduced		50	30	61		53
		Absent		10	15	39		11
	Auto.	Normal	100	46	24	4	100	39
		Reduced		27	43	74		61
		Absent		-				
		Excess		27	33	22		
NUMBER OF FETUSES			34	30	29	24	35	44

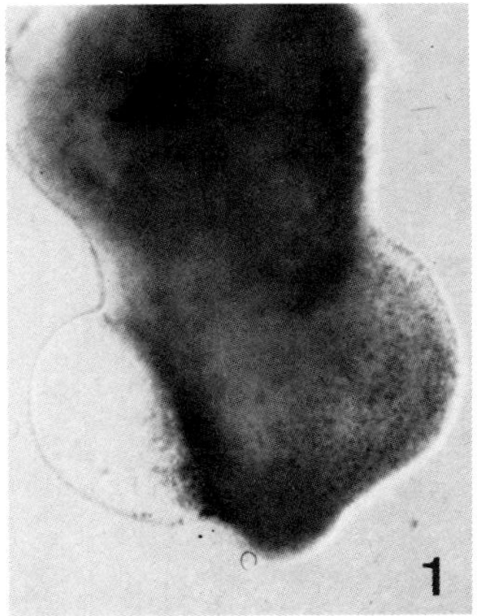

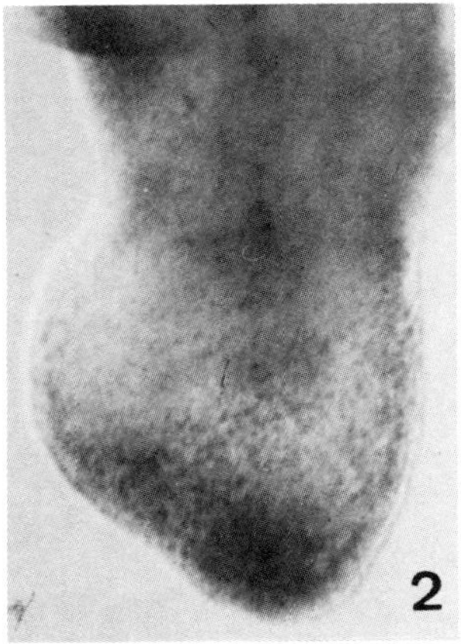

Right hinlimb buds Ara-C treated on day 11(10mg/K)

Fig.1 : dorsal view—30h after treatment : note the postaxial bleb.

Fig.2 : plantar view—54h after treatment.

Bibliography

Amprimo R, Ambrosi G (1973) Experimental analysis
    of the chick embryo limb bud growth. Arch.Biol.
    (Bruxelles) 84:35.
Chaube S, Murphy ML (1963) Teratogenic effect of
    hadacidin(a new growth inhibitory chemical) on
    the rat fetus. J.Exp.Zool. 152:67.
Chaube S, Kreis K, Uchida, Murphy ML (1968)
    The teratogenic effect of 1-B-D-arabinofusano-
    sylcytosine in the rat. Protection by deoxycy-
    tidine. Biochem.Pharm. 17:1213.
Janners MY, Searls RL (1970) Changes in rate of
    cellular proliferation during the differencia-
    tion of cartilage and muscle in the mesenchyme
    of the embryonic chick wing.Develop.Biol.23:136.
Karnovsky DA, Lacon CR (1966) The effects of
    1-B-D-arabinofuranosylcytosine on the developing
    chick embryo. Biochem.Pharm. 15:1435.
Kochhar DM, Penner JD, McDay JA (1978) Limb
    development in mouse embryos. II.Reduction
    defects, cytotoxicity and inhibition of DNA
    synthesis produced by cytosine arabinoside.
    Teratology 18:71.
Milaire J (1969) Etude morphogénétique de la
    syndactylie postaxiale provoquée chez le rat
    par l'hadacidine. I.Analyse des anomalies chez
    l'adulte,le foetus à terme et les embryons de
    15,16 et 17 jours. Arch.Biol.(Liège) 80 : 167.
Milaire J (1971) Etude morphogénétique de la
    syndactylie postaxiale provoquée chez le rat
    par l'hadacidine. II.Les bourgeons de membres
    chez les embryons de 12 à 14 jours. Arch.Biol.
    (Liège) 82:253.
Milaire J (1976) Rudimentation digitale au cours
    du développement normal de l'autopode chez les
    Mammifères. Colloques internationaux C.N.R.S.
    266:221.
Milaire J (1978) Approches morphologiques,
    histochimiques et expérimentales de la genèse
    des malformations des membres. Bull.Acad.Méd.
    Belg. 133:402.
Rooze M (1980) The effect of cytosine-arabinoside
    on limb morphogenesis in the mouse.in Merker HJ
    Nau H,Neubert D(Eds) Teratology of the limbs.
    W.de Gruyter & Co. Berlin,New York. p.355.

Rooze M. Action tératogène de l'hadacidine sur le développement des membres postérieurs chez la souris. Arch.Biol.(Bruxelles): in preparation.

Scott WS, Ritter ES, Wilson JG (1975) Studies on induction of polydactyly in rats with cytosine arabinoside. Devel.Biol. 45:103.

Stark RS, Searls RL (1973) A description of chick wing bud development and a model of limb morphogenesis. Devel.Biol. 33:138.

Limb Development and Regeneration
Part A, pages 377–385
© **1983 Alan R. Liss, Inc., 150 Fifth Avenue, New York, NY 10011**

BIOACTIVATION OF THALIDOMIDE BY A MONKEY LIVER FRACTION IN
A RAT LIMB CULTURE SYSTEM*

Thomas H. Shepard, M.D.[*] and Kohei Shiota, M.D.[**]

Central Laboratory for Human Embryology
University of Washington School of Medicine RD-20
Seattle, Washington 98195

The impact of the thalidomide epidemic in humans needs
no repeating. Thalidomide is selectively teratogenic in
humans, monkeys and rabbits but not in the rat (Cahen 1966;
Fabro 1981; Scott et al 1977). Variation in the sensitivity
of different species to teratogens is one of the basic principles of teratology, and without a better understanding of
the underlying molecular mechanisms the extrapolation of
animal experimental findings to human risk are very difficult.
Thalidomide may offer a good study model to this problem.

Schumacher et al (1968) observed that rabbit liver homogenates enhanced the rate of disappearance of thalidomide
from incubation mixtures whereas rat liver homogenates did
not. In addition, more thalidomide metabolites were bound
to liver macromolecules in the rabbit than in the rat. Their
studies suggested to Gordon et al (1981) that thalidomide or
one of its metabolites might undergo oxidative metabolism
via an aryl oxide intermediate and that this might be the
ultimate teratogen. This form of bioactivation is common
in carcinogenesis and mutagenesis as well as teratogenesis
(Juchau 1981; Fantel et al 1979). Gordon et al (1981) using
an assay composed of human lymphocytes were able to show that
a liver monooxygenate systems from human, monkey and rabbit
enhanced thalidomide toxicity but rat liver did not. Their
work stimulated this present study of bioactivation of thalidomide in an embryonic system. Since the underlying morpho-
logic alteration found in the embryo is thought to be failure
of condensation of the precartilaginous skeleton (Vickers
1967), it seemed logical to study early limb buds in culture.

METHODS

Sprague-Dawley rats were obtained from a local vendor
(Tyler Laboratories, Bellevue, Wa.). The morning following
copulation was termed the start of day zero of pregnancy.
After maternal sacrifice with ethyl ether embryos were carefully examined and staged by measuring crown-rump length as
well as the outline of the limb buds. The forelimb buds
were dissected away from embryos on days 12, 13 or 14 using
watchmaker's forceps and dissecting needles. The limb buds
were mounted on lens paper supported by wire grids at the
surface of the medium (Moscona et al 1965).

The liver enzyme sources were from pregnant Sprague-
Dawley rats or three year old male Macaca nemestrina. None
of the animals was pretreated with an inducing agent. The
method of preparation was similar in each species and has
been described (Fantel et al 1981). The supernatant of the
homogenates (S-9) obtained after centrifugation at 9,000Xg
contained components of the mixed-function monooxygenase
system. It was stored under liquid nitrogen. After thawing
the preparation was diluted in Hanks balanced salt solution
before addition to the culture plates. The protein contents
of the S-9 for the rat and monkey were 88 and 104 mg per ml
respectively.

Culture medium consisted of BJG_b medium (Gibco Laboratories) with 0.8 ml fetal calf serum and 0.1 ml of antibiotics (penicillin, streptomycin and mycostatin at concentrations
of 100 units, 50 µg and 10 units per ml of medium respectively). Cofactors were NADPH and glucose-6-phosphate to give
a final concentration of 0.1 mM for both. The thalidomide
was obtained from Grünenthal Company, West Germany and had
a melting point of 271-273°C. The thalidomide was dissolved
in redistilled dimethyl sulfoxide just prior to addition to
medium. The final concentration was 70 µg per ml. The total
volume of the medium was adjusted to 3.5 ml per petri dish.
The medium was changed every two or three days and the limb
buds harvested on the 7th day. Incubation was carried out
at 38°C in an incubator with 5% CO_2 in air. For some experiments 300 µl of the reaction mixtures were added to a
dialysis bag containing 100 µl of monkey S-9. The dialysis
bag measured 2.55 mm in diameter and had a 12-14,000 molecular weight cutoff. Concentrations of monkey S-9 were 0.6,
0.3 or 0.03 µl and for rat S-9 3.0 or 0.3 µl per ml of
medium.

The growth of limb buds was assessed by projecting and tracing their outline on paper and measuring the surface area with a planimeter. (Keuffel & Esser Compensating Polar Planimeter). Protein content was measured after sonication using the method of Bradford (1976). The limbs were fixed in Bouin's and stained by alcian blue (Blankenburg, 1982). Statistical analysis was performed using student's t test (Edwards 1955).

RESULTS

The results given in Figure 1 and Table 1 demonstrate that the addition of S-9 prepared from monkey liver reduces the growth of rat limb buds in culture. The complete system, that is, with thalidomide, S-9 and cofactors markedly reduced growth. The S-9 with thalidomide but without cofactors retarded growth to some extent as did S-9 with cofactors but without thalidomide. Concentrations of cofactors higher than o.1 mM were toxic as were amounts of monkey S-9 over 0.3 µl per ml but neither thalidomide nor cofactors alone in concentrations used inhibited growth. The findings were repeated in four experiments which used either 0.03 µl or 0.3 µl of S-9 per ml of medium. The addition of 100 µl of S-9 to the inside of a dialysis bag was not associated with toxicity to the limbs and, with the complete system, significant retardation of limb growth was measured.

The findings with rat S-9 are illustrated in Table 2. Limb buds in the media containing S-9 were smaller than the controls but neither thalidomide nor thalidomide with cofactors for the monooxygenase system enhanced toxicity. Concentration of S-9 used was 0.3 µl per ml of medium.

A very close correlation between the traced surface area and the protein content of limbs was found. This is illustrated in Figure 1. The variability of growth in the different experimental groups is shown. The surface area of upper limb buds on day 12 was 0.5-0.6 mm^2 and between 2-3 mm^2 after six days in vitro. The protein content increased about four-fold.

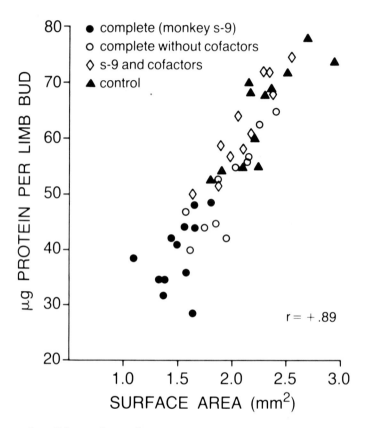

Figure 1. Plot of surface areas versus protein content of day 12 rat limb buds grown in vitro for six days. The complete medium (●) contained monkey S-9 (0.03 µl per ml); NADPH and glucose-6-phosphate (0.1 mM); and thalidomide (70 µg per ml). The other culture conditions were thalidomide and S-9 without cofactors (o), S-9 and cofactors without thalidomide (◇) and medium alone (▲). Correlation coeficient was +0.89.

Table 1. Growth of Rat Limb Buds in the Presence of
Monkey S-9

	A	B	C	D
S-9	+	+	+	−
Cofactors	+	−	+	−
Thalidomide	+	+	−	−
Surface area % of control	66.2*	86.6	92.8	100
S.D.	12.0	12.7	11.6	14.1
n	12	12	13	12
Protein % of control	61.0*	81.8	94.8	100
S.D.	16.0	13.4	13.4	14.1
n	12	13	13	12

* Significantly less than corresponding figure in Column B, C and D (p<0.01). B did not differ significantly from C nor was C significantly less than D. B was less than D at borderline significance, p<0.05.
Medium: BJG_b 2.6 ml, fetal calf serum 0.8 ml, antibiotics 0.1 ml.
Monkey S-9 0.03 µl per ml of medium. Day 12 limb buds grown for 6 days.
Cofactors glucose-6-phosphate (0.1 mM) and NADPH (0.1 mM)
Thalidomide 70 µg per ml of medium.

Table 2. Growth of Rat Limb Buds in the Presence of Rat S-9

	A	B	C	D
S-9	+	+	+	−
Cofactors	+	−	+	−
Thalidomide	+	+	−	−
Surface area % of control	87.3	85.0	77.5*	100
S.D.	10.5	9.7	10.9	17.6
n	10	10	10	10
Protein % of control	80.2**	76.2	61.9	100
S.D.	17.9	30.9	21.0	17.5
n	10	10	10	10

Conditions the same as in Table 1 except the final
concentration of S-9 was 0.3 μl per ml of medium.
* Surface area was significantly reduced from control D
(p<0.01).
** Values for column A, B and C were significantly less
than control D.

DISCUSSION

These preliminary results suggest that the mechanism
of action of thalidomide on embryonic tissue requires metabolic activation via the mixed function monooxygenase system.
Gordon et al, 1981 have previously shown that thalidomide
toxicity in human lymphocytes can be enhanced by adding
liver microsomes from species that are sensitive to thalidomide teratogenicity. Other attempts to study thalidomide
teratogenicity in the embryo have been made (Lash & Saxen
1972; Klein et al 1982 and Neubert 1982). Lash and Saxen
(1972) showed in organ culture that thalidomide inhibited
the activation of human limb chondrogenesis by human
embryonic mesonephros. Klein et al (1981) added serum from
monkeys receiving thalidomide to rat embryo culture and
found variable effects including lethality and some malformations (but no limb defects). In organ culture studies of
mouse limb buds after maternal pretreatment with monooxygenase
stimulators Neubert (1981) has presented preliminary studies
indicating that a thalidomide analog, 3-(1,3-dihydra-1-oxo-
2H-Isoindol-2-yl-2.6-dioxopeperidine (EM-12), inhibited growth.

Studies reported have suggested further experiments are needed to determine the true nature of the bioactivation. Inhibitors of the mixed function monooxygenase system such as carbon monoxide and metapyrone should be tested. Epoxide hydrolase (EC 3,3,2,3) an enzyme which hydrolyzes epoxides would be expected to decrease the thalidomide toxicity while two inhibitors of the enzyme reaction, cyclohexene oxide and 1,2-eopxy-3,3,3-trichloropropane should enhance toxicity. Teratogenic and non-teratogenic analogs of thalidomide should be tested.

Studies using rabbit and human liver sources are in progress and are needed to help prove the species specificity of the reaction system.

There are several problems with this culture system. Considerable variability of the surface area and protein content of cultured limbs grown for six days was present even when taken from the same litter. The variability may be due to interlitter limb bud size differences, or differences in surgical removal or nutrition in culture. The S-9 preparations were found to be toxic in concentrations over 0.3 μl per ml of medium. The use of more highly purified enzyme preparations or other dialysis bag systems are being tried. Further drawbacks to this assay are the time and care of explant preparation and six day culture period.

It is of significance that although limb defects cannot be produced in vivo in the rat fetus by thalidomide, limb buds are affected in vitro when a liver enzyme system from a thalidomide sensitive animal (monkey) is present in the system. One could hypothesize that the embryo is at the mercy of maternal metabolites of the agent. The ability to interchange metabolizing systems from different species may be useful to extrapolate teratologic activity between species and in particular between small laboratory animals and humans.

SUMMARY

The effect of thalidomide on a rat limb bud system has been studied in vitro. Thalidomide and adult monkey liver supernatants activated with cofactors for the mixed function monooxygenase system inhibited growth. Similar preparations from the rat did not inhibit growth.

Planimetric measures of the surface area of limb buds were found to correlate closely with their protein contents. The ability to compare metabolizing systems from different mammals may be useful in extrapolating teratologic data between different species.

Cahen RL (1966). Experimental and clinical chemoteratogenesis. Adv Pharmacol 4:263-349.

Fabro S (1981). Biochemical basis of thalidomide teratogenicity. In Juchau MR (ed): "The Biochemical Basis of Chemical Teratogenesis," New York: Elsevier/North-Holland, p 159-178.

Scott WJ, Fradkin R, and Wilson JG (1977). Non-confirmation of thalidomide induced teratogenesis in rats and mice. Teratology 16:333-336.

Schumacher HJ, Wilson JG, Terapane JF and Rosedale SL (1970). Thalidomide: disposition in rhesus monkey and studies of its hydrolysis in tissues of this and other species. J Pharmacol Exp Ther 173:265-269.

Gordon GB, Spielberg SP, Blake DA and Balasubramanian V (1981). Thalidomide teratogenesis: evidence for a toxic arene oxide metabolite. Proc Natl Acad Sci 78:2545-2548.

Juchau MR (1981). "The Biochemical Basis of Chemical Teratogenesis," New York: Elsevier/North-Holland.

Fantel AG, Greenaway JC, Juchau MR and Shepard TH (1979). Teratogenic bioactivation of cyclophosphamide in vitro. Life Sci 25:67-72.

Vickers TH (1967). Concerning the morphogenesis of thalidomide dysmelia in rabbits. The Br J of Exp Path 48:580-591.

Moscona A, Trowell OA and Willmer EN (1965). "Cells and Tissues in Culture," Willmer EN (ed) London and New York: Academic Press p 77-87.

Fantel AG, Greenaway JC, Shepard TH, Juchau MR and Selleck SB (1981). The teratogenicity of cytochalasin D and its inhibition by drug metabolism. Teratology 23:223-231.

Bradford M (1976). A rapid and sensitive method for the quantitation of microgram quantities of protein utilizing the principle of protein-dye binding. Anal Biochem 72: 248-254.

Blankenburg G (1981). Some methods for culturing embryonic tissues. In Neubert D and Merker HJ (eds): "Culture Techniques," Berlin and New York: Walter de Gruyter, p 590.

Edwards AL (1955). Statistical Methods for the Behavorial Sciences, New York: Rinehart.

Lash JW and Saxen L (1972). Human teratogenesis: in vitro studies on thalidomide-inhibited chondrogenesis. Dev Biol 28:(1)61-70.

Klein NW, Plenefisch JD, Carey SW, Chatot CL and Clapper L (1981). Evaluation of serum teratogenic activity using rat embryo cultures. In Neubert D and Merker HJ (eds): "Culture Techniques," Berlin and New York: Walter de Gruyter p 67-74.

Neubert D and Bluth U (1981). Limb bud organ cultures from mouse embryos after apparent induction of monooxygenase in utero: Effects of cyclophosphamide, dimethylnitrosamine and some thalidomide derivatives. In Neubert D and Merker HJ (eds): "Culture Techniques," Berlin and New York: Walter de Gruyter p 175-195.

*Supported by NIH Grants HD00836, HD12717 and HD02843
**PRESENT ADDRESS: Kohei Shiota, M.D.
                 Congenital Anomaly Research Center
                 School of Medicine
                 Kyoto University
                 Sakyo-ku, Kyoto 606, Japan

Limb Development and Regeneration
Part A, pages 387–397
© **1983 Alan R. Liss, Inc., 150 Fifth Avenue, New York, NY 10011**

EFFECT OF THALIDOMIDE-DERIVATIVES ON LIMB
DEVELOPMENT IN CULTURE

Diether Neubert and Ralf Krowke

Institut für Toxikologie und Embryopharmakologie
der Freien Universität Berlin,
Garystr. 9, D-1000 Berlin 33

Introduction:

The mechanism of the teratogenic action of the most noto-
rious human teratogen - thalidomide - is still rather obscure. There are 3 questions which especially require
answering:

1. Is it the original substance itself which is responsi-
    ble for the teratogenic effect, or is it a hydrolysis
    product or an active metabolite?

2. What is the reason for the pronounced "species-spe-
    cificity" (i.e. rat or mouse vs. primates) of this te-
    ratogen?

3. What is the reason for the pronounced "phase-specifi-
    city" during organogenesis and the pattern of abnor-
    malities induced in primates?

One of the major drawbacks in revealing the mode of the
teratogenic action of thalidomide is the lack of a conveni-
ent model for studying embryotoxic effects. The most widely used animal species - rats and mice - do not respond
properly to the action of this embryotoxic substance.

Within the last years we have established culture systems - "whole-embryo" cultures and predominantly limb
bud cultures - which permit a closer study of the mode of
embryotoxic actions of teratogens. These tests are now
routinely used in our laboratory and more than 50,000
limb buds have been cultured.

One of the main disadvantages of such systems (using
limb buds from mouse or rat embryos) is, that they - as

may be expected - do not respond to thalidomide.

Since our general interest was to study the possibility of drug activation in embryonic tissues (cf. NEUBERT and TAPKEN, 1978), we also attempted to add a drug metabolizing capacity to our culture system. All of these approaches have not been very satisfactory - even those attempting the addition of a purified cytochrome-P-$_{450}$ system (reconstituted system) to the cultures. We have, therefore, tried another approach - that of charging the explants to be cultured with a drug-activating capacity - by inducing drug-metabolizing enzymes in the explants. Such studies have now been performed for more than a year and we have obtained results which appear promising (NEUBERT and BLUTH, 1981).

Here, we present some data on the effect of thalidomide and some thalidomide derivatives on in vitro development of limb buds from mouse embryos and on some additional studies aimed at an elucidation of biochemical properties of thalidomide.

Experimental Conditions

Pregnant NMRI mice were pretreated with various substances known to be inducers of xenobiotic-metabolizing enzymes. The inducing agent was given orally on three consecutive days. The animals were sacrificed 24 hours after the last dose on days 10, 11 or 12 of pregnancy.

Limb buds were cultured using a chemically defined medium and a suspension culture technique (cf. BLANKEN-BURG, 1981). About 15 limb buds were cultured in a "rotating bottle" system; 10 to 12 cultures were initiated in one experimental series per day. The limbs were allowed to differentiate in culture for 3 or 6 days. Thalidomide or its derivatives were dissolved in the culture medium without the addition of solubilizers or solvents. For evaluation the explants were fixed and the developed cartilage was stained with alcian blue.

For studies on the binding of thalidomide on cellular components, either a microsomal fraction of mouse, rat or marmoset liver (either normal or after pretreatment with an "inducer") or a homogenate of whole 11-day-old mouse embryos (either normal or after pretreatment with rifampicin) were used. The homogenates or cell fractions were incubated with 45 µg/ml ^{14}C-thalidomide (sp. act. 5.68

µCi/mmol) or 100 µg/ml ^{14}C-EM-12 (sp. act. 2.44 µCi/ mmol) in the presence of NADP (0.45 mM), glucose-6-phosphate (4 mM), glucose-6-phosphate dehydrogenase (0.5 µ/ml), 4 mM nicotinamide, 50 mM $KH_2 PO_4$, 7.5 mM $MgCl_2$ and 50 mM Tris-HCl pH 7.4 for 15 to 30 min at 37°C. Final concentration: 300 to 500 µg microsomal protein/ml. The reaction was stopped by the addition of 1 ml of dioxane to the 1 ml incubation sample and the sediment formed was washed 8 times with dioxane. This was followed by solubilization in TS1 (tissue solubilizer, RPI-Corp.), the radioactivity was counted in a liquid scintilation counter (LKB 1215 Rack-β-II) and the protein content of the samples was determined (briuret-method).

Experimental conditions for measuring prolyl hydroxylase are given in the legend to Table 2.

Results and Discussion

Three kinds of experiments were performed: (1) studies with limb bud cultures, (2) binding studies with ^{14}C-thalidomide and some thalidomide derivatives, and (3) studies on prolyl hydroxylase of mouse embryos.

1.   Studies with limb bud cultures

The mouse limb buds were either dissected from normal 10-, 11- or 12-day-old mouse embryos or from embryos of mice pretreated with inducers of monooxygenases.

Within the last years, we have cultured hundreds of limb buds from 10.5- to 12-day-old mouse embryos in the presence of thalidomide or its teratogenic derivative EM-12 under various experimental conditions, without seeing any effect on differentiation in culture. Abnormal development could be induced in vitro under certain experimental conditions with some of the hydrolysis products of thalidomide (NEUBERT et al., 1978), at high concentrations. We have also performed some orienting studies culturing the limb buds of about 60-day-old marmoset embryos. This stage roughly corresponds to 11- to 12-day-old mouse embryos. Differentiation in culture may be observed but it is much less satisfactory than that obtained with mouse embryos. Again, we never obtained any interference with development in culture in the presence of 100 µg/ml EM-12 or 50 µg/ml thalidomide.

Some clues have become available which indicate that thalidomide might require metabolic activation in order to exert its teratogenic effect (GORDON et al., 1981; NEUBERT and BLUTH, 1981). We attempted to increase the capacity for metabolic conversion of certain xenobiotics in the explants used for culturing. Two approaches appear feasible: (1) the addition of a drug-metabolizing system to the culture, or (2) the attempt to "induce" drug-metabolizing enzymes within the explants. Here we would like to report on the second approach.

The chance of inducing monooxygenases in early fetal or even embryonic rodent tissues does not appear too great when considering the reports which have accumulated over the last decades. These reports state that the activity of such cytochrome-P_{450}-dependent monooxygenases was found to be inmeasurable and attempts to induce these monooxygenase reactions failed, unless they were induced during the perinatal period. But, more recently, it was found that aryl hydrocarbon hydroxylase (AHH) can be induced during the embryonic stage (e.g. day 11) in the mouse (NEUBERT and TAPKEN, 1978) or possibly even at the preimplantation stage (PEDERSEN, 1981). Another clue that leads us to believe, that the statement cited above may have to be modified, comes from a publication from NAU and GANSAU (1981), indicating that metabolic conversion of diazepam is detectable with sensitive methods as early as on day 11 or 13 of gestation in the mouse. The attempt to induce monooxygenases in embryonic or early fetal tissues is connected with two major difficulties: (1) a variety of different monooxygenases with rather varying substrate specificities exist. In the (hypothetical) case of a metabolic conversion of thalidomide no clue exists as to which type of cytochrome-P_{450} may be involved (if monoxygenases are involved at all!); (2) a given "inducing agent" will only induce a certain type of monooxygenase activity and a large variety of inducers with different specificities are known today. A negative outcome of an induction experiment may either mean that the enzyme has not been induced under the experimental conditions chosen or that the substrate used is not converted by the enzyme induced.

In our experimental set-up - pretreatment of pregnant animals with an inducer and culturing of the limbs buds of embryos exposed in utero - we found a drastic interference with cartilage formation using the teratogenic thalidomide derivative EM-12. We chose EM-12 because it is

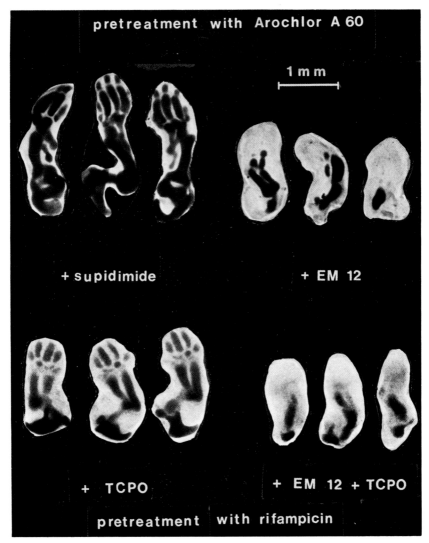

Fig. 1:   Effect of EM-12 on limb development in culture.

upper part: Hindlimbs of 12-day-old mouse embryos exposed in utero to Arochlor A 60 were cultured for 3 days in the presence of 100 µg/ ml supidimide or EM-12.

lower part: Forelimbs of 11-day-old mouse embryos exposed in utero to rifampicin were cultured for 3 days in the presence of 3 µg/ml TCPO or EM-12 + TCPO.

The explants on the left correspond to controls.

thalidomide

EM 12

EM 87

better water-soluble than thalidomide. The substance was added to the culture medium at a concentration of 100 µg/ ml. An apparently non-teratogenic (cf. HENDRICKX and HELM, 1980) thalidomide derivative - EM-87 = supidimid - did not interfere with the morphogenetic differentiation in culture. These results favour the argument that a metabolic conversion of EM-12 is a prerequisite for its teratogenic action. It appears that a species, intrinsically insensitive to the action of thalidomide, may be charged with at least some degree of sensitivity by increasing its ability for drug metabolism. But, we have to realize that we are still faced with a series of difficulties and our findings so far, by no means allow an explaination for the specificity of the teratogenic action of thalidomide. There are a number of reasons for this:

(a) so far, we have not been able to change the susceptibility in vivo by pretreatment.

(b) only certain (early) stages of limb development seem to respond to the action of EM-12 in culture. After cartilage has started to form, no effect can be induced - even in explants of marmoset limb buds. This corresponds to the findings in vivo and indicates an interference with - until now completely unknown - specific differentiation processes.

(c) high concentrations of EM-12 are needed in vitro. Apparently the enzyme activity induced is low and only at high substrate concentrations are sufficient (but small) amounts of an inhibitor formed.

(d) the effect of the induction is variable and most, but not all, experiments are met with success.

(e) in many cases, the presence of an additional inhibitor (e.g. TCPO) is required for an effect of EM-12 on development. Although this additional compound is ineffective itself at the concentration used, (but very effective alone at higher concentrations), it seems to produce a "predamage" and increases the susceptibility of the limb tissue to

EM-12. Although we are favouring the hypothesis that TCPO blocks some SH-groups or -receptors in the limbs, we are still lacking evidence on the mode of this amplification of the EM-12 effect.

(f) our studies have so far not proven that tissues from primates (marmosets, which are very sensitive to the action of thalidomide in vivo) - including tissues at the late embryonic stage (day 60 of gestation) - have a higher capacity to "activate" EM-12 or thalidomide than mouse tissue after induction.

Six "inducers" have been used in our studies so far. From these (cf. Table 1), 3 have been found to be "positive", i.e., the limb buds from pretreated animals responded to the action of EM-12 in vitro. The remaining three, especially those known to induce AHH, and in addition, phenobarbital, the first and most thoroughly studied inducer, were found to be "negative".

Table 1:

Potency of "inducers" of drug-metabolizing enzymes to activate the effectiveness of EM-12 in vitro

EM-12 is	
active in vitro	inactive in vitro
after pretreatment with:	
rifampicin	phenobarbital
Arochlor A 60	benzo(a)pyrene
β-naphthoflavone	TCDD

2. Binding studies with ^{14}C-thalidomide

In order to obtain some further information on the possible mode of the thalidomide action and on the neccessity for this teratogen to be metabolically activated, we performed additional biochemical studies with embryonic and non-embryonic mammalian tissues. We investigated whether thalidomide or EM-12 may be bound to cell components and whether a metabolic activation is the prerequisite for such a binding. The studies were performed with

^{14}C-labelled EM-12 or thalidomide in vitro (Table 2). Furthermore, in vivo studies are in progress.

A binding of the radioactive label to (protein) components of the microsomes is found when ^{14}C-EM-12 or ^{14}C-thalidomide are incubated with such cell fractions in the presence of the typical cofactors required for a monooxygenase reaction. We like to avoid the term "covalent-binding", since it is little defined. The radioactive label bound to the (presumably) protein is resistant to treatment with either acid (1 n perchloric acid) or alkali (0.3 n NaOH) at 37°C (2 hrs).

Table 2:

Binding of ^{14}C-thalidomide to mouse tissue components

tissue preparation	pretreatment	dpm/mg protein	
		15 min.	30 min.
liver microsomes	none	3 200	3 700
	rifampicin	17 500	22 500
embryo (day 11) homogenate	none	100	---
	rifampicin	1 200	---

(Experimental conditions are given in the text.)

The data obtained so far indicate the following aspects:

- the extent of binding, using mouse liver microsomes, is greatly enhanced when microsomes of animals after a pretreatment with "inducers" are used;

- ^{14}C-EM-12, as well as ^{14}C-thalidomide, are bound under the experimental conditions used;

- the binding is not pronounced when liver microsomes from untreated marmosets are used. This thalidomide-sensitive primate species does not activate thalidomide better than mouse embryos;

- very little (if any) binding is found when homogenates of (un-pretreated) 11-day-old mouse embryos are used;

- a low but measurable binding is found when homogenates of 11-day-old mouse embryos are used after a pretreatment of the mothers with, e.g. rifampicin.

3. Inhibition of prolyl hydroxylase

Preliminary results suggest that preincubation of a mouse embryo homogenate with thalidomide or EM-12, may form a metabolite which inhibits prolyl hydroxylase. Again, pretreatment with, e.g. rifampicin, greatly enhances this effect (Table 3).

Table 3:

Effect of thalidomide on prolyl hydroxylase

(mouse embryo (day 11) homogenate; pretreatment with rifampicin; preincubation with the inhibitor)

inhibitor	activity of prolyl hydroxylase (dpm x min $^{-1}$ x mg protein $^{-1}$
none	$273 \stackrel{\wedge}{=} 100$ %
20 µg/ml thalidomide	$160 \stackrel{\wedge}{=} 59$ %
25 µg/ml EM-12	$119 \stackrel{\wedge}{=} 44$ %

The homogenate was preincubated with the inhibitors for 3 min in the presence of 0.45 mM NADP, 4 mM glucose-6-phosphate, 5 µg/ml G-6-P-DH, 7.5 mM $MgCl_2$ and 100 mM Tris-HCl pH 7.5. The prolyl hydroxylase reaction was then started by adding ^{14}C-protocollagen substrate (120 000 dpm), 100 µg/ml ascorbate, 15µg/ml α-ketoglutarate and 14 µg/ml ferroammonium sulfate. The reaction was stopped after 5 min (37°C) by immersing the reaction vessels into liquid nitrogen. Method of HUTTON et al., 1966.

Summary and Conclusions

(a) Explants of mouse embryos are rendered sensitive to the action of thalidomide and its teratogenic derivatives by pretreatment in vivo with some agents known as "inducers" of drug-metabolizing enzymes. Nothing is known so far as to the enzymes involved and to the chemical structure of the metabolite(s) formed.

(b) We do not know whether the same type of metabolite

is responsible for the other two reactions (binding and prolyl hydroxylase), which we studied.

(c) Although a formation of an arene oxide may be suggested by some of the information available - and such a metabolite may even exist - we do not consider it a likely activation product for the teratogenic effect of thalidomide. This is because of known relationships between the chemical structure of thalidomide derivatives and teratogenicity and because of other factors known on the specific teratogenicity of thalidomide. We rather favour an "activation" which occurs at or around the phthalimide-N-atom.

(d) It appears feasible that more than one metabolic converstion may take place at the thalidomide molecule. Not all of these may be responsible for the teratogenic action in primates. The easiest way to distinguish between specific and "unspecific" effects in this respect is to prove that structurally related non-teratogens - such as EM-82 - come out negative in the model used. This has not been accomplished in the studies by GORDON et al. (1981), nor - so far - in our binding studies (2) nor in those on prolyl hydroxylase (3). This prerequisite is, however, fulfilled in the case of our limb bud studies. This may indicate that we are not studying "unspecific" effects or artefacts.

(e) Finally, it should be remembered that the "inducers" used in our studies not only induce cytochrome-P_{450} but also other enzymes, e.g. reductases!

ACKNOWLEDGEMENTS

These studies were carried out in collaboration with Ursula Bluth, Gudrun Blankenburg and Ursula Jacob-Müller. Grants from the Deutsche Forschungsgemeinschaft supported these investigations. We are indebted to Dr. Frankus, Chemie Grünenthal, for providing us with the labelled and unlabelled thalidomide derivatives and for many valuable discussions and to Jane Klein-Friedrich for preparing the manuscript.

References

Blankenburg, G (1981). Limb bud cultures. In Neubert D, Merker HJ (eds): "Culture Techniques", Berlin: de Gruyter, pp 590.

Gordon GB, Spielberg SP, Blake DA, Balasubramanian V (1981). Thalidomide teratogenesis: Evidence for a toxic arene oxide metabolite. Proc Natl Acad Sci, 78 : 2545.

Hendrickx AG, Helm FCh (1980). Nonteratogenicity of a structural analog of thalidomide in pregnant baboons (Papio cynocephalus). Teratology 22 : 179.

Hutton JJ, Tappel AL, Udenfriend S (1966). A rapid assay for collagen proline hydroxylase. Anal Biochem 16 : 384.

Nau H, Gansau C (1981). Development of cytochrome P_{450}-dependent drug metabolizing enzyme activities in mouse and human tissues in vitro. In Neubert D, Merker HJ (eds): "Culture Techniques" Berlin: de Gruyter, pp 495.

Neubert D., Bluth U (1981). Limb bud organ cultures from mouse embryos after apparent induction of monooxygenases in utero. Effects of cyclophosphamide, dimethylnitrosamine and some thalidomide derivatives. In Neubert D, Merker HJ (eds): "Culture Techniques," Berlin: de Gruyter, pp 175.

Neubert D, Tapken S (1978). Some data on the induction of monooxygenases in fetal and neonatal mouse tissue. In Neubert D, Merker HJ, Nau H, Langman J (eds): "Role of Pharmacokinetics in Prenatal and Perinatal Toxicology," Stuttgart: Thieme Publishers, pp 69.

Neubert D, Tapken S, Baumann I (1978). Influence of potential thalidomide metabolites and hydrolysis products on limb development in organ culture. In Neubert D, Merker HJ, Nau H, Langman J (eds): "Role of Pharmacokinetics in Prenatal and Perinatal Toxicology", Stuttgart: Thieme Publ, pp 359.

Pedersen RA (1981). Benzo(a)pyrene metabolism in early mouse embryos, In Neubert D, Merker HJ (eds): "Culture Techniques," Berlin: de Gruyter, pp 447.

Limb Development and Regeneration
Part A, pages 399–405
© 1983 Alan R. Liss, Inc., 150 Fifth Avenue, New York, NY 10011

THALIDOMIDE AND THE NEURAL CREST

Janet McCredie, M.D., F.R.C.R.

Associate Professor of Radiology,
University of Sydney
N.S.W. Australia. 2006

Our present state of knowledge on the subjects of
development and regeneration of the limbs relies very
substantially on evidence gained by experimentation with
lower forms of life. The avian embryo holds pride of
place in studies of limb development, while amphibia have
revealed much basic information about limb regeneration.
For some workers, the prime motivation in investigating
limb development is a simple quest for truth, or curiosity.
Other investigators pursue research on these topics in the
hope of ultimately preventing errors in development,
particularly human malformations. It is easy for individual
researchers to cone down their interest, and their reading,
to their chosen model or species. Thus it is possible
that a specialist in apes will neglect to note facts which
have been established in sea urchins, while the specialist
in cockroaches fails to absorb facts which have been
established in human beings. At a conference such as this,
we should each strive to avoid the quagmires of super-
specialisation, and to address the task of grappling with
the knowledge derived from models other than our own. We
all grope after the natural laws of limb growth. It is
quite possible that the evidence for these laws is
fragmented. Therefore we must have the courage to borrow
from one another any fragments of information which appear
to fit into the jigsaw. The riddle of D.N.A. was finally
solved, we are told, by tinkering with a cardboard mock-up
of the molecular structure, the facts having been borrowed
from a variety of investigative sources.

The present paper juxtaposes several concepts, derived

from several species, in an attempt to explain why limbs fail to develop normally. Conversely, through studying developmental failure, some insight into factors necessary for the normal development of limbs should emerge.

The model chosen for study of failure of limb development was thalidomide embryopathy in the human. Failure of limb development ranged from total absence of the limb (amelia) to absence, hypoplasia, fusion, or other disorganisation of a single digit. Between these extremes, a confusing variety of reduction defects was recorded. The general pattern of skeletal reduction was the stripping out of a longitudinal band of tissue, such as loss of radius and thumb, or loss of tibia and hallux (Henkel and Willert 1969). This embryopathy was first recognised in the human, but it was subsequently reproduced in rabbits, rats, mice, guinea pigs, dogs, cats, monkeys and other sub-human primates (Cahen 1966). Not only did the drug prevent normal development of the limb, but it also prevented normal formation of internal organs, resulting in congenital heart disease, atresia and stenosis of sections of the gastro-intestinal tract, and aplasia and hypoplasia of various organs such as the eye, ear, spleen, kidney, gallbladder, and appendix (Lenz 1962,1963).

The two antagonists in this embryological combat are the drug and the embryo. Let us examine properties of each of these.

THALIDOMIDE IN RODENTS AND HUMANS.

From pharmacological studies in rabbits and rats, Schumacher, Blake, Gurian and Gillette (1968), and Williams, Schumacher, Fabro and Smith (1965) established three important facts about thalidomide:

The intact molecule is the teratogen.

The thalidomide molecule hydrolyses rapidly in plasma in normal conditions, with a steep rise and fall, within 3 hours of oral intake, to half the peak plasma level at one hour.

The products of hydrolysis are not teratogenic. Thus the teratogenic action is a short, pulsed insult, maximal at one hour after ingestion.

A fourth property of thalidomide was that it caused a profound and intractable sensory peripheral neuropathy in healthy human adults. Many people who took this sedative complained of burning sensations, tingling and numbness in the hands and feet. Neurophysiological and neuropathological studies on these patients eight years after the drug had been withdrawn, showed that they were still suffering the symptoms of sensory nerve damage, which had the characteristics of an axonal degeneration affecting mainly the longest nerve fibres in the body (Fullerton and Kremer, 1961, Fullerton and O'Sullivan, 1968).

Since animals cannot describe sensations such as tingling and numbness, sensory neuropathy in animals is almost impossible to diagnose, especially in its early stages. It is easy therefore, for teratologists to overlook the fact that thalidomide attacked the sensory neurons in humans. Could the sensory neuron be the target of thalidomide in the embryo?

THE EMBRYO AND THE NEURAL CREST

The above question directs attention to a primordial tissue in the embryo, the neural crest. It is well-established in vertebrates that this narrow strip of tissue is the source of the sensory and autonomic nerves, and that by extensive migration and multiplication, crest cells partake in formation of facial and spinal cartilages, melanoblasts, odontoblasts, Schwann cells and meninges (Horstadius 1950, Johnston 1966, Weston 1970, Le Douarin 1980). The majority of laboratory studies have used avian embryos, since mammalian embryos are inaccessible to experimental intervention. It is accepted practice to apply the major principles extrapolated from chick embryology to assist in understanding the morphogenesis of higher vertebrates.

The neural groove folds up and forms the neural tube in a progressive fashion from cephalic to caudal end of the embryo. At any one time, an embryo will contain a range of stages of neural tube development, most advanced at the cephalic end. Neural crest development parallels this cephalo-caudal gradation, in a bilateral symmetrical sequence, so that maturation of the paired neural crest derivatives in any embryo is found to be more advanced at the cranial than at the caudal end of the neuraxis. This would provide a spectrum of stages and a potential spectrum of vulnerability to an invading neurotoxin. Such an agent might temporarily inhibit the normal function of neural crest cells at that level, sparing the more mature crest above, and the more primitive crest caudal to that site. Growth deficits ensuing in tissues dependant upon the inhibited crest segments would tend to be bilateral and symmetrical. This anatomic pattern of deformities was documented by clinicians in several large series of thalidomide cases reported in the literature (Henkel and Willert 1969, Pfeiffer and Kosenow 1962, Smithells 1973).

Neural crest is first visible in the human embryo at day 18 of gestation. Over the next three weeks, the crest cells divide and migrate, forming paired spinal sensory ganglia and some cranial nerve ganglia, and the autonomic ganglia, down to the intrinsic ganglion cells around organ primordia. Human cranial and cervical dorsal root ganglia are well-formed by 28 days gestation (Langman, 1969). Neurites emerge from crest cells as soon as their migration commences, in chick (Weston 1970), and in rabbit (Tennyson 1965). Nerve axons have been demonstrated within the upper third of forelimb buds of rabbit embryos prior to chondrogenesis (McCredie, Cameron and Shoobridge 1978, Cameron and McCredie 1982). Thus the peripheral nervous system in the early embryo is more mature than the limb structures which it is destined to innervate.

Against this background of precocious development of neural crest, the sensitivity of the embryo to thalidomide deserves close scrutiny. The human embryo is sensitive to thalidomide from 21 to 42 days gestation (Knapp, Lenz and Nowack, 1962, Lenz 1963). Early in this "sensitive period", the drug causes head and neck deformities. Taken between 24 and 36 days gestation, the drug induces upper limb defects; when taken after 35 days, abnormalities of the legs result. Upper limb buds are first visible in the human embryo at 28 days gestation (Nishimura 1973), and the earliest mesenchymal condensation for humerus is not present until 42 days gestation (Hamilton and Mossman, 1972). Thus the "sensitive period" coincides with neural crest development, and pre-dates the earliest differentiation of skeletal structures within the limb buds (McCredie 1981).

NEUROTROPHISM

Does a relationship exist between sensory nerves and limb growth? There is established evidence of a biological dependance of limb regeneration upon sensory innervation in adult newts. Using Triturus and Amblystoma as the experimental models, Marcus Singer and his colleagues have recorded a series of classical experiments which have verified and amplified previous concepts of neurotrophism.

Singer sectioned either motor or sensory cervical nerve roots, amputated the forelimbs, and showed that intact sensory innervation was essential for normal regeneration of the limb. He concluded that the sensory nerve possesses a growth-stimulating property. Later experiments demonstrated some degree of trophic function in motor and autonomic nerves, but this is quantitatively much less than that exhibited by sensory neurons.

Translation of the biological principle of sensory neurotrophism from adult newt to human embryo lacks justification from some points of view, yet it has support from others. Many scientists regard the newt as an adult form which has retained the embryonic property of limb generation, a capacity lost by the adults of higher vertebrates. Bearing in mind the common ancestry of all vertebrates, it is not impossible that there may be a common neurotrophic mechanism underlying regeneration in amphibia and limb morphogenesis in all vertebrates (McCredie 1981).

Experiments on mammalian nerves are technically difficult at very early stages of embryogenesis. Zalewski (1979) has shown that sensory neurons in the immature foetus of rat possess a trophic ability over the formation of taste buds. Mizell and Isaacs (1970) used the opossum fetus to show that a local implantation of neural tissue provoked regeneration of amputated limbs in that embryo. Human teratology records monsters with well-formed limbs in the absence of brain (anencephalus), spinal cord (amyelia), or heart (acardius). In such cases, peripheral nerves and ganglion cells have been recorded. This is at least circumstantial evidence of a link between neural crest elements and limb generation.

Cahen RL, (1966) Experimental and clinical chemoteratogenesis. Adv Pharmacol.Garattini S, Shore PA (eds) 4:263

Cameron J, McCredie J (1982) Innervation of the undifferentiated limb bud in rabbit embryo. J Anat. (in press)

Fullerton PM, Kremer M (1961) Neuropathy after intake of thalidomide (Distaval). Brit med J. 2: 855

Fullerton PM, O'Sullivan DJ (1968) Thalidomide neuropathy. A clinical, electrophysiological and histological follow-up study. J Neurol Neurosurg Psychiat. 31: 543.

Hamilton WJ Mossman HW (1972) Human Embryology. Heffer ambridge p 542

Henkel H-L Willert H-G (1969) Dysmelia: A classification and a pattern of malformation in a group of congenital defects of the limbs. J Bone Joint Surg 51: 399

Horstadius S (1950) The Neural Crest Oxford University Press London

Inman V Saunders JB de C (1944) Referred pain from skeletal structures J Nerv Ment Dis 99: 660

Johnston MC (1966) A radioautographic study of migration and fate of cranial neural crest cells in the chick embryo Anat Rec 156:143

Knapp K Lenz W Nowack E (1962) Multiple congenital abnormalities Lancet 2: 725

Langman J (1969) Medical Embryology Williams and Wilkins Baltimore p 292

Le Douarin N (1980) The migration and differentiation of neural crest cells. Current Topics in Dev Biol 16: 31

Lenz W (1962) Thalidomide and congenital abnormalities Lancet 1 :45

Lenz W (1963) Das thalidomid syndrom Fortschr Med 81 :148

McCredie J Cameron J Shoobridge R (1978) Congenital malformations and the neural crest Lancet 2 : 761

McCredie J (1975) Segmental embryonic peripheral neuropathy Pediat Radiol 3 : 162

McCredie J (1976) Neural crest defects : A neuroanatomic basis for classification of multiple malformations related to phocomelia J. Neurol Sci 28 : 373

McCredie J (1977) Sclerotome subtraction. A radiologic interpretation of reduction deformities of the limbs. Birth Defects Original Article Series XIII, 3D,65

McCredie J (1981) Thalidomide and the neural crest. Assessment of a hypothesis. In Hetzel BS, Smith RM (eds) :"Fetal Brain Disorders: Recent approaches to the problem of mental deficiency", Elsevier N.Holland p 426

Mizell M. Isaacs JJ (1970) Induced regeneration of hind limbs in the newborn opossum Amer Zool 10 : 141

Nishimura H (1973) Personal communication

Pfeiffer RA Kosenow W (1962) Thalidomide and congenital abnormalities Lancet 1 : 45

Schumacher H Blake DA, Gurian J Gillette JR (1968) A comparison of the teratogenic activity of thalidomide in rabbits and rats J Pharmacol exp Ther 162 : 189

Singer M (1943) The nervous system and regeneration of the forelimb of adult Triturus. II The role of the sensory nerve supply J Exp Zool 92 : 297

Singer M (1946) The nervous system and regeneration of the forelimb of adult Triturus. V The influence of number of nerve fibres including a quantitative study of limb innervation J exp Zool 101 299: 337

Singer M (1947) The nervous system and regeneration of the forelimb of adult Triturus VI A further study of the importance of nerve number including quantitative measurements of limb innervation J exp Zool 104 223-249

Singer M (1964) The trophic quality of the neuron: Some theoretical considerations In M Singer and JP Schade (eds) Mechanisms of Neural Regeneration, Progress in Brain Research Vol 13 Elsevier Amsterdam pp 228-232

Singer M (1974) Neurotrophic control of limb regeneration in the newt; Ann NY Acad Sci 228 308-322

Williams RT Schumacher H Fabro S Smith RL (1965) The chemistry and metabolism of thalidomide In Robson Sullivan Smith (eds) : "A sumposium on embryopathic activity of drugs" London : Churchill

Zalewski A (1974) Trophic function of the neurons in transplanted neonatal ganglia Exp Neurol 45 : 189

Limb Development and Regeneration
Part A, pages 407–412
© **1983 Alan R. Liss, Inc., 150 Fifth Avenue, New York, NY 10011**

POSTAMPUTATIONAL HEALING OF MOUSE DIGITS MODIFIED BY TRAUMA

Daniel A. Neufeld, Ph.D.

Department of Anatomy
University of South Dakota School of Medicine
Vermillion, S.D.   57069

The postamputational healing process in mice has been
altered (Neufeld, 1980) such that several days after treatment the amputation site resembled a blastema which exists
on a regenerating newt limb. To accomplish this, skin was
stripped from a stump several weeks after amputation, and
stump tissues were exposed to a saturated solution of sodium
chloride. The rationale for the above experiment came from
experiments in which regeneration was induced in postmeta-
morphic frogs by exposing amputation stumps to altered salt
concentrations (Rose, 1945).

Young anurans can also be made to initiate regeneration
by a variety of chemical stimuli or, in the absence of
chemicals, by simple mechanical trauma (review by Polezhaev,
1972). The present experiment was performed to determine
whether trauma was an effective stimulant in mammals, and
whether it might be used instead of, or added to, salt
treatments in future attempts to induce blastema formation
in mice. Results indicate that trauma is an effective
stimulus but that the effect is dependent upon details of
the trauma technique.

Materials and Methods

Adult male white mice fed standard laboratory chow and
each weighing more than 21 grams were used. Following
intraperitoneal injection of Nembutol, the middle (third)
toe of the left posterior manus was swabbed with 70 percent
ethanol and amputated by large iridectomy scissors through
the mid diaphysis of the proximal phalanx. A tourniquet

was used to minimize blood loss, but no hemostasis or anti-
biotic was applied to the wound surface. A variety of treatments were employed with a minimum of 2 animals per treatment. Digits of control animals were either amputated without additional intervention, or were amputated and on a
designated postamputation day (p.d.) the skin was removed
from the apex of the stump. Trauma was administered to
experimental animals on the day of amputation and, or, once
or twice thereafter (see details below). Under Nembutal
anesthesia, the skin which covered the end of the stump was
excised by circumferential incision ("circumcision") approximately even with the end of the bone. A sterile 30 gauge
needle or fine-tipped tungsten wire was used to tease apart
the stump tissue. Typically, this required approximately
50 sharp strokes. Fourteen to 16 days postamputation,
depending upon the time of last treatment and appearance of
wound, animals were sacrificed. The left foot was amputated
by scalpel through the metatarsals and fixed by immersion in
Bouins. The stump of the middle toe was dissected free, and
its hair was removed by forceps under a dissecting microscope. Tissues remained in Bouins for a total of 7 days,
and in a decalcifying solution of sodium citrate and formic
acid for 21 days. They were processed in paraplast, serially
sectioned at 7 micrometers, and stained with hematoxylin and
eosin. Several slides were stained with Alcian Blue 8GX,
pH 5.8, in varying concentrations of MgCl (method of Scott
and Dorland, 1965). In addition, blastemas of regenerating
newt limbs were processed as above and used for comparison
with mouse toe stumps.

Results

Effectiveness of the treatments was proportional to
the intensity of the treatments. One group of animals
(represented by Figures 1A and 1B) received no treatment
after amputation. Their amputation site was characterized
by a covering of full-thickness skin, which contained hair
follicles and sweat glands, separated from bone by a bursa.
Simple circumcision at 8 p.d. (Figure 1C) eliminated
epidermal appendages and reduced collagen and the bursa.
However, fibroblasts reappeared in lamellae and skin con-
tracture again reduced the subepithelial area. Circumcision with trauma on 6 p.d. and recircumcision on 12 p.d.
(Figures 1D and 1E) eliminated the bursa and disrupted the
subepithelial cells. More intensive treatment, circumcision
and trauma at least twice after amputation (Figures 1F and

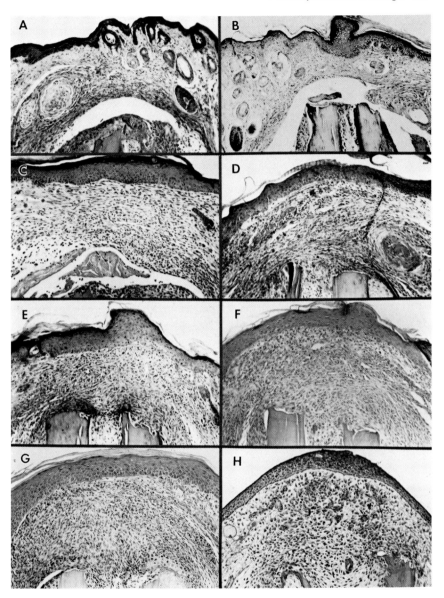

Figure 1. Histological Effects of Trauma Treatment. Amputation sites at 14 to 16 days postamputation. Mouse toes (A–G) are arranged in sequence of increasing treatment intensity. A newt limb blastema (H) is also included.

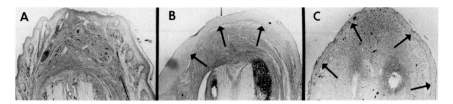

Figure 2. Control (A) and Experimental (B) Mouse Toes and
Newt Limb (C) Stained with Alcian Blue. An alcianophilic
line can be seen in the newt limb and the treated mouse toe
at the epithelial-mesenchymal interface (indicated by arrows).

1G), increased the disrupted cell population causing a
resemblance to the blastema of a regenerating newt limb
(Figure 1H).

    Alcian blue stained sections revealed that the pattern
of alcianophilia of maximally treated mice resembled that of
a newt blastema (Figure 2). In both instances maximal
staining was observed near the bone and a thin alcianophilic line was observed at the epithelial-mesenchymal
interface. In control mice and differentiated newt skin the
staining pattern was correlated with differentiated tissues,
and the subepithelial alcianophilic line was absent or
diminished.

Discussion

    Reports have emphasized the histological criteria,
including the absence of dermis but the presence of numerous
mesenchymal cells covered by a thickened epithelium, which
characterize the growing or regenerating limb bud (see recent
reviews by Hinchliffe and Johnson, 1980, and Wallace, 1981).
An initial attempt to generate blastemas on stumps of mouse
toes succeeded in eliminating dermis, but was unable to
generate a mass of subepithelial cells (Schotté and Smith,
1961). Recently, blastema-like features were at least
partially generated (Neufeld, 1980), and the present work
gives evidence that such features can be generated by at
least one other method.

Several authors have proposed that extracellular matrix components influence limb morphogenesis by inhibiting chondrogenesis (Toole, 1971), determining cell aggregation patterns (Newman and Frisch, 1979), and ultimately increasing cAMP during differentiation (Kosher et al., 1981). The use of alcian blue to localize polyanionic extracellular materials in this study represents a preliminary attempt to investigate differences in the distribution of matrix components in treated mouse toes and in regenerating newt limbs. The similarity in staining pattern is interpreted as a positive sign in an attempt to generate mammalian blastemas. The specific distribution of polyanionic moieties is unknown, however.

The potential for mammalian regeneration is not clear. Induced partial regeneration in mammals has been reported by several authors (reviewed by Neufeld, 1980, and recent work by Sisken and Fowler, 1982, and Smith, 1981). However, the accuracy of these claims is difficult to assess. All fail to show either a photograph of the gross outgrowth, a histological section of the outgrowth, or a midsagittal section of some intermediate stage of development to indicate from where the outgrowth arose. The absence of a stable marker at the amputation level also can result in ambiguous data, as pointed out by Fleming and Tassava (1981).

The current series of experiments which attempt to generate a blastema in mammals are conducted on the premise that if a blastema forms, then a variety of questions can be pursued. Will mammals initiate a regenerative response? What constitutes a regenerative response? Can such a response be sustained? Furthermore, if growth can be initiated, questions about establishment of positional information and functions of extracellular matrix can be addressed in this system, as they are currently being investigated elsewhere in developing and regenerating limbs.

References

Fleming, M. and R. Tassava (1981). Preamputation and postamputation histology of the neonatal opossum hindlimb: Implications for regeneration experiments. J Exp Zool 215:143.
Hinchliffe, J. and D. Johnson (1980). The Development of the Vertebrate Limb. Clarendon Press, Oxford, p. 268.
Kosher, R., M. Savage, and K. Walker (1981). A gradation

of hyaluronate accumulation along the proximodistal axis of the embryonic chick limb bud. J Embryol Exp Morph 63:85.

Neufeld, D. (1980). Partial blastema formation after amputation in adult mice. J Exp Zool 212:31.

Newman, S. and H. Frisch (1979). Dynamics of skeletal pattern formation in developing chick limb. Science 205:662.

Polezhaev, L. (1972). Loss and Restoration of Regenerative Capacity in Tissues and Organs of Animals. Harvard Univ. Press, Cambridge, Massachusetts, p. 385.

Rose, S. (1945). The effect of NaCl in stimulating regeneration of limbs of frogs. J Morphol 77:119.

Schotté, O. and C. Smith (1961). Effects of ACTH and of cortisone upon amputational wound healing processes in mice digits. J Exp Zool 146:209.

Scott, J. and J. Dorling (1965). Differential staining of acid glycosaminoglycans by alcian blue salt solutions. Histochemistry 5:221.

Sisken, B. and I. Fowler (1982). Effects of augmented nerve supply on amputated rat limbs. Anat Rec 202:177.

Smith, S. (1981). The role of electrode position in the electrical induction of limb regeneration in subadult rats. Bioelectrochemistry and Bioenergetics 8:661.

Toole, B. and J. Gross (1971). The extracellular matrix of the regenerating newt limb: synthesis and removal of hyaluronate prior to differentiation. Develop Biol 25:57.

Wallace, H. (1981). Vertebrate Limb Regeneration. John Wiley & Sons, New York, p. 276.

Limb Development and Regeneration
Part A, pages 413–422
© **1983 Alan R. Liss, Inc., 150 Fifth Avenue, New York, NY 10011**

LIMB ANOMALIES AS A CONSEQUENCE OF SPATIALLY-RESTRICTING
UTERINE ENVIRONMENTS

John M. Graham, Jr., M.D., ScD.

Department of Maternal and Child Health
Dartmouth Medical School
Hanover, NH   03755

Most limb anomalies occur in otherwise normal children.
Often, no cause for the problem is evident, and the lack of
symmetrical involvement or a broader pattern of defects may
indicate that such problems are not due to mutant genes or
teratogenic drug exposures. Recent evidence suggests that
some limb anomalies might be associated with spatially-restricting uterine environments (Graham et al., 1979). The
purpose of this study is to update and summarize additional
clinical findings that support this hypothesis and to suggest
how early limb compression within such an environment might
result in a variety of limb anomalies.

Vascular disruption, with subsequent loss of previously
normally formed tissue is suggested as the underlying pathogenetic mechanism. Such disruption can result from either
extrinsic vascular compression or intrinsic vascular occlusion, and proof of such an etiology may require specialized
studies which are not routinely performed. It is a second
purpose of this study to encourage the routine use of specific
studies that might foster further understanding of the causes
for such limb anomalies.

CASE MATERIAL

The pertinent clinical findings for 30 instances of
human limb reduction defects are set forth in Tables 1 and 2.
These findings include 3 instances associated with enlarged
uterine myomas, 8 instances associated with maternal uterine
malformation, and 19 instances associated with early amnion
rupture sequence.

TABLE 1:   LIMB ANOMALIES ASSOCIATED WITH UTERINE STRUCTURAL ABNORMALITIES.

Case Number	Type of Uterine Structural Abnormality	Type of Limb Anomaly and Associated Findings
1*	10 cm. myoma at cervical os	Hypoplastic right lower leg with absent 5th ray.
2	Multiple rapidly enlarging myomata	Hypoplastic left lower leg with absent 4th and 5th rays. Intrauterine fracture of tibia.
3	Multiple myomata in cul de sac	Hypoplastic right hand with absent 5th ray; hypoplastic right leg with 2 digit foot.
4*	Bicornuate Uterus	Hypoplastic left arm and hand with syndactyly and absent 4th and 5th digital rays.
5*	Bicornuate Uterus	Hypoplastic right 5th finger. Hypoplastic right leg and foot.
6*	Bicornuate Uterus	Distal hypomelia of right arm with absent hand and wrist.
7*	Bicornuate Uterus (Hand-foot-uterus Syndrome)	Distal hypomelia of left arm and leg (Hand-foot-uterus Syndrome; postaxial brachydactyly).
8	Bicornuate Uterus	Asymmetric distal hypomelia of left and right legs.
9	Bicornuate Uterus	Hypoplastic left arm and hand with absent 4th and 5th rays.
10	Bicornuate Uterus	Rudimentary left leg with 2 toes. Deformed right leg with 4 toes. Left-sided abdominal wall defect. Retroflexed buttocks, short umbilical cord.
11	Bicornuate Uterus	Hypoplastic left arm with rudimentary wrist and absent hand.

*Previously published by Graham et al., J. Pediatr., 96:1052, 1980.

TABLE 2: LIMB ANOMALIES ASSOCIATED WITH EARLY AMNION RUPTURE SEQUENCE

Case Number	Body Wall Defect	Scoliotic Convexity	Band-related Findings	Limb Anomalies
1*	———	———	Cleft lip & palate Constrictive rings on fingers	Hypoplastic left arm; absent 1st digital ray. Syndactyly right hand.
2	———	———	Constrictive rings on fingers	Hypoplastic left arm and hand.
3*	Left thorax and abdomen	Left	Cleft lip & palate Constrictive rings on digits	Hypoplastic left pelvis & leg; absent fibula & lateral rays. Syndactyly left hand; absent first 2 rays.
4**	Left abdomen	Left	———	Absent left leg.
5	Left thorax and abdomen	Left	Craniofacial cleft	Absent left arm.
6**	Left abdomen	Left	———	Rudimentary left leg.
7	Left thorax and abdomen	Left	Craniofacial cleft	Hypoplastic left arm; 3-digit hand.
8**	Right thorax and abdomen	?	Craniofacial cleft	Absent right arm.
9**	Left thorax and abdomen	Head flexed to left	Craniofacial cleft	Absent left arm and scapula.
10**	Anterior abdomen	Left with 180° pelvic rotation	———	Rudimentary right leg
11	Right thorax and abdomen	Right	Craniofacial cleft	Rudimentary right arm
12**	Left thorax and abdomen	Left	———	Absent left arm. Hypoplastic right leg; 3 digit foot.
13**	Left thorax and abdomen	Left	Craniofacial cleft	Rudimentary left arm.
14**	Left abdomen	———	———	Absent left leg and hemi-pelvis.
15**	Left abdomen	Left	———	Rudimentary left leg.
16**	Anterior abdomen	———	———	Rudimentary right foot.
17	Left abdomen	Left	———	Absent left arm. Hypoplastic right leg with 3-digit foot.
18	Left thorax and abdomen	Left	Craniofacial cleft	Absent left arm.
19	Right thorax and abdomen	?	Craniofacial cleft	Absent right arm. Anteriorly-placed left arm.

*Previously published by Graham et al., J. Pediatr., 96:1052, 1980.

**Previously published by Miller et al., J. Pediatr., 98:292, 1981.

DISCUSSION

The concept that some congenital anomalies might arise from spatially-restricting uterine environments is not new. In fact, Hippocrates was one of the first to suggest that "...infants become crippled in the following way: where in the womb there is a narrowness at the part where in fact the crippling is produced, it is inevitable that the body moving in a narrow place shall be crippled on that part." The following sections will summarize evidence that confirms the association between spatially-restricting uterine environments and limb anomalies and conclude with a section summarizing the evidence for vascular disruption as one possible underlying pathogenetic mechanism.

Ectopic Tubal Gestation

One type of embryo that might be examined in an attempt to gain some understanding of the adverse impact of early uterine restrictive influences consists of embryos which implant ectopically in a fallopian tube. This kind of an abnormal implantation occurs about 1 in 50 human pregnancies and results in a high frequency of structural anomalies in such embryos. One study of 44 ectopic conceptuses by Stratford (1970) demonstrated structural abnormalities in 67 percent of these specimens. A second study of 76 embryos obtained from pregnancies occurring within a fallopian tube demonstrated abnormalities in over 50 percent of the embryos (Poland et al., 1976). More recently, Matsunaga and Shiota (1980) conducted a study of 3,614 well-preserved human embryos derived from artificial termination of pregnancy. They were interested in determining whether ectopic implantations or enlarged myomas could enhance the prevalence of localized malformations of the embryo. Among 43 ectopic conceptuses, 11.6 percent were malformed (versus 3.3 percent among the normally implanted pregnancies). Unilateral amelia was significantly increased among the ectopic cases, and it was concluded that this was a reflection of the teratogenic impact of early spatial restriction.

Uterine Structural Anomalies

Table 1 describes instances of limb reduction anomalies associated with enlarged uterine myomas in 3 cases and with a maternal bicornuate uterus in 8 cases. The placenta was

not examined in any of these cases, and no data are available
on the vast majority of normal infants born to women with such
uterine structural anomalies. Uterine malformations occur
with varying degrees of severity in 1-2 percent of women.

Two previous studies lend support to the hypothesis that
myomatous uteri may be associated with a currently undefined
risk for limb anomalies in offspring delivered from such uterine cavities. Among 97 conceptuses from myomatous pregnancies,
6.2 percent were malformed and caudal dysplasia was significantly increased among this group (Matsunaga and Shiota,
1980). The authors suggested that this might be a consequence
of the positional effects of mechanical distress. In addition
to the cases reported herein, Hedenstedt (1944) described an
infant with limb deficiency affecting all four limbs, who was
delivered from a uterus containing a myoma which was approximately 16 cm. in diameter. Figure 1 demonstrates a hypoplastic left leg with an intrauterine fracture of the left
tibia and absence of 2 postaxial digital rays. The mother
had a 10-year history of uterine myomata and these enlarged
rapidly during this pregnancy. The fetus was in a prolonged
breech position (like the other 2 instances of lower limb
deficiency associated with uterine myomata in Table 1).

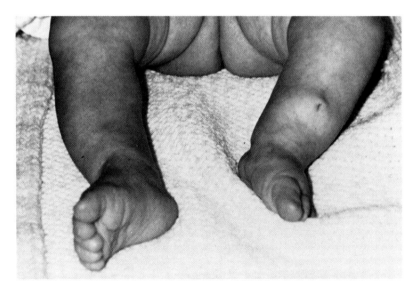

Figure 1. Hypoplastic left leg associated with maternal
uterine myomata (Case 2, Table 1).

Though data concerning infants born with limb reduction anomalies to women with uterine structural abnormalities are biassed by the retrospective nature of their ascertainment, there appears to be enough evidence to support an association between the 2 findings. The total lack of any placental studies in such cases limits speculation about a possible cause-and-effect relationship. Since in most instances, the structurally abnormal uterus may not result in significant fetal compression until the latter half of pregnancy, it is quite possible that such abnormal uterine cavities may pre-dispose toward placental problems that may result in limb reduction anomalies through mechanisms other than fetal limb compression.

The association between uterine structural abnormalities and limb reduction defects in the offspring needs further evaluation through good prospective studies before reliable risk figures can be generated. A retrospective study by Miller et al. (1979) indicates that the range of problems associated with a maternal bicornuate uterus can include fetal loss and prematurity, as well as structural defects in the offspring. Preliminary indications suggest that as succes-sive pregnancies fill out the malformed uterine cavity, the volume available for a developing fetus can enlarge. This may diminish the risk for problems associated with uterine restrictive influences, as the number of pregnancies success-fully carried to term increases.

Early Amnion Rupture Sequence

Normally, the fetus is surrounded by a protective cushion of amniotic fluid throughout gestation, with rupture of the fetal membranes occurring at or shortly before the moment of delivery. When the amniotic membranes rupture earlier than this time, it can be associated with a variety of structural defects in the offspring. Since the publication of Torpin's initial studies (1968), amputation defects of limbs and digits have been associated with early amnion rupture, and Baker and Rudolph (1971) extended this association to include distal syndactyly. These defects are thought to relate to the attachment of amniotic bands to fetal parts, particularly the limbs and digits, and they are often associated with positional foot deformations.

More recently, Jones et al. (1974) and Higginbottom et al. (1979) described craniofacial defects associated with amniotic bands and noted that the time at which the amnion rupture occurred was an important determining factor in the kind of structural defect produced. Not all of the structural defects associated with amnion rupture can be ascribed to the impact of amniotic bands alone. Graham et al. (1980) recently implicated intrauterine compression from very early amnion rupture as one cause for limb deficiency that went beyond the constricting influence of amniotic bands. They noted that in previously-reported instances of limb deficiency, associated with insilateral body wall defects, band-related craniofacial clefts, and severe scoliosis, the regional association of the limb and body wall deficiency on the outer surface of the scoliotic convexity was consistent with the disruptive effects of early embryonic compression resulting from the loss of amniotic fluid.

This constellation of defects is highly similar to that produced in rat fetuses subjected to amniotic puncture at 14 to 16 days' gestation (Kennedy and Persaud, 1977). Such puncture resulted in loss of amniotic fluid, followed rapidly by vascular disruption and necrotic loss of tissue in the limbs, head and body wall. The defects noted in affected fetuses included limb deficiency, body wall deficiency, craniofacial defects, neural tube defects, scoliosis, postural deformations, growth deficiency, band-related defects, and a shortened umbilical cord.

Miller et al. (1981) recently described 10 instances of early amnion rupture in the human and analyzed an additional 17 cases reported previously in the literature. They noted the following nonrandom regional associations in these 27 cases: 1.) total concordance between the side of the major limb deficiency and the side of body wall deficiency, 2.) a striking relationship between the direction of scoliotic convexity and the side of the limb/body wall deficiency (73 percent of cases), 3.) total concordance between craniad neural tube defects and upper limb deficiency, 4.) total concordance of major upper limb deficiency and upper body wall deficiency, and 5.) total concordance of major lower limb deficiency and lower body wall deficiency. These unusual regional associations between defects appear to be due to the disruptive effects of early embryonic compression. Table 2 includes the cases reported by Graham et al. (1980) and Miller et al. (1981) as well as 7 additional cases. A recent study by Graham and Smith (1981) suggests that the range of limb defects associated with early amnion rupture can range from limb deficiency of all gradations to preaxial polydactyly.

Figure 2 demonstrates some of the limb anomalies that have been associated with early amnion rupture.

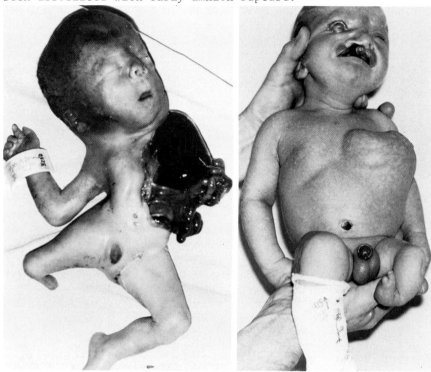

Figure 2.   Limb deficiency associated with early amnion rupture: (Left) Case 12, Table 2 from Miller et al. (1981), (Right) Case 3, Table 2.

Possible Pathogenesis for Limb Reduction Anomalies
and Suggestions for Clinical Evaluation

The studies by Kennedy and Persaud (1977) suggest that experimental amnion rupture in rats resulted in limb reduction defects through a process of vascular engorgement, hemorrhage and necrosis. A layer of reformed periderm separated the distal necrotic portion of the limb and eventually amputated it, leaving only the stump of a previously well-formed limb. In other instances, edematous blebs appeared to amputate the distal segment of the limb. The earliest changes consisted of engorgement, dilitation and disruption of the blood vessels of the limbs and heads of treated fetuses, followed by massive hemorrhage into the surrounding tissues.

Kennedy and Persaud (1977) suggested that these changes might be the consequence of umbilical compression and/or placental constriction, mediated by the spatial orientation of the fetus at the time of amnion puncture. These findings suggest that intrauterine alterations in placental and/or umbilical vascular pressure could result in limb reduction anomalies.

Recently, clinical studies in humans have suggested that other types of vascular compromise can result in limb reduction anomalies. Carey et al. (1982) described 10 instances of limb reduction anomalies associated with urethral obstruction malformation sequence. In these cases, limb reduction defects were hypothesized to have resulted from iliac vessel compression due to progressive enlargement of the fetal bladder. Hoyme et al. (1982) described 4 patients with unilateral transverse limb reduction defects due to in-utero vascular accidents. Microscopic evidence of fetal vascular occlusive disease was present on multiple sections of placenta in 3 instances, suggesting that occlusion of the brachial artery was secondary to embolization from the placental vascular thrombi. In the fourth instance, a massive thrombus was observed to be occluding the brachial artery, following loss of the fetus due to placental abruption.

These studies suggest that children born with sporadic limb reduction defects deserve a careful gross and histologic examination of the placenta for evidence of possible vascular problems or early amnion rupture. Furthermore, some consideration should also be given to the shape of the mother's uterine cavity and to whether any predisposing factors exist for abnormal placental attachments. It has recently been recognized that short umbilical cords may be associated with spatially-constrictive uterine environments (Miller et al., 1981). Long umbilical cords may also be at risk for vascular compromise (Hoyme et al., 1982). Thus umbilical cord length should also be routinely recorded in children born with limb reduction anomalies.

ACKNOWLEDGEMENTS

The expert technical assistance of Ms. Karen McKenney is much appreciated. Case materials were contributed by Drs. D.W. Smith, M.E. Miller, M.C. Higginbottom, K.L. Jones, B.D. Hall, J.W. Hanson, J. Zonana, T.H. Shepard, M.J. Stephan, and L.B. Holmes, and their help is gratefully acknowledged.

REFERENCES

Baker CJ, Randolph AJ (1971). Congenital ring constrictions and intrauterine amputations. Am J Dis Child 121:393.
Carey JC, Eggert LD, Curry CJ (1982). Lower limb deficiency and the urethral obstruction sequence. Birth Defects Original Article Series, in press.
Graham JM, Jr, Higginbottom MC, Smith DW (1981). Preaxial polydactyly of the foot associated with early amnion rupture: Evidence for mechanical teratogenesis? J Pediatr 98: 943.
Graham JM, Jr, Miller ME, Stephan MJ, Smith DW (1980). Limb reduction anomalies and early in utero limb compression. J Pediatr 96:1052.
Hedenstedt S (1944). Zwie Falle von totaler amelie. Acta Obstet Gynecol Scand 24:271.
Higginbottom MC, Jones KL, Hall BD, Smith DW (1979). The amniotic band disruption complex: Timing of amniotic rupture and variable spectra of consequent defects. J Pediatr 95:544.
Hoyme EH, Jones KL, Van Allen MI, Saunders BS, Benirschke K (1982). The vascular pathogenesis of transverse limb reduction defects. J Pediatr, in press.
Jones KL, Smith DW, Hall BD, Hall JG, Ebbin AJ, Massoud H, Golbus M (1974). A pattern of craniofacial and limb defects secondary to aberrant tissue bands. J Pediatr 84:90.
Kennedy LA, Persaud TVN (1977). Pathogenesis of developmental defects induced in the rat by amniotic sac puncture. Acta Anat 97:23.
Matsunaga E, Shiota K (1980). Ectopic pregnancy and myomatous uteri: Teratogenic effects and maternal characteristics. Teratology 21:61.
Miller ME, Dunn PM, Smith DW (1979). Uterine malformation and fetal deformation. J Pediatr 94:387.
Miller ME, Graham JM, Jr, Higginbottom MC, Smith DW (1981). Compression-related defects from early amnion rupture: Evidence for mechanical teratogenesis. J Pediatr 98:292.
Miller ME, Higginbottom MC, Smith DW (1981). Short umbilical cord: Its origin and relevance. Pediatrics 67:618.
Poland BF, Dill FJ, Stybo C (1976). Embryonic development in ectopic human pregnancy. Teratology 14:315.
Stratford BF (1970). Abnormalities of early human development. Amer J Obstet Gynec 107:1223.
Torpin R (1968). Fetal Malformations Caused by Amnion Rupture During Gestation. Charles C. Thomas Publisher, Springfield IL.

Limb Development and Regeneration
Part A, pages 423–429
© **1983 Alan R. Liss, Inc., 150 Fifth Avenue, New York, NY 10010**

UNIQUE PATTERN OF LIMB MALFORMATIONS ASSOCIATED WITH CARBONIC ANHYDRASE INHIBITION.[1]

William J. Scott, Claire M. Schreiner and
Kenneth S. Hirsch[2]

Children's Hospital Research Foundation and
Department of Pediatrics, University of Cincinnati
College of Medicine, Cincinnati, Ohio 45229

In 1965 Layton and Hallesy made a remarkable discovery. They found that when the drug acetazolamide was incorporated into the diet of pregnant rats or mice many of the offspring were missing postaxial digits on the right forelimb. This phenomenon has been repeated many times in different laboratories so that there can be no doubt that acetazolamide has a predilection to induce distal, postaxial right fore-limb deformity. The left forelimb is also sensitive to this effect but at higher dose levels and proximal and preaxial forelimb structures can also be affected at higher dose levels (Wilson et al. 1968).

Attention to the mechanism by which acetazolamide produces this unique effect has centered on carbonic anhydrase inhibition. The reasons for this focussed effort are: (1) the major pharmacodynamic effect of acetazolamide is inhibition of this enzyme (Maren 1967) and (2) other carbonic anhydrase inhibitors induce the same unusual pattern of deformity (Wilson et al. 1968; Hallesy and Layton 1967; Maren and Ellison 1972).

Detracting from this hypothesis has been the inability to detect carbonic anhydrase within the embryo during the period of development when acetazolamide is an effective teratogen. Wilson et al. (1968) were unable to detect carbonic anhydrase activity in rat embryos until day 13 of

[1]Supported by NIH grants HD09951 and ES07051.
[2]Present address: Eli Lilly & Co., Greenfield Laboratories, Greenfield, Indiana 46140.

gestation, two days after the end of the sensitive period in this species.

Recently, using a different method of assay, we have been able to detect carbonic anhydrase activity in rat and mouse embryos during the teratogenically sensitive phase of development (Hirsch et al. in press). Studies in mice employed two inbred strains, CBA/J which are quite sensitive to acetazolamide-induced ectrodactyly and SWV mice which are relatively resistant to this effect (Biddle 1975a,b). The specific activity of carbonic anhydrase in whole embryo homogenates was greater in the teratogenically sensitive CBA/J mouse but the difference was not statistically signigicant. More importantly the sensitivity to inhibition by acetazolamide was greater in CBA/J embryo homogenates (Hirsch et al. in press), providing a rational explanation for the resistance of SWV embryos to acetazolamide teratogenesis.

Next we attempted to define the site within the embryo at which carbonic anhydrase inhibition initiated abnormal limb development. Using the histochemical technique of Hansson (1967) we found carbonic anhydrase activity in many embryonic structures (Figure 1). Particular attention was paid to enzyme distribution within the forelimb bud. Staining, indicative of enzyme activity was absent from the ectoderm, including the apical ectodermal ridge (Figure 2). Enzyme activity was greatest in the area between the ectoderm and adjacent mesenchyme. From these studies we reasoned that carbonic anhydrase in the forelimb is not the primary site of action for acetazolamide because: (1) carbonic anhydrase activity was not localized postaxially where the malformation occurs; (2) there was no observable difference in enzyme activity between right and left forelimbs, although the right side is predominantly affected by malformation; (3) forelimb buds of resistant SWV mouse embryos stained essentially the same as sensitive rat and CBA/J mouse forelimbs; and (4) carbonic anhydrase activity in the limbs of sensitive and resistant embryos was equally inhibited by incorporation of acetazolamide in the incubation medium.

The only consistent difference observed between susceptible and resistant embryos occurred in the primitive, nucleated erythrocytes. With the standard incubation time of 6 minutes carbonic anhydrase activity was seen in almost

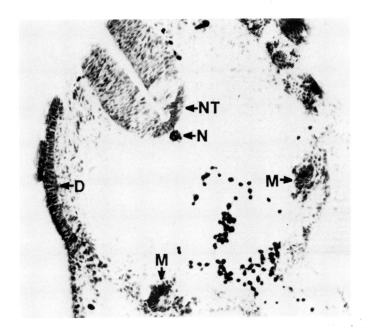

Figure 1. Day 11 rat embryo. NT = neural tube; N = notochord; M = mesonephros; D = dermatome. Sections were incubated for 6 min. and counterstained lightly with eosin. Original magnification X 190. Bar = 50 μm (Reprinted from Schreiner et al., 1981.)

all erythrocytes of the susceptible day 11 rat and day 10 CBA/J embryos (Figure 3A). Under identical conditions, nearly half of the nucleated erythrocytes of the resistant day 10 SWV embryo remained unstained. (Figure 3B). This latter population of unstained erythrocytes does contain carbonic anhydrase since when incubation time was increased to 9 minutes nearly all the erythrocytes in SWV embryos were stained.

At least two possibilities could account for the low carbonic anhydrase activity of this second population of SWV nucleated erythrocytes. They could contain a lower amount of the same isozyme found in the nucleated

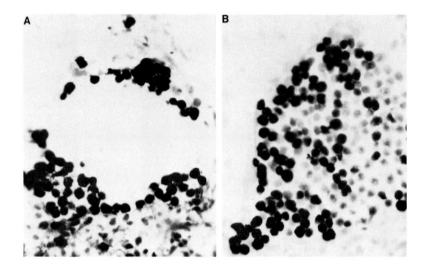

Figure 2. Histochemical staining of nucleated erythrocytes
in the dorsal aorta: (A) day 11 rat; (B) day 10 SWV mouse.
Sections were incubated for 6 min. Original magnification
X 590. Bars = 10 μm. (Reprinted from Schreiner et al.,
1981.)

erythrocytes that stained at 6 minutes of incubation.
Alternatively, the carbonic anhydrase activity in the
nucleated erythrocytes that do not stain until 9 minutes
of incubation could be due to an isozyme of carbonic
anhydrase with lower catalytic activity.

To help distinguish between these alternatives we
examined the inhibition of carbonic anhydrase activity by
acetazolamide histochemically in nucleated erythrocytes
from SWV, CBA/J and rat embryos (Table 1). At 10^{-5} M
acetazolamide, carbonic anhydrase activity was completely
inhibited in the rat and both mouse strains. At 5×10^{-6} M
acetazolamide, carbonic anhydrase activity in rat and CBA/J
mouse erythrocytes was completely inhibited for 20 minutes,
but the SWV erythrocytes stained after 12-15 minutes of
incubation. At 10^{-6} M acetazolamide, nucleated erythrocytes

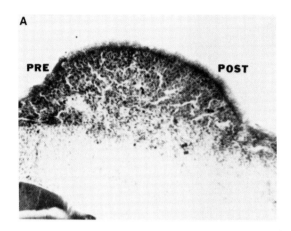

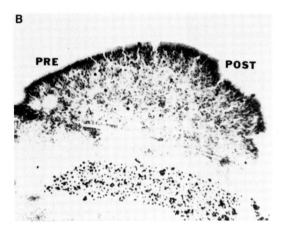

Figure 3. Histochemical staining of forelimb buds: (A)
day 11 rat; (B) day 10 SWV mouse. Sections were incubated
for 6 min and counterstained lightly with eosin. Note
absence of staining in ectoderm. Also note the differential staining of the nucleated erythrocytes below limb
in Figure 3B. Original magnification X 160. Bars = 50 μm.
(Reprinted from Schreiner et al., 1981.)

stained at 9 minutes in the SWV mouse, at 12 minutes in the
rat and at 15 minutes in the CBA/J mouse. It is well known
that the ability of acetazolamide to bind with carbonic
anhydrase varies with the form of enzyme tested. High

Table 1. Inhibition of carbonic anhydrase activity by acetazolamide in nucleated erythrocytes of the SWV mouse, rat, and CBA/J mouse.

	Acetazolamide concentration[a] (M)									
	10^{-5}		5 x 10^{-6}					10^{-6}		
	Time[b] (min)									
RBC source	20	9	12	15	20	6	9	12	15	20
SWV	–	–	(+)	+	+	–	(+)	+	+	+
Rat	–	–	–	–	–	–	–	(+)	+	+
CBA/J	–	–	–	–	–	–	–	–	(+)	+

[a]Ranking is relative to staining of controls: + = no inhibition; (+) = partial inhibition; – = complete inhibition.
[b]For each concentration, sections were floated on the incubation media and removed at various time intervals. Each point represents a minimum of three sections from three different incubations.
(Reprinted from Schreiner et al., 1981).

activity carbonic anhydrase (CA II) has a greater binding affinity for acetazolamide than does the low activity form (CA I) (Tashian 1977). The fact that a five to ten-fold greater amount of acetazolamide was needed to completely inhibit enzyme activity in SWV nucleated erythrocytes than in CBA/J or rat embryos supports the explanation that a low activity isozyme exists in SWV nucleated erythrocytes.

Thus we hypothesize that the existence of low activity carbonic anhydrase with low acetazolamide binding affinity in the erythrocytes of SWV mouse embryos plays an important role in resistance to acetazolamide terato-genesis. We expect that one of the physiological effects of carbonic anhydrase inhibition, such as hypercapnia or lowered pH, might be the basis of abnormal limb development and that SWV embryos are spared because a portion of their nucleated erythrocytes possess a form of carbonic anhy-drase that is not inhibited at usual teratogenic levels of acetazolamide. Accordingly, objectives of future studies are threefold: (1) elucidation of the isozymic

profiles of the nucleated erythrocytes of sensitive and
resistant embryos; (2) elucidation of vascular changes
resulting from carbonic anhydrase inhibition; and (3)
clarifying the relationship between these changes and
the limb tissue where the malformation manifests itself.

References

Biddle FG (1975a). Teratogenesis of acetazolamide in the
  CBA/J and SWV strains of mice. I. Teratology.
  Teratology 11:31.
Biddle FG (1975b). Teratogenesis of acetazolamide in the
  CBA/J and SWV strains of mice. II. Genetic control
  of the teratogenic response. Teratology 11:37.
Hallesy DW, Layton WM (1967). Forelimb deformity of
  offspring rats given dichlorophenamide during pregnancy.
  Proc Soc Exp Biol Med 26:6.
Hansson HPJ (1967). Histochemical demonstration of
  carbonic anhydrase activity. Histochemie 11:112.
Hirsch KS, Wilson JG, Scott WJ, O'Flaherty EJ (in press).
  Acetazolamide teratology and its association with
  carbonic anhydrase inhibition in the mouse. Teratogen
  Carcinogen Mutagen.
Layton WM, Hallesy D (1965). Deformity of forelimb in
  rats; association with high doses of acetazolamide.
  Science 149:306.
Maren TH (1967). Carbonic anhydrase: chemistry,
  physiology, and inhibition. Physiol Rev 47:595.
Maren TH, Ellison AC (1972). The teratological effect
  of benzolamide, a new carbonic anhydrase inhibitor.
  Johns Hopkins Med J 130:116.
Schreiner CM, Hirsch KS, Scott WJ (1981). Carbonic
  anhydrase distribution in rodent embryos and its
  relationship to acetazolamide teratogenesis.
  J Histochem Cytochem 29:1213.
Tashian RE (1977). Evolution and regulation of the
  carbonic anhydrase isozymes. In Rattazzi MC, Scandalios
  JG, Whitt GS (eds): "Isozymes: Current Topics in
  Biological and Medical Research," New York: Alan R.
  Liss, vol. 2, p 21.
Wilson JG, Maren TH, Takano K, Ellison A (1968).
  Teratogenic action of carbonic anhydrase inhibitors
  in the rat. Teratology 1:51.

SECTION FOUR
REGENERATION

Limb Development and Regeneration
Part A, pages 433–443
© **1983 Alan R. Liss, Inc., 150 Fifth Avenue, New York, NY 10011**

POSITIONAL MEMORY IN VERTEBRATE LIMB DEVELOPMENT AND
REGENERATION

Bruce M. Carlson. M.D., Ph.D.

Departments of Anatomy and Biological Sciences
University of Michigan
Ann Arbor, Michigan 48109 USA

During the past decade, the interpretations of many
morphogenetic phenomena in limb development and regeneration
have been based upon the concept of positional information,
as formulated by Wolpert (1969). In embryonic development,
cells of the early limb bud are assumed to be exposed to
some form of positional cue(s) or signal(s) by which they
are in some manner able to determine their relative positions in the limb field. On the basis of this information
they subsequently take part in pattern formation and undergo
appropriate cytodifferentiation. It is commonly assumed
that, having been exposed to positional cues, the cells acquire positional values, which represent their unique ad-
dresses within the three-dimensional matrix of the developing limb.

A number of studies on limb regeneration in both vertebrates and invertebrates have provided evidence for the retention of some expressible record of their original position
by certain tissues of the limb. This is known as positional
memory. The remainder of this article will summarize the
main characteristics of positional memory in the limbs of
vertebrates capable of epimorphic regeneration and also the
evidence for positional memory in the developing amniote
limb. Emphasis will be placed upon positional memory along
the transverse axes of the limb.

POSITIONAL MEMORY IN LIMBS CAPABLE OF EPIMORPHIC REGENERATION

A number of investigators working on the limbs of urodele amphibians (Droin 1959; Settles 1967; Lheureux 1972; Carlson 1974) have reported the formation of supernumerary regenerates after the axial rotation of some component (usually the skin) of the limb and amputation through the level of the rotated tissue. Early interpretations of this phenomenon were various, but implicit in most was the idea that the rotation disrupted some lines of morphogenetic communication within the regenerating limb. Further refinements of the interpretations of rotation experiments have been based upon the assumption that the rotated tissues retain a memory of their original position in the limb and that when tissues with disparate positional values are juxtaposed, they interact. There is evidence in both insects and amphibians (See reviews by French et al. 1976; Bryant et al. 1981) that the interaction takes the form of intercalary regeneration.

Very little is known about positional memory in the amphibian limb. Experimental studies have defined the following characteristics of positional memory along the transverse axes of the amphibian limb:

1. Positional memory resides principally, if not exclusively in dermis and muscle. There is little or no memory in epidermis, skeletal tissues or nerve (Carlson 1974, 1975a; Lheureux 1972, 1975a). It is felt by a number of investigators that in muscle positional memory is located mainly, or entirely, in the connective tissue component, but this assumption awaits experimental confirmation.

2. In the stylopodium of the axolotl forelimb, positional memory of one type is distributed along the anteroposterior axis, but not along the dorsoventral or the proximodistal axes (Carlson 1974). In the zeugopodial segment of the axolotl forelimb (Carlson, unpublished) and in forelimbs of other urodeles tested such a clear linear localization is not seen.

3. The positional memory in muscle and dermis persists throughout most, if not all of life in urodeles. This is indicated by the formation of supernumerary limbs after the rotation of stump tissues or blastemas in adult wild-caught newts, which are usually of indeterminate age (Settles 1967; Iten and Bryant 1975) and in 5-6-year old axolotls (Carlson, unpublished).

4. Positional memory in 180°-rotated muscle and dermis is very stable, persisting for as long as two years after rotation within the limb field (Carlson 1974, 1975a; Lheureux 1975a).

5. Positional values (and the memory of them) are estab-

lished very early in amphibian limb development. The formation of supernumerary limbs in Harrison's (1921) limb disc
rotations is in all likelihood an example of the same type
of positional effect that one sees after the rotation of
tissues in adult limbs.

6. Positional effects due to the apposition of tissues
with disparate positional values are not made manifest in the
absence of appropriate conditions for epimorphic regeneration
(Carlson 1974, 1975b). Limbs bearing rotated tissues show no
gross evidence of a response until or unless other conditions
leading to epimorphic regeneration are present. Despite the
absence of a grossly evident response, local intercalation of
missing positional values may occur according to some models
(French et al. 1976; Bryant et al. 1981). In insects, local
lateral intercalation between tissues with different positional values is readily apparent.

7. In order to affect the morphology of an epimorphic regenerative process, the tissues bearing positional memory
must be at or in the site of epimorphic activity. In skin
rotation experiments even a thin band of normally oriented
skin interposed between the rotated tissue and the wound
surface is sufficient to block the morphogenetic effect of
the rotated tissue (Carlson 1975a; Lheureux 1972).

8. In axolotls, positional memory is not expressed when a
cross-transplanted muscle is minced and allowed to degenerate
and then regenerate as an individual muscle (Carlson 1975b).
Yet positional memory persists, because when a limb bearing
cross-transplanted muscles in either the peak of the degenerative phase (5 days) or well into the regenerative phase
(30 days) is amputated, a complex multiple regenerate forms.

9. In the axolotl, positional memory does not depend upon
the integrity of the extracellular matrix, for multiple regenerates form after the implantation of positionally dislocated, minced dermis (Tank 1981; Carlson, unpublished) or
muscle (Carlson, 1975b).

10. The expression of positional memory (i.e. the formation of multiple regenerates) in tissue rotation experiments
is not related to axial rotation per se, but is rather due
to positional dislocation alone (Carlson 1975a).

11. The positional memory of stump tissues can influence
the morphogenesis of a regeneration blastema. This has been
demonstrated both by blastema rotation experiments (Iten and
Bryant 1975; Tank 1978) and by rotating stump skin proximal
to a regeneration blastema (Tank 1977).

12. Whether or not positional memory or its expression
are affected by x-radiation is presently unclear. In exper-

iments involving the irradiation and rotation of stump tissues before amputation, normal rather than multiple regenerates form (Rahmani and Kiortsis 1961; Carlson 1974; Lheureux 1975b; Tank 1981). Yet when non-irradiated blastemas are rotated about irradiated stumps, supernumerary regenerates consistently appear (Holder et al. 1979; Maden 1979), indicating the preservation and expression of positional memory in stump tissues.

Discussion of the various models of amphibian limb morphogenesis in which positional memory is implicit is beyond the scope of this article, but virtually all assume that the cells of the stable, non-regenerating limb contain some sort of positional values in a covert or overt form. Another common assumption is that if conditions are appropriate for distal transformation in an epimorphic system, abundant cell division, leading to the formation of a regeneration blastema, occurs at the interface between tissues or cell masses possessing different positional values.

POSITIONAL MEMORY IN THE LIMBS OF AMNIOTES

Less attention has been given to the question of positional memory in the limbs of vertebrates not capable of epimorphic regeneration, if for no other reason than there has been no readily available means of assaying for its presence or absence. The assay system used in urodeles, namely the occurrence of abnormal epimorphic regeneration, is by definition not available in these forms. Yet, it is important to learn whether or not the tissues of amniote limbs contain a memory of their positional values because, according to our present understanding, epimorphic regeneration would be theoretically impossible in limbs not possessing a functional set of positional values in some of the component tissues.

The principal questions that must be posed are 1) Do the tissues of developmentally stable amniote limbs possess an organized set of positional information, or is it absent? 2) If present, are the positional values readily expressible or are they unavailable because of some sort of repression or internalization? 3) If a system of expressible positional information cannot be demonstrated in the tissues of stable limbs, can it be demonstrated in developing limbs? 4)If expressible positional information is present within the tissues of non-regenerating limbs, is the inability to regenerate due to a deficiency of permissive or coordinating factors?

In investigating positional memory in the limbs of amniotes, a new assay system is clearly necessary. One alternative is to use the embryonic limb bud as the reactive system. For this, the avian wing bud is the obvious choice. A recent series of papers from Iten's laboratory (Iten 1982; Iten and Murphy 1980; Javois and Iten 1981, 1982) has produced evidence which has been interpreted to demonstrate a system of positional information in the developing wing of the chick embryo. The evidence has been suggested to explain several aspects of limb morphogenesis on the basis of the polar coordinate model (French et al. 1976). This model (Iten et al. 1981) has been considered to be opposed to the ZPA model of digital specification (Tickle et al. 1975; Summerbell and Tickle 1977). Of relevance to the present discussion is the validity of the evidence for and against the presence of positional memory in the avian limb bud. During the past year I have been investigating the question of positional memory in amniote limbs and, where relevant, I shall summarize my yet unpublished findings as they relate to the central issues. The major focus of my experiments has been to investigate certain aspects of positional memory in limb bud mesoderm cultured in vitro, but such experiments are based upon the definition of an adequate model for demonstrating positional memory in vivo. To do so, I have conducted experiments with donor tissue from quail wing buds and chick wing buds as hosts. The results of these experiments strongly suggest that a form of positional memory does, indeed, exist in the mesodermal cells of the amniote limb bud.

Iten and Murphy (1980) removed wedges of mesoderm + overlying ectoderm from anterior locations in stage 21 wing buds of chicks and grafted them into posterior slits at the level of the somite 19-20 junction. They reported a high incidence of growth of supernumerary structures with recognizable digits. As the disparity between levels of origin and insertion of the grafts decreased, the percentage of supernumerary structures decreased.

Both Fallon and Thoms (1979) and Iten and Murphy (1980) cited poor results when pieces of anterior chick mesoderm were grafted into posterior locations of chick host wing buds. My experiments on grafting chick anterior mesoderm also produced dramatically poorer results than those obtained with grafts of mesoderm + ectoderm, but a 45% incidence of super-

numerary structures (mainly nodules of cartilage) was obtained. Experiments with grafts of anterior quail mesoderm
into posterior slits in chick wing buds have produced a higher incidence of supernumerary structures (85% in preliminary
experiments), but in hundreds of grafts I have never seen an
identifiable digital structure. Instead, the structures were
either irregular nodules or rods (sometimes jointed) of cartilage. In my control series with grafts of mesoderm alone
(anterior-to-anterior, posterior-to-posterior or flank-to-
posterior slits) the overall incidence of supernumerary
structures was less than 5%.

In anterior-to-posterior mesodermal grafts the bulk of
the supernumerary structures are comprised of graft cells
(Fallon and Thoms 1979; Carlson, unpublished), but in larger
structures, intercalary regions are mosaics of graft and
host cells (Iten, 1982; Carlson, unpublished). With this
knowledge, it is not surprising that high doses of x-radiation to the donor tissue in anterior-to-posterior grafts
suppress the formation of supernumerary structures (Honig
1980). This is in sharp contrast to the results of experiments in which supernumerary structures form after the grafting of x-radiated ZPA´s into anterior host sites, for the
proposed mechanism of induction of supernumerary growth is
quite different in these experiments.

One of the most difficult criticisms to answer in anterior-to-posterior grafts is the possibility that the morphogenesis of supernumerary structures is due to the effect of
the putative ZPA morphogen upon the cells of the grafted anterior tissue that has been placed close to the ZPA. Iten
and Murphy (1980) felt that if this were the case, the percentages and morphology of supernumerary structures after
grafts into posterior slits of somite 17-18-level grafts
should be the same as that of 16-17-level grafts because both
are exposed to the same level of ZPA morphogen. Little enough
is known about normal ZPA mechanisms to allow indirect arguments based upon possible exposure to different levels of
morphogen either before or after grafting to be considered
definitive on either side of the question. Nevertheless
this is an important point to be considered, but its solution
depends upon devising ways of experimentally dissociating
high levels of ZPA morphogen from extreme posterior positional values.

Javois and Iten (1981) grafted posterior wedges from

various levels into standard anterior slits and found decreased percentages of supernumerary structures with decreasing distances between the levels of graft origin and the anterior slit. In my recent experiments, grafts of cultured
posterior quail wing mesoderm placed into homologous posterior slits running parallel to the longitudinal axis of the
chick host wings produced no supernumerary structures. However, when cultured posterior mesoderm was placed into transverse slits running from the posterior margin to the middle
of the wing bud, a 34% incidence of supernumerary structures,
mainly cartilaginous nodules, occurred, indicating a positional
effect on cartilage differentiation from posterior mesoderm.
Of greater significance was the finding that the supenumerary
cartilage was of graft origin. If the supernumerary structures had been formed as the result of the activity of ZPA
morphogen from the graft (for the donor mesoderm was obtained from areas of high ZPA activity), then one would have anticipated that the supernumerary structures would have been
of host origin. Additional evidence in favor of local positional interactions in the developing avian wing bud has been
reported by Javois et al. (1981), who found that for any
given level of anteroposterior disparity between graft and
host, the supernumerary response was greater if the dorsoventral orientation of the graft v s. host was also reversed.

The accumulated evidence from various grafting experiments provides considerable evidence for the existence of
some form of positional memory in the avian wing bud. The
existence of positional memory need not, however, be incompatible with a morphogenetically active ZPA, since the ZPA
hypothesis is concerned principally with the specification
of structures. Demonstrating the existence of a memory of
position does not in itself explain how the memory of positional values was originally established. Present experimental evidence is not incompatible with a morphogenetically
active (actual or potential) ZPA present in the limb bud concurrently with the memory of position . by mesodermal cells
that may have been exposed to the influences of the ZPA or
other positional cues.

Very little is known about the properties of positional
memory in the avian wing bud. In vivo grafting experiments
(Carlson, unpublished) have shown that it can be demonstrated
as early as HH stage 17, and it is not present, or if present,
not expressible in simple grafts, in late quail embryos. The
expression of positional memory in quail anterior mesoderm

declines sharply from the 4th through the 7th day of embryonic development. A large number of tissue culture experiments (Carlson, unpublished) have shown that positional memory can be maintained in pieces of anterior mesoderm cultured in vitro for up to 3 days outside the limb field. Thus it appears that the system of memory does not require constant input from the active limb field for its maintenance over that time.

The culture experiments of myself and many others (e.g. Ahrens et al. 1977; Gumpel-Pinot 1980) leave no doubt that pieces or aggregates of wing bud mesoderm possess the ability to form cartilage in isolation. In relating this to in vitro-in vivo grafting operations as well as the results of purely in vivo grafts of limb bud mesoderm, it might be just as important to consider the positionally related suppression of the tendency to self-differentiate into cartilage when dealing with homologous level control grafts as it is to consider the positionally related stimulation of differentiation and morphogenetic outgrowth in heterologous level grafts.

COMPARISON OF POSITIONAL MEMORY BETWEEN ANIMALS CAPABLE OF EPIMORPHIC LIMB REGENERATION AND HIGHER VERTEBRATES

There is now considerable evidence for the existence of a system of long-term positional memory in the limbs of animals capable of epimorphic regeneration. Positional memory also appears to exist in the amniote limb, but it can only be expressed during the early formative periods of limb development. Whether such memory is later erased or is tightly repressed is not known.

In urodeles, positional memory does not seem to be expressed in the morphogenetically stable limb, but it can be demonstrated by confronting tissues with the memory of different positional values. The role of positional memory in the normally developing limb is obscure, although according to some morphogenetic models it may be related to distal transformation. In animals with long-term positional memory, the confrontation of tissues with different positional values leads to morphogenetic activity at the interface, in the form of intercalary regeneration and, if conditions are right, distal transformation. The nature of the activity at the interface of juxtaposed tissues with short-term positional memory of differing positional values is less well documented. In demonstrations of either form of positional memory, the response is limited by the inherent capacity of the system to change its state of differentiation and the direction of morphogenesis.

ACKNOWLEDGMENT

Original research was supported by grants from the MDA and while a Faculty Scholar, Josiah Macy Jr. Foundation. My sincere thanks are due to Prof. Lauri Saxén at the University of Helsinki for providing the facilities of his laboratories to me for my experiments on the developing avian wing.

REFERENCES

Ahrens PB, Solursh M, Reiter RS (1977). Stage-related capacity for limb chondrogenesis in cell culture. Devel Biol 60:69.
Bryant SV, French V, Bryant PJ (1981). Distal regeneration and symmetry. Science 212:993.
Carlson BM (1974). Morphogenetic interactions between rotated skin cuffs and underlying stump tissues in regenerating axolotl forelimbs. Devel Biol 39:263.
Carlson BM (1975a). The effects of rotation and positional change of stump tissues upon morphogenesis of the regenerating axolotl limb. Devel biol 47:269.
Carlson BM (1975b). Multiple regeneration from axolotl limb stumps bearing cross-transplanted minced muscle regenerates. Devel Biol 45:203.
Droin A. (1959). Potentiailtès morphogenès dans la peau du triton en régénération. Rev Suisse Zool 66:641.
Fallon JF, Thoms SD (1979). A test of the polar coordinate model in the chick wing bud. Anat Rec 193:534.
French V, Bryant PJ, Bryant SV (1976). Pattern regulation in epimorphic fields. Science 193:969.
Gumpel-Pinot M (1980). Ectoderm and mesoderm interaction in the limb bud of the chick embryo studied by transfilter cultures: Cartilage differentiation and ultrastructural observations. J Embryol Exp Morph 59:157.
Harrison RG (1921). On relations of symmetry in transplanted limbs. J Exp Zool 32:1.
Holder N, Bryant SV, Tank PW (1979). Interactions between irradiated and unirradiated tissues during supernumerary limb formation in the newt. J Exp Zool 208:303.
Iten LE (1982). Pattern specification and pattern regulation in the embryonic chick limb bud. Am Zool 22:117.
Iten LE, Bryant SV (1975). The interaction between the blas-

tema and stump in the establishment of the anterior-posterior organization of the limb regenerate. Devel Biol 44:119.

Iten LE, Murphy DJ (1980). Pattern regulation in the embryonic chick limb: Supernumerary limb formation with anterior (non-ZPA) limb bud tissue. Devel Biol 75:373.

Iten LE, Murphy DJ, Javois LC (1981). Wing buds with three ZPAs. J Exp Zool 215:103.

Javois LC, Iten LE (1981). Position of origin of donor posterior chick wing bud tissue transplanted to an anterior host site determines the extra structures formed. Devel Biol 82:329.

Javois LC, Iten LE (1982). Supernumerary limb structures after juxtaposing dorsal and ventral chick wing bud cells. Devel biol 90:127.

Javois LC, Iten LE, Murphy DJ (1981). Formation of supernumerary structures by the embryonic chick wing depends on the position and orientation of a graft in a host limb bud. Devel Biol 82:343.

Lheureux E (1972). Contribution à l'étude du rôle de la peau et des tissus axiaux du membre dans la déclenchement de morphogenèses régénératrices anormales chez le triton Pleurodeles waltlii Michah. Ann Embry. Morph 5:165.

Lheureux E (1975a). Nouvelles données sur les rôles de la peau et des tissus internes dans la régénération du membre du triton Pleurodeles waltlii Michah. Roux' Arch 176:285.

Lheureux E (1975b). Régénération des membres irradiés de Pleurodeles waltlii Michah. Roux'Arch 176:303.

Maden M (1979). Regulation and limb regeneration: The effect of partial irradiation. J Embryol Exp Morph 52:183.

Rahmani T, Kiortsis V (1961). Le rôle de la peau et des tissus profonds dans le régénération de la patte. Rev Suisse Zool 68:91.

Settles HE (1967). Supernumerary regeneration caused by ninety degree skin rotation in the adult newt, Triturus viridescens. Ph.D. Thesis, Tulane University, New Orleans.

Summerbell D, Tickle C (1977). Pattern formation along the anteroposterior axis of the chick limb bud. In Ede DA, Hinchliffe JR and Balls M (eds): "Vertebrate Limb and Somite Morphogenesis," Cambridge, Cambridge Univ. Press, p. 41.

Tank PW (1977). The timing of morphogenetic events in the regenerating forelimb of the axolotl, Ambystoma mexicanum. Devel Biol 57:15.

Tank PW (1978). The occurrence of supernumerary limbs following blastemal transplantation in the regenerating forelimb of the axolotl, Ambystoma mexicanum. Devel Biol 62:143.

Tank PW (1981). The ability of localized implants of whole or minced dermis to disrupt pattern formation in the regenerating forelimb of the axolotl. Am J Anat 162:315.

Tickle C, Summerbell D, Wolpert L (1975). Positional signalling and specification of digits in chick limb morphogenesis. Nature (London) 254:199.

Wolpert L (1969). Positional information and the spatial
  pattern of cellular differentiation. J Theoret Biol 25:1.

Limb Development and Regeneration
Part A, pages 445–454
© **1983 Alan R. Liss, Inc., 150 Fifth Avenue, New York, NY 10011**

VITAMIN A AND THE CONTROL OF PATTERN IN REGENERATING LIMBS

Malcolm Maden

National Institute for Medical Research,
Mill Hill, London, U.K.

The normal process of limb regeneration ensures that an
exact copy of the elements removed by amputation are replaced.
This simple observation implies that blastemal cells at the
level of the cut must have some knowledge of their position
in the limb in order to know from which level to commence
regeneration. Previous experiments have revealed that this
knowledge of position is not reflected in relatively simple
parameters such as the number of cells in the early blastema
or the cell-cycle time of blastemal cells (Maden, 1976). We
are therefore tempted to resort to more intangible concepts
such as positional information (Wolpert, 1969) in order to
explain the phenomenology of regeneration.

To investigate in what form positional information may
be encoded in the limb we need a reliable method of altering
the normal course of pattern formation during regeneration.
Recent experiments with Vitamin A (Maden, 1982) have revealed
the great potential that this molecule could have in
contributing to our understanding of the nature of positional
information. This work is briefly reviewed below, along
with further data on the effects of Vitamin A on the
regenerating limbs of young axolotls, Ambystoma mexicanum.

The Effect of Vitamin A

When 50-70mm axolotls were immersed in a solution of
retinol palmitate (15 IU/ml) for 12 days after bilateral
amputation through the radius and ulna, various abnormalities
in the proximodistal sequence of regenerated elements were

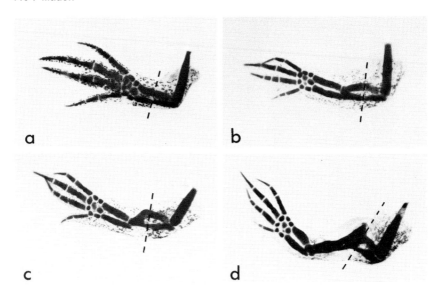

Fig. 1. Cleared whole mounts of regenerated limbs stained
with Victoria blue. All limbs were amputated through the
radius and ulna, the broken line marking the amputation plane.
a, Control. Distal radius and ulna, carpals and digits
replaced. b, After 12 days in 15 IU/ml retinol palmitate
this limb regenerated an extra radius and ulna in tandem.
c, Same treatment as b, but here an elbow joint and distal
end of the humerus has also been intercalated. d, After
15 days in 75 IU/ml retinol palmitate a complete new limb
has regenerated from the amputation plane. Magnifications:
a, x13 b, x11 c, x10 d, x9.

apparent. Instead of replacing only those elements removed
(Fig. 1a), extra tissue was regenerated ranging from extra
carpals, an extra length of radius and ulna, an extra
complete radius and ulna in tandem (Fig. 1b) or an extra
distal humerus and radius and ulna (Fig. 1c). If the length
of time that the animals were immersed in retinol palmitate
was increased to 15 days and the concentration increased to
75 IU/ml, then complete new limbs were regenerated from the
cut ends of the radius and ulna (Fig. 1d) in 7 out of 8
cases.

A clear concentration and time effect was apparent : as the concentration of Vitamin A was increased and its time of administration lengthened, so the level at which the limb was repeated became more proximal (Table 1 of Maden, 1982).

It is important to emphasize that, as can be seen from the Figures, the anteroposterior axis (digit and carpal sequence) of the regenerates are perfectly normal. Several limbs have also been serially sectioned to study their muscle patterns and these too were normal. Therefore Vitamin A seems to uniquely affect the proximodistal axis of regenerating axolotl limbs.

Other Amputation Levels

Similar concentration and time effects were apparent after amputation through limb levels other than the radius and ulna. After amputating either through the humerus or carpals a complete limb can be regenerated from the cut (Fig. 2a). However, the concentration of retinol palmitate needed to produce this result was lower from the humerus level than from the carpal level. This suggests that the amount of Vitamin A needed to induce the regeneration of a complete limb decreases as the amputation plane becomes more proximal.

These experiments have also been performed on hindlimbs at all three amputation levels. Here too similar time and concentration effects are found. At high concentrations complete limbs can be regenerated from amputation through the tibia and fibula (Fig. 2b) or tarsals (Fig 2c).

The Effect of Other Retinoids

The effect of retinol acetate, retinol palmitate, retinol and retinoic acid on regeneration after amputation through the radius and ulna have been compared. At a concentration of 0.12mM (equivalent to 15 IU/ml retinol palmitate) there was a clear difference in the degree of repetition caused by the 4 retinoids. Retinoic acid was the most potent, causing the regeneration of complete limbs from the amputation plane in 3 out of 15 cases (Fig 3a). Retinol was the second most effective and 1 out of 16 cases regenerated a complete limb. Retinol palmitate only

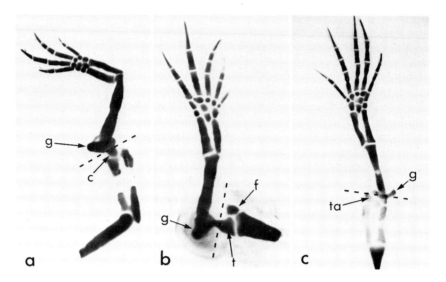

Fig. 2. Regenerates from limbs placed in 75 IU/ml (a,b) or
150 IU/ml retinol palmitate for 15 days. Broken lines mark
the amputation planes. a, Complete limb regenerated from an
amputation through the carpals. One remaining carpal (c)
can still be seen. b, Full hindlimb regenerate from amputation
through the tibia and fibula (t), (f). c, Full hindlimb
regenerate from amputation through the tarsals (ta). In all
these cases, in addition to the complete limb a part of the
girdle (g) has also regenerated. Magnifications : a, x9
b, x10 c, x7.

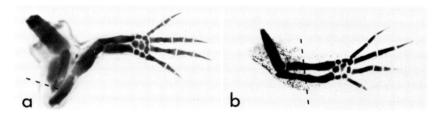

Fig. 3. Regenerates from limbs placed in the same
concentration (0.12mM) of retinoic acid (a) or retinol
palmitate (b) for 15 days. Broken lines mark the amputation
plane. In a, a complete limb has regenerated from the radius
and ulna, but in b only an abnormally long radius and ulna
can be detected. Magnifications : a, x8 b, x12.

produced limbs with abnormally long radii and ulnae (5 out of
20 cases) (Fig. 3b), no proximal elements being regenerated
and retinol acetate was the least effective with abnormally
long radii and ulnae in 2 out of 20 cases.   By recording
the degree of reduplication on a 0 (normal) to 5 (complete
limb) score and summing the values the following relative
efficacy was obtained: retinoic acid 80, retinol 68,
retinol palmitate 10, retinol acetate 4.

The Effect of Other Vitamins

     To determine whether the effect of Vitamin A was
unique, other vitamins have been tested for their effect on
regeneration after amputation through the radius and ulna
or tibia and fibula.   The concentration of solutions used
was 0.6 mM (equivalent to 75 IU/ml of retinol palmitate)
and the vitamins used were B1, B2, B6, C and E.   The only
noticeable effect of these compounds was an occasional
reduction in the number of digits regenerated, particularly
in the hindlimb and in Vitamin E phalanges tended to fuse.
Otherwise normal regenerates were produced.   Thus the
effect of Vitamin A is unique.

Time Course of Action

     In order to define the phase of regeneration during
which Vitamin A acts, the following two experiments were
performed.   Firstly, after amputation through the radius
and ulna or tibia and fibula (zeugopodium) the animals were
placed in 0.6 mM retinol palmitate (75 IU/ml) for varying
periods from 2 to 20 days.   The degree of reduplication of
the regenerates was scored as before on a 0 (normal) to
5 (complete limb from amputation plane) score and the
result is shown in Fig. 4a.   In both fore and hindlimbs
the degree of reduplication increased to a maximum at 12
days and remained there until 18 days.   But, if the
amputated limbs were kept in retinol palmitate for as long
as 20 days then regeneration was totally inhibited.

     In the second experiment the limbs were amputated at
the same level and varying times (0-20 days) allowed to
elapse before placing them in 0.6 mM retinol palmitate for
15 days.   The results are shown in Fig. 4b, scored as
before.   This reveals that if amputated limbs are placed

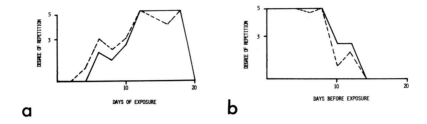

Fig.4.   Average degree of reduplication of regenerates placed in 75 IU/ml retinol palmitate.   Broken lines – forelimbs amputated through the radius and ulna, solid lines – hindlimbs amputated through the tibia and fibula.   See text for details of degree of reduplication. a, limbs placed in Vitamin A immediately after amputation for increasing periods of time.   b, limbs allowed to regenerate for increasing periods of time before placing in Vitamin A.

in Vitamin A during the first 8 days of regeneration then a high degree of reduplication will result, but after longer delays the effectiveness decreases.

Therefore, these two complementary experiments demonstrate that in order to have the maximum effect, Vitamin A has to be present during the early phases of regeneration.   These include dedifferentiation and the establishment of the blastema, the time during which the important events of pattern formation are taking place.

Cellular Effects of Vitamin A

Observations during the course of these experiments revealed that in high concentrations of Vitamin A regeneration is inhibited, yet dramatic effects on pattern formation are manifest after removal from Vitamin A.   This is shown in Fig. 5a with camera lucida drawings. After 15 days, experimental limbs (lower set) have only a pimple of epidermis at the tip, whereas control limbs (upper set) have virtually completed regeneration.   A preliminary investigation of

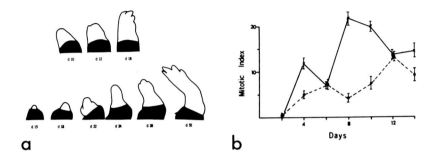

Fig. 5a. Camera lucida drawings of the external form of
control (upper set) and experimental (lower set) regenerating
limbs after amputation through the radius and ulna. The
hatched area marks the stump, the clear area the regenerate.
The time elapsed, in days, since amputation is recorded
under each drawing. Regeneration in control limbs is
virtually complete by 18 days (top right), whereas after
15 days in 75 IU/ml retinol palmitate (bottom left) no
regeneration has taken place. Only upon removal from
Vitamin A does regeneration then commence. Fig. 5b. Mitotic
indices (per thousand) of control (solid lines) and
experimental (broken lines) regenerates at various times
after amputation. Experimental limbs were placed in 75 IU/ml
retinol palmitate for the duration of the experiment (14
days). Although mitosis is not totally inhibited in
experimental limbs, the two large bursts of mitotic activity
shown by control limbs are prevented. Bars mark standard
errors.

the cellular effects of Vitamin A has indicated that mitosis
is severely reduced, but not abolished (Fig. 5b). This
reveals the paradoxical nature of Vitamin A - it suppresses
cell division yet once released from this inhibition, extra
pattern is then regenerated.

Is Time Delay the Cause of the Effect?

It is possible that the delay in regeneration caused by the suppression of mitosis (Fig. 5) could be responsible for the profound effects on pattern formation. Fortunately, a simple test of this proposition can be performed because denervated axolotl limbs do not regenerate and the mechanism of action, via the inhibition of cell division, is the same (Singer, 1978). When the denervated limb becomes reinnervated, regeneration commences, thus the delay before regeneration can be manipulated at will.

Forelimbs were denervated by crushing the nerves at the brachial plexus and then amputated through the radius and ulna. They were redenervated on days 6 and 12 post amputation. After 15 days no regeneration had taken place, as in the Vitamin A treated limbs, but upon reinnervation, regeneration commenced. Such denervated limbs, as expected, only regenerated the missing elements - distal radius and ulna and hand (Fig. 6). Therefore time delay per se is not responsible for the effects of Vitamin A on pattern formation.

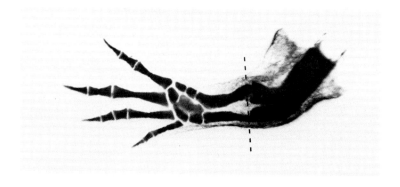

Fig. 6. Limb regenerated after amputation through the radius and ulna (broken line), having been denervated for 15-20 days. Only those elements removed by amputation were replaced, in contrast to Vitamin A treatment (cf Figs 1,2 & 3). Magnification x 12.

Discussion

The dramatic effects of Vitamin A on the regenerating
axolotl limb are to cause tandem repetition of elements
specifically in the proximodistal axis. It is most potent
in its action during the early phases of regeneration, that
is dedifferentiation and the establishment of the blastema.
What seems to be occurring is that the proximal boundary
value of the cells of the early blastema is being changed
to a more proximal value by Vitamin A. The more Vitamin A,
the more proximal the altered positional value becomes,
leading eventually to the regeneration of a complete limb
(including part of the girdle - Fig. 2) from a distal
amputation site. The anteroposterior and dorsoventral axes
of the extra pattern elements are perfectly normal. What
this means in terms of molecular changes occurring in the
blastemal cells is, of course, entirely speculative as no
biochemical experiments have yet been performed on amphibian
tissues. For the present, a more fruitful approach is to
compare the results described above with those in other
systems.

Experiments on developing chick limbs have revealed
that Vitamin A affects this system in a very different
manner (Summerbell and Harvey, this Symposium; Tickle,
Alberts, Wolpert and Lee, 1982). Here, duplications in the
anteroposterior axis are produced and no effects on the
proximodistal axis have been detected. These contrasting
results raise several important questions. Is this
difference in behaviour due to differences in the mechanisms
of development and regeneration or are they due to differences
between birds and amphibians ? To answer these questions we
need to perform the same experiment on the developing limbs
and then the regenerating limbs of the same species.
Preliminary evidence from Rana and axolotls indicates that
the answer to both the above questions is yes, thereby
considerably complicating the situation.

Vitamin A, applied either in the water, as in the
regeneration experiments, or by local application, as in
the chick experiments, causes deletions in the proximodistal
axis of developing limb buds of Rana and axolotls (Maden,
unpublished). This compliments the known teratogenic
effects of Vitamin A on mammalian limb development.
No effect on the anteroposterior axis could be detected,
revealing a difference in developmental behaviour between

amphibians (and presumably mammals also) and chicks. On the
other hand the regenerating limbs of axolotls produce tandem
repetitions of elements revealing a difference between the
developmental and regenerative behaviour of the same species.
Furthermore, experiments on regenerating Rana limbs have
revealed that both the anteroposterior and proximodistal
axes can be duplicated after Vitamin A treatment. Clearly,
the situation is rather complicated and many more experiments
need to be performed to unravel the phenomenology of
Vitamin A effects before a fruitful biochemical approach
can be made. When the time is ripe for such a course an
understanding of the molecular basis of pattern formation
will hopefully ensue.

I thank Katriye Mustafa for excellent technical
assistance during the course of this work.

Maden M (1976). Blastemal kinetics and pattern formation
    during amphibian limb regeneration. J.Embryol exp Morph
    36: 561-574.
Maden M (1982). Vitamin A and pattern formation in the
    regenerating limb. Nature 295: 672-675.
Singer M (1978). On the nature of the neurotrophic
    phenomenon in urodele limb regeneration. Amer Zool
    18: 829-841.
Summerbell D, Harvey F (1982). Vitamin A and the control
    of pattern in limb development. This Symposium.
Tickle C, Alberts B, Wolpert, L, Lee J (1982). Local
    application of retinoic acid to the limb bond mimics the
    action of the polarizing region. Nature 296: 564-566.
Wolpert L (1969). Positional information and the spatial
    pattern of cellular differentiation. J theor Biol
    25: 1-47.

Limb Development and Regeneration
Part A, pages 455–465
© **1983 Alan R. Liss, Inc., 150 Fifth Avenue, New York, NY 10011**

THE ORIGIN OF TISSUES IN THE X-IRRADIATED REGENERATING LIMB
OF THE NEWT *PLEURODELES WALTLII*

Emile Lheureux

Université des Sciences et Techniques de Lille
Laboratoire de Morphogenèse Animale
59655 Villeneuve d'Ascq Cedex. France

Limb regeneration in newts is inhibited by appropriate
doses of X-irradiation. Umanski (1937) and many workers
after him demonstrated that transplantation of a variety of
non irradiated tissues to irradiated limb stumps reversed
this inhibition. Grafted tissues were thought to provide
all the cells of regenerates (Umanski 1937, 1938 ; Thornton
1942 ; Wallace, Wallace 1973 ; Wallace, Maden, Wallace 1974 ;
Maden 1979 b). According to this hypothesis, a study of
metaplastic conversion of grafted tissues could be performed by the observation of regenerate histological sections.
However, several investigations demonstrated that both irradiated and non irradiated cells participated to regenerating tissues (Liosner 1947 ; Sidorova 1949 ; Lagan 1961 ;
Desselle 1968 ; Desselle, Gontcharoff 1978). Then, several
authors have traced cells derived from labelled grafts in
irradiated as well as non irradiated regenerating limbs
(Hay 1952 ; Barr 1964 ; Steen 1968 ; Namenwirth 1974 ; Dunis,
Namenwirth 1977 ; Desselle, Gontcharoff 1978). Cells derived
from non irradiated triploid tissues transplanted to irradiated diploid limb stumps were traced among regenerating
tissues by the presence of three nucleoli instead of two
(Namenwirth 1974 ; Dunis, Namenwirth 1977). Metaplasia was
observed but the histological studies were ineffective to
detect the possible presence of irradiated cells among regenerating tissues. On the other hand, labelling of the
stump tissues by means of tritiated thymidine, prior to
irradiation, allowed to detect a few irradiated cells in
the blastema , but the fate of these cells was not studied
(Desha 1974). Although the use of tritiated thymidine may
be criticized, the presence of irradiated cells in the

blastema cannot be excluded. Another method has been used
to study the fate of both irradiated and non irradiated tissues (Desselle, Gontcharoff 1978). It consisted in transplanting non irradiated cartilage from phalanges of *Desmogna-
thus fuscus* into irradiated limb stump of *Triturus vulgaris*
and calculating the percentage of cells from each species
by means of cytophotometric analysis of regenerate cell nuclei. Desselle and Gontcharoff showed that only 5 % to 15 %
of the cells derived from non irradiated grafts.

In order to approach the problem of the participation
of irradiated cells to regenerating tissues and to determine the ability of tissues to undergo metaplasia, we transplanted in *Pleurodeles waltlii* non irradiated triploid tissues to irradiated diploid limb stumps and performed cytophotometric analysis of regenerating tissue cell nuclei.

MATERIALS AND METHODS

The operations were performed on 60-70 mm larval newt
Pleurodeles waltlii bred in our laboratory.

X-irradiation.

The entire right arm and shoulder of host specimen were
irradiated with 2000 R of X-rays. The apparatus used was a
Transfoleix 90.20 Massiot, set at 85 kV, 2.5 mA, with no
added filtration. The target distance was 5.5 cm. The dose
rate of irradiation was 160 R per minute.

Triploid Animals.

Triploid donor animals were obtained by a cold treatment of freshly spawned newt eggs (Beetschen, 1960). Fertilized eggs were collected from a spawning female
and placed in a 0°C water bath for 10 hours, then returned
to room temperature to develop. Among surviving larvae,
85 % were triploid.

Transplantation of grafts.

The triploid tissues used as grafts consisted of either

cartilage, muscle or skin. Pure cartilage from humerus diaphysis was inserted under the skin of the host-irradiated
limb stump. Skin cuff grafts were removed from the upper
arm of triploid donor animals. Pieces of adhering muscle
were carefully removed from the grafts. Prior to transplantation, the corresponding irradiated skin of the host's
upper arm was removed and discarded. Muscle grafts were
composed of 2 pieces of muscle excised from ventral and
dorsal musculature of the upperarm, which were inserted under the skin of the host upper arm. Operated limbs were
amputated through the grafts two days later.

Histology.

A few regenerates resulting from skin graft, were
fixed in Bouin's fluid and then decalcified, embedded in
tissuemat, and serially sectioned at 7 μm. The sections were
stained with azan.

Cytophotometry.

In order to separate epidermis from dermis, pieces of
skin were soaked for 20 minutes in 0.1 % EDTA in modified
Steinberg's solution, lacking calcium and magnesium and
adjusted to pH 8.3 with Tris buffer. Dermis, muscle and
cartilage samples underwent an enzymatic treatment for half
an hour in 2 % trypsin in modified Steinberg's solution.
Every sample was then fixed in Carnoy's fixative. Squashes
of different tissue samples were prepared, then hydrolysed
in 1 N H Cl at 60°C for 7 minutes and stained in Feulgen
for one hour.

Measurements were made by the two wavelenghts method
(560 nm, 498 nm) using a Leitz MPV cytophotometer. From
the values (in arbitrary units) obtained by measurements
of 100 diploid cell nuclei, a mean value m 1 was calculated.
It represents a relative value of diploid cell DNA content.
Then a second value m2 was calculated from m1 value :
m2 = 1.5 m1. m2 value represents a relative value of triploid cell DNA content. Measurements of DNA content of
regenerated tissue cell nuclei were compared with m1 and
m2 values. Histograms were drawn so that diploid and triploid
cell populations of the regenerating tissues could be easily
distinguished (Fig. 1, 2).

RESULTS

It was shown in a previous paper (Lheureux 1975) that regeneration of an amputated limb which had been irradiated with 2OOO R of X-rays was inhibited. Moreover, non irradiated skin cuffs or muscle fragments were able to induce limb regeneration, when transplanted to an irradiated limb stump.

No regenerate resulted from Pure Cartilage Grafts.

Cartilage from humerus diaphysis was unable to induce regeneration of irradiated limbs. In a few cases (8 out of 23) very little spikes were observed from contralateral diploid humerus cartilage. Transplantation of triploid cartilage were attempted (10 cases) but no regeneration occured.

Epidermis.

The dynamics of replacement of irradiated epidermis by non irradiated epidermis have been studied by means of cytophotometric analysis in order to distinguish diploid cells of the host animal and triploid cells from a triploid skin cuff grafted at the most proximal level of an irradiated forelimb. Detailed experiments and results will be published in a separate paper. The main results are the following ones. (1). All the irradiated epidermis cells of an irradiated limb are replaced by non irradiated triploid cells migrating distally. This replacement is completed within 3 to 4 weeks. (2). The regenerates resulting from non-irradiated muscle grafts are always covered with non irradiated epidermis cells. (3). When the growth of regenerates has ended, epidermis as a whole continues to migrate slowly distally so that the epidermis of the limb is replaced within six months.

Regenerates resulting from Grafts of Triploid skin.

The regenerates resulting from skin grafts seemed to be devoid of muscle so that a histological study was performed.
Histological data. Four regenerates from irradiated limb stumps bearing non irradiated cleaned skin cuffs were cut into serial sections and stained for a histological

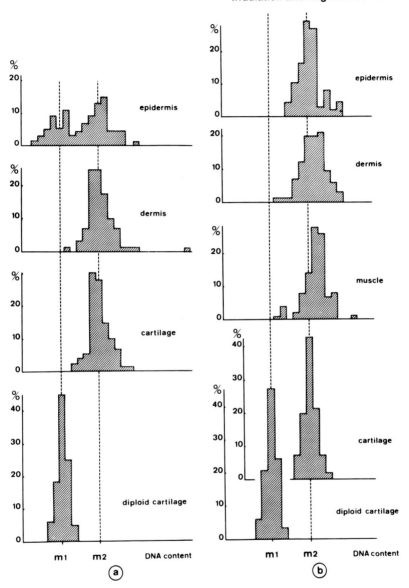

Fig. 1. Two regenerates from triploid dermis graft. m1 and m2 respectively represent DNA content mean values of diploid and triploid cell nuclei. Lower histogram : control diploid cartilage. Upper histograms : regenerated tissues. a: regenerate without muscle. b: regenerate with muscle.

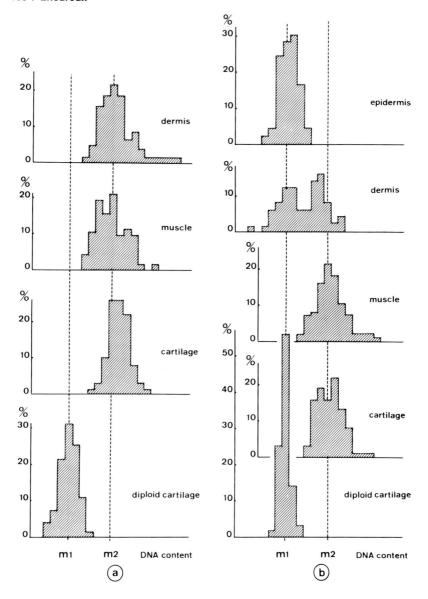

Fig. 2. Two regenerates from triploid muscle graft. For explanation see Fig. 1. a: All the mesodermal regenerated tissues are triploid (3 cases). b: Cartilage and muscle are triploid and many diploid cells are present in dermis (4 cases). Epidermis is always diploid (5 cases were analysed).

examination. Skin and skeleton were present but muscle was
seen in only one regenerate. In the three other regenerates
connective tissue filled the space between epidermis and
skeleton.

Cytophotometric data. Nine regenerates were used to
study the distribution of diploid and triploid cells among
regenerating tissues. 7 limbs were operated and the regenerates were analysed. Two limbs were reamputated and the second regenerates were also studied. All the regenerates but
one corresponding to a second regeneration, did not show
any muscle when dissected. In every regenerate, samples of
dermis and cartilage showed only triploid cells (Fig. 1 a).
In the regenerate provided with muscle all the mesodermal
tissues also derived from triploid graft (Fig. 1 b). In 4
regenerates,epidermis was studied too. It was triploid in
3 cases (Fig. 1 b) and both triploid and diploid in the last
case (Fig. 1 a). This result pointed out the slow migration
of the whole epidermis towards the distal tip.

Finally dedifferentiated cells from dermis could at
least redifferentiate into dermis and cartilage.

Regenerates resulting from Grafts of Triploid Muscle.

All the normal limb tissues were present in the 7 studied regenerates. Epidermis, analysed in 5 limbs was diploid
(Fig. 2 b). Cartilage and muscle samples were only constituted by triploid cells. However dermis showed either only
triploid cells (Fig. 2 a) or both triploid and diploid cells
(Fig. 2 b).

These results demonstrated that the cells dedifferentiated from muscle can redifferentiate into muscle, cartilage and dermis cells. Presence of diploid cells will be discussed later.

DISCUSSION

From these results three main questions arise (1). Why
cartilage implants did not allow irradiated limb stumps to
regenerate ? (2). Did irradiated cells actually participate
to regenerated tissue formation ? (3). What might be the
limits of transdifferentiation during limb regeneration ?

Cartilage and absence of regeneration.

According to the model of pattern regulation in epimorphic fields (French, Bryant, Bryant 1976) it is easy to
understand that regenerates can develop from either a skin
cuff or a graft composed of two fragments of muscle from
dorsal and ventral upper arm. If cartilage grafts cannot re-
initiate regeneration of irradiated limb, it may be thought
that cartilages do not have positional values or more likely
cannot regenerate appropriate values to produce a regenerate.
This conclusion explains the failure of regeneration of an
irradiated limb stump bearing a non irradiated dorsal or
ventral skin strip (Lheureux 1975). Eggert (1966) transplanted scapula cartilage to irradiated limb stumps of *Triturus
viridescens* and did not obtain any regenerate. However, in
other species regenerates were produced from irradiated limb
stumps bearing cartilage grafts (Desselle, 1968 ; Wallace,
Maden, Wallace 1974 ; Desselle, Gontcharoff 1978). Thus,
different species seem to behave differently facing the
same experiments. The age of the animals could be taken in
consideration (Wallace, Maden, Wallace 1974). This experiment
will have to be repeated on younger *Pleurodeles* larvae.

Do Irradiated Cells actually participate to Regeneration ?

Our results showed that diploid cells did not participate to cartilage and muscle constitution. Dermis originated
from triploid grafts but, in a few regenerates developed
from stumps bearing triploid muscle grafts, it was composed
of both triploid and diploid cells. Further investigations
will have to be carried out to know whether irradiated cells
actually participated to regenerated dermis or whether the
analysed samples were contaminated with epidermis cells
uncompletely separated from dermis. Nevertheless, it is not
undue to suppose that non irradiated tissues were the entire
source of the regenerating tissues inasmuch as X-rays were
shown to block blastemal cell cycle (Maden 1979 a). Likewise,
Namenwirth (1974) indicated that the percentage of cells
showing three nucleoli in all-triploid limbs was comparable
with the percentage of cells showing three nucleoli in
regenerates resulting from triploid grafts transplanted to
diploid irradiated limb stumps. However, our results were
inconsistent with those of Desselle and Gontcharoff (1978)
who demonstrated that 5 % to 15 % of the regenerating tissue
cells proceeded from grafts and the rest from irradiated

stumps. So, similar experiments led to different results.
This points out the need to irradiate in standard conditions.

Transdifferentiation in regenerating tissues.

Our results showed that cells originating from dermis
or skeletal muscle could express other phenotypes than their
own phenotypes. Dermis gave rise to dermis and cartilage
while muscle was the origin of new muscle dermis and cartilage. However, there is no evidence that regenerated muscle could originate from dermis. When we obtained muscle
from skin graft, a few muscle fibers could have adhered the
skin and escaped detection when skin cuffs were cleaned.
Our results are in line with those of Namenwirth (1974) and
Dunis and Namenwirth (1977). From our data it was not possible to determine which cells underwent a metaplastic conversion. Since cartilage derived from muscle as well as dermis
mesenchymatous cells might give rise to chondrocytes. However
we must keep in mind that *in vitro* experiments showed that
cartilage can arise from either muscle cells or fibroblasts
of rat skeletal muscle (Nathanson and Hay, 1980). Since
muscle resulting from dermis grafts (Namenwirth 1974 ;
Dunis and Namenwirth, 1977 ; this report) was only observed
in a few cases, transdifferentiation of mesenchymatous cells
into myoblasts cannot be affirmed. Conversion of chondrocytes
into myoblasts was not yet accurately established. Eggert
(1966) and Steen (1968) showed a relative stability of chondrocyte phenotype. However when triploid cartilage was transplanted with diploid muscle, a few chondrocytes could undergo a transdifferentiation into myoblasts (Namenwirth, 1974).

The use of pure tissues associated with a heritable
label and a reliable detection of cells among regenerating
tissues should contribute to the progress of our studies.

REFERENCES

Barr HJ (1964). The fate of epidermal cells during limb re-
    generation in larval *Xenopus*. Anat Rec 148:358.
Beetchen JC (1960). Recherches sur l'hétéroploidie expéri-
    mentale chez un Amphibien Urodèle *Pleurodeles waltlii*
    Michah. Bull Biol Fr Belg 96:12.
Desha DL (1974). Irradiated cells and blastema formation in
    the adult newt *Notophthalmus viridescens*. J Embryol exp
    Morph 32:405.

Desselle JC (1968). Restauration de la régénération par des implants de cartilage dans les membres irradiés de *Triturus cristatus*. C R Acad Sci 267:1642.

Desselle JC, Gontcharoff M (1978). Cytophotometric detection of the participation of cartilage grafts in regeneration of X-rayed Urodele limbs. Biol Cell 33:45.

Dunis DA, Namenwirth M (1977). The role of grafted skin in the regeneration of X-irradiated axolotl limbs. Develop Biol 56:97.

Eggert RC (1966). The response of X-irradiated limbs of adult Urodeles to autografts of normal cartilage. J Exp Zool 161:369.

French V, Bryant PJ, Bryant SV (1976). Pattern regulation in epimorphic fields. Science 193:969.

Hay ED (1952). The role of epithelium in Amphibian limb regeneration, studied by haploid and triploid transplants. Amer J Anat 91:447.

Lagan M (1961). Regeneration from implanted dissociated cells II. Regenerates produced by dissociated cells derived from different organs. Folia Biol 9:3.

Lheureux E (1975). Régénération des membres irradiés de *Pleurodeles waltlii* Michah (Urodèle). Influence des qualités et orientations des greffons non irradiés. Wilhelm Roux Arch Devel Biol 176:303.

Liosner LD (1947). Concerning the restoration of the regenerative capability of axolotl irradiated limbs (in Russian) Dokl Akad Nauk S S S R 57:633.

Maden M (1979 a). Neurotrophic and X-ray blocks in the blastemal cell cycle. J Embryol Exp Morph 50:169.

Maden M (1979 b). The role of irradiated tissue during pattern formation in the regenerating limb. J Embryol Exp Morph 50:235.

Namenwirth M (1974). The inheritance of cell differentiation during limb regeneration in the axolotl. Develop Biol 41:42.

Nathanson MA, Hay ED (1980). Analysis of cartilage differentiation from skeletal muscle grown on bone matrix. I Ultrastructural aspects. Develop Biol 78:301.

Sidorova VF (1949). The regeneration of radiografted limb after transplanting unradiografted skin. Transl Dokl Akad Nauk S S S R 68:973.

Steen TG (1968). Stability of chondrocyte differentiation and contribution of muscle to cartilage during limb regeneration in the axolotl (*Siredon mexicanum*) J Exp Zool 167:49.

Thornton CS (1942). Studies on the origin of the regeneration blastema in *Triturus viridescens*. J Exp Zool 89:375.

Umanski E (1937). An investigation of the regenerative process in amphibians by the method of exclusion of individual tissues from Roentgen beams.(in Russian). Biol Zh 6:739.

Umanski E (1938). Regenerative potencies of the skin of the axolotl.(in Russian). Bull Biol Med Exp URSS 6:142.

Wallace BM, Wallace H (1973). Participation of grafted nerves in Amphibian limb regeneration. J Embryol Exp Morph 29:559.

Wallace H, Maden M, Wallace BM (1974). Participation of cartilage grafts in Amphibian limb regeneration. J Embryol. Exp Morph 32:391.

Limb Development and Regeneration
Part A, pages 467–476
© 1983 Alan R. Liss, Inc., 150 Fifth Avenue, New York, NY 10011

AMPHIBIAN LIMB REGENERATION: DISTAL TRANSFORMATION

David L. Stocum

Department of Genetics
and Development
University of Illinois
Urbana, IL   61801

A major tenet about pattern regulation and regeneration is that cells monitor their position within a morphogenetic field (Wolpert, 1969, 1971)and that fields are stable only when cells are in contact with normal neighbors. Creating a discontinuity in a field renders it unstable, and evokes adjustments in the remaining parts which restore normal neighbors (Mittenthal, 1981). The restoration can take place by proportionally adjusting the positional values of remaining cells in the absence of growth to make a miniature whole (morphallaxis), or by the addition of cells by proliferation, assigning to them the missing positional values (epimorphosis).

A good system for the study of pattern regeneration is the amphibian limb, which regenerates epimorphically by dedifferentiation of stump cells under a wound epidermis to form a growing blastema. A major question directed to this system is: what is the nature of the cellular interactions governing the restoration of missing positional values during regeneration? Blastemal morphogenesis is independent of any stump inductive influence; hence, the interactions responsible for restoration of normal cell neighbors take place strictly within the blastema itself (Stocum, 1980a; Stocum, 1982).

One model for these interactions which has been considerably successful in accounting for the asymmetric circumferential pattern of regenerating limbs is the two-dimensional polar coordinate model (Bryant, French, Bryant, 1981). In this model, the cross-section of the limb is represented by a circle. Cells have two positional values, one on the circle which specifies position on the limb circumference, and one on a radius of the circle, which specifies position along the proximal-distal (PD) axis. The model postulates that non-neighboring blastema cells migrate under the wound epidermis, in any of several possible patterns, from the circumference toward the center of the amputation plane. Confrontation of the disparate circumferential positional values carried by these cells triggers their division and the intercalary re-generation of a new circle of circumferential values. Two rules govern the restoration of pattern. The shortest in-tercalation rule states that intercalation of cirumferential values takes place via the shortest route on the circle be-tween non-neighboring cells. The distalization rule states that, if the circumferential values of cells in the new cir-cle are the same as those of adjacent cells in the old cir-cle, the new cells must adopt a more central (distal) value. Distal transformation is thus dependent upon, and is the automatic outcome of, circumferential intercalation. If no opportunity for circumferential intercalation exists, cell division and distal transformation will be halted.

An important observation in formulating the distaliza-tion rule was that double posterior stylopodia regenerate fore or hindlimbs with mirror-image sets of digits, which can be either convergent or divergent in the anterior- posterior (AP) plane, but double anterior stylopodia regenerate only a truncated extension of the femur or humerus (Bryant, 1976; Stocum, 1978; Holder, Tank, Bryant, 1980). This observation suggested that the posterior half of the limb carries more than half the sequence of circumferential positional values, and indicated a third rule of regeneration, that the extent of distal transformation is proportional to the fraction of the circumferential sequence represented in the stump (Stocum, 1978; Bryant, French, Bryant, 1981). This rule and the distalization rule can account for the regenerative non-equivalence of double anterior and posterior limbs. In both, blastema cells with like circumferential values will sooner or later (depending on the pattern of cell migration) confront each other during regeneration, thus denying the opportunity for further circumferential intercalation and, hence, distal transformation. This will occur sooner in dou-ble anterior limbs than in double posterior ones, since the former have fewer positional values that can interact, re-sulting in the production of truncated regenerates.

Intercalary regeneration occurs in the PD axis when a discontinuity is made in this axis by grafting a distal blastema to a more proximal stump level, the stump being the source of cells for the intercalated structures (Pescitelli, Stocum, 1980). Double anterior thigh stumps also intercalate a symmetrical femur and tibia when either normal or double anterior wrist blastemas are grafted to them (Stocum, 1980b, Stocum, 1981). In terms of the polar coordinate model, this observation suggests that circumferential intercalation is not involved in PD intercalary regeneration, and that blastema cells distalize solely in response to the PD discontinuity. The polar coordinate model, therefore, views the distal transformation of terminal and PD intercalary regeneration as the result of two different patterns of cell interaction. Essentially, grafting a distal blastema proximally is viewed as creating a discontinuity, whereas cutting off a distal segment of the limb is not.

I have modified the polar coordinate model so that a uniform set of rules can be applied in three dimensions to both "intercalary" and "terminal" regeneration (Stocum, 1980b). My model treats the loss of any portion of a limb as a discontinuity, with restoration of missing positional values always occurring by intercalary regeneration within the blastema. Blastema cells have three coordinates of positional information. Two of these (circumferential and radial) specify position in the transverse plane, and the third specifies position on the PD axis. The wound epidermis everywhere on the limb is postulated to have a single positional value of zero, which signifies external boundary (see Maden, 1977). When a free surface is created by simple amputation, dedifferentiated and dividing cells will come into contact with the wound epidermis. Distalization of blastema cells is then the result of intercalary regenration to eliminate the discontinuity between the positional values represented by the wound epidermis and the amputation level, just as intermediate positional values are intercalated after grafting a distal blastema to a proximal limb stump. Intercalation in either case could proceed by an averaging mechanism (Maden, 1977) or by a serial mechanism in which positional value changes distally by only one unit at a time.

In this model, circumferential intercalation is not in itself sufficient for distal transformation, nor is confrontation of cells with like positional values sufficient in itself to halt distal transformation. This brings up the

interesting possibility that stylopodial double anterior regenerates may not be truly truncated (i.e., ending in a stylopodial positional value), but might instead include more
distal segments. The apparent truncation would be an illu-
sion created by the convergence resulting from progressive
elimination of midline values.

To test this notion, double anterior thighs were made on
the right hindlimbs of large (90-110 mm length) and small
(40-50 mm length) white and dark axolotls. After a 10-21 day
healing period, the double anterior hindlimbs were amputated
through mid-thigh. When the resulting blastemas reached
early to medium bud stages (staging according to Stocum,
1979) the following experiments were done (Fig. 1).

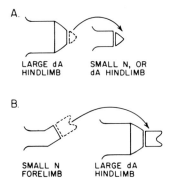

Fig. 1. Diagram of operations.(A) Distal tip of double anterior thigh blastema grafted to normal or double anterior
thigh stump. (B) Normal wrist blastema grafted to distal tip
of double anterior thigh blastema.

    1. The distal 25-50% of double anterior blastemas from
large animals was homografted to the proximal thigh stumps of
either normal or double anterior hindlimbs of small animals.

    2. The distal 0-50% was removed from the double an-
terior thigh blastemas of large animals. Normal wrist blastemas or the distal 30-50% of normal upper arm blastemas
(medium bud to 4-fingerbud stages) from small animals were
then autografted or homografted to the cut surfaces of the
double anterior blastema.

If distal positional values are represented anywhere in

the distal half of a double anterior thigh blastema, the
first operation creates a PD discontinuity and intercalary
regeneration is expected to ensue, whereas the second operation creates no PD discontinuity and no intercalation is expected. If the PD pattern in the double anterior blastemas
is truncated, the reverse results are expected.

The results are summarized in Tables 1,2, and Figs. 2,3.
Intercalation occurred in 3 of 8 cases (37.5%) after grafting
the tip of a double anterior thigh blastema to a double anterior thigh stump (Table 1). In two of these cases the

Table 1. Results of grafting the distal 25-50% of early or
medium bud (10-15 day) double anterior thigh blastemas to
double anterior or normal thigh stumps. Partial intercalation means that the complete femur or femur and knee joint
regenerated; complete intercalation means that a whole zeugopodium regenerated. dA = double anterior; N = normal.

		Intercalation			
Host	Cases	None	Partial	Complete	Normal Regeneration
da	8	5(62.5%)	1(12.5%)	2(25%)	0
N	4	1(25%)	0	0	4 (75%)

distal 25% of medium bud blastemas was grafted, and the femur
and a symmetrical zeugopodium were regenerated, followed by a
single digit (Fig. 2A). In the third case, the distal 50% of
an early bud blastema was grafted, but only the femur and a
knee joint were completed (Fig. 2B). The remaining 5 cases
exhibited no intercalation (Fig. 3A). Three of these grafts
were from early bud blastemas and two from medium bud blastemas. The results of this experiment thus suggest that at
least the distal 25% of double anterior medium bud blastemas
contains distal positional values and that double anterior
regenerates therefore are not truly truncated. Hence, medium
bud double anterior thigh blastemas are as complete in their
PD axial pattern as normal medium bud blastemas, whose
pattern is specified at least up through the first row of
basipodial elements (Stocum and Dearlove, 1972). Resorption
of the graft (which contains few cells) could account for the
failure to intercalate in the majority of cases, a notion
that is strengthened by the fact that double anterior

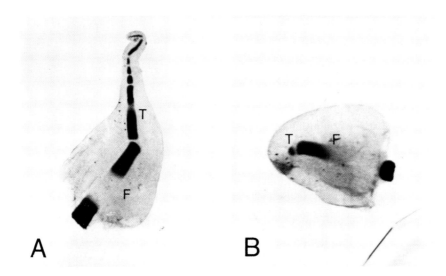

Fig. 2 (A) Regeneration of symmetrical femur (F) and tibia (T) after grafting the distal 25% of a medium bud double anterior thigh blastema to a double anterior thigh stump. (B) Regeneration of femur and small piece of tibial cartilage after grafting the distal 50% of an early bud double anterior thigh blastema to a double anterior thigh stump. X 60.

blastema tips obviously resorb when grafted to normal stumps (Table 1). Three of 4 cases (75%) regenerated normally with no sign of the graft at the tip of the regenerate. However, it is also possible that the lack of regeneration was due to failure to produce a PD discontinuity, since all of the cases that exhibited no intercalation were ones in which the distal 40-50% of the double anterior donor blastema were grafted. This experiment is being repeated using the tissue markers afforded by heterografts.

The results of grafting normal blastemas to the tips of double anterior thigh blastemas appear to contradict the results and interpretation of the experiment just described (Table 2). Only 1 of 15 cases exhibited the predicted failure to intercalate. The remaining cases intercalated the femur (2 cases) or the femur plus a symmetrical tibia (12 cases) (Fig. 3B). Five of these cases might be explained by

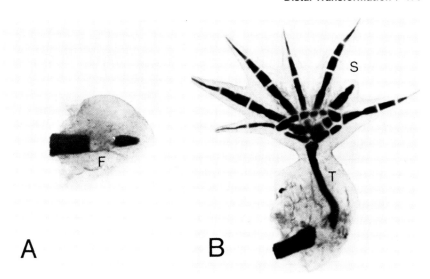

Fig. 3. (A) Apparently truncated regenerate formed after grafting the distal 50% of a medium bud double anterior thigh blastema to a double anterior thigh stump. This is also the typical appearance of regenerates produced after simple amputation of double anterior thigh stumps. F = femur. (B) Regenerate produced after grafting a normal wrist blastema to the distal tip of a double anterior thigh blastema. The hand developed normally, and a mirror-image supernumerary hand (S) developed on the posterior side of the primary. A symmetrical tibia (T) was intercalated. X 60.

the fact that the donor blastemas were grafted to host levels (40–50%) which might have positional values proximal to the autopodium. However, the single case which did not intercalate was also one in which the donor blastema was placed at the 50% level of the host. Furthermore, the remaining 9 donor blastemas were grafted to the 0–30% level of the host blastema, yet all intercalated. These results therefore suggest the opposite conclusion, that cells in the apex of double anterior blastemas have stylopodial positional values and the regenerates made by these blastemas are really truncated.

Table 2. Results of grafting normal wrist blastemas or the distal 30-50% of upper arm blastemas to the 0-50% level of host double anterior thigh blastemas. The donor normal blastemas ranged from early bud to 4-fingerbud; the host double anterior blastemas ranged from early bud to late bud. Partial intercalation means that a complete femur regenerated; complete intercalation means that a complete tibia also regenerated.

	Intercalation		
Cases	None	Partial	Complete
15	1(6.7%)	2(13.3%)	12(80%)

Which interpretation is correct? A definitive answer cannot yet be given, but the data can be reconciled in favor of the idea that distal positional values are represented in double anterior regenerates. It is possible that host blastema cells with distal positional values are injured and die when a normal blastema is grafted to the tip of a double anterior blastema. Prospective structures distal to the stylopodium might thus be eliminated, creating a PD discontinuity which would be filled in by intercalation of a symmetrical zeugopodium. If so, this particular experiment would not be a good test of the hypothesis. Histological studies are being done to determine whether such cell death does occur.

It should be emphasized that, even if it be shown that double anterior regenerates are not truly truncated, this will not strictly disprove the distalization mechanism postulated by the polar coordinate model. For example, more circumferential intercalation than is currently assumed possible by the model may occur during the regeneration of double anterior limbs; the whole PD pattern might be established early in regeneration by the distalization rule (or some other mechanism, including a morphallactic one) and require extensive cell division only to generate a cell mass sufficient to expand the pattern – hence, the complete PD pattern may be present in a double anterior regenerate, but cannot be discerned because, in addition to being severely hypomorphic in the AP plane, it is telescoped due to lack of PD growth. Regardless of the mechanism of distal transformation, pattern expansion might be impossible in double anterior blastemas for geometric and mechanical reasons. Anterior-posterior pattern convergence might not allow the formation of a mass of cells at the tip of the blastema sufficient to support the continued distal extension of blood vessels. Cell division to expand the PD pattern may therefore be halted not because normal neighbors have been satisfied in the circumference, but because systemic support for growth cannot be extended. We are currently conducting additional experiments to test the notion that regeneration from a free surface occurs by intercalation between the wound epidermis and more proximal positional values.

REFERENCES

Bryant SV (1976). Regenerative failure of double half limbs in Notophthalmus viridescens. Nature 263:676.

Bryant SV, French V, Bryant PJ (1981). Distal regeneration and symmetry. Science. 212:993.

Holder N, Tank PW, Bryant SV (1980). Regeneration of symmetrical forelimbs in the axolotl, Ambystoma mexicanum. Dev Biol 74:302.

Maden M (1977). The regeneration of positional information in the amphibian limb. J Theoret Biol 69:735.

Mittenthal J (1981). The rule of normal neighbors: A hypothesis for morphogenetic pattern regulation. Dev Biol 88:15.

Pescitelli MJ, Stocum DL (1980). The origin of skeletal structures during intercalary regeneration of larval Ambystoma limbs. Dev Biol 79:255.

Stocum DL (1978). Regeneration of symmetrical hindlimbs in larval salamanders. Science 200:790.

Stocum DL (1979). Stages of forelimb regeneration in Ambystoma maculatum. J Exp Zool 209:395.

Stocum DL (1980a). Autonomous development of reciprocally exchanged regeneration blastemas of normal forelimbs and symmetrical hindlimbs. J Exp Zool 212:361.

Stocum DL (1980b). Intercalary regeneration of symmetrical thighs in the axolotl, Ambystoma mexicanum. Dev Biol 79:276.

Stocum DL (1981). Distal transformation in regenerating double anterior axolotl limbs. J Embryol Exp Morph 65:3.

Stocum DL (1982). Determination of axial polarity in the urodele limb regeneration blastema. J Embryol Exp Morph (in press).

Stocum DL, Dearlove GE (1972). Epidermal—mesodermal
   interaction during morphogenesis of the limb regeneration
   blastema in larval salamanders. J Exp Zool 181:49·
Wolpert, L (1969). Positional information and the spatial
   pattern of cellular differentiation. J Theor Biol
   25:1·
Wolpert, L (1971). Positional information and pattern
   formation. Curr Top Dev Biol 6:183·

Limb Development and Regeneration
Part A, pages 477–490
© 1983 Alan R. Liss, Inc., 150 Fifth Avenue, New York, NY 10011

MORPHOGENESIS OF THE REGENERATING LIMB BLASTEMA OF THE
AXOLOTL: SHAPE, AUTONOMY AND PATTERN

Nigel Holder and Susan Reynolds

Department of Anatomy,
King's College University of London,
Strand, London WC2R 2LS.

INTRODUCTION

In this paper we attempt to leave the mainstream studies of the control of pattern regulation in amphibian limbs
and concentrate on morphogenesis of the blastema. The focus
of attention will be the description and possible controls
of blastemal shape. This particular facet of blastemal
morphogenesis is of crucial importance to our understanding
of pattern formation yet has been almost disregarded experimentally. The two key questions we address are: 1. Does
the shape of the blastema alter at different levels of
amputation in the proximal to distal (p-d) axis and at
different temporal stages of regeneration at any given
level? 2. If so, how is this change in shape controlled and
to what degree are the controlling factors properties of the
stump upon which the blastema develops or the blastema
itself as it begins to form?

It is difficult to see why shape as a property of
limb morphogenesis has been neglected for so long because it
affects our interpretation and understanding of all of the
modern theories for the creation of spatial patterns. Indeed, the present set of experiments was initially stimulated by results from grafting experiments designed to examine the cellular interactions controlling pattern formation.
The recent results of experiments where the regenerative
ability of symmetrical limb stumps was tested (Holder et
al., 1980) strongly suggested that symmetrical tissues regenerated basically different patterns depending on the proximal to distal limb level at which they were amputated (see

also Bryant et al., 1982; Holder, 1981; Tank and Holder, 1981). For example, double posterior upper arm stumps regenerated double posterior mirror image symmetrical limbs with between three and six digits if amputated immediately after grafting; whereas subsequent amputation of such symmetrical structures in the lower arm resulted in significant increases in pattern elements about the line of symmetry (Holder et al., 1980). One interpretation of how these patterns are formed is based on the local short range cell-cell interactions and specific rules of cell behaviour which are suggested by the polar-coordinate model (French et al., 1976; Bryant et al., 1981). Short range interaction models focus attention on the likely contacts between cells during blastemal outgrowth, and, amongst other important considerations (see Bryant et al., 1982; Holder, 1981), these contacts will be greatly affected by blastemal shape and healing modes (see also French, 1981). In the specific case of the experiment mentioned above the appearance of anterior pattern elements in the midline of a double posterior symmetrical limb can be brought about only by contact between dorsal and ventral cells in the midline of the outgrowing blastema. The results of the experiment suggest that the likelihood of this event occurring is much higher in the forearm than in the upper arm. In terms of the polar-coordinate model, therefore, one would predict that the dorsal and ventral sides of the limb are much closer together in the forearm than in the upper arm. That is, the shape of the blastemas formed in the two regions must be different. The same deduction can be drawn from other experiments performed upon the upper and lower arms which considered the effect of wounding on cell contact and distal outgrowth. The effects of wounding are markedly different in the two regions with a clear healing time effect being seen in the upper arm following amputation of symmetrical stumps but not in the lower arm (cf. Tank and Holder, 1978; Krasner and Bryant, 1980). We can state at the outset that preliminary results have demonstrated that a clear difference in shape exists between forearm and upper arm blastemas of comparable stage (Holder, 1981 and figure 1). In this paper we extend this description of blastemal shape and begin to examine how it is controlled.

The regenerating amphibian limb blastema is well suited to studies of morphogenesis and pattern formation because stump tissues are easily manipulated surgically. The present studies take advantage of this fact and the relative influence of three major stump tissues – dermis, muscle and cartilage, on blastemal shape are examined by removing each in turn from the stump and subsequently amputating the limb at the appropriate level. The results of these experiments strongly suggest that the skin, and particularly the dermal component, plays a significant role in shaping the blastema. It also becomes apparent from this study that the blastemal shape is under blastemal control from an early stage and that the stump which initially provides the blastemal cell population, plays a minimal role. In short, the blastema is autonomous in terms of the control of its own shape in much the same way as it is in controlling the pattern of limb parts it will produce (see Holder and Tank, 1979; Stocum, 1980a and b).

Description of blastemal shape.

In order to describe blastemal shape limbs were amputated at different proximal to distal levels and were allowed to regenerate for different lengths of time to specific stages. Six categories were established. The forelimb was amputated at either mid–upper arm, mid–lower arm or wrist levels. Upper arm blastemas were allowed to regenerate to the stage of medium bud (MB) or Palette (see Tank et al., 1976); lower arm blastemas to MB and wrist blastemas to late bud. The remaining two categories involved amputations of legs at either mid upper leg (thigh) or mid lower leg (shank) level and the resulting blastemas were fixed at MB.

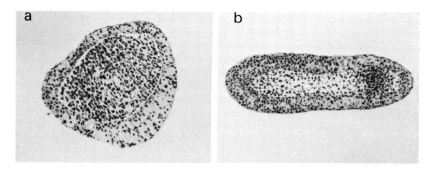

Fig. 1. Transverse sections of MB blastemas from an upper axon stump (a) and a lower arm stump (b). Note the clear difference in shape.

At the required stage blastemas were removed together with a large part of the remaining stump, fixed in Bouin's fluid and processed for was embedding. In all cases 10 serial transverse sections were cut of stump and blastema and these were stained with either haematoxylin and eosin or Mallory's trichrome. In order to assay for shape every tenth section (100 μm levels) was drawn with a camera lucida so that the outline of the ectoderm and mesoderm was apparent. A simple assessment of shape was then obtained by measuring the length of orthogonal axes (dorsal-ventral and anterior-posterior) and dividing the length of the long axis into that of the short axis. The ratio thus obtained (r - see figure 2) approached 1 as a perfect circle and 0 as the shape became increasingly elliptical (see Holder, 1981).

The results of these analyses for each of the six categories are shown in the first set of graphic representations (figures 1 and 2a-d). In each category examined the 95% confidence limits were calculated and are shown on the appropriate graphs. The range of individual measurements which were averaged at each 100 μm step from the distal tip can therefore be assessed. In addition, where it was deemed useful, t tests were carried out between the means of pooled data from specific categories (see table 1, 2 and 3). The means themselves give a rough indication of the shape of the blastemas when early regeneration stages, such as MB blastemas, are compared. Thus the upper arm MB mean value of r is 0.68 which gives the impression of the blastema being much more rounded than the elliptical lower arm blastema which has a mean r value of 0.26 (table 1). However, the comparison of the means at later stages becomes misleading because the shape range within the whole blastema becomes large. For example, a palette-shaped blastema has the range of both the upper arm and lower arm MB blastemas combined (figure 2b). The means for upper and lower arm MB and upper and lower leg MB blastemas are shown in table 1. In each case within the respective limbs each mean is significantly different from the other at the .001 level.

When blastemas are allowed to regenerate to the later stages of late bud (LB) and palette (Pal) it becomes clear that even though a blastema may arise from an upper limb stump it becomes progressively more elliptical distally as it reaches levels where distal structures (forearm and hand) are to be formed. This observation is shown in figure 2b, where it can be seen that the distal regions of palette

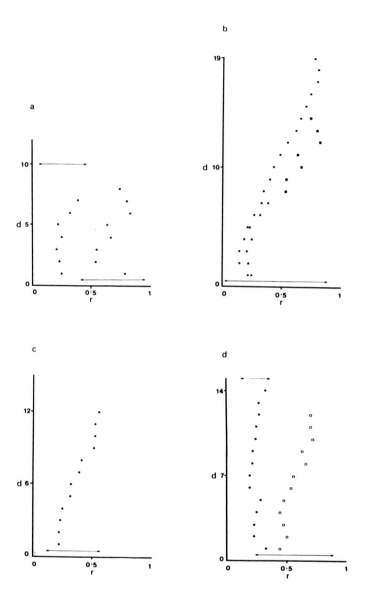

Fig. 2. Graphic representations of measurements taken from serial section sets of blastemas of different regenerative stage and different proximo–distal levels. r = shape ratio (see text) and d is the distance from the distal tip in 100 steps. The arrows on each graph represent the 95% confidence

limits for the range of each particular population. The mean of ranges of a and d are given in Table 1.
a. Upper (•) and lower (▲) arm MB. Upper arrow refers to lower arm values.
b. Palette staged upper arm blastema (•) with 95% confidence limits. The remaining values are transferred from the adjacent graph to compare palette with MB's from upper (■) and lower (▲) arms. When combined into one series the MG ranges are closely similar to that of the palette stage.
c. Wrist LB staged blastema.
d. Upper (□) and lower (•) leg M-LB amputations. The upper arrow refers to lower leg values.

---

staged blastemas have an r value of about 0.25. Also shown on this graph are superimposed values of MB blastemas of lower arms and upper arms demonstrating that a palette-staged blastema is equivalent to a MB of a forearm developing distal to a MB of an upper arm stump. As the transition to a more elliptical appearance occurs gradually from the stump towards the blastemal tip this observation is the first suggestion that shape is an autonomous function of blastema cells with a particular proximal to distal positional value, and may not be influence directly by the stump upon which the blastema sits. This conclusion is reinforced by the results of experiments where stump tissues are manipulated prior to amputation, which are discussed below.

Blastemal shape as an autonomous feature of proximo-distal level

Following amputation of a normal limb dedifferentiation of the stump occurs and the dedifferentiated mesodermal cells stream forward to form a blastema which is bounded by a thin epidermal layer. It seems possible, therefore, that a MB blastema would take on a shape determined mechanically by the shape and form of the stump upon which it forms. For example, the shape of the flattened forearm MB blastema could be determined by two skeletal elements in the stump (radius and ulna) as compared with the circular form of the upper arm medium bud which forms round a single skeletal element, the humerus. This possibility, along with the possible influence of other stump tissues, the dermis and the muscle, was tested by the removal of each of these

TABLE 1:

NORMAL MB SHAPES

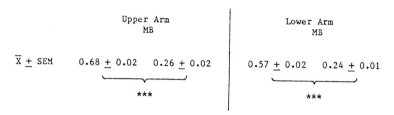

	Upper Arm MB		Lower Arm MB	
$\overline{X} \pm SEM$	0.68 ± 0.02	0.26 ± 0.02	0.57 ± 0.02	0.24 ± 0.01
	_____		_____	
	***		***	

tissues from the stump prior to amputation and the subsequent examination of shape of the MB blastemas which form
following amputation.

Three basic operations were performed: (a) In the
lower arm either the radius or the ulna was removed and the
limb allowed to heal for two to three weeks before amputation.   (b) In the leg the skin was removed as a cuff in
either the shank or thigh region and amputation was performed a week later, by which time the epidermis from surrounding tissues had grown over the exposed soft tissue (see
Carlson, 1975).   Thus the dermis at the amputation plane
was much reduced.   (c) Specific muscles were removed from
the thigh region: the large ventral musculus puboischio
tibialis (mpit, 3 cases), two small dorsal muscles (musculus
extensor ilio tibialis pars anterior and pars posterior:
meilt′ and meilt′′) plus the ventral mpit (1 case), and
three dorsal muscles (meilt′ and meilt′′) and musculus ilio-
fibularis plus the ventral mpit (1 case). Because the shape
of the blastema derived from each of these types of operation, in the case of the bone and muscle removals control
operations were performed in which the skin was opened and
then sutured closed exactly as in the experimental operations;   the underlying tissues, however, were left un-
touched.   In all three operational categories between 3 and
5 blastemas were allowed to regenerate to the stage of MB-LB
when they were fixed, sectioned and analysed as before.   The
results are presented in graphic form in figure 3a–d, and
the mean r values and their statistical significances presented in table 2.

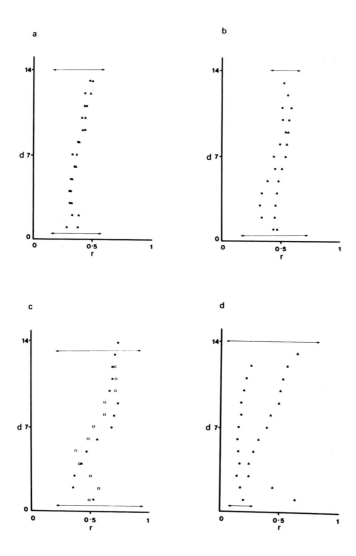

Fig. 3. Graphic representations of measurements taken from serial section sets of sections of M/LB blastemas derived from stumps to which various tissues have been removed. See Table 2 for mean value comparisons and significance. Desig-

nations as in figure 2.

a. Removal of the ulna (•) compared with sham operated controls (▲). The upper arrow refers to sham control values.

b. Removal of the radius (•) compared with sham operated controls (▲). The upper arrow refers to sham control values.

c. Muscle removals from the upper leg (thigh) level (•) compared with sham operated controls (□). The upper arrow refers to sham control values.

d. Removal of dermis from the upper leg (▲) and lower leg (•). The upper arrow refers to the upper leg values.

The most striking outcome of these results is that major surgical alteration of the stump by removal of muscle (figure 3c) and cartilage (figure 3a-b) has remarkably little effect upon the shape of the blastema. This is particularly clear in the case of the bone removals (figure 3a-b) where removal of either the radius or the ulna does not alter the shape of the blastema significantly as compared with values for control operations. The same is true of the muscle removals even though when ventral and dorsal muscles were removed together the operated thigh prior to amputation appeared substantially flatter than normal.

When the dermis was removed the resulting MB leg blastemas appeared significantly flattened at both upper and

TABLE 2:

TISSUE REMOVAL AND AMPUTATION

Operation	Position of amputation	Mean ± SEM	Significance at 0.001 (t-test)	Comment
Dermis removal	Upper leg	0.43 ± 0.02	***	Flattens blastema.
	Lower leg	0.17 ± 0.01	***	Flattens blastema.
Muscle removal	Upper leg	0.58 ± 0.02	−	No effect as compared to sham.
Ulna removal	Lower arm	0.36 ± 0.02	−	No effect as compared to sham.
Radius removal	Lower arm	0.44 ± 0.02	−	No effect as compared to sham.

lower arm levels (figure 4d and table 2). Although the reasons for this are unclear, it suggested to us that the dermis may be playing a significant role in establishing blastemal shape, at least at the early stages of regeneration. This influence is also consistent with other types of experiments involving positional mismatches of dermis with respect to the remaining stump tissues which have suggested that the dermis is an important tissue in pattern regulation (Carlson, 1974, 1975; Tank, 1981).

To examine further the possible role in morphogenesis of the dermis in particular and the skin in general we completed an additional set of experiments in which whole cuffs of skin were exchanged between proximal (thigh) and distal (shank) levels of the leg. After amputation through the graft MB/LB blastemas were fixed, sectioned and analysed as before. The striking result of these experiments (figure 4) was that the distal skin around a proximal stump caused the forming blastema to flatten significantly, and proximal skin around a distal stump caused the blastema to become significantly more rounded (table 3 and figure 4). It seems reasonable to assert therefore that the skin, and most probably the dermal component, plays a significant role in controlling blastemal shape.

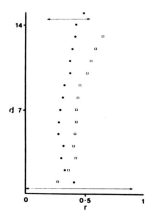

Fig. 4. Section measurements from M/LB blastemas derived from upper leg (•) and lower leg (□) stumps bearing skin from the opposing limb region. Note that from the upper leg stump bearing lower leg skin the blastema is flatter than the blastema derived from the reciprocal combination. The

designations are as for figure 2 and the upper arrow refers to upper leg amputations. The means for these distributions appear in Table 3.

Conclusions: Shape, autonomy and pattern.

The results of the experiments reported here demonstrate that blastemas derived from different levels of the proximo-distal axis are of different shape. Furthermore, these shape changes seem to be a property of level specific information in the limb. The autonomous control of blastemal shape is suggested by the fact that a palette staged blastema derived from an upper arm amputation begins "spontaneously" to flatten as the appropriate proximo-distal level is reached during outgrowth, even though the distal end of the blastema has derived from more rounded proximal blastemal tissue (see figure 2b). The tissue removal experiments involving muscle and cartilage indicate that the stump is unlikely to play any crucial mechanical role in shaping the blastema. In addition, consistent with a level specific autonomous control system is the observation that proximo-distally misplaced skin cuffs can significantly alter blastema shape in a level specific manner. It seems likely therefore that shape can be thought of in the same way as spatial pattern regulation of anatomical features such as the skeleton and muscle during regeneration. In short, it seems that positional information somehow controls shape of the blastema as a whole in addition to the spatial arrangements of cell types that is occurring during outgrowth.

TABLE 3:

SKIN TRANSPLANTATIONS

Graft	Position of amputation	Mean ± SEM	Significance at 0.001 (t-test)	Comment
Distal to proximal	Upper leg	0.35 ± 0.01	***	Flattens blastema.
Proximal to distal	Lower leg	0.44 ± 0.03	***	Rounds the blastema.

The autonomous control of cell patterning within the blastema has been demonstrated many times. For example, when muscle tissues (Carlson, 1972) and cartilagenous skeletal elements (Goss, 1958) are removed from the stump they are reformed distal to the level of amputation by the blastema. Similarly, MB blastemas derived from symmetrical or normal stump behave autonomously when they are reciprocally exchanged (Holder and Tank, 1979; Stocum, 1980a). It seems that once a MB blastemal cell population has been established it behaves as an autonomous unit with respect to pattern and form and the stump upon which it sits does little more than maintain a supply of trophic factors and oxygen.

The description of blastemal shape and the experiments involving the dermis which show that shape can be altered in predictable ways open the path for further study of the relationship between shape, form and pattern formation during limb regeneration. It has already been outlined in the introduction that shape is crucial for an interpretation of pattern regulation in terms of a short range cell-cell interaction system such as the polar coordinate model (see Bryant et al., 1981, 1982; Holder, 1981). However, shape is also critical to the formulation of models involving long range signalling systems such as the zpa model for the developing chick limb bud (see Tickle et al., 1975 and Smith and Wolpert, 1981) or prepattern models such as that derived by Newman and Fritsch (1979) also for the establishment of the skeletal pattern in the chick limb bud. The application of these models should now be testable in a novel way; by altering blastemal shape following specific surgical manipulation of stump tissues and examining pattern regulation in the resultant blastema. Such experiments, coupled with a histological study of blastema formation should shed more light on the elusive but crucial relationship between pattern formation and morphogenesis.

REFERENCES

Bryant SV, Holder N, Tank PW (1982). Cell-cell interactions and distal outgrowth in amphibian limbs. Amer Zool 22: 143.
Bryant SV, French V, Bryant PJ (1981). Distal regeneration and symmetry. Science 212:993.
Carlson BM (1972). Muscle morphogenesis in axolotl limb regenerates after removal of stump musculature. Dev Biol 28:487.

Carlson BM (1974). Morphogenetic interactions between rotated skin cuffs and underlying stump tissues in regenerating axolotl forelimbs. Dev Biol 39:263.

Carlson BM (1975). The effects of rotation and positional change of stump tissues upon morphogenesis of the regenerating axolotl limb. Dev Biol 47:269.

French V (1981). Pattern regulation and regeneration. Phil Trans Roy Soc Lond B 295:601.

French V, Bryant PJ, Bryant SV (1976). Pattern regulation in epimorphic fields. Science 193:969.

Goss RJ (1958). Skeletal regeneration in amphibians. J Embryol Exp Morph 6:638.

Holder N (1981). Pattern formation and growth in the regenerating limbs of urodelean amphibians. J Embryol Exp Morph 65(supp):19.

Holder N, Tank PW (1979). Morphogenetic interactions occurring between blastemas and stumps after exchanging blastemas between normal and double-half forelimbs in the axolotl, Ambystoma mexicanum. Dev Biol 68:271.

Holder N, Tank PW, Bryant SV (1980). Regeneration of symmetrical forelimbs in the axolotl, Ambystoma mexicanum. Dev Biol 74:302.

Krasner GN, Bryant SV (1980). Distal transformation from double-half forelimbs in the axolotl Ambystoma mexicanum. Dev Biol 74:315.

Newman SA, Frisch, HL (1979). Dynamics of skeletal pattern formation in developing chick limb. Science 205:662.

Smith JC, Wolpert L (1981). Pattern formation along the anteroposterior axis of the chick wing: the increase in width following a polarising zone graft and the effect of X-irradiation. J Embryol Exp Morph 63:127.

Stocum DL (1980a). Autonomous development of recoprocally exchanged regeneration blastemas of normal forelimbs and symmetrical hindlimbs. J Exp Zool 212:361.

Stocum DL (1980b). Intercalary regeneration of symmetrical thighs in axolotl Ambystoma mexicanum. Dev Biol 79:276.

Tank PW (1981). The ability of localised implants of whole or minced dermis to disrupt pattern formation in the regenerating forelimb of the axolotl. Am J Anat 162:315.

Tank PW, Carlson BM, Connelly TG (1976). A staging system for forelimb regeneration in the axolotl, Ambystoma mexicanum. J Morphol 150:117.

Tank PW, Holder N (1978). The effect of healing time on the proximo-distal organisation of double-half forelimb regenerates in the axolotl, Ambystoma mexicanum. Dev Biol 66:72.

Tank PW, Holder N (1981). Pattern regulation in the regenerating limbs of urodele amphibians. Q Rev Biol 56:113.

Tickle C, Summerbell D, Wolpert L (1975). Positional signalling and specification of digits in chick limb morphogenesis. Nature 254:199.

Limb Development and Regeneration
Part A, pages 491–500
© **1983 Alan R. Liss, Inc., 150 Fifth Avenue, New York, NY 10011**

REGENERATION OF SKELETAL MUSCLE IN NOTOPHTHALMUS VIRIDESCENS

Jo Ann Cameron

Department of Anatomical Sciences
University of Illinois
Urbana, Illinois 61801

Urodele limb regeneration is an epimorphic process
which follows the same sequence of pattern formation and
tissue differentiation as does the limb bud. A striking
difference between the embryonic limb bud and the regeneration blastema is the origin of the blastema cells. These
cells originate through dedifferentiation of mature stump
tissues in an area which extends several millimeters proximal to the level of amputation. All of the tissues of the
stump, except nerve axons, blood vessels, and epidermis,
undergo dedifferentiation. Through the aggregation of
these dedifferentiated cells beneath the epidermis of the
distal tip of the limb stump, the blastema is formed
(Thornton, 1938; and Hay, 1959).

The connective tissue elements which dedifferentiate
are cartilage or bone, in the case of adult salamanders,
and fibroblasts. Schwann cells associated with the nerve
axons also dedifferentiate. Through the application of
tritiated thymidine, nucleolar cell markers, and pigment
cells, the involvement of these tissues in the formation of
a blastema and eventual redifferentiation of the limb has
been documented (Steen, 1968; Namenwirth, 1974; and Maden,
1977). The limitation of most of these studies is that the
labeled tissues contained a mixture of several cell types.
The only tissue for which a pure population of cells can be
obtained, short of cloning, is cartilage. The other
tissues, bone and nerve sheath, are intimately associated
with fibroblasts which are impossible to separate.

The role of dedifferentiated skeletal muscle in the

formation of a blastema and redifferentiation of the limb
tissues has not been demonstrated convincingly. Muscle is
permeated with fibroblasts and Schwann cells, which dedifferentiation affects as well as the multinucleated fibers
and the satellite cells which may be present. The myogenic
capability of satellite cells, which lie within the basal
laminae of the muscle fibers, has been demonstrated in
vitro (Konigsberg et al, 1975; and Bischoff, 1980) and in
vivo (Snow, 1978). Despite the problems involved with
tracing the fate of cells which are liberated when
amphibian skeletal muscle dedifferentiates, the initial
phases of the process can be described for two species.

Muscle fiber dedifferentiation has been examined at
the fine structural level for regenerating larval Ambystoma
punctatum forelimbs (Hay, 1959) and adult Notophthalmus
viridescens forelimbs (Lentz, 1969). Briefly, muscle
dedifferentiation involves the gradual loss of organelles
related to contractile function of the muscle fiber and the
acquisition of structures associated with protein synthesis, a major activity of the dividing blastema cell. In
particular, there is a decrease in the myofilaments and
sarcoplasmic reticulum, and zones of amorphous material,
presumably disorganized myofilament proteins, are associated with intact filaments. The fiber fragments into
nucleated and non-nucleated cytoplasmic compartments. The
non-nucleated compartments degenerate. The nuclei of the
newly formed cells contain prominent nucleoli, much
euchromatin, and a thin deposition of heterochromatin at
the nuclear envelope. The cytoplasm of these cells
contains rough endoplasmic reticulum and Golgi; the
presence of scattered myofilaments is the last cytological
indication of their origin. Beyond this point in dedifferentiation it is impossible to identify the blastema cells
originating from the multinucleated fibers from those which
arose from stimulation of fibroblasts, Schwann cells, or
satellite cells if they are present in the muscle.

The case of Notophthalmus viridescens muscle regeneration presents several interesting opportunities for
investigating the possible sources of myoblasts during limb
regeneration. Newts are one of the few tetrapod species
which can regenerate a limb during the adult phase of their
life cycle. In addition, the adult limb muscle tissue does
not contain satellite cells (Popiela, 1976). Therefore,
the contribution of muscle fiber dedifferentiation and the

possibility of metaplasia of other mononucleated cells in
muscle redifferentiation can be investigated in this
system.

Since muscle fiber dedifferentiation is considered a
major component of newt limb regeneration, one would
predict that if the muscle fibers are destroyed, muscle
regeneration will be reduced. We have investigated this
possibility in vivo, using a minced muscle system, and in
vitro, culturing muscle explants and cells derived from the
explants. Minced muscle regeneration has been studied
extensively in mammalian and amphibian species (Carlson,
1972). These systems are characterized by two features:
1) the mincing process destroys virtually all myonuclei by
terminally injuring the fibers, and 2) in all cases studied
thus far, there is a surviving population of myogenic cells
which form myoptubes with varying degrees of efficiency
depending upon the species. Analysis of the minced muscle
in rat and frog at various times during regeneration
indicates that the myogenic cells come from the satellite
cells. These cells are first apparent within the remaining
basal laminae of the injured fibers and actually form a
cuff of basophilic cells around the injured fibers. Later,
as these fibers' contents are cleared away by macrophages,
the cells of the basophilic cuff begin to fuse forming new
myotubes. The resulting muscle has been influenced by the
bed in which it has developed: it makes the correct
tendinous attachments, and although smaller, it resembles
the original muscle in form and function (Carlson, 1972).

EXPERIMENTS
Adult newts were obtained from Lee's Newt Farm (Oak
Ridge, TN) and housed in aquaria containing 1% Holtfreter's
solution. For all experiments the newts were anesthetized
in 0.075% chloretone (Kodak). Post-operatively, the newts
were housed in a container with moss and a little 1%
Holtfreter's solution for the duration of the experiment.

Regeneration of Minced Muscle In Vivo
A flap of skin was peeled back exposing the extensor
cruris and tibialis fibularis muscles of both hindlimbs in
45 newts. With iridectomy scissors this muscle was minced
in situ into pieces approximately 1mm^3. The skin flap
was repositioned over the muscle in each limb and the
muscle was allowed to regenerate for varying periods of
time up to 50 days. In a control group, the entire muscle

was removed, the skin flap repositioned over the empty
muscle bed, and regeneration was allowed for 45 days. The
progress of regeneration was examined at the macroscopic
and histological levels during the 50 day interval.

Macroscopic observations of exposed, fixed muscle
were made at weekly intervals. The pattern of minced
muscle regeneration followed closely that observed in
another amphibian, the axolotl (Carlson, 1972). The
original shape of the mass of minced fragments was molded
by the bone, muscle, and skin which delimit the muscle bed.
Progressive reduction in the size of the minced mass began
around 7 days post-operatively and continued until 5 weeks.
The accumulation of extravascular blood around the fragments gradually decreased, and a regular pattern of circulation was established through the minced fibers. The
fragments decreased in size, and by 4 weeks the regenerate
appeared as a miniature version (approximately one-third
the width) of the original muscle, with normal tendinous
insertions.

Histological examination of the minced muscle at 2
days revealed degenerative changes in the sarcoplasm and
invasion of macrophages within the fibers. The sarcoplasm
of the degenerating muscle became vacuolated and lost
striations. The nuclei of the degenerating fragments were
irregularly shaped with spherically clumped chromatin. By
day 4 the sarcoplasm of the minced fibers was completely
removed in some peripheral areas. Basophilic cells with
little cytoplasm were present almost entirely outside the
fiber fragments, which were still apparent and arranged
randomly. The long axis of these cells was usually aligned
with the axis of the adjacent fragment at this stage of
regeneration. By day 16 fusion of the myoblasts had
occurred. The nuclei of these newly formed myotubes were
centrally located and the cytoplasm was basophilic.

During weeks 4 and 5 the myogenic area continued to
expand through proliferation of myoblasts, their fusion,
and by increase in the volume of cytoplasm of the myotubes.
As the cytoplasm increased, it gradually became eosino-
philic and striations appeared. The 8 week regenerating
muscle fibers were approximately the same size as unminced
fibers. The muscle fibers were arranged more irregularly
and more connective tissue was present between these muscle
fibers than in the unminced. The control series of muscle

removed completely did not regenerate at all.

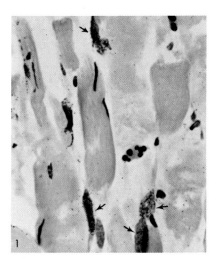 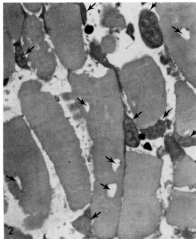

Fig 1 Section through a 4 day minced muscle in vivo.
Arrows point to tritiated thymidine labeled cells found
outside the degenerating fibers. Cells from this
population later fuse to form myotubes. X600.

Fig 2 Section through a 4 week minced muscle explant.
Arrows point to myonuclear ghosts within the injured fibers
and to mononucleated cells adjacent to these fibers. X600.

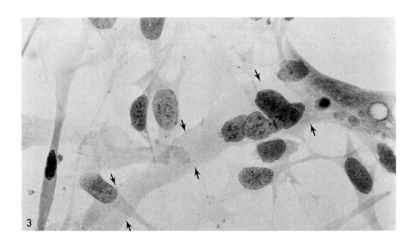

Fig 3   Myotube formed within 4 week secondary culture of
newt muscle mononucleated cells.   X600.

Primary Cultures of Newt Skeletal Muscle
      In order to investigate further the possibility that
newt muscle regenerates from a population of cells which
occurs within the muscle tissue, we developed a culture
system which supports newt muscle regeneration (Schrag and
Cameron, in preparation).   The culture medium was tailored
to the special needs of amphibian cells which include an
osmolarity of 260 mOs, 1% pCO_2, and pH 7.2-7.4 (Balls et
al, 1976).   Minimal essential medium (MEM) was used at 89%
strength and buffered with bicarbonate (0.32mg/ml) and 1%
CO_2.   The MEM was supplemented with 5% Fetal Bovine Serum
(Flow), 100 U/ml Penicillin, 50 µg/ml Streptomycin, 0.125%
Gentamycin Sulfate Solution (Sigma), 1 µg/ml thymidine
(Sigma), 2.5 µg/ml Fungizone (Gibco), and 292 µg/ml L-
glutamine (Sigma).   Aqueous solutions of four hormones were
also added at the following concentrations: 28 U bovine
insulin (Sigma), 20 µg somatotropin (ICN), 20 µg hydrocortisone (ICN), and 1 ng L-thyroxine (Sigma) added to each
100 ml.   Dibutyryl cyclic AMP (Sigma) was added at 50
µg/ml. The calcium ion concentration was 1.8 mM to promote
fusion of the myoblasts.   The complete medium was filtered
through a 0.45 µm Millipore filter, stored at 8°C, and used
within one week.   The cultures were grown at ambient temperature and fed every three days for periods up to 160
days.

      The forelimbs were removed below the distal humerus
and sterilized in 1% sodium hypochlorite solution followed
by several rinses in sterile Earles Balanced Salt Soltuion
(EBSS) (Gibco), pH 7.4.   The muscle dissections were done
in sterile EBSS in petri dishes.   After the forearm was
skinned, muscles were cut at their attachments, removed
carefully from the bones, and cut into 3 mm^3 pieces.   In
some experiments, the pieces were then further minced to
about 1 mm^3.   The muscle pieces were placed in sterile
0.05% crystalline trypsin (Gibco) made with Pucks calcium
and magnesium free salt solution at pH 7.6, soaked for 12-
24 hours at 8°C, then stirred in the same solution at room
temperature for 30 minutes, transferred to 0.03% soybean
trypsin inhibitor (Gibco), and placed on collagen-coated
35mm plates with 1 ml medium.

      Histological analysis of the explants at days 1 to 3,

and at weekly intervals thereafter until 4 weeks, revealed intact fibers and healthy myonuclei within the explants which had not been minced prior to culturing. Size of the fibers and peripheral location of the myonuclei indicated that these fibers survived explantation and were not newly regenerated. There was no evidence of dedifferentiation of the muscle fibers.

Explants which had been minced contained pyknotic myonuclei by day 3 and pyknotic myonuclei and nuclear ghosts thereafter. Although the fibers gradually lost their striations, they never completely degenerated by the end of 10 weeks. The mononucleated cell types within the explant were not distinguishable from each other at the light microscopic level. Mitotic figures were present in less than 1% of the cells in the explants. The number of mononucleated cells in the explants diminished by 5 weeks, while those which migrated onto the plate proliferated; presumably most of the cells migrated to the plate soon after they began to divide.

Cellular outgrowth from the explants began on days 5 through 10. Outgrowth continued and the cells were studied from day 5 to day 160, although the cultures remained viable and continued to grow after 160 days. Three cell morphologies were present on the plate: flat epithelial, stellate, and bipolar. These resembled closely the blastema cell morphologies observed by Jabaily et al (1982), who demonstrated that single blastema cells can assume all three configurations. We have not determined whether these shapes represent modulations of a single cell type or three separate cell phenotypes in our culture system. After 2 to 3 weeks the cells on the plate increased markedly in number. At 3 weeks many of the cells formed aggregations and by 4 weeks many cells within the aggregations lined up and fused. Electron micrographs of these tubes revealed developing myofibrils. No degeneration of the myotubes was observed once they had formed.

Secondary Cultures of Newt Skeletal Muscle

In order to more closely observe the fusing cells, to determine if a particular cell configuration was more frequently associated with fusion, we grew the cells at clonal density. At 3 to 4 weeks cells that had migrated onto the primary culture plates were treated with 0.05% trypsin in Puck's until they rounded up and then

transferred to 0.03% soybean trypsin inhibitor. After
washing in medium the cells were plated in secondary
cultures at 200 cells per 16mm well. The secondary
cultures consisted of cells which displayed the same three
configurations. These cells proliferated and migrated
together until they formed prefusion aggregates. The
aggregates were composed of cells which displayed more than
one type of configuration. By 3 to 4 weeks myotubes formed
within the aggregates, and not all of the cells within the
aggregate fused into myotubes. It was not possible to
predict myogenic cells on the basis of morphology, as can
be done with accuracy in quail muscle cultures. A similar
sequence of prefusion aggregation was also observed in newt
blastema cell cultures by Jabaily et al (1982).

DISCUSSION
    The in vivo studies of minced muscle regeneration in
newts demonstrate that their skeletal muscle regenerates in
a pattern resembling one which has been characterized in
other vertebrates. Mincing triggers degeneration of the
muscle fibers and their myonuclei. One difference between
this and other minced muscle systems is that cells which
resemble myoblasts are not found within the basal laminae
of the degenerating fragments, but are found in the extracellular space. Basophilic cuffs of cells which surround
degenerating fibers in species which contain satellite
cells are not found in regenerating newt muscle. This
suggests that the myogenic cells originate outside the
basal laminae of the minced fibers. The cells which lie
outside the basal laminae of the fibers in this muscle are
fibroblasts, pericyte cells, Schwann cells, and endothelial
cells. The occurrence of pyknotic myonuclei and macrophage
invasion of the minced fragments is evidence that normal
fiber dedifferentiation did not occur. Therefore, dedifferentiation is not prerequisite for minced muscle regeneration in the newt. Although the source of myogenic cells
was not identified, the great majority clearly came from
mononucleated cells present within the muscle tissue prior
to mincing. The possibility of neogenic satellite cell
formation by immediate fragmentation into nucleated
cytoplasmic compartments from the injured fibers cannot be
ruled out completely, but seems unlikely.

    Our observations of cultured muscle explants demon-
strate that newt skeletal muscle can remain in a healthy-
appearing state for several weeks following explanation.

Although muscle fiber dedifferentiation was not observed under these culture conditions, the possibility exists for further _in vitro_ investigations of the factors which may induce fiber dedifferentiation. Again we cannot rule out the formation of satellite cells and their contribution to the mononucleated cell population; however, this is not a major source of myogenic cells in our system. The observation that mononucleated cells from unminced cultures, which contain myonuclei after several weeks, give rise to myogenic cells is evidence that extensive fiber dedifferentiation does not occur.

The mononucleated cell population within the muscle tissue gives rise to cells which resemble blastema cells in culture. These cells are stimulated to proliferate in the absence of fiber dedifferentiation. Some of these cells fuse to form myotubes in the same manner as has been described for blastema cells. Therefore it seems entirely possible that mononucleated cells from muscle tissue contribute to the myogenic cells of the blastema.

The potentially powerful technique of culturing amphibian regeneration cells has opened the door to many experiments in pattern formation and differentiation at the cellular and molecular levels. Previous attempts to culture amphibian myogenic cells had met with little success except for those of Steen (1973) who cultured _Xenopus laevis_ hindlimb myoblasts in a medium which promoted proliferation and differentiation into myotubes at clonal density. We have repeated this result with a modification of the presently described medium. We have also achieved culture conditions which support _Ambystoma mexicanum_ blastema cell proliferation, and their differentiation into cartilage and muscle within time intervals which are comparable to regeneration in the animal (Hinterberger and Cameron, in preparation).

The research was supported in part by NSF PCM 79-19338. I would especially like to thank Dr. Irwin R. Konigsberg in whose laboratory I had the opportunity to develop the goal of culturing amphibian myogenic cells. My students and I continue to benefit from his seminal contributions to biology.

Balls M, Brown D, Fleming N (1976) Long term amphibian
 organ culture. In Prescott D (ed): "Methods in Cell
 Biology," New York: Academic Press, p. 213.
Bischoff R (1980) Plasticity of the myofiber satellite
 cell complex in culture. In Pette D (ed): "Plasticity
 of Muscle," Berlin: Walter de Gruyter, p. 119.
Carlson BM (1972). "The Regeneration of Minced Muscles."
 Basel: S Karger.
Hay ED (1959) Electron microscopic observations of muscle
 dedifferentiation in regenerating Amblystoma limbs.
 Develop Biol 1:555.
Hinterberger TJ, Cameron JA (1982) Muscle and cartilage
 differentiation in axolotl limb blastema cell cultures.
 In preparation.
Jabaily JA, Blue P, Singer M (1982) The culturing of
 dissociated newt forelimb regenerate cells. J exp Zool
 219:67.
Konigsberg UR, Lipton BH, Konigsberg IR (1975) The
 regenerative response of single mature muscle fibers
 isolated in vitro. Develop Biol 45:260.
Lentz TL (1969) Cytological studies of muscle dedifferentiation and differentiation during limb regeneration of
 the newt Triturus. Amer J Anat 124:447.
Maden M (1977) The role of Schwann cells in paradoxical
 regeneration in the axolotl. J Embryol exp Morph 41:1.
Namemwirth MR (1974) The inheritance of cell differentiation during limb regeneration in the axolotl. Develop
 Biol 41:42.
Popiela H (1976) Muscle satellite cells in urodele
 amphibians: facilitated identification of satellite
 cells using ruthenium red staining. J exp Zool 198:57.
Schrag JA, Cameron JA (1982) Regeneration of adult newt
 skeletal muscle in vitro. In preparation.
Snow MH (1978) An autoradiographic study of satellite cell
 differentiation into regenerating myotubes following
 transplantation of muscles in young rats. Cell Tiss Res
 186:535.
Steen TP (1968) Stability of chondrocyte differentiation
 and contribution of muscle to cartilage during limb
 regeneration in the axolotl (Siredon mexicanum). J exp
 Zool 167:49.
Steen TP (1973) The role of muscle cells in Xenopus limb
 regeneration. Amer Zool 13:1349.
Thornton CS (1938) The histogenesis of muscle in the
 regenerating forelimb of larval Amblystoma punctatum.
 J Morph 62:17.

Limb Development and Regeneration
Part A, pages 501–512
© **1983 Alan R. Liss, Inc., 150 Fifth Avenue, New York, NY 10011**

GROWTH FACTORS FROM NERVES AND THEIR ROLES DURING LIMB
REGENERATION

Anthony L. Mescher

Anatomy Section, Medical Sciences Program
Indiana University School of Medicine
Bloomington, Indiana 47405

Limb regeneration in amphibians has long been known to
represent a developing system in which cell proliferation
is dependent upon an influence emanating from nerves, a
phenomenon which Singer (1952) referred to as the nerve's
"trophic" effect. Initial attempts to characterize the
factor responsible for this growth-promoting effect involved infusing partially purified extracts of peripheral
and central nervous tissues into denervated regeneration
blastemas in vivo and assaying the results in terms of
enhanced macromolecular synthesis in the blastemal cells
(Singer, 1974). Results of these studies indicate that the
active agent is probably a basic protein of relatively low
molecular weight (Singer et al., 1976; Singer, 1978).

The evidence from Singer's (1978) work that neurotrophic factors are not species-specific, together with the
difficulties in quantifying effective doses in these in-
fusion studies, have encouraged workers to use extracts of
nervous tissue from larger vertebrates and simpler in vitro
bioassays in efforts to purify growth-promoting or mitogenic substances in brain and peripheral nerves. The
observation that aqueous extracts of adult brain are approximately twice as mitogenic for cultured fibroblasts as any
other tissue extract tested was reported from two laboratories independently over 40 years ago (Hoffman et al.,
1940; Trowell and Willmer, 1939). Only in recent years,
however, has some degree of success been achieved in the
characterization of factors responsible for such effects.
A basic protein with an approximate molecular weight of
13,000 and having potent mitogenic activity for fibroblasts

was purified from bovine brain or pituitary by Gospodarowicz
(1974). Purification of a similar polypeptide from bovine
spinal cord has also been reported (Jennings et al., 1979).
Testing extracts of chick embryo brain in an assay based
on ^{14}C-amino acid incorporation in cultured regeneration
blastemas, Choo et al. (1981) found most of the activity
associated with a basic protein with a molecular weight of
13,500, although several somewhat larger proteins also
showed some activity. Lentz and his co-workers (1981) have
partially purified from rat brain a protein capable of
maintaining acetylcholinesterase activity of newt muscle
in culture. Though not completely characterized, this
factor also appears to be a small basic protein. Neurally
derived proteins with a wide range of molecular weights
have been reported by investigators in many laboratories
to promote the proliferation and/or differentiation of
embryonic skeletal muscle cells in vitro. Noteworthy
among these is the protein sciatin, purified with an
apparent molecular weight of 84,000 from chicken peripheral
nerve by Markelonis et al., 1980 (possibly the same as that
reported by Popiela and Ellis, 1981), which Oh and Markelonis
(1980) assert is the active principle in the chick embryo
extract widely used as a medium supplement for in vitro
studies of myogenesis.

Although differences in purification techniques and
bioassays make direct comparisons difficult, it has become
apparent from these recent studies that a variety of
factors with "neurotrophic" properties may exist in
nervous tissue. Since several of these reports, as well as
the in vivo work of Singer et al. (1976), indicate the
presence of relatively small, basic proteins in the active
fraction, the possibility that the fibroblast growth factor
(FGF) purified by Gospodarowicz (1974) may be involved in
the observed biological responses should be considered.
This polypeptide has been shown to be an extremely potent
mitogen for many types of mammalian cells in vitro
(Gospodarowicz et al., 1978a) and is active in all of the
assays for "neurotrophic" factors mentioned above. Infused
FGF promotes the resumption of mitotic activity in denervated newt forelimb blastemas in vivo (Mescher and
Gospodarowicz, 1979) and when added to cultured blastemas,
FGF at a concentration of 10 ng/ml is as effective as the
optimal dose of brain extract in promoting ^{3}H-thymidine
incorporation (Mescher and Loh, 1981), ^{14}C-amino acid
incorporation and mitotic activity (Carlone et al., 1981).

Similar concentrations of FGF have also been shown to maintain total acetylcholinesterase activity in cultured newt triceps muscle above that in untreated contralateral controls after one week (Carlone et al., 1981). Moreover, FGF at 0.1 µg/ml, without additional chick embryo extract, has an effect on the proliferation and development of fetal bovine muscle cells (Gospodarowicz et al., 1976; Gospodarowicz and Mescher, 1977) comparable to that reported by Oh and Markelonis (1980) using sciatin at 25 µg/ml with chick embryo muscle cells.

Amino acid analyses have shown that brain FGF is a family of three polypeptides, identical to residues 44-153, 44-166, and 91-153 of myelin basic protein (MBP) (Westall et al., 1978). This protein, which is an important constituent of both central and peripheral nervous system myelin and whose amino acid sequence is highly conservative (Gregson, 1976), has been studied extensively since its recognition as the antigen responsible for the autoimmune disease experimental allergic encephalomyelitis, a model for multiple sclerosis (Einstein, 1972). Westall and co-workers (1978) reported that FGF crossreacted antigenically with MBP and was encephalitogenic in guinea pigs and rats. These authors emphasized the likelihood that fragments with FGF activity are generated proteolytically from MBP during the purification procedure, since brain contains an endogenous proteinase which cleaves the basic protein between residues 43-44 and 89-90 (Whitaker and Seyer, 1979).

The assertion that brain FGF is derived from the basic protein of myelin has been challenged, however, by other workers who suggest that the basic protein fragments are not mitogenic and that the FGF activity is due to other, more acidic, polypeptides present as only 1 or 2% of the protein in the preparation. It has been reported (1) that peptides corresponding to those identified as FGF but produced by in vitro proteolysis of purified MBP are not mitogenic (Thomas et al., 1980; Chiang et al., 1980), (2) that the mitogenic activity in FGF is resolvable by isoelective focusing at a more acidic pH than that of the MBP fragments (Thomas et al., 1980; Kellett et al., 1981), and (3) that the FGF activity is not retained on a column of Sepharose coupled with chicken antibodies to bovine MBP (Thomas et al., 1980). These results by themselves however do not immediately refute the suggestion that FGF is derived from MBP. Since standard methods of MBP

purification (Dunkley and Carnegie, 1974) involve prolonged
extraction at a pH of approximately 3, which causes loss
of mitogenic activity from FGF (Gospodarowicz et al., 1978b),
it is not surprising that peptides prepared from purified
MBP are inactive. Isoelectric focusing of brain FGF has
been done by Lennon and Gospodarowicz (pers. com.), who
find the mitogenic activity at pH 9.6, coincident exactly
with that of the MBP fragments, rather than at pH 4.8-5.8
(Thomas et al., 1980) or pH 7.2-7.4 and 8.1-8.6 (Kellett
et al., 1981). Nor does the failure of antibodies against
intact MBP to bind FGF disprove the derivation of FGF from
MBP, since the fragments that represent FGF may not be
recognized by such antibodies (Hashim et al., 1979; Whitaker
et al., 1977).

The controversy concerning the identity of FGF is
important since it involves the possibility that a growth
factor purified from nervous tissue may be derived from
the myelin component of the tissue rather than from the
neurons themselves. Determination of the actual source of
the extracted mitogenic activity is necessary before one
can speculate on the physiological or developmental role(s)
of the factors. We have continued to test the hypothesis
that FGF is derived from MBP, not only because of its relevance to the concept of "neurotrophic" processes, but also
because of its utility in explaining the cell proliferation
which accompanies demyelination in both central and peripheral nervous tissue. Selective release and partial
degradation of the basic protein are important early events
in the demyelination process and occur at the same time as
the onset of proliferation among involved astrocytes in the
brain or Schwann cells, vascular and connective tissue cells
in peripheral nerves (Hallpike, 1976; Einstein, 1972).
Westall (1980) has shown that cerebrospinal fluid from
multiple sclerosis patients, which contains MBP and its
fragments released during demyelination, is approximately
5 times more mitogenic for cultured fibroblasts than control
cerebrospinal fluid.

Complete antiserum to bovine MBP, rather than purified
antibodies selected by their ability to bind intact MBP as
used by Thomas et al. (1980), might be more heterogeneous
with regard to the binding of MBP fragments. Using standard
methods of immunoabsorption (Kwapinski, 1972), we have
tested the effects of such antiserum on the biological
activity of brain FGF. MBP antiserum removed essentially

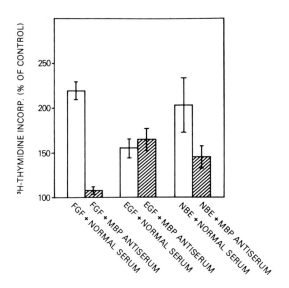

Fig. 1. Effects of rabbit preimmune serum and MBP antiserum on the mitogenic activity of FGF, EFG and NBE for
cultured blastemas. Aliquots of the growth factors and
brain extract were immunoabsorbed by the method of Kwapinski
(1972) and added to cultures at 10 ng FGF/ml, 10 ng EGF/ml,
and 100 µg NBE/ml as described previously (Mescher and Loh,
1981). Each bar indicates the mean ± S.E. of three blastemas.

all of the mitogenic activity of FGF, as indicated by [3]H-
thymidine incorporation in cultured regeneration blastemas
(Fig. 1). Preimmune serum did not reduce FGF's activity
and the antiserum had no effect on the activity of another
mitogen, epidermal growth factor (EGF), in the same assay.
These results are consistent with the report by Westall et
al. (1980) that brain FGF is identical with a fragment of
MBP and suggest that the purified antibodies used by Thomas
et al. (1980) were ineffective because they failed to recognize isolated FGF. To determine whether MBP fragments contribute to the mitogenic activity of crude extracts of
brain, we investigated the effects of the antiserum on the
newt brain extract (NBE) studied previously in the cultured

blastema assay (Mescher and Loh, 1981). Similar experimental approaches have been used to demonstrate the presence of nerve growth factor in extracts of salivary glands (Cohen, 1960) and the presence of EGF in human milk (Carpenter, 1980). As shown in Figure 1, MBP antiserum reduced the activity of the NBE by 50% as compared to the activity of extract treated with preimmune serum. The effect of the antiserum on NBE indicates that a significant portion of the growth-promoting activity from blastema cells in brain extracts may be due to FGF, a conclusion consistent with the physical properties of several partially purified "neurotrophic" factors discussed earlier.

It has been known for some time that proliferation of Schwann cells, fibroblasts and vascular endothelial cells increases greatly during degeneration of injured myelinated nerves, but not unmyelinated nerves (Abercrombie and Johnson, 1946; Joseph, 1947; Romine et al., 1976). Abercrombie (1957) discussed two possible mechanisms for this phenomenon: diffusion of "wound hormones" from the disintegrating myelin or the creation of space into which the surrounding cells grow when the large myelinated fibers, but not the much smaller unmyelinated fibers, degenerate. If the first alternative is correct, FGF released proteolytically from MBP could be one of the factors responsible for the observed effect. We have recently shown (Yachnis and Mescher, 1982) that the release of mitogens rather than the creation of tissue space is the explanation for the proliferative response. When segments of murine sciatic nerve, which is completely myelinated, and of abdominal vagus nerve, which is essentially unmyelinated, were co-cultured with quiescent serum-deprived fibroblasts, the degenerating sciatic nerve fascicles produced significantly higher levels of DNA synthesis in the fibroblasts than the abdominal vagus nerve fascicles of similar size (Fig. 2). This effect was dose-dependent and inversely proportional to the distance from the nerve (Yachnis and Mescher, 1982). Although the identity of the released mitogen(s) remains unknown, experiments preliminary to the characterization of the active factors are underway. Extracts of myelinated and unmyelinated nerves were prepared by a method similar to that of Oh (1975) from the trigeminal and olfactory nerves respectively of the long-nose garfish (Lepisosteus osseus), a source from which quantities of unmyelinated nerve adequate for biochemical analyses can be obtained. Testing these preparations in the murine fibroblast assay showed

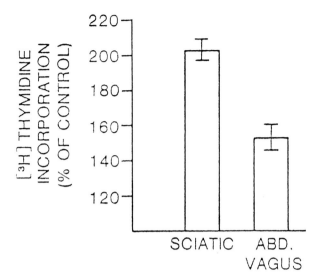

Fig. 2. Mitogenic effects of degenerating myelinated sciatic
nerve and unmyelinated abdominal vagus nerve. The Balb/c
3T3 cell assay (Yachnis and Mescher, 1982) was used and
results are expressed as percent of control cultures without
added nerve. Each bar represents the mean ± S.E. of three
cultures and the difference is significant (p< 0.005).

that myelinated nerve extract was more mitogenic than that
of unmyelinated nerve at concentrations ranging from 25 to
250 µg/ml (Fig. 3), thus providing further evidence for the
existence of myelin-derived growth factors.

Efforts to purify mitogenic or myogenic factors from
brain or peripheral nerves are usually motivated by the
search for neurotrophic factors which are presumed to be
present in the axoplasm. The possibility that polypeptides
with such activities may be present in other, nonneuronal
cells of nervous tissue is seldom considered in discussions
of this work. The experiments cited here, together with
the reports of Westall (1980; Westall et al., 1978) and the
correlation between demyelination and localized cell pro-

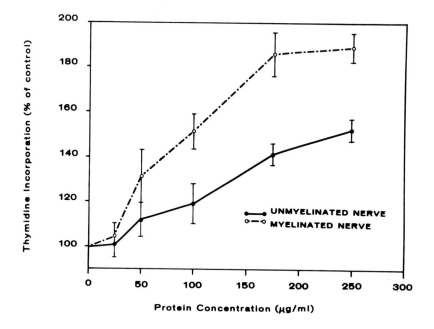

Fig. 3. Dose response curves for mitogenic activity of garfish nerve extracts. The Balb/c 3T3 cell assay (Yachnis and Mescher, 1982) was used and results are expressed as percent of control cultures without nerve extract. Each point represents the mean ± S.E. of three cultures. Significant differences were seen at concentrations greater than 50 µg/ml.

liferation (Abercrombie, 1957; Henson, 1965), suggest that factors derived from myelin are mitogenic both in vivo and in vitro. Such factors may be completely unrelated to the axoplasmic agents being sought, but still have "neurotrophic" activity in the bioassays used in the purification procedures. As discussed above, FGF purified from bovine brain is highly active in various assays similar to those used for neurotrophic factors. The reduction in the mitogenic activity of brain extract by immunoabsorption with antiserum against MBP, the putative precursor to FGF, indicates that this factor may indeed be one of the active agents in the extracts used in studies on the neurotrophic control of limb regeneration. While it is clear that

blastemal cells respond to FGF at very low concentrations
in vitro (Mescher and Loh, 1981; Carlone et al., 1981),
the physiological significance of this factor in limb regeneration remains speculative. As we have discussed more
fully elsewhere (Gospodarowicz and Mescher, 1980), two
possible roles for FGF in the regeneration process present
themselves in light of its derivation from myelin: as one
of the factors promoting dedifferentiation upon its release
from the myelin of the injured nerves or as a mitogen released distally at the nerve fiber endings after transfer
into the axoplasm from the Schwann cell.

The author's research reported here was supported by
NIH grants GM 27735 and GM 31080.

References:

Abercrombie M (1957). Localized formation of new tissue
    in an adult mammal, Symp Soc Exp Biol 11:235.
Abercrombie M, Johnson ML (1946). Quantitative histology
    of Wallerian degeneration. I. Nuclear population in
    rabbit sciatic nerve. J Anat 80, 37.
Carlone RL, Ganagarajah M, Rathbone MP (1981). Bovine
    pituitary fibroblast growth factor has neurotrophic
    activity for newt limb regenerates and skeletal
    muscle in vitro. Exp Cell Res 132: 15.
Carpenter G (1980). Epidermal growth factor is a major
    growth-promoting agent in human milk. Science
    210: 198.
Chiang TM, Whitaker JN, Seyer JM, Kang AH (1980). Effect
    of peptides of bovine myelin basic protein on dermal
    fibroblasts. J Neurosci Res 5:439.
Choo AF, Logan DM, Rathbone MP (1981). Nerve trophic
    effects: partial purification from chick embryo
    brains of proteins that stimulate protein synthesis
    in cultured newt blastemata. Exp Neurol 73:558.
Cohen S (1960). Purification of a nerve-growth promoting
    protein from the mouse salivary gland and its neuro-
    cytotoxic antiserum. Proc Nat Acad Sci 46:302.
Dunkley PR, Carnegie PR (1974). Isolation of myelin basic
    proteins. In Marks N, Rodnight R (eds): "Research
    Methods in Neurochemistry", Vol 2, New York: Plenum
    Press, p 219.

Einstein ER (1972). Basic protein of myelin and its role in experimental allergic encephalomyelitis and multiple sclerosis. In Lajtha A (ed): "Handbook of Neurochemistry", Vol VII, New York: Plenum Press, p 107.

Gospodarowicz D (1974). Localisation of a fibroblast growth factor and its effect alone and with hydrocortisone on 3T3 cell growth. Nature 249:123.

Gospodarowicz D, Mescher AL (1977). A comparison of the responses of cultured myoblasts and chondrocytes to fibroblast and epidermal growth factors. J Cell Physiol 93:117.

Gospodarowicz D, Mescher AL (1980). Fibroblast growth factor and the control of vertebrate regeneration and repair. Ann NY Acad Sci 339:151.

Gospodarowicz D, Weseman J, Moran JS, Lindstrom J (1976). Effect of a fibroblast growth factor on the division and fusion of bovine myoblasts. J Cell Biol 70:395.

Gospodarowicz D, Moran JS, Mescher AL (1978a). Cellular specificities of fibroblast growth factor and epidermal growth factor. In Papaconstantinou J, Rutter WJ (eds): "Molecular Control of Proliferation and Differentiation, New York: Academic Press p 33.

Gospodarowicz D, Bialecki H, Greenberg G (1978b). Purification of the fibroblast growth factor activity from bovine brain. J Biol Chem 253:3736.

Gregson NA (1976). The chemistry and structure of myelin. In Landon DN (ed): "The Peripheral Nerve," London: Chapman and Hall, p 512.

Hallpike JF (1976). Histochemistry of peripheral nerves and nerve terminals. In Landon DN (ed): "The Peripheral Nerve," London: Chapman and Hall, p 605.

Hashim GA, Sharpe RD, Carvalho EF (1979). Experimental allergic encephalomyelitis: sequestered encephalito-genic determinant in the bovine myelin basic protein. J Neurochem 32:73.

Henson RA (1965). The demyelinations. In Cumings JN, Kremer M (eds): "Biochemical Aspects of Neurological Disease", Philadelphia: FA Davis Co., p 214.

Hoffman RS, Tenenbaum E, Doljanski L (1940). The growth activating effect of extracts of adult tissue on fibroblast colonies in vitro. Growth 4:207.

Jennings T, Jones RD, Lipton A (1979). A growth factor from spinal cord. J Cell Physiol 100:273.

Joseph J (1947). Absence of cell multiplication during degeneration of non-myelinated nerves. J Anat 81:135.

Kellett JG, Tanaka T, Rowe JM, Shiu RPC, Friesen HG (1981). The characterization of growth factor activity in human brain. J Biol Chem 256:54.

Kwapinksi JBG (1972). "Methodology of Immunochemical and Immunological Research." New York: Wiley-Interscience, p 286.

Lentz TL, Addis JS, Chester J (1981). Partial purification and characterization of a nerve trophic factor regulating muscle acetylcholinesterase activity. Exp Neurol 73:542.

Markelonis GJ, Kemerer VF, Oh TH (1980). Sciatin: purification and characterization of a myotrophic protein from chicken sciatic nerves. J Biol Chem 255:8967.

Mescher AL, Gospodarowicz D (1979). Mitogenic effect of a growth factor derived from myelin on denervated re-generates of newt forelimbs. J Exp Zool 207:497.

Mescher AL, Loh JJ (1981). Newt forelimb regeneration blastemas in vitro: cellular response to explanation and effects of various growth-promoting substances. J Exp Zool 216:235.

Oh TH (1975). Neurotrophic effects: characterization of the nerve extract that stimulates muscle development in culture. Exp Neurol 46:432.

Oh TH, Markelonis GJ (1980). Dependence of in vitro myogenesis on a trophic protein present in chicken embryo extract. Proc Nat Acad Sci 77:6922.

Popiela H, Ellis S (1981). Neurotrophic factor: charac-terization and partial purification. Dev Biol 83:266.

Romine JS, Bray GM, Aquayo AJ (1976). Schwann cell multi-plication after crush injury of unmyelinated fibers. Arch Neurol 33:49.

Singer M (1952). The influence of the nerve in regeneration of the emphibian extremity. Quart Rev Biol 27, 169.

Singer M (1974). Neurotrophic control of limb regeneration in the newt. Ann NY Acad Sci 228:308.

Singer M (1978). On the nature of the neurotrophic phenomenon in urodele limb regeneration. Amer Zool 18:829.

Singer M, Maier CE, McNutt WS (1976). Neurotrophic activity of brain extracts on forelimb regeneration in the urodele, Triturus. J Exp Zool 96:131.

Thomas KA, Riley MC, Lemmon SK, Baglan NC, Bradshaw RA
(1980). Brain fibroblast growth factor: nonidentity
with myelin basic protein fragments. J Biol Chem
255:5517.
Trowell OE, Willmer EN (1939). Studies on the growth of
tissue in vitro. J Exp Biol 16:60.
Westall FC (1980). Demyelinating disease and mitogens of
myelin origin. Ann NY Acad Sci 339:139.
Westall FC, Lennon VA, Gospodarowicz D (1978). Brain-
derived fibroblast growth factor: identity with a
fragment of the basic protein of myelin. Proc Nat
Acad Sci 75:4675.
Whitaker JN, Seyer JM (1979). The sequential limited
degradation of bovine myelin basic protein by bovine
brain cathepsin D. J Biol Chem 254:6956.
Whitaker JN, Jen Chou CH, Chou FCH, Kibler RF (1977).
Molecular internalization of a region of myelin
basic protein. J Exp Med 146:317.
Yachnis AT, Mescher AL (1982). Stimulation of DNA synthesis
in Balb/c 3T3 cells by peripheral nerve degenerating
in vitro. Exp Neurol 76:139.

Limb Development and Regeneration
Part A, pages 513–524
© **1983 Alan R. Liss, Inc., 150 Fifth Avenue, New York, NY 10011**

ROLES OF NEURAL PEPTIDE SUBSTANCE P AND CALCIUM IN BLASTEMA
CELL PROLIFERATION IN THE NEWT <u>NOTOPHTHALMUS</u> <u>VIRIDESCENS</u>

Morton Globus, Swani Vethamany-Globus,
    Agnes Kesik, and Guy Milton
Department of Biology
University of Waterloo
Waterloo, Ontario, N2L 3G1

The proliferation of animal cells unquestionably
involves a multifactorial regulatory chain of events
involving both extracellular initiators and intracellular
mediators in order to achieve overt mitosis. Neural, hormonal and epidermal influences, which we have referred to as
the "tripartite control" of proliferation (Globus, 1978)
are all external signals. The recent discovery that many
neural peptides seem to act as mitogens has evoked considerable interest but also raises the question of the
diversity of factors required to yield full expression of
trophic stimulation. According to Berridge (1975),
external mitogens regulate internal signals such as
calcium, which he considers is the primary regulator of
cell division. This study was undertaken to deal with some
aspects of both external and internal mitogenic signals.

SUBSTANCE P AS A CANDIDATE FOR MEDIATOR OF NEUROTROPHIC
EFFECTS.

It is generally accepted that the primary role of
nerves in limb regeneration is the promotion of blastema
cell proliferation in the regenerate (Singer and Craven,
1948; Tassava and Mescher, 1975; Globus and Vethamany-
Globus, 1977). Although a neurotrophic factor has not as
yet been identified, the available evidence indicates that
trophic activity resides in peptides of relatively low
molecular weight, possibly less than 5000 daltons (Jabaily
and Singer, 1978; Singer, 1978). Although attention is
currently being given (Gospodarowicz et al., 1978) to

polypeptide growth factors such as FGF (M.W. 13,400), which has been shown to be mitogenic for a variety of mammalian cells and appears to stimulate blastema cell proliferation, and to EGF (M.W. 6,045), very low molecular weight neurally derived peptides have not as yet been investigated for neurotrophic activity during limb regeneration.

Substance P, first extracted from equine brain (von Euler and Gaddum, 1931) is widely but unequally distributed throughout the CNS (Brownstein et al. 1976; Takahashi and Otsuka, 1975), with high concentrations of immunoreactive substance P in the substantia nigra and hypothalamic regions of the brain (Powell et al., 1973; Leeman and Hammerschlag, 1967). It is contained in high concentrations within sensory neurons, localized to neuronal perikarya in dorsal root ganglia, to neuronal processes within the dorsal horn of the spinal cord (Hökfelt et al., 1975) and within terminal axons of sensory neurons (Pickel, Reis and Leeman, 1977). Interestingly, the concentration of bioassayable substance P (or that estimated by RIA) in dorsal nerve roots is considerably greater than that in the ventral roots of the spinal cord (Takahashi and Otsuka, 1975) and an analysis of subcellular fractions of rat hypothalamic tissue has shown that it is present in highest concentration in the synaptosomal fraction (Whittaker et al., 1964). Substance P has been isolated in pure form from bovine hypothalamus (Chang and Leeman, 1970) and the structure elucidated by Chang et al. (1971) was found to have the amino acid sequence H-Arg-Pro-Lys-Pro-Gln-Gln-Phe-Phe-Gly-Leu-Met-NH$_2$, and an estimated molecular weight of 1340.

Substance P - Stimulated Increase in Blastema Cell Mitotic Activity.

In order to assess the effect of substance P on proliferation in the limb regenerate, mid-cone stage blastemata were cultured on ultrathin Millipore filters according to the method of Globus and Vethamany-Globus (1977) and divided into 3 groups; (a) ganglionated, (b) non-ganglionated and (c) substance P - treated explants. A very low concentration of substance P (10 picograms SP per ml of medium) was employed. The explants were cultured for 48 hours at 28± 0.5°C, histological sections were prepared, mitotic indices were calculated and a Chi^2 test for equality of binomial proportions was performed. The results (Figure 1) show that the overall mitotic index of SP - treated explants (1.31) was significantly greater (p < 0.001) than controls (MI = 0.71) and fell high within the range of mitotic activity observed in ganglionated explants.

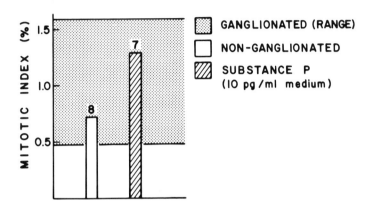

Fig. 1. Limb blastema explants cultured in the absence of nerves with substance P in the medium at a concentration of 10 picograms per ml medium. Mitotic index was compared with ganglionated and non-ganglionated explants. The numbers over bars refer to the number of blastemata evaluated.

Suppression of Mitotic Activity by Substance P Antiserum

Having observed that a peptide, substance P, in pure form has mitogenic activity, one would expect a suppression of mitosis if the peptide were immunologically inactivated. Accordingly, dorsal root ganglia, reportedly rich in SP, were pretreated for 24 hours with SP antiserum prepared in rabbits. Ganglionated blastemata were subsequently cultured for 72 hours (21 ± 1°C) in the presence of SP antiserum at 1:100, 1:1000, and 1:10000 dilutions in the medium. Two control groups served for comparative purposes; (a) a non-treated ganglionated group and (b) one which had been treated with normal rabbit serum (serum control, ganglionated). The results (Figure 2) show that SP antiserum at 1:100 dilution suppressed mitotic activity to a level of approximately 57% of that in untreated controls. At dilutions of 1:1000 and 1:10,000, mitotic activity remained suppressed to levels of 64% and 69%, respectively, however, a progressive recovery of mitotic activity was evident as antiserum concentration was reduced. The results indicate that substance P is sensitive to the antiserum at extremely low concentrations and suggests that SP may participate in the neurotrophic promotion of blastema cell proliferation.

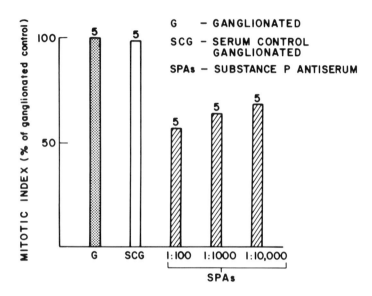

Fig. 2. Ganglionated blastemata were cultured in the presence of substance P antiserum. The overall mitotic index was compared with untreated ganglionated explants and normal serum-treated, ganglionated explants. Three dilutions of the antiserum were employed, 1:100, 1:1000 and 1:10,000 parts in the medium. The numbers over bars refer to the number of blastemata evaluated.

The general localization of substance P in nervous tissues, the presence of high concentrations of it in dorsal root ganglia and terminal axons of sensory nerves, and the finding that it is concentrated in the synaptosomal fraction, is consistent with the localization of neurotrophic activity reported by Marcus Singer (1952, 1978; Singer Maier and McNutt, 1976). Interestingly, electrical stimulation of small diameter afferent neurons leads to a release of substance P (Yaksh et al., 1980). The present study shows that substance P stimulates mitosis in the limb blastema in vitro and that activity is suppressed by antiserum to substance P. The undecapeptide is, therefore, a logical candidate for mediator of neurotrophic effects. An extensive investigation of localization and activity of other low molecular weight peptides is underway in our laboratory.

ROLE OF CALCIUM IN THE INTERNAL REGULATION OF BLASTEMA CELL MITOSIS

Many cellular responses to external stimuli are at least partly mediated by changes in ion flux. Cell proliferation is known to be an ion-mediated response involving calcium (Rasmussen, 1975; Whitfield et al., 1979). An increase in intracellular calcium may be the primary intracellular signal for division in many cells (Berridge, 1975; Nagle and Egrie, 1981). According to Berridge, the calcium signal, although generated by external mitogens, may be regulated in combination with internal factors. Calcium is now thought to play a role in assembly and function of the mitotic apparatus (Nagle and Egrie, 1981). Although its importance in limb regeneration has been recognized (Vethamany-Globus et al., 1978) it has received little attention in the literature.

Changes in Extracellular Calcium Concentration

These experiments were performed in vitro using ganglionated mid-cone stage blastemata cultured for 48 hours according to the transfilter method of Globus and Vethamany-Globus, (1977). The culture medium utilized for this purpose was calcium-free Spinner medium (Gibco), adjusted to amphibian tonicity (225 ± 5 m osmol). Because of a previously established requirement for serum in the medium, fetal calf serum was included at a concentration of 2%; this medium was the zero standard, ie. no additional calcium added other than that contained in the serum. In the other experimental groups, additional Ca^{++} was added to give 0.5, 0.75 and 1.8 mM Ca^{++} concentrations, in excess of the zero standard value. 1.8 mM Ca^{++} corresponds to the values we have used in previous experiments and therefore served as the control in these studies. The results shown in figure 4 reveal that no substantial change in mitotic activity was recorded, however, at 0.5 mM appeared to be optimal. Additional attention is presently being given to the ionic composition of tissue culture media used in regeneration studies.

Effects of Calcium Flux on Mitosis

Ionophore A23187 is known to transfer calcium across biologic membranes and cause an increase in intracellular calcium levels, ie. a net influx (Rasmussen, 1975). It has been suggested (Luckasen et al., 1974) that the ionophore behaves as a mitogen in this regard. To test this hypothesis in the limb regeneration system, animals at an early cone stage of regeneration were injected intraperitoneally (I.P.) with 0.1 ml of 5×10^{-7} M ionophore A23187; (Calbiochem) which was dissolved in 10 mM dimethylsulfoxide (DMSO) and administered twice daily for a period of 3 days. Control animals received 0.1 ml of 10 mM DMSO per injection. The results (Figure 3) show that ionophore A23187 induced a significant ($p < 0.001$) increase (221%) in mitotic index, when compared with the control group.

These experiments were repeated in vitro. Ionophore A23187 (5×10^{-7} M) was added to the medium at a concentration of 0.33 µl per ml of medium, in which cone stage blastemata were cultured transfilter to sensory ganglia for 48 hours. A similar significant ($p < 0.001$) increase (> 200%) in mitotic index was recorded when compared with control explants (see Figure 4). The data suggest that ionophore-induced Ca^{++} influx, resulting in an increase in intracellular Ca^{++}, is mitogenic; these results are consistent with other reports (Luckasen et al., 1974).

Effect of Chlorpromazine

An increase in intracellular Ca^{++} concentration is known to lead to the formation of an active Ca^{++} - calmodulin complex (Means and Dedman, 1980, review). The phenothiazine antipsychotic drug, chlorpromazine, which has been reported to bind to the intracellular calcium receptor, calmodulin, causes inhibition of several enzyme systems regulated by calmodulin (Levin and Weiss, 1979; Weiss and Wallace, 1980). When newts at an early cone stage of forelimb regeneration were injected I.P. with 0.1 ml of 5 mM chlorpromazine HCl (Sigma), twice daily for 3 days, the

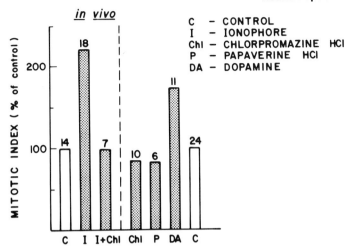

Fig. 3. Newts at an early cone stage of regeneration were injected I.P. with experimental compounds, twice daily for 3 days. 1½ hours after the last injection, the blastemata were assessed for mitotic activity. The number of blastemata evaluated is shown over each bar.

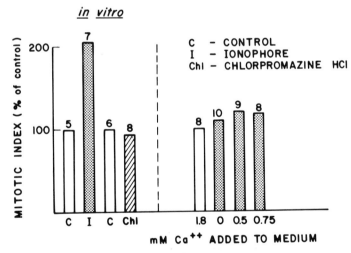

Figure 4. Midcone stage blastemata were cultured transfilter to sensory ganglia for 48 hours, in the presence of experimental compounds (left) or different concentrations of calcium (right). The blastemata were assessed for mitotic activity. The numbers over the bars refer to the number of blastemata evaluated.

mitotic index of blastemal cells was significantly (p < 0.005) decreased when compared with controls (Figure 3). Although a more than twofold increase in mitotic index was observed using Ca^{++} ionophore A23187 alone, the results show that when Ca^{++} influx was followed by chlorpromazine inhibition of the active Ca^{++} - calmodulin complex, the mitotic activity was markedly lower than ionophore treatment alone, and in fact, was similar to control values. These results extend the earlier findings of Sicard and Dinicola (1974) who have shown that chlorpromazine causes retardation of limb regeneration when injected into newts.

Chlorpromazine treatment was employed in vitro as well, using ganglionated blastemata cultured in medium containing 0.33 µl chlorpromazine HCl per ml. Preliminary findings indicate that this dosage may have been too low since a significant effect was not achieved see (Figure 4).

Chlorpromazine is also known to be an antagonist to dopamine and thereby mimics the effect of dopamine depletion. We treated newts at an early cone stage of limb regeneration with dopamine HCl (100 picograms/ .1 ml per injection) twice daily for a period of 3 days, and then assessed its effect on mitotic activity in the blastema. Preliminary results (Figure 3) show that the treatment promoted a significant (p < 0.001) increase (175%) in mitotic activity when compared with controls.

Promotion of Calcium Efflux

Experiments were also performed in vivo with papaverine which promotes a net calcium efflux, lowering intracellular Ca^{++} levels (Duprat and Kan, 1981). Animals which had reached an early cone stage of regeneration were injected I.P. twice daily for 3 days with 0.1 ml of papaverine HCl (Sigma), prepared in amphibian saline. Controls were injected with 0.1 ml of saline. Under these conditions, a significant (p < 0.001) decrease in mitotic index was recorded (Figure 3), when compared with controls, which lends support to the notion that changes in intracellular Ca^{++} concentration may trigger the internal regulation of mitosis during amphibian limb regeneration.

Calmodulin Levels in the Blastema

Upon Ca^{++} binding, conformational changes in calmodulin occur, allowing it to associate with calmodulin-regulated protein kinases which lead to the initiation of mitosis. Nagle and Egrie (1981) localized calmodulin in the mitotic apparatus of dividing tissue culture cells, and many others including Berridge (1975), Cheung (1980) and Chafouleas (1982), have ascribed an important role for calmodulin in the cell cycle. Having recorded a large stimulatory effect of ionophore A23187 on blastema cell mitotic activity and having observed an inhibitory effect by chlorpromazine, we questioned whether there may be a correlation between calmodulin levels and stimulatory/inhibitory treatments. Newts at an early cone stage of forelimb regeneration were treated with either ionophore A23187 or chlorpromazine HCl as described above. 1½ hours after the last injection, blastemata (now mid-cone stage) were excised and individually weighed on a microbalance. The blastemata were homogenized and centrifuged, and calmodulin levels in the supernatant were determined by radioimmunoassay. The data (Table 1), expressed in

Table 1. Radioimmunoassay of calmodulin levels in mid-cone stage blastemata of the newt, Notophthalmus viridescens.

Treatment	Nanograms Calmodulin per mg blastema tissue (mean ± SD)	Difference from Control
Control	1.48 ± 0.55	–
Ionophore	1.24 ± 0.47	not signif.
Chlorpromazine	1.50 ± 0.90	not signif.

Newts were injected I.P. with 0.1 ml of either Ionophore A23187 (5×10^{-7}M), chlorpromazine HCl (5×10^{-3}M) or dimethylsulphoxide (DMSO control, 0.5%), twice daily for a period of 3 days. Blastemata were homogenized in buffer, centrifuged and the supernatant was assayed for calmodulin. Data were subjected to a student's t test; p values > 0.05 were taken as not statistically significant.

nanograms of calmodulin per mg wet weight of blastema tissue, indicate that calmodulin levels were not significantly altered from control values (close to 1.5 ng per mg of blastema tissue) in response to the treatments employed. Chafouleas et al. (1982) have similarly reasoned that the regulation mediated by such a multifunctional regulatory protein must be related to the effective concentration of the protein in the cell at any given time. They have shown that calmodulin levels are elevated twofold at late G_1 and/or early S and are maintained throughout the remainder of the cell cycle until cytokinesis. The G_1 daughter cells then contain half the intracellular calmodulin level found prior to cell division.

It is clear that neural peptides are involved as external mitogens in blastema cell proliferation and that intracellular calcium ion concentrations may mediate the actions of extracellular stimuli, however, it remains to be determined whether neurotrophic influences on the blastema are mediated by internal calcium signals.

Acknowledgements

This work was supported by NSERC Grant No. A6933 to M. Globus and No. A9753 to S. Vethamany-Globus. The portion of this work dealing with calcium represents part of Mrs. Kesik's Master's dissertation. Mr. Guy Milton carried out the substance P experiments.

Berridge MJ (1975). Control of cell division: A unifying hypothesis. J Cyclic Nucleotide Res 1:305.
Brownstein MJ, Mroz EA, Kizer JS, Palkovits M, Leeman SE (1976). Regional distribution of substance P in the brain of the rat. Brain Res 116:299.
Chafouleas JG, Bolton WE, Hidaka H, Boyd AE III, Means AR (1982). Calmodulin and the cell cycle. Cell 28:41.
Chang MM, Leeman SE (1970). Isolation of a sialogogic peptide from bovine hypothalamic tissue and its characterization as substance P. J Biol Chem 245:4784.
Chang MM, Leeman SE, Niall HD (1971). Amino acid sequence of substance P. Nature New Biol 232:86.
Cheng WY (1980). Calmodulin plays a pivotal role in cellular regulation. Science 207:19.
Duprat AM, Kan P (1981). Stimulating effect of the divalent cation ionophore A23187 on in vitro neuroblast differentiation. Experientia 37:154.

Euler US von, Gaddum JH (1931). An unidentified depressor substance in certain tissue extracts. J Physiol (Lond) 72:74.

Globus M, Vethamany-Globus S (1977). Transfilter mitogenic effect of dorsal root ganglia on cultured regeneration blastemata, in the newt, Notophthalmus viridescens. Develop Biol 56:316.

Globus M (1978). Neurotrophic contribution to a proposed tripartite control of the mitotic cycle in the regeneration blastema of the newt, Notophthalmus viridescens. Amer Zool 18:855.

Gospodorowicz D, Moran JS, Mescher AL (1978). Cellular specificities of fibroblast growth factor and epidermal growth factor. In Papaconstantinou J, Rutter WJ (eds): "Molecular Control of Proliferation and Differentation", New York: Academic Press, p.33.

Hökfelt T, Kellerth JO, Nilsson G, Pernow B (1975). Experimental immunohistochemical studies on the localization and distribution of substance P in cat primary sensory neurons. Brain Res 100:235.

Jabaily JA, Singer M (1978). Neurotrophic and hepatotrophic stimulation of embryonic chick muscles in vitro. Devel Biol 64:189.

Leeman, SE, Hammerschlag R. (1967). Stimulation of salivary secretion by a factor extracted from hypothalamic tissues. Endocrinol 81:803.

Levin RM, Weiss B (1979). Selective binding of antipsychotics and other psychoactive agents to the calcium – dependent activator of cyclic nucleotide phosphodiesterase. Pharmacol Exp Ther 208:454.

Luckasen JR, White JG, Kersey JH (1974). Mitogenic properties of a calcium ionophore, A23187. PNAS 71: 5088.

Means AR, Dedman JR (1980). Calmodulin – an intracellular calcium receptor. Nature 285:73.

Nagle BW, Egrie JC (1981). Calmodulin and ATPases in the mitotic apparatus. In Zimmerman AM, Forer A (eds): "Mitosis/Cytokinesis", New York: Academic Press, p.337.

Pickel VM, Reis DJ, Leeman SE (1977). Ultrastructural localization of substance P in neurons of rat spinal cord. Brain Res 122:534.

Powell D, Leeman SE, Tregear GW, Niall HD, Potts JT Jr (1973). Radio immuno-assay for substance P. Nature (Lond) 241:252.

Rasmussen H (1975). Ions as 'second messengers'. In Weissmann G, Claiborne R (eds): "Cell Membranes", HP Publishing p. 203.

Sicard R, Dinicola AF (1974). Retardation of progressive forelimb regeneration by treatment with neuropharmacological agents. Oncology 30:442.

Singer M (1952). The influence of the nerve in regeneration of the amphibian extremity. Q Rev Biol 27:169.

Singer M (1978). On the nature of the neurotrophic phenomenon in urodele limb regeneration. Amer Zool 18: 829.

Singer M, Craven L (1948). The growth and morphogenesis of the regenerating forelimb of adult Triturus following denervation at various stages of development. J Exp Zool 108:279.

Singer M, Maier C, McNutt W (1976). Neurotrophic activity of brain extracts on forelimb regeneration in the urodele, Triturus. J Exp Zool 196:131.

Tassava RA, Mescher AL (1975). The role of injury, nerves and the wound epidermis during the initiation of amphibian limb regeneration. Differentiation 4:23.

Takahashi T, Otsuka M (1975). Regional distribution of substance P in the spinal cord and nerve roots of the cat and the effect of dorsal root section. Brain Res 87:1.

Vethamany-Globus S, Globus M, Tomlinson B, (1978). Neural and hormonal stimulation of DNA and protein synthesis in cultured regeneration blastemata, in the newt, Notophthalmus viridescens. Develop Biol 65:183.

Weiss B, Wallace TL (1980). Mechanisms and pharmacological implications of altering calmodulin activity. In Cheung WY (ed): "Calcium and Cell Function", New York: Academic Press, p. 329.

Whitfield JF, Boynton AL, Macmanus JP, Sikorska M, Tsang BK, (1979). The regulation of cell proliferation by calcium and cyclic AMP. Molec Cell Biochem 27:155.

Whittaker VP, Michaelson IA, Kirkland RJ (1964). The separation of synaptic vesicles from nerve-ending particles ('Synaptosomes'). Biochem J 90:293.

Yaksh TL, Jessel TM, Gamse R, Mudge AW, Leeman SE (1980). Intrathecal morphine inhibits substance P release from mammalian spinal cord in vivo. Nature (Lond) 286:155.

Limb Development and Regeneration
Part A, pages 525–536
© 1983 Alan R. Liss, Inc., 150 Fifth Avenue, New York, NY 10011

METHOD FOR 3-DIMENSIONAL ANALYSIS OF PATTERNS OF THYMIDINE
LABELING IN REGENERATING AND DEVELOPING LIMBS

Thomas G. Connelly and Fred L. Bookstein

Department of Anatomy and Center for Human Growth
The University of Michigan
Ann Arbor, Michigan 48109

Recently the regenerating amphibian limb has become a
highly favored model for the study of morphogenesis.
Experiments on regenerating limbs have contributed greatly
to the development of a theoretical framework for
understanding mechanisms of morphogenetic control (Bryant et
al, 1982; Tank and Holder, 1981). Regeneration studies
have augmented those carried out on developing avian limbs
and together, the two systems have yielded much of what we
know about epimorphic morphogenesis and pattern formation /
regulation in vertebrate development. It is clear from the
data on regenerating and developing limbs that positional
information is not uniformly distributed throughout the
limb. One important problem then becomes the identification
of patterns within the developing structure which could
reflect the nature of the material(s) or processes
responsible for the basis of the positional information
itself.

Faber (1960) proposed that regenerate outgrowth occurs
due to the existence of an "apical proliferation center".
This particular concept is attractive from the standpoint of
viewing the longitudinal outgrowth of the limb in much the
same was as growth of the root tip occurs via the apical
meristem. However, this hypothesis has contributed little
to our understanding of the relationship between
longitudinal outgrowth and the morphogenesis which occurs as
the blastema changes shape from the Late Bud to Palette and
Early Digits stages. A significant amount of shape change
occurs which suggests to us that there are regional

differences in growth which cannot be adequately explained by a single "apical proliferation center". With only a few exceptions the studies devoted to analysis of growth in the regenerating or developing limb have focused on that structure in only a 2-dimensional sense. The majority of the papers published have examined proliferative activities primarily along the proximodistal axis, (Chalkley, 1954; Hearson, 1966; Shuraleff, 1968; Smith and Crawley, 1977; Stocum, 1980). The reason for a lack of adequate data on 3-dimensional patterns of proliferation as an indicator of growth appears to be the absence of an efficient method for accumulating the massive amounts of information such studies would entail. With this in mind we set about investigating the possibility of using computer-assisted methods for obtaining and analyzing data on the 3-dimensional distribution of proliferative activity in regenerating and developing amphibian limbs. We chose to analyze the pattern of ^{3}H-thymidine (^{3}H-TdR) as an indicator of such activity for 2 main reasons. First, there is no computer-assisted method for automatically identifying mitoses in histological sections. Secondly, most studies of mitosis in regenerating limbs have shown the index to be between 0 and 3%. With such low levels of division the probability of error is very high. Thus, even if one were to use manual methods for counting mitotic figures, reliable counts with low variance would be difficult to obtain. We recognize that there may be some difference between patterns of DNA synthesis and patterns of actual cell division. However, it would appear that following ^{3}H-TdR patterns is a good means of narrowing the regions or stages to be studied when actual mitotic counts could be made. When this project was begun there was no available means of automatically or semiautomatically doing even thymidine labeling indices. A method has now been devised which allows this to be done.

SEMIAUTOMATIC METHODS

Newts (Notophthalmus viridescens) are subjected to bilateral forelimb amputation just proximal to the wrist. At various time intervals after amputation the animals are injected with 10 microcuries per gram body weight of ^{3}H-TdR. The precursor is allowed to incorporate for 4-6 hours and the limbs are then fixed in 2.5% glutaraldehyde in 0.1 M phosphate buffer. The limbs are embedded in methacrylate (Sorval, JB-4) and serial 2 micron thick cross sections are cut from tip to base. Sections are saved only at 30 micron intervals. The base of the blastema is defined as the first section containing bony tissue. In order to maintain the serial order of the sections each one is mounted individually on a drop of water placed on a subbed microscope slide. The subbing medium causes the water to bead so that each section will stay in a particular place on the slide. The sections are then treated with an aldehyde blocking agent (Dimedone) for 22 hours and subsequently stained with the Feulgen reaction. Thus, only nuclear profiles are stained. The stained sections are then coated with liquid Kodak NTB-2 emulsion and processed for autoradiography

The resultant autoradiograms are digitized on a TV-based image analyzer (Quantimet 720). This instrument has 64 grey level sensitivity and its camera is relatively insensitive in the red region of the spectrum. In white light the pink nuclei appear very pale to the system while the dark profiles of heavily labeled cells appear very black . If a green filter (530 nm) is then placed in the illumination path the pink nuclei absorb heavily and appear dark to the system . By digitizing the sections first in white light, and then in filtered light it is possible to distinguish labeled (at least heavily labeled) cells from unlabeled (or lightly labeled) cells.

The resultant digital image pairs must then be processed by a second computer in order to actually extract the location of labeled or unlabeled nuclear profiles from the image. This has been accomplished in collaboration with Dr. Stanley Sternberg using a spatial processor (Cytocomputer) and algorithms of his design. To avoid the necessity for making digital mosaics of the sections each section is digitized at a magnification low enough to encompass the largest section encountered, preferably within a 512 X 512 picture element (pixel) matrix. The initial magnification used on the sections studied thus far is 25.2 X. The digital images are processed by the Cytocomputer in such a way that all operations are performed on the entire image at once. We have trained the system using large photomosaics or camera lucida drawings to corroborate the machine counts and have been able to count to within 3% of manual counts. The algorithm is in error but the error is consistent. We feel that this is acceptable error since manual counts of sections containing large numbers of nuclear profiles usually have high error rates due to operator boredom or fatigue. It requires 10-20 minutes of machine time to actually acquire and process the image pairs for each section (depending on the image size). The results of the processing are in machine-readable form so another potential source of error, human transcription of the resultant counts, is eliminated. All the data may be directed to other programs for contour plotting or reconstruction. In the digitizing step the operator can edit out unwanted areas. At this point in the project only patterns of mesenchymal cell activity are being analyzed. Thus a method has been developed which allows the investigator to look at patterns of ^{3}H-TdR incorporation in sections with large numbers of nuclear profiles, a task which would be prohibitively tedious by manual methods. We currently have a problem with maintaining count accuracy due to changes in threshold grey level values resulting from variability in the digitizer from run to run. The problem is being attacked in collaboration with Dr. Edward Delp using methods of adaptive thresholding. This approach will allow us to compare grey level ranges between a test image and the training image so that an appropriate threshold level may be set.

MANUAL METHODS

In order to continue to generate some data which can be used in developing 3-dimensional reconstructions we have begun to use manual digitization methods. Because cross-sections of newt limbs are large and contain huge numbers of cells we have chosen to use the developing hindlimb of the Mexican axolotl, Ambystoma mexicanum as a model.

Axolotl embryos injected with ^{3}H-TdR have been kindly provided by Dr. David Stocum. These embryos are embedded in paraffin and serially cross sectioned at 5 microns through the developing hindlimb region. Following autoradiography sections spaced 25 microns apart are traced with a camera lucida and then the tracings are digitized manually using a Summagraphics tablet linked to a Tektronix 4054 graphics computer. Points defining the boundary of the sections and the locations of labeled and unlabeled nuclear profiles are entered into computer programs written by one of us (FLB). These programs generate 2-dimensional maps of the labeling index or cell density within the sections (Fig. 1). The data are also used to generate 3-dimensional images of labeling intensity or cell density (Fig. 2). In order to enhance the visual quality of the 3-dimensional displays we are now in the process of adapting programs written for reconstruction of computed axial tomograms (CAT's) and for synthetic animation to generate surfaces within surfaces.

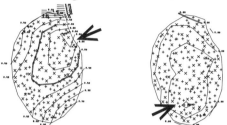

Fig. 1. Two-dimensional plots of labeling index (A) and cell density (B) in one cross section from a limb of the pair in Fig. 3. The section is taken from the region of highest labeling intensity approximately 170 micrometers from the tip of the limb. The regions of highest activity are indicated by arrows. The X's indicate labeled cells and the +'s indicate unlabeled features.

From the above discussion it can be seen that it is possible to analyze the 3-dimensional patterns of cell density and thymidine labeling index in a developing structure. Data are not available from many limbs yet but we have been able to determine (as did Smith and Crawley) that patterns of 3H-TdR incorporation are not uniform throughout the blastema. We have also looked at the distribution of ^{3}H-TdR incorporation along the 3 axes of 6 embryonic axolotl hindlimbs. In all of those limbs the peak labeling index along the P-D axis occurs some distance proximal to the tip. In 4 of the 6 limbs (the longest) the peak of label appears to be found in sections between 150 and 220 microns proximal to the tip. Compare the patterns of relative labeling index for a right and left limb pair from this group (Fig. 3). Notice too that the labeling index in the sections nearest the tip is significantly lower than the peak index in the limb. The distribution of labeling index and cell density in the cross sectional dimensions for one such section is shown in Fig. 1. Notice that those patterns are not random but show regions of higher activity. The distribution of these peaks seems to change between proximal and distal levels. At more proximal levels the highest ^{3}H-TdR incorporation occurs in the future posterior region of the limb and the highest cell densities occur centrally or in the future anterior regions. At more distal levels the trend appears reversed with greater ^{3}H-TdR

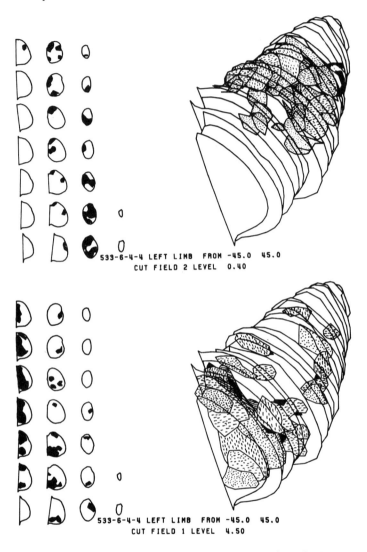

533-6-4-4 LEFT LIMB FROM -45.0 45.0
CUT FIELD 2 LEVEL 0.40

533-6-4-4 LEFT LIMB FROM -45.0 45.0
CUT FIELD 1 LEVEL 4.50

Fig. 2. A 3-dimensional display of labeling index (A) and cell density (B) in the left hindlimb of a developing axolotl. The stippled regions in the drawings on the right correspond in 3-D to the blackened areas in the 2-D tracings on the left. The display is intended to show boundaries of regions with 40% labeling index or 4.5 cells / sample box.

the tip and forms an acute angle. This indicates that at
the time of branching, growth is occurring primarily in the
anterior or posterior regions and not at the distal tip.
Because of the pronounced dorsal curvature of the regenerate
it appears that that region is the major axis of growth.
This means that the first digit forms by a reduction in the
rate of outgrowth in that area relative to more dorsal
(posterior) regions. This results in the formation of a
prominent bulge posteriorly and a notch between that mass
and the first digit. The rest of the digits may then form
by a continued shift in growth posteriorly. Elongation of
the regenerate does not occur due to growth at the middistal
tip of the blastema. Since the distance between the two
major branch points in the medial axis (the proximal one
being an artefact of the line indicating the amputation
plane) increases, yet the distance from the most distal
branch point to the tip remains relatively constant it
appears that longitudinal growth is occurring from some
region proximal to the tip. These results suggest that the
change in form occurring during digit formation is due to
differential growth of the regenerate along the antero-
posterior axis. In fact, examination of the 3-dimensional
reconstructions bears this out. When taken with the data
described for developing embryonic limbs it would appear
feasible now to concentrate on growth changes occurring in
the distal one-third of the regenerate, primarily on
differences along the cross sectional axes of the limb.

We recognize that patterns of 3H-TdR incorporation may
not be equated with growth of the regenerate and agree that
they probably represent only one component in growth as a
whole. We must certainly look at other parameters including
matrix density patterns etc. to fully understand the
process. However, with further adaptations of the computer-
assisted methods described in this paper such analysis is
definitely feasible.

It is frustrating to present such preliminary data
without sufficient evidence to draw firm conclusions about
how it may be integrated into one or more of the currently
favored models for pattern formation and regulation. Yet we
believe that our method of analysis will allow us to
generate sufficient data to carefully analyze such questions
in the future. There is evidence from other lines which
suggests that it is worthwhile to pursue the task of
describing the relation between internal and external growth

parameters. Evidence from Bryant and others suggests that positional values are not uniformly distributed around the circumference of the limb but is weighted more heavily towards the anteroventral region of the limb. Tank and Holder (1979) have demonstrated differential patterns in absolute cell number across the crossectional axes of the salamander limb, but it is difficult to relate their result to the actual density of blastema cells in the developing regenerate. Their methods are also not the best since they do not take into account the varying amounts of cellular material (relative size differences) in the various tissue types which make up the limb. Virtually nothing is known about the spatial distribution of important components of the extracellular matrix in regenerates.

Two reports concerning distribution of proliferative activities in the regeneration blastema have looked at axes other than the proximo-distal. Shuraleff (Ph.D. thesis 1968) confirmed Chalkley's finding of a lowered mitotic index directly beneath the apical epithelium, described a dense core of mesenchymal cells within the blastema, and suggested a tendency toward higher mitotic activity in the lateral regions of the blastema. His sampling methodology was unfortunately designed in such a way as to minimize the probability of finding any regional differences in the cross-sectional axes. Smith and Crawley (1977) examined the thymidine labeling patterns in axolotl limb regenerates with manual methods similar to those described above. However, the limited sample size and their use of longitudinal sections creates a problem in accurately estimating activity along the long axis of the limb. They did, however, demonstrate complicated patterns of labeling intensity throughout the blastema.

The advantages to the methods described in this paper are that they are not dependent upon a "sampling methodology" and are really an unbiased means of estimating distribution of labeling in all axial dimensions of the limb.

In his Ph.D. thesis, Shuraleff proposed a model for regenerate outgrowth which looked at the blastema in a more biomechanical sense. His hypothesis was that distal outgrowth occurred because the apical cap provided a reserve of cells which could slide proximally to allow for stretching from mesenchymal cell proliferation. When the incorporation anteriorly and the cell density higher centrally or more posterior(Fig.2). These data are from larvae with limbs at a stage equivalent to the medium to late bud stage. Clearly we need more data for more limbs of a wider range of stages before any firm conclusions can be drawn. The preliminary results are especially intriguing particularly in light of the data presented below on analysis of the gross growth patterns of regenerating limbs in the axolotl.

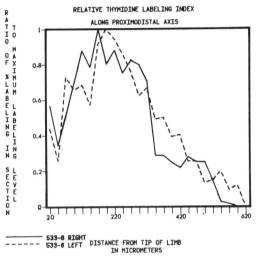

Fig. 3. A plot of the ratio of labeling index in a section /maximum sectional labeling index in the limb along the proximodistal axis of two developing axolotl hindlimbs. The left is reproduced in 3-D in Fig. 2. Note that the peak labeling level is proximal to the tip. Two-dimensional plots of labeling index and cell density for a section of the right limb at the peak labeling level are shown in Fig. 1.

GROSS MEASUREMENTS OF FORM CHANGE

In order to determine if there is an objective means of evaluating regenerate growth and morphogenesis from serial observations of growing limbs we have been using the method of the "medial axis transform" (MAT; Blum, 1973), which is a means of describing a shape based on the shape itself. This procedure computes a "skeleton" of the shape and displays branches in the "skeleton" where major shape changes are present. We have looked at the MATs of digitized camera lucida tracings of 8 axolotl forelimb regenerates which were part of the original data set used by Tank et al (1976) in devising a staging system for axolotl limb regeneration. An analysis of the MATs shows that there is a remarkable consistency from limb to limb in the points along the longitudinal axis at which branches indicating the formation of digits appear (Fig. 4). There is likewise a remarkable consistency in the angles between the branches once they form.

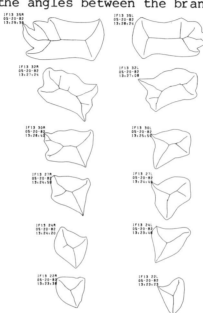

Fig. 4. Medial axes for two axolotl regenerates. Note the consistency in branch formation, angle and distance from distal tip to the first branch . The proximal branch is an artifact caused by forcing the algorithm to consider the amputation point as a surface. Frontal view of regenerates from 22 days (bottom) to 35 days (top) .

The most informative results from this study are, however, related to the way in which the branches seem to appear. Because of the method of computing the MAT we can determine how the shape is changing to give rise to branch points in the "skeleton". In order for the first branch to apear there must be a change in the radius of curvature of the distal region of the regenerate such that the tip flattens with respect to the future anterior and posterior regions. The branch always occurs some distance proximal to "stretchiness" of the apical cap was exhausted, distal outgrowth slowed, and hand morphogenesis ensued. Given the fact that we know that the basement membrane (adepidermal membrane) of the regenerate epidermis is absent, and that it reforms in a proximo-distal fashion, and that such structures have been shown to have an important effect on morphogenesis in salivary glands (Bernfield, 1981) this hypothesis deserves further consideration. The MAT method shows (at least circumstantially) that by reducing growth in the mid-distal region one can actually control morphogenesis along the A-P axis. There may be no need to postulate a dual mechanism for control of morphogenesis along the A-P and P-D axes. Perhaps a refinement of the progress zone hypothesis is all that is necessary. Perhaps too, we should investigate more carefully the possible mechanical functions of the regenerate epidermis as well as its possible funtions as a source of mitogen or as a current generator.

(Supported by grants NSF-PCM77-04643 to T.G.C. and DE-05410 to F.L.B.)

Bernfield M (1981). Organization and remodeling of the extracellular matrix in morphogenesis. In Connelly TG, Brinkley LL, Carlson BM (eds): "Morphogenesis and Pattern Formation," New York: Raven Press, p 139.

Blum H (1973). Biological shape and visual science. J Theor Biol 38:205.

Bryant SV, Holder N, Tank PW (1982). Cell-cell interactions and distal outgrowth in amphibian limbs. Amer Zool 22:143.

Chalkley DT (1954). A quantitative histological analysis of forelimb regeneration in Triturus viridescens. J Morphol 94:21.

Faber J (1960). An experimental analysis of regional organization in the regenerating forelimb of the axolotl (Ambystoma mexicanum). Archiv de Biol 71:1.

Hearson L (1966). An analysis of apical proliferation in the forelimb regeneration blastema of the axolotl, Ambystoma mexicanum. Ph D Thesis, Michigan State University.

Shuraleff N (1968). The effect of nerve augmentation on the cellular activity of mesenchymatous cells within the regeneration blastema of the hindlimb of the axolotl, Ambystoma mexicanum. Ph D Thesis, Michigan State University.

Smith AR, Crawley AM (1977). The pattern of cell division during growth of the blastema of regenerating newt limbs. J Embryol Exp Morphol 37:33.

Stocum DL (1980). The relation of mitotic index, cell density and growth to pattern regulation in regenerating Ambystoma maculatum forelimbs. J Exp Zool 212:233.

Tank PW, Holder N (1981). Pattern regulation in the regenerating limbs of urodele amphibians. Quart Rev Biol 56:113.

Tank PW, Carlson BM, Connelly TG (1976). A staging system for forelimb regeneration in the axolotl, Ambystoma mexicanum. J Morphol 150:117.

Limb Development and Regeneration
Part A, pages 537–545
© 1983 Alan R. Liss, Inc., 150 Fifth Avenue, New York, NY 10011

EFFECTS OF PARTIAL DENERVATION ON THE NEWT BLASTEMA CELL
CYCLE

Roy A. Tassava, David P. Treece, Cherie L. Olsen

Department of Zoology
The Ohio State University
Columbus, Ohio   43210

The injury of amputation causes the mesodermal cells of
a salamander limb to leave their G-0 differentiated state
and enter the G-2 phase of the cell cycle (Tassava and
Mescher, 1975). After this dedifferentiation process, the
resultant mesenchymal-like cells normally progress through
several rounds of the cell cycle, including the S phase
(DNA synthesis), G-2, and M (mitosis). The daughter cells
accumulate to form the blastema from which the new limb
tissues ultimately redifferentiate.

In a limb which is denervated at or near the time of
amputation, cell proliferation is minimal and a blastema
does not form (Singer, 1978; Mescher and Tassava, 1975). In
fact, nerves are needed throughout the early so-called nerve-
dependent stages of blastema formation. It is not until
about 8 days post-amputation in larval Ambystoma (Butler and
Schotté, 1944) and 17 days post-amputation in adult newts
(Singer and Craven, 1948) that blastemas become nerve-independent and are able to complete development into small, but
complete limbs after denervation.

To attempt to assess the cellular mechanisms by which
nerves influence regeneration, we initiated studies designed
to compare dedifferentiation, RNA synthesis, DNA synthesis,
and mitosis in denervated versus innervated limb stumps. Of
all these activities, mitosis appeared to be the most
adversely affected, occurring at levels far below those of
regenerating limbs (Kelly and Tassava, 1973; Mescher and
Tassava, 1975; Tassava et al., 1974). Because H^3-thymidine
labeling invariably showed dedifferentiated cells to be

synthesizing nuclear DNA, we hypothesized that nerves acted in the G-2 phase of the cell cycle to "set the conditions for mitosis" (Mescher and Tassava, 1975; Tassava and Mescher, 1975). Others have since suggested that nerves act throughout the cell cycle (Globus, 1978) or in the G-1 phase (Maden, 1979).

Tassava and McCullough (1978) measured the cell cycle, (CCT) by use of the percent labeled mitoses method (Mitchison, 1971; McCullough and Tassava, 1976), of nerve-dependent axolotl blastemas after denervation but obtained inconsistent results. In one group of blastemas, in which limbs were denervated 8 days after amputation and the CCT was measured through days 9-12, the cell cycle increased in length by 13.5 hrs compared to contralateral control blastemas. The increase was mainly in the G-1 phase. In a second group, in which limbs were denervated on day 8 but the CCT was measured through days 10-13, the CCT increased by only 3.5 hrs and included increases in both G-1 and G-2. One of the difficulties with measuring the CCT after complete denervation is that the mitotic index decreases rapidly to zero (Singer and Craven, 1948; Tassava and McCullough, 1978). It is therefore difficult to obtain sufficient mitotic figures to complete the labeled mitoses curve through the 3rd 50% intercept (McCullough and Tassava, 1976).

We have been able to circumvent this problem and to investigate the cell cycle role of nerves without the usual complicating features of blastema resorption or delayed and/or abnormal morphogenesis (pattern formation).

The system we have developed is to partially denervate 14 day adult newt forelimb blastemas.

The initial experiment was designed to compare the G-2 and S phases of the cell cycle in control and partially denervated blastemas.

METHODS

Forelimbs of adult newts (Notophthalmus viridescens) were bilaterally amputated through the mid radius/ulna. On day 14 post-amputation, limbs were either completely denervated (8 limbs), partially denervated (4th and 5th nerves cut; 17 limbs), or sham-operated (24 limbs). Limbs were

observed through time and compared for stages and size.
Limbs of a second group of newts were amputated as above and
left limbs were partially denervated on day 14. On day 17,
3 days after partial denervation, each newt was given an
injection of 10 μci of (H3-methyl) thymidine (62 ci/mM) in
0.1 ml of sterile distilled water. After a 2 hr incorporation period, each newt was given a chase of unlabeled
(cold) thymidine (1000 x by weight) (Wallace and Maden, 1976).
At 2, 8, 16, 24, 32, and 40 hrs post-injection, limbs were
sampled and prepared for histology and autoradiography.
Alternate longitudinal sections of both control and partially
denervated blastemas were sampled for labeled and unlabeled
mitotic figures. Percent labeled mitoses curves were constructed through 40 hrs from which the lengths of $G_2 + \frac{1}{2}M$
and the S phase were determined.

RESULTS

Complete denervation at 14 days post-amputation resulted
in inhibition of blastema outgrowth and subsequent resorption
in every case (Table 1). It can be concluded that these
young, early bud stage blastemas were nerve-dependent.
Removing 2/3 of the limb nerve supply by partial denervation
clearly had a much less adverse effect since 13 of 17 (77%)
(Table 1) continued to progress through the various stages
of regeneration.

Table 1. A comparison of the effects of partial versus
complete denervation on nerve-dependent newt regenerates
(observations on day 30 post-amputation). Those which did
not complete regeneration resorbed back to the amputation
surface.

Nerve(s) severed	Stage	Total #	# Completing regeneration	Resorbed
Sham	Day 14	24	24	0
4,5	Day 14	17	13	4
3,4,5	Day 14	8	0	8

A comparison of the blastema stages (according to Iten and Bryant, 1973) on day 30 post-amputation (day 16 post-partial denervation) can be seen in Table 2. These data show that blastemas of partially denervated limbs did not lag behind those of sham-operated (control) limbs in terms of stages. Comparable stages in the groups were also seen before day 30 and after day 30 (data not included).

Table 2. A comparison of stages of partially denervated (left) adult newt limb regenerates with normally innervated (right) regenerates (observations on day 30 post-amputation).

		# of regenerates at:					
Limb	No Regeneration	EB	MB	LB	Pal	Dig	Total
L	4	0	1	2	4	6	17
R	0	0	1	1	8	7	17

When blastema sizes were compared, it was immediately clear that partially denervated blastemas were always shorter and narrower (anterior/posterior axis) than sham-operated control blastemas, at all stages examined. When these dimensions were compared for digit stage regenerates on day 30, the control regenerates were 36% longer and 21% wider than the partially denervated regenerates.

Tabulation of the percent labeled mitoses from Feulgen stained sections after autoradiography (at least 25 mitotic figures/limb) showed very similar values for the two groups of blastemas. No labeled mitoses were seen at 2 hrs post-injection of H-3 thymidine, but by 8 hrs post-injection about 90% of the mitotic figures (prophase, metaphase, anaphase, telophase) were labeled (over background) (Table 3). The labeling percent stayed near 90% through 16 and 24 hrs. At 40 hrs, the percent labeled mitoses was 34% in each group of blastemas. After constructing a percent labeled mitoses curve (not shown) the lengths of $G-2 + \frac{1}{2} M$ (time zero to the 1st 50% intercept; McCullough and Tassava, 1976) and the S phase (1st 50% intercept to the 2nd 50% intercept) could be calculated (Table 4). Cells of the left, partially denervated blastemas traversed the $G-2 + \frac{1}{2} M$ distance in 6 hrs compared to 5 hrs for the control right limbs. The S phase of cells of the partially denervated blastemas was

31.1 hrs in length compared to 32.4 hrs for the control blastemas (Table 4).

Table 3. Tabulation of the percent labeled mitotic figures of partially denervated (left) regenerates and control (right) through 40 hrs post-labeling. At least 3 regenerates were sampled at 2, 8, 32, and 40 hrs. Only one regenerate was sampled at 16 and 24 hrs.

		% labeled mitotic figures at:				
Limb	2 hrs	8 hrs	16 hrs	24 hrs	32 hrs	40 hrs
Left	0	86±5	92	90	82±6	34±3
Right	0	92±2	91	92	81±4	34±6

Table 4. Comparison of the time required for cells of partially denervated and control regenerates to traverse the G_2 and S phases of the cell cycle.

	Length (hrs) of:	
Limb	$G_2+\frac{1}{2}M$	S phase
Left	6.0	31.1
Right	5.0	32.4

The total time required for cells to traverse G-2, $\frac{1}{2}$M, and the S phase was equal for the 2 groups of blastemas.

DISCUSSION

The 14 day post-amputation newt blastemas of the present study were clearly nerve-dependent, since after complete denervation outgrowth stopped and resorption occurred. Cellular effects of complete denervation on nerve-dependent blastemas have been examined recently by Tassava and McCullough (1978) with axolotls and by Loyd and Connelly (1981) with newts. When nerve-dependent blastemas of axolotls were completely denervated, the mitotic index

decreased from 2% to zero during the subsequent 7 days whereas
the H-3 thymidine labeling index decreased from 70% to 30%
(Tassava and McCullough, 1978). The cells which block in the
cell cycle after denervation are apparently rapidly removed
because no accumulation of G-1 or G-2 cells was detected by
microspectrophotometry (Loyd and Connelly, 1981). Removal
of these blocked cells, by whatever mechanism, undoubtedly
accounts for blastema resorption (Tassava and McCullough,
1978). Completely denervated blastemas are not well-suited
for cell cycle studies because of the complications due to
resorption and decreases in mitotic index.

To avoid these complications we initiated studies to
examine the effects of partial denervation. By delaying
partial denervation to 14 days post-amputation, it has been
possible to examine effects of limited innervation (1/3 of
the normal supply) on blastema outgrowth without delaying
morphogenesis. It is of interest that nearly 80% of these
14 day nerve-dependent blastemas continued to develop after
partial denervation. This is in contrast to the more adverse
effects of partial denervation (4th + 5th nerves) at the time
of amputation (Singer and Egloff, 1949). In these latter
experiments over 50% of the limbs regenerated but were delayed
by 3 to 8 days. In our own experiments (Tassava, unpublished),
partial denervation (4th + 5th nerves) at one day post-amputation resulted in non-regeneration in 80% of the cases.

By 14 days after amputation, not only has dedifferentiation been completed but cellular proliferation is well underway, as exemplified by continued increases from day 7 to day
14 in H-3 thymidine labeling (Hay and Fischman, 1961; Mescher
and Tassava, 1975) and mitotic index (Chalkley, 1954; Mescher
and Tassava, 1975). The 3rd nerve alone is sufficient for
the continuation of regeneration providing proliferation has
gotten this 14 day "head start." What is significant is that
growth of the regenerate is adversely affected, as compared
to controls, as shown by measurements of the distal/proximal
(length) and anterior/posterior (width) axes. Even if one
assumes the dorsal/ventral (height) axis of partially denervated and control regenerates to be equal, it can be estimated
that the average volumes of the partially denervated 30 day
(digit stage) regenerates is, at the most, only 60% that of
the controls.

This decreased size of partially denervated regenerates
cannot be accounted for by a greater cellular density and/or

smaller cells (Tassava, unpublished). Presumably then, partial denervation has an adverse effect on proliferation and therefore fewer cells are produced. This type of an effect could result from two possible mechanisms. (1) After partial denervation some cells might block completely at some point in the cell cycle. For example, for every 10 cells which are traversing the cell cycle, only 7 might complete mitosis. This would result in only 14 daughter cells instead of the potential 20. Any cells blocking before mitosis would not contribute to a labeled mitoses curve. The labeled mitoses curve would show no change in the CCT if the non-blocked cells traversed the cycle in the same time as cells in control blastemas. Such a mechanism has been hypothesized to result from fewer blastema-nerve axon contacts with limited innervation. Those cells not contacted by nerve axons are blocked (Tassava and McCullough, 1978). (2) An alternative mechanism could involve an adverse effect of partial dener-vation on all blastema cells such that the average cell cycle time would be increased. Such a mechanism is consistent with the view expressed by Singer and Egloff (1949). When the number of nerve fibers is reduced their chemical contribution of a regeneration-evoking chemical to the general milieu of the limb is also reduced. In a given period of time, for example 90 hrs, control blastema cells might complete 2 cell cycles and therefore the population would increase 4 times (assuming a 45 hr CCT). If the CCT is increased to 55 hrs in partially denervated blastemas, then a period of 110 hrs would be required for the same population increase. The increase in the cell cycle could presumably be in any one phase of the cell cycle or proportionately distributed throughout the cycle.

The present results do not distinguish between these 2 alternatives. However, the lengths of G-2 + ½M and the S phase are not different in partially denervated blastemas compared to controls. Therefore there is clearly no increase in the G-2 phase or the S phase (alternative 2 above) due to partial denervation. Extension of the labeled mitoses curve through 72 hrs will make it possible to determine the length of the G-1 phase and also the entire cell cycle time. The height of the 2nd labeled mitoses peak may also indicate whether or not cells are completely blocked in G-1 or, alternatively, whether the G-1 phase is increased in length.

If after extension of the labeled mitoses curve the CCT of partially denervated blastemas is found to be equal to that of control blastemas, then alternative #1 above would be supported.

REFERENCES

Chalkley DT (1954). A quantitative histological analysis of forelimb regeneration in Triturus. J Morph 94:21.

Globus M (1978). Neurotrophic contribution to a proposed tripartite control of the mitotic cycle in the regeneration blastema of the newt. Amer Zool 18:855.

Hay ED, Fischman DA (1961). Origin of the blastema in regenerating limbs of the newt Triturus viridescens. An autoradiographic study using tritiated thymidine to follow cell proliferation and migration. Devel Biol 3:26.

Iten LE, Bryant SV (1973). Forelimb regeneration from different levels of amputation in the newt: length, rate, and stages. Wilhelm Roux Arch 173:263.

Kelly DJ, Tassava RA (1973). Cell division and ribonucleic acid synthesis during the initiation of limb regeneration in larval axolotls (Ambystoma mexicanum). J Exp Zool 185:45.

Loyd RM, Connelly TG (1981). Microdensitometric analysis of denervation effects on newt blastema cells. Experientia 37:967.

Maden M (1979). Neurotrophic and X-ray blocks in the blastemal cell cycle. J Embryol Exp Morph 50:169.

McCullough WD, Tassava RA (1976). Determination of the blastema cell cycle on regenerating limbs of the larval axolotl. Ohio J Sci 76:63.

Mescher AL, Tassava RA (1975). Denervation effects on DNA replication and mitosis during the initiation of limb regeneration in adult newts. Devel Biol 44:187.

Mitchison JM (1971). "The Biology of The Cell Cycle." Cambridge, England: Cambridge University Press.

Schotte OE, Butler EG (1944). Phases in regeneration of the urodele limb and their dependence upon the nervous system. J Exp Zool 97:95.

Singer M (1978). On the nature of the neurotrophic phenomenon in urodele limb regeneration. Amer Zool 18:829.

Singer M, Craven L (1948). The growth and morphogenesis of the regenerating forelimb of adult Triturus following denervation at various stages of development. J Exp Zool 108:272.

Singer M, Egloff FRL (1949). The nervous system and regeneration of the forelimb of adult Triturus. VIII. The effect of limited quantities on regeneration. J Exp Zool 111:295.

Tassava RA, Bennett LL, Zitnik GD (1974). DNA synthesis without mitosis in amputated denervated forelimbs of larval axolotls. J Exp Zool 190:111.

Tassava RA, McCullough W (1978). Neural control of cell cycle events in regenerating salamander limbs. Amer Zool 18:843.

Tassava RA, Mescher AL (1975). The roles of injury, nerves, and the wound epidermis during the initiation of amphibian limb regeneration. Differen 4:23.

Wallace H, Maden M (1976). The cell cycle during amphibian limb regeneration. J Cell Sci 20:539.

Supported by National Science Foundation.

Limb Development and Regeneration
Part A, pages 547–555
© **1983 Alan R. Liss, Inc., 150 Fifth Avenue, New York, NY 10011**

DNA SYNTHESIS AND CELLULAR PROLIFERATION IN CULTURED NEWT
BLASTEMAS

Bernard Lassalle

Université des Sciences et Techniques de Lille
Laboratoire de Morphogenèse Animale
59655 Villeneuve d'Ascq Cédex   France

Regeneration of the adult newt forelimb is
characterized by two successive stages. The first one leads
to an accumulation of dedifferentiated cells which constitute the cone-shaped blastema. Blastemal cell proliferation is
greatly influenced by the presence of nerves. A limb denervated at this stage does not regenerate. The second one,
the morphogenesis, leads to the reconstruction of the
amputated part by differentiation of blastemal cells. This
stage is nerve-independent (Singer and Craven, 1948). The
precise nature of nervous control upon the first stage is
still unknown. Denervation produces a cellular proliferation arrest and a general decrease of macromolecular synthesis (Dresden, 1969 ; Lebowitz and Singer, 1970 ; Singer
and Caston, 1972).

Infusions of the limb with substances such as
brain extracts reestablish macromolecular synthesis near its
original level (Jabaily and Singer, 1977). However, the study of nervous control system *in vivo* is difficult : denervation is never complete, furthermore, numerous other incon-
trollable agents (hormonal state of animals, quantification
of infused materials...) complicate experimental results.
Thus, *in vitro* blastema culture appears as a good approach
to the study of blastemal cellular proliferation and its
control by nervous system. We used this method to estimate
the effects of several agents (nervous system, hormones,
epidermal migration) on DNA synthesis and cellular proliferation of newt blastemas.

MATERIALS AND METHODS

       Adult newts (*Pleurodeles waltlii*) reared in
laboratory were kept in tap water (24°C) and fed chironomid
larvae and beef liver twice a week. After anesthetizing
the animals in MS 222, limbs were amputated bilaterally at
midhumeral level and allowed to regenerate to cone stage.
At the time of explantation, blastemas were excised, sterilized in a 1 % chloramine solution for 1 min. and then
rinsed in sterile medium. Individual explants were cultured
on supporting grids 24 to 96 hours in Leibovitz's L 15
medium diluted to a tonicity appropriate for amphibian cells
(260 mOsm) and containing penicillin (100 U/ml) and streptomycin (100 μg/ml). In some experiments, MEM, CMRL 1066,
Ringer supplemented with fetal calf serum were used. Cultures were maintained at 25° C in a normal atmosphere.

       Sterile dorsal root ganglia (16 th and 17 th
crural, 3 rd and 4 th brachial) were excised from donor
animal either at the time of blastemas explantation and
then cultured next to the blastema or 3 days before blastemas explantation and then grafted in the blastema (one
ganglion per blastema).

       To determine the levels of DNA synthesis, each
blastema was cultured for a 24 hr period in the presence
of 3 μ Ci/ml 3H-thymidine (specific activity, 1 Ci/mM).
During this time, the incorporation of 3H-thymidine into
DNA is linear (see Fig. 1). The cultures were rinsed with
fresh medium and then homogenized individually with a
potter-Elvehjem homogenizer in 1.0 ml of cold distilled
water. Each homogenate was transferred to a tube to which
was added 1.0 ml distilled water rinses of the homogenizer ,
yielding a total volume of 2.0 ml. 1.6 ml of the homogenate
was precipitated 15 min. with 0.2 ml trichloracetic acid at
0°C. The homogenate was then filtered through glass micro-
fibre filter (Whatman GF/C) and the radioactivity in the
collected precipitate determined by methods of liquid
scintillation counting. Protein content was determined on
the remaining homogenate (0.4 ml) by modified Lowry method
(Markwell *et al.*, 1978).

       Insulin, Concanavalin A or dibutyryl cAMP
were added in the culture medium just before the time
of explantation.

To determine mitotic index, the cultures were fixed in Bouin's fluid, embedded in paraffin and serially sectioned at 7 μm in a longitudinal plane. The sections were stained with Groat's hematoxylin and counterstained with Nuclear Red. Mitotic figures were counted in every second section of the entire blastema and the total number of blastema cells was counted in every fourth section. The mitotic index (%) was then calculated on each blastema.

RESULTS

Culture Conditions.

Several culture media adjusted to amphibian tonicity and containing 0.14 IU/ml insulin (Globus, Vethamany-Globus, 1977) were evaluated for their ability to maintain DNA synthesis. Leibovitz's L 15 medium alone had this ability. In other culture media as Ringer supplemented with fetal calf serum, MEM supplemented or not with fetal calf serum, CMRL 1066, DNA synthesis declined more or less rapidly. For these reasons, Leibovitz's L 15 medium was used in further experiments.

Although media tonicity (ranged between 240 to 340 mOsm) did not modify DNA synthesis of blastemas (Fig. 2), Leibovitz's L 15 medium was always diluted to 70 % to obtain an osmotic concentration of 260 ± 10 mOsm.

Insulin Effects.

Without insulin in the medium, DNA synthesis declined rapidly (Fig. 3 A). Three and 4 days after explantation, levels of 3H-thymidine incorporation had decreased respectively to a baseline of 48 % and 22 % of those observed 24 hr after explantation.

With 0.035 IU/ml insulin in the medium, DNA synthesis declined less rapidly (respectively 60 % and 50 %, 3 and 4 days after explantation ; fig. 3B). With 0.14 IU/ml insulin in the medium, the decrease of DNA synthesis was prevented (Fig. 3C). Furthermore, the stimulation of DNA synthesis obtained with insulin was proportional to its concentration (Fig. 4).

Nervous System Effects.

A series of experiments was performed in order
to determine the effects of nerves on 3H-thymidine incorporation into DNA. The results were different according as
the ganglion was put next to the blastema at the time of
explantation or grafted into the blastema 3 days before
explantation. In the first case, the increase of DNA synthesis was 59 % over non-ganglionated explants (Fig. 5).
In the second case, the increase seemed much higher (more
than 100 %), however in this case, the determination of DNA
synthesis per proteins unit was difficult to estimate because grafted ganglion homogenized with the blastema overesti-
mated protein content. Estimation of mitotic index on blastemas cultured in the same conditions gave a more accurate
value. Non-ganglionated blastemas had a mitotic index of
0.742. Mitotic index of blastema with a grafted ganglion
(1.4) was higher than that of blastemas with ganglion put
next to them (1.205).

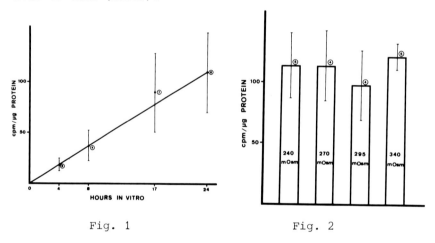

Fig. 1   Fig. 2

Fig. 1 - A time course study for 3H-thymidine incorporation
into DNA in newt forelimb blastemas 3 days after explantation. Each point represents the mean ± standard deviation.
The number of blastemas is shown in the circles.

Fig. 2 - A comparaison of the effects of culture medium
osmolarity on the 3H-thymidine incorporation into DNA in
newt forelimb blastemas 3 days after explantation. Means
are indicated by bar values ; the number of blastemas
is shown in the circles and standard deviations are given.

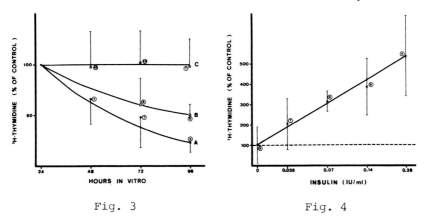

Fig. 3   Fig. 4

Fig. 3 - Levels of 3H-thymidine incorporation in newt forelimb blastemas after explantation to organ culture in Leibovitz's L 15 medium without insulin (A), with 0.035 IU/ml
insulin (B), with 0.07 IU/ml insulin (C). Each point represents the mean ± standard deviation. The number of blastemas
is shown in the circles.

Fig. 4 - Effects of insulin on 3H-thymidine incorporation in
cultured blastemas 3 days after explantation. Each point
represents the mean ± standard deviation. The number of
blastemas in shown in the circles.

Effects of Epidermal Migration Inhibitors.

            In some experimental series standard deviations
were very high especially because some blastemas did not incorporate 3H-thymidine. After histological observations, we
assumed that blastemas with low DNA synthesis were those in
which epidermis had migrated on the cut surface after explantation. Numerous pycnotic cells were observed in these
blastemas. In order to ascertain this assumption, blastemas
were cultivated with several epidermal migration inhibitors
(Donaldson, Mason, 1978 ; Dunlap, Donaldson, 1980) : cAMP,
Concanavalin A. The effects of these inhibitors were to
increase DNA synthesis and to decrease standard deviation
values (Fig. 6).

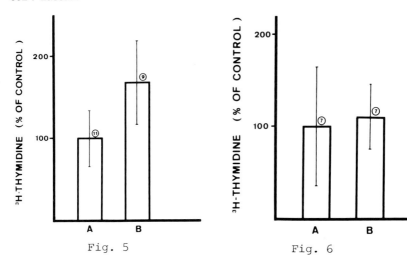

Fig. 5   Fig. 6

Fig. 5 - Effects of nervous system (sympathetic newt ganglion) on 3H-thymidine incorporation into DNA in blastemas.
Blastemas were cultured 48 hr without (A) or with ganglion
(B) in Leibovitz's L 15 medium. Levels of incorporated radioactivity were determined after an additional 24 hr of culture with 3H-thymidine. Means are indicated by bar values,
number of blastemas is shown in the circles and standard
deviations are given. Nervous system produced a significant
increase in 3H-thymidine incorporation (B) over control value
(A) with P < 0.01.

Fig. 6 - Effects of concanavalin A (50 μg/ml) on 3H-thymidine
incorporation in blastemas 3 days after explantation. Means
are indicated by bar values, number of blastemas is shown
in circles and standard deviations are given. Difference
between means is not significant. Standard deviations are
reduced. Concanavalin A (25 μg/ml) cAMP (10^{-5} M) gave similar
results.

DISCUSSION

          Culture media must be able to sustain mitosis
and DNA synthesis in cultured blastemas at a level close to
the one observed *in vivo*. Numerous culture media were used
for this purpose : CMRL 1415 (Globus, Liversage, 1975 ;
Globus, Vethamany-Globus, 1977), MEM (Vethamany-Globus *et al.*,
1978 ; Globus *et al.*, 1980 ; Tomlinson *et al.*, 1981),

Leibovitz L 15 (Carlone, Foret, 1979 ; Mesher, Loh, 1981).
Conn *et al.* (1979) have tested several culture media for
their ability to sustain proliferation of newt blastemal
tissue for a long time (5 to 7 days) contrary to other authors who, except Globus and Vethamany-Globus (1977), maintained their cultures during only 24 to 48 hr. On the basis
of gross histological observations Conn *et al.* (1979) concluded that CMRL 1066 adjusted to a tonicity of 250 mOsm supports
the continued maintenance and proliferation of newt limb
blastemal cells. In opposition to these results, our data
show that the best DNA synthetic capacity is provided by
Leibovitz's L15 medium. Moreover, variations in a large
range of osmotic concentration do not affect DNA synthesis.
All culture media used by the authors were supplemented with
growth factors such as serum, insulin or both. Serum has
the disadvantage of adding unknown constituents in the culture medium. For this reason, we used insulin (Globus, Vethamany-Globus, 1977) as growth factor. The insulin dose-
response data showed a linear increase in the incorporation
levels of 3H-thymidine with insulin concentration ranging
from 0.035 - 0.28 IU/ml which is in line with the results of
similar experiments carried out by Mesher and Loh (1981).
The insulin concentration required to stimulate DNA synthesis in cultured blastemas is not physiological since the
insulin concentration in serum from both intact and regenerating newts averages 20 µU/ml (Liversage, Banks, 1980).
Insulin probably acts as an analogue of somatomedins
(Ewton, Florini, 1981) on cultured blastemas. The insulin
concentration (0.035 IU/ml) used in our experiments supported a DNA synthesis to a level similar to the one achieved
in blastemas denervated *in vivo* (Singer, Caston, 1972).

        The increase of DNA synthesis in blastemas
cultured with ganglion previously shown by Vethamany-Globus
et al. (1978) was questionable "due to the presence of many
unknown hormones, antiproteases and low-molecular-weight
components in serum" (Conn *et al.*, 1979). Our results clearly
show that in a chemically defined medium, nervous system
increases DNA synthesis and cellular proliferation. The
precise nature of neurotrophic factor is still unknown.

Both concanavalin A and cAMP were ineffective in promoting DNA synthetic activity. Their effect is to prevent epidermal migration. Close contact between mesodermis and the culture medium is necessary to maintain cytological integrity of newt blastemal cells.

REFERENCES

Carlone RL, Foret JE (1979). Stimulation of mitosis in cultured limb blastemata of the newt, *Notophthalmus viridescens*. J Exp Zool 210:245.

Conn ME, Dearlove GE, Dresden MH (1979). Selection of a chemically defined medium for culturing adult newt forelimb regenerates. In Vitro 15:409.

Donaldson DJ, Mason JM (1978). Inhibition of protein synthesis in newt epidermal cells : Effects on cell migration and concanavalin A-mediated inhibition of migration *in vivo*. Growth 42:243.

Dresden MH (1969). Denervation effects on newt limb regeneration : DNA, RNA and protein synthesis. Develop Biol 19:311.

Dunlap MK, Donaldson DJ (1980). Effect of cAMP and related compounds on newt epidermal cell migration both *in vivo* and *in vitro*. J Exp Zool 212:13.

Ewton DZ, Florini JR (1981). Effects of the somatomedins and insulin on myoblast differentiation *in vitro*. Develop Biol 86:31.

Globus M, Liversage RA (1975). *In vitro* studies of limb regeneration in adult *Diemictylus viridescens* : neural dependence of blastema cells for growth and differentiation. J Embryol Exp Morphol 33:813.

Globus M, Vethamany-Globus S (1977). Transfilter mitogenic effect of dorsal root ganglia on cultured regeneration blastemata, in the newt, *Notophthalmus viridescens*. Develop Biol 56:316

Globus M, Vethamany-Globus S, Lee YCI (1980). Effect of apical epidermal cap on mitotic cycle and cartilage differentiation in regeneration blastemata in the newt, *Notophthalmus viridescens*. Develop Biol 75:358.

Jabaily JA, Singer M (1977). Neurotrophic stimulation of DNA synthesis in the regenerating forelimb of the newt, *Triturus*. J Exp Zool 199:251.

Lebowitz P, Singer M (1970). Neurotrophic control of protein synthesis in the regenerating limb of the newt, *Triturus*. Nature, 225:824.

Liversage RA, Banks BJ (1980). Insulin levels in the serum of intact and regenerating adult newts (*Notophthalmus viridescens*). J Exp Zool 211:259.

Markwell MAK, Haas SM, Bieber LL, Tolbert NE (1978). A modification of the lowry procedure to simplify protein determination in membrane and lipoprotein samples. Anal Biochem 87:206.

Mesher AL, Loh JJ (1981). Newt forelimb regeneration blastemas *in vitro* : cellular response to explantation and effects of various growth-promoting substances. J Exp Zool 216:235.

Singer M, Caston SD (1972). Neurotrophic dependence of macromolecular synthesis in the early limb regenerate of the newt, *Triturus*. J Embryol Exp Morphol 28:1.

Singer M, Craven L (1948). The growth and morphogenesis of regenerating forelimb of adult *Triturus* following denervation at various stages of development. J Exp Zool 108:279.

Tomlinson BL, Globus M, Vethamany-Globus S (1981). Promotion of mitosis in cultured newt limb regenerates by a diffusible nerve factor. In Vitro 17:167.

Vethamany-Globus S, Globus M, Tomlinson B (1978). Neural and hormonal stimulation of DNA and protein synthesis in cultured regeneration blastemata in the newt *Notophthalmus viridescens*. Develop Biol 65:183.

Limb Development and Regeneration
Part A, pages 557–563
© **1983 Alan R. Liss, Inc., 150 Fifth Avenue, New York, NY 10011**

REGIONAL DIFFERENCES OF PROTEIN SYNTHESIS IN THE LIMB
REGENERATION BLASTEMA OF THE AXOLOTL

Jonathan M.W. Slack

Imperial Cancer Research Fund,
Mill Hill Laboratories,
London, NW7 1AD, England

This study was initiated with the hope of finding
biochemical correlates of "epigenetic codings". These are
the numbers which those who work on regeneration believe are
written on the cells of the limb and which control the
spatial pattern of differentiation. They are often called
"positional values" although I prefer the more neutral term
"epigenetic coding" since I believe that there need not
necessarily be a one-to-one relationship between coding and
position within the organ. The evidence from morphological
experiments suggests that the codings form one or more serial
hierarchies (Slack, 1980). They are stable in the
differentiated tissues but can change in the regeneration
blastema, the changes tending to restore the continuity of
a sequence of states (Lewis, 1981).

If one has no idea about the molecular basis of the
codings then a study of two dimensional protein gels seems a
good place to start since this technique can separate more
protein species than any other. It is true that the coding
substances need not themselves be proteins but whatever they
are they should at least be made by enzymes which are
proteins. The method adopted was to characterise the protein
synthesis patterns of the different cell types found in the
limb and then to look for differences between regions which
have the same cellular composition.

The details of the technique used are described elsewhere (Slack, 1982). Briefly, blastemas were dissected in a
physiological saline and then labelled with ^{35}S-methionine
in tissue culture medium for six hours. The total proteins

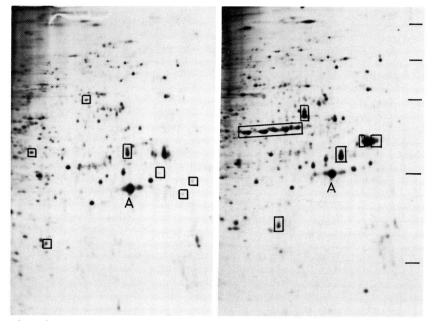

Fig. 1. Two dimensional gels of proteins synthesised by
isolated mesenchyme (left) and epidermis (right) from cone
stage proximal forelimb blastemas. The boxes show mesenchyme
specific and epidermis specific proteins. The bars on the
right show the positions of molecular weight markers (top
to bottom : 200, 92, 67, 46 and 30 kd). The pH range is
about 7.5 to 4.5 from left to right. "A" is actin.

were fractionated by two dimensional gel electrophoresis,
which involves an initial separation by isoelectric focussing
and a second separation by molecular weight. The newly
synthesised proteins were visualised by fluorography.

Results
        The regeneration blastema consists of a mesenchymal core
surrounded by an epidermis. These tissues can easily be
separated by microdissection and a number of comparisons were
made between them. Fig. 1 shows gels of protein made by
forelimb proximal cone stage mesenchyme and epidermis. Seven
mesenchyme-specific and thirteen epidermis-specific proteins
are shown. These are not only found in these blastemas but
also in later stages, in hindlimb blastemas and in forelimb

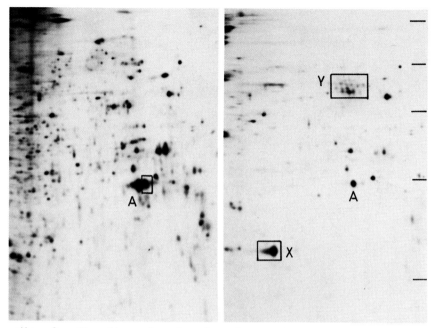

Fig. 2. Two dimensional gels of proteins synthesised by muscle (left) and cartilage (right) dissected from early digit stage regenerates. A:actin.

blastemas from distal amputation levels. They are therefore reproducible markers of mesenchyme and epidermis respectively.

The general pattern of spots remains much the same until overt cytodifferentiation begins. By the early digit stage the muscle and cartilage have diverged from one another and from the earlier mesenchyme. The muscle displays a prominent α-actin spot and the cartilage shows two groups of spots labelled X and Y on Fig. 2.

So this part of the study confirms what we already know : that different histological cell types make different proteins. These differences are easy to find since they exist among the major species, but is it possible to find more minor differences between samples of the same cellular composition taken from different places ?

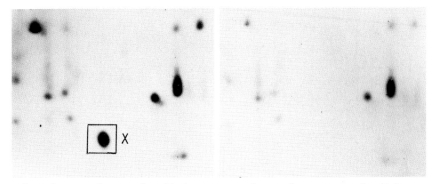

Fig. 3. At the early digit stage there is a reproducible gradient of the cartilage specific protein "X" from anterior to posterior. Left : digit 1, right : digit 4.

Comparisons were made between forelimb and hindlimb, between proximal amputations and distal amputations and between anterior and posterior regions within a blastema. In many cases the epidermis and mesenchyme were compared separately. These experiments are only significant if repeated since small differences are found between samples perhaps due to genetic variation among the animals, or to slight variations in extraction and preparation. No significant differences were found in any of these comparisons. The differences which were found were either not reproducible, or they were associated with visible differences in cellular composition or they were associated with differences in developmental stage.

These latter points deserve some emphasis since this type of experiment has recently been carried out on other systems where such factors are not entirely controlled (Jäckle and Kalthoff 1981, Greenberg and Adler 1982). For example, when gels are run of proteins synthesized by each of the four digits from an early digit blastema a few differences are apparent. Among these is a gradient of a 35,000 mw protein from anterior to posterior (Fig. 3). But this is actually the X protein which is specific to cartilage (Fig. 2b). Since the anterior digits differentiate earlier than the posterior ones it is possible to find a time at which there is a gradient of synthesis. But of course this is a most unlikely candidate for a coding substance :

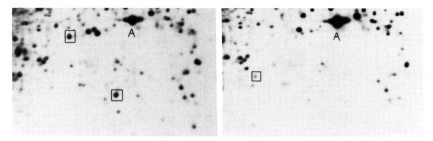

Fig. 4. Effect of vitamin A treatment on protein synthesis. On the left is a portion of a 2-D gel of blastema mesenchyme from a distal amputation level. On the right is the pattern produced by distal blastema mesenchyme after two weeks exposure of the animals to 75 iu/ml retinol palmitate. Animals were treated by Dr. M. Maden.

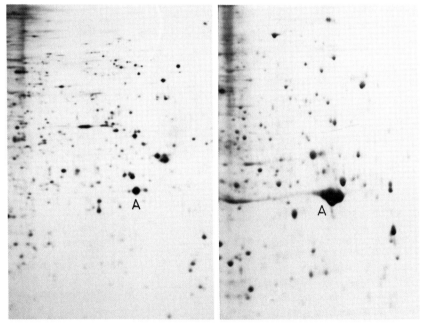

Fig. 5. Two dimensional gels of proteins synthesized by pleurodele limb regeneration blastemas (left) and cells grown from larval limb buds (right).

its abundance and tissue specificity suggests that it is
probably a component of the cartilage matrix.

Fig. 4 shows gels of distal blastema mesenchyme with
and without prior treatment of the animals by vitamin A.
This has been shown by Maden (1982) to have a remarkable
"proximalising" effect and therefore probably affects the
coding system. The three differences shown are reproducible.
Unfortunately we cannot say that they represent a shift
to a more proximal pattern since there is no significant
difference between normal proximal and distal blastemas. It
is therefore difficult to assess whether or not the observed
changes are important.

The last figure shows a comparison between protein
synthesis in a blastema and in mesenchyme cells grown from
larval limb buds which had been through ten passages in
vitro. It is striking how different the patterns are; in
particular the cultured cells are making a lot more actin
than the blastemas although do not appear to be
differentiating into muscle. At present we do not know
whether the difference is due to the in vitro environment
or to the possibility that the cultures are derived from
cells which are a minority in the limb bud itself. What we
call "mesenchyme" may not be homogeneous and pluripotent
but may consist of a mixture of differently committed cell
types (eg, Chevallier et al. 1977).

Conclusions

The study described here has proved negative with
regard to its main objective. But this has not stopped
us believing in the reality of the codings, it simply
means that finding them is not going to be easy. We now
know that there are no vast biochemical differences
between cells of the same histological type in different
regions. Any differences that do exist will be rather
subtle and it will presumably be necessary to make a
correct guess about the class of molecule to examine
before finding the coding substances themselves. I have
previously suggested that the carbohydrate chains of cell
surface glycoproteins and glycolipids would lend themselves
to encoding a serial hierarchy of states (Slack 1980) and
am now proceeding to examine them in the first instance in
early embryos and eventually in secondary fields such as
the limb.

References

Chevallier, A., Kieny, M., Mauger, A. and Sengel, P. (1977)
  Developmental fate of the somitic mesoderm in the chick
  embryo. pp421-432 in Vertebrate Limb and Somite
  Morphogenesis. ed. Ede, D.A., Hinchliffe, J.R. and
  Balls, M. Cambridge University Press
Greenberg, R.M. and Adler, P.N. (1982) Protein synthesis
  and accumulation in Drosophila melanogaster imaginal
  discs: identification of a protein with a non-random
  spatial distribution. Dev. Biol. 89:273
Jäckle, H. and Kalthoff, K. (1981) Proteins foretelling head
  or abdomen development in the embryo of Smittia spec.
  (Chironomidae, Diptera). Dev. Biol. 85:287
Lewis, J. (1981) Simpler rules for epimorphic regeneration.
  The polar coordinate model without polar coordinates.
  J. Theor. Biol. 88:371
Maden, M. (1982) Vitamin A and pattern formation in the
  regenerating limb. Nature 295:672
Slack, J.M.W. (1980) A serial threshold theory of
  regeneration. J. Theor. Biol. 82:105
Slack, J.M.W. (1982) Protein synthesis during limb
  regeneration in the axolotl. J. Embryol. Exp. Morph.
  in press

Limb Development and Regeneration
Part A, pages 565–575
© 1983 Alan R. Liss, Inc., 150 Fifth Avenue, New York, NY 10011

SKIN OF NON-LIMB ORIGIN BLOCKS REGENERATION OF THE NEWT
FORELIMB

Patrick W. Tank, Ph.D.

Department of Anatomy, Slot 510
University of Arkansas for Medical Sciences
Little Rock, Arkansas 72205

The epidermis covering the amputation surface of the
amphibian limb plays an essential role in the accumulation
of a regeneration blastema. This has been shown by
experiments in which limb stumps that have been deprived of
an epidermal covering have failed to regenerate (reviewed by
Goss, 1969; Carlson, 1974a). It is not clear whether
epidermis plays a role in pattern formation during
regeneration. There are two schools of thought on this
subject. The first school contends that the epidermis
provides a permissive stimulus for regeneration but does not
perform an instructive role in the process. Most of the
recent models for pattern formation during regeneration fall
under this school of thought (reviewed by Tank and Holder,
1981). The second school of thought considers the epidermis
to play an instructive role during pattern formation
(reviewed by Rose, 1970). Experiments in which tail
epidermis was grafted to limbs (Glade, 1963) or in which
epidermis overlying the blastema was rotated (Tank, 1977) or
grafted eccentrically (Thornton, 1960) suggest an
instructive role for epidermis.

One intriguing example of the influence epidermis may
have on pattern formation comes from the work of Thornton
(1962) in which forelimb skin was replaced by skin from the
head in Ambystoma talpoideum. When the limb was amputated
through the graft no regeneration occurred. Thornton
interpreted this result to mean that epidermis of head
origin would not support a regenerative response. An
alternative interpretation can be offered for Thornton's
results. The dermal component of head skin was also present

at the amputation surface and may have influenced the
results. The purpose of the present study is to examine the
influence of epidermis and dermis in the regenerative
failure of limb stumps whose skin has been replaced by skin
of non-limb origin.

MATERIALS AND METHODS

Several experimental designs were utilized in the
present study. A description of the surgical and technical
methods common to all experimental groups will be presented
first. Specific surgical techniques will be described with
the results.

These experiments utilized male adult newts
(Notophthalmus viridescens) purchased from Lee's New Farm,
Oak Ridge, TN. Experimental animals were fed three times
weekly on beef liver and were maintained in aged tap water
in individual plastic containers. During the course of the
experiments, animals were maintained at 23°C in a low
temperature incubator. Prior to surgery, all animals were
anesthetized by immersion in ethyl m-aminobenzoate
methanesulfonate (Eastman) diluted 1:1000 in aged tap water.

The basic surgical design was to graft skin of flank
origin to the left upper forelimb. Skin incisions were made
in the transverse plane of the limb just distal to the
shoulder and just proximal to the elbow. A mid-dorsal
incision was made to connect the first two incisions and the
skin was peeled off with the aid of watchmaker's forceps.
The forelimb skin was discarded and the upper forelimb was
wrapped in skin or dermis obtained from the left flank of
the same animal. The graft was supported by a wrapping of
lens tissue during the early healing phase. Within 24 hrs
the graft was sufficiently healed so that lens tissue was no
longer necessary. Sham surgery was performed on the right
forelimb immediately after surgery was completed on the left
forelimb.

Following surgery animals were placed on gauze in 100%
Steinberg medium (Hamburger, 1960) to which had been added
0.02% streptomycin sulfate (Sigma) and 0.02% penicillin G
(1670 U/mg, Sigma) to retard infection. They were
maintained in this solution for two days in a refrigerator
(5°C) to reduce their mobility while the grafts healed.

Animals were then removed from the cold and transferred to
20% Steinberg medium containing antibiotics at 23°C for 5
days. Animals were then transferred into aged tap water at
23°C for the remainder of the study.

Grafts were permitted to heal for 30 days prior to
amputation. The level of amputation (except were noted) was
through the distal 1/3 of the graft of flank tissue. Bone
and soft tissues were trimmed to produce a flat amputation
surface. Approximately 3 mm of grafted tissue remained
proximal to the plane of amputation after trimming of the
stump.

Limbs were permitted to regenerate for 60 days after
amputation. They were then drawn with the aid of a camera
lucida, harvested and preserved in Bouin's fluid. Whole
mounts of regenerated limbs were prepared by the method of
Iten and Bryant (1973).

RESULTS

Experiment I was designed to answer the question "Will
replacement of limb skin by a cuff of flank skin inhibit
regeneration of the newt forelimb?". The methods and
results are summarized in Figure 1. A rectangular piece of
skin was removed from the flank region dorsal to the lateral
line. The long dimension of this piece of skin was in the
anteroposterior axis (Fig. 1). The skin was grafted to the
left upper forelimb so that the dorsal edge of the graft was
proximal on the limb and the posterior end of the graft was
on the anterior surface of the limb (Fig. 1). Limbs were
amputated 30 days after grafting. After 60 days of
regeneration, 96% of these limbs failed to regenerate
externally visible outgrowths (Fig. 1). One limb
regenerated a hypomorphic single digit and one limb
regenerated a normal four-digit hand. Sham-operated
surgical controls were prepared by removing the skin from
the upper forelimb and replacing it in harmony with the
stump. A piece of skin equivalent in size to a graft was
then removed from the flank and discarded. Sham-operated
controls were amputated after 30 days of healing. Normal
four-digit hands were regenerated in 93% of the control
limbs. One control regenerated a hypomorphic three-digit
hand of normal asymmetry.

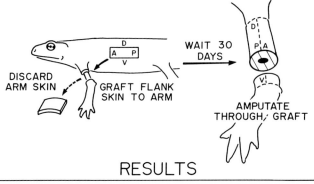

	TOTAL	NON-REGENERATES	REGENERATES		
			HYPO-MORPHIC	NORMAL	HYPER-MORPHIC
EXP.	51	49 (96 %)	1 (2%)	1 (2%)	0
CONT.	16	0	1 (7%)	15 (93%)	0

Fig. 1. Methods and results of Experiment I.

Experiment II was designed to answer the question "Does
the orientation of the flank skin graft alter its ability to
prevent forelimb regeneration?". The methods and results
are summarized in Figure 2. A rectangular piece of flank
skin was removed from the region dorsal to the lateral line.
The long dimension of this piece of skin was in the
dorsoventral axis (Fig. 2). The skin was grafted to the
upper forelimb in two ways; either the anterior edge or the
posterior edge was proximal. The dorsal edge or the ventral
edge of the flank skin contacted anterior limb tissues,
respectively (Fig. 2). Following amputation 30 days later,
100% of these limbs failed to regenerate externally visible
structures (Fig. 2).

Experiment III was designed to answer the question
"Will epidermis of flank-skin origin support limb
regeneration?". The methods and results may be seen in
Figure 3. A rectangular piece of flank skin was grafted to
the upper forelimb in exactly the same way as described in
Experiment I (Fig. 1). The graft was permitted to heal for
30 days at which time the forelimb skin was removed distal
to the graft. The limb was then amputated 3 mm below the

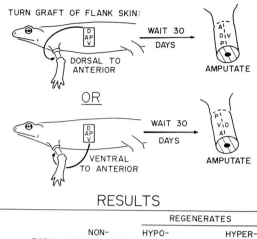

Fig. 2. Methods and results of Experiment II.

graft (Fig. 3). The denuded portion of the limb stump was
epithelialized by epidermis of flank skin origin. The
result was a stump of limb tissues covered by flank
epidermis (Fig. 3). After 60 days of regeneration 86% of
these limbs had regenerated normal four-digit hands (Fig.
3). One limb failed to regenerate, one limb formed a
hypermorphic five-digit hand, and two limbs regenerated
hypomorphic single digits. Sham surgical controls were
prepared by removing and replacing limb skin in normal
orientation, and discarding a piece of flank skin equivalent
in size to a graft. After 30 days of healing, forelimb skin
was removed distal to the sham-operated graft and the limb
was amputated 3 mm distal to the graft. Epidermis arising
from limb skin migrated over deep limb tissues at the
amputation surface. A total of 93% of these sham-operated
limbs regenerated normal four-digit hands after 60 days of
regeneration (Fig. 3). Only one limb regenerated a
hypomorphic two-digit hand.

Experiment IV was designed to answer the question "Will
a limb stump bearing flank dermis regenerate?". The methods
and results are presented in Figure 4. A piece of flank
skin was removed from the region dorsal to the lateral line.

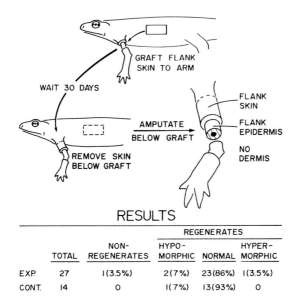

Fig. 3. Methods and results of Experiment III.

The skin was immersed in 0.1% versene in 0.6% NaCl.
After 15 minutes at room temperature (18°C) the piece of
skin was transferred to 0.6% NaCl. The epidermis was then
mechanically removed under a dissecting microscope with aid
of watchmaker's forceps. Following a 3-minute rinse in 0.6%
NaCl the epidermis-free piece of dermis was grafted to the
upper forelimb. Orientation of the graft was identical to
that described for Experiment I (Fig. 1). During the
process of healing, epidermis from limb skin migrated over
the flank dermis. After 30 days of healing, the limbs were
amputated through the grafts. Limb epidermis covered flank
dermis at the plane of amputation (Fig. 4). After 60 days
of regeneration 66% of the limbs had failed to produce an
externally visible outgrowth (Fig. 4). Two limbs
regenerated normal four-digit hands, one limb regenerated a
three-digit hypomorphic hand and one limb regenerated a
one-digit hypomorphic hand. Sham-operated surgical controls
were prepared by removing the skin from the upper forelimb
and treating it with versene as described above. Following
mechanical removal of the epidermis, the limb dermis was
returned to its original location and orientation. A piece
of flank skin equivalent in size to a graft was then removed

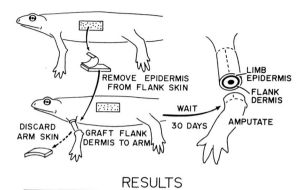

RESULTS

	TOTAL	NON-REGENERATES	REGENERATES		
			HYPO-MORPHIC	NORMAL	HYPER-MORPHIC
EXP.	12	8 (66%)	2 (17%)	2 (17%)	0
CONT.	5	0	0	5 (100%)	0

Fig. 4. Methods and results of Experiment IV.

and discarded. Following 30 days of healing each control limb was amputated through the distal 1/3 of the graft. All sham-operated control forelimbs regenerated normally (Fig. 4).

DISCUSSION

Several conclusions may be drawn from the results presented here. First, replacement of limb skin by skin of flank origin prevents regeneration of the forelimb (results of Experiment I). Second, the orientation of the graft of flank skin does not affect its ability to prevent regeneration (results of Experiments I and II). Third, flank epidermis is capable of supporting a regenerative response in the limb (results of Experiment III). Finally, flank dermis grafted to the limb prevents limb regeneration (results of Experiment IV).

Many experiments have demonstrated that skin of forelimb origin has a profound effect on pattern formation (reviewed by Tank and Holder, 1981). Most notably, rotation of a cuff of forelimb skin by 180° around the stump causes multiple regenerates following amputation through the rotated skin (Carlson, 1974b; Lheureux, 1972, 1975; Tank, 1977). Epidermis-free skin grafts yield similar results, indicating that dermis is the component of skin that is active in stimulating multiple regenerates (Carlson, 1975). Forelimb epidermis rotated in a similar manner does not cause multiple regenerates (Carlson, 1975). This evidence and the results of the present study on flank skin strongly suggest a permissive (not instructive) role for both limb and flank epidermis. Limb and flank dermis, on the other hand, appear to share the ability to influence pattern formation in the regenerating limb.

The above arguments stand in contrast to the results of several experiments cited earlier which indicated an instructive role for epidermis (Glade, 1963; Tank, 1977; Thornton, 1962). One criticism that applies to all three of these earlier studies is that grafts of epidermis could have been contaminated by blastemal cells or cells of mesodermal origin. There was no way to prevent this contamination and these results must be viewed with caution. At this writing it is safe to say that a permissive function of epidermis has been clearly demonstrated. An instructive function of epidermis during pattern formation has not been clearly demonstrated.

Failure of limbs to regenerate after replacement of limb skin with skin from other body regions is not a new finding. In addition to the work of Thornton (1962), Polejaieff (1935) and Effimov (1931, 1933) have reported similar findings after grafting head skin to the forelimbs of axolotls. Also, Trampusch (1958a,b) performed a series of experiments testing the ability of skin from various body regions to either support forelimb regeneration or restore regenerative ability to X-rayed hind limbs of axolotls. These investigators agree that head skin does not support forelimb or hind limb regeneration. Body skin does not rejuvenate the regenerative ability of an X-rayed limb (Trampusch, 1958a). These results lend support to the concept of limb territories, since tissues obtained from unirradiated limbs do rejuvenate the regenerative ability of X-rayed stumps.

The results of the present study indicate that flank epidermis supports regeneration of limb stumps. The cells of the dermis must be considered the cause of regenerative inhibition. To explain this phenomenon based on dermal cell contribution requires a departure in thinking from the type of interactions thought to take place between ectoderm and mesoderm in developing limbs of other vertebrates. It can be assumed that cells of non-limb regions do not possess positional information in a format that is normally present in the limb field. When a graft of tissue (such as head or flank skin) is placed in the limb field followed by amputation, dedifferentiation of cells from the graft could lead to accumulation of non-limb cells in the regeneration blastema. If the non-limb cells are present but cannot participate in meaningful communication, pattern formation could be disrupted. Stated differently, cells of flank skin origin might occupy space in the early blastema and effectively (albeit passively) prevent contacts between cells that have dedifferentiated from limb tissues. If one thinks in terms of current models of pattern formation [such as polar coordinates (Bryant et al., 1981) or averaging (Maden, 1977; Stocum, 1980)] the inability of cells to make contact could spell the difference between success and failure to organize the limb pattern.

The hypothesis that cells of non-limb origin act as mechanical barriers to cellular communication between limb cells fits well with most of the current models of pattern formation. It is particularly compatible with the idea that local cell-to-cell interactions are essential to the process of pattern formation (Bryant et al., 1982). Direct evidence of the dedifferentiation of non-limb cells and visualization of these cells entering the blastema would provide support for this hypothesis. This can be accomplished with 3N cell markers in axolotls. These projects await future study in our lab.

ACKNOWLEDGEMENTS

The author wishes to express his gratitude to Janet Hayden for technical assistance and to Denise Wells for performing preliminary experiments with flank skin grafts. A special thanks to Nancy Stone for typing the manuscript. Supported by grant No. PCM 79-20503 awarded by the National Science Foundation.

REFERENCES

Bryant SV, French V., Bryant PJ (1981). Distal regeneration and symmetry. Science 212:993.

Bryant SV, Holder N., Tank PW (1982). Cell-cell interactions and distal outgrowth in amphibian limbs. Amer Zool 22:143.

Carlson BM (1974a). Factors controlling the intiation and cessation of early events in the regenerative process. In Sherbet GV (ed): "Neoplasia and Cell Differentiation," Karger, Basel, p 60.

Carlson BM (1974b). Morphogenetic interactions between rotated skin cuffs and underlying stump tissues in regenerating axolotl forelimbs. Dev Biol 39:263.

Carlson BM (1975). The effects of rotation and positional change of stump tissues upon morphogenesis of the regenerating axolotl limb. Dev Biol 47:269.

Effimov MI (1931). Die Materialien zur Erlernung der Gesetzmassigkeit in den Erscheinungen der Regeneration. Z Exp Biol 7:352.

Effimov MI (1933). Die Rolle der Haut in Prozess der Regeneration eines Organs beim Axolotl. Z Biol 2: (reviewed by Polejaieff, 1935).

Glade RW (1963). Effects of tail skin, epidermis and dermis on limb regeneration in Triturus viridescens and Siredon mexicanum. J Exp Zool 152:169.

Goss RJ (1969). "Principles of Regeneration". Academic Press, New York.

Hamburger V. (1960). "A Manual of Experimental Embryology". 2nd edition. The University of Chicago Press, Chicago.

Iten LE, Bryant SV (1973). Forelimb regeneration from different levels of amputation in the newt, Notophthalmus viridescens: length, rate and stages. Wilhelm Roux' Arch Entwicklungsmech Org 173:263.

Lheureux E (1972). Contribution a l'etude du role de la peau et des tissus axiaux du membre dans le declenchment de morphogenese regeneratrices abnormales chez le triton Pleurodeles waltlii Michah. Ann Embryol Morphog 5:165.

Lheureux E (1975). Nouvelles donnees sur les roles de la peau et des tissus internes dans la regeneration du membre du triton Pleurodeles waltlii Michah (Amphibien Urodele). Wilhelm Roux' Arch Entwicklungsmech Org 176:285.

Maden M (1977). The regeneration of positional information in the amphibian limb. J Theor Biol 69:735.

Polejaieff LW. (1935). Uber die Rolle des Epithels in den anfanglichen Entwicklungsstadien einer Regenerationsanlage der Extremitat beim Axolotl. Wilhelm Roux' Arch Entwicklungsmech Org 133:701.

Rose SM (1970). "Regeneration". Appleton Century Crofts, New York.

Stocum DL (1980). Intercalary regeneration of symmetrical thighs of the axolotl, Ambystoma mexicanum. Dev Biol 79:276.

Tank PW (1977). The timing of morphogenetic events in the regenerating forelimb of the axolotl, Ambystoma mexicanum. Dev Biol 57:15.

Tank PW, Holder N (1981). Pattern regulation in the regenerating limbs of urodele amphibians. Q Rev Biol 56:113.

Thornton CS (1960). Influence of an eccentric epidermal cap on limb regeneration in Ambystoma larvae. Dev Biol 2:551.

Thornton CS (1962). Influence of head skin on limb regeneration in urodele amphibians. J Exp Zool 150:5.

Trampusch HAL (1958a). The action of X-rays on the morphogenetic field I: Heterotopic grafts on irradiated limbs. Proc Kon Ned Akad V Wetensch 61:417.

Tampusch HAL (1958b). The action of X-rays on the morphogenetic field II: Heterotopic skin on irradiated tails. Proc Kon Ned Akad V Wetensch 61:531.

Limb Development and Regeneration
Part A, pages 577–586
© **1983 Alan R. Liss, Inc., 150 Fifth Avenue, New York, NY 10011**

PATTERN REGULATION DURING URODELE LIMB REGENERATION:
THE EFFECTS OF ATYPICAL BASES

Charles E. Dinsmore, Ph.D.

Department of Anatomy
Rush Medical College
Chicago, Illinois 60612

Recent interest in the analogies between limb development and regeneration has broadened the base of insight
into problems in both areas (reviews by Faber 1971, 1976;
Stocum, 1975). Nevertheless, urodele limb regeneration
differs significantly from limb development in that its
base of origin is a stump comprised of several differentiated tissues which have been subjected simultaneously to
the severe trauma of amputation. The concisely orchestrat-
ed sequence of post-amputation responses including the
dedifferentiation of local stump tissues and the subsequent formation of the blastema have been described and
reviewed elsewhere (Goss, 1961, 1969; Thornton, 1968).
It is important, however, to recall that limb regeneration
normally replaces only those structures distal to the level
of amputation. The obvious role of the stump is therefore
to provide not only material (cells and nutrients) and
type specificity (e.g., limb stumps produce limb regenerates), but also level specific positional information, as
proposed in the conceptual framework introduced by Wolpert
(1969).

The purpose of this report is to examine three experimental procedures which address stump-blastema interactions
and, from these studies, to consider how atypical bases
may interact with and affect limb regenerate morphogenesis.

Interactions between Tissue and Epimorphic Regeneration
The definitions of and differences between tissue regeneration and epimorphic regeneration have been well documented by Carlson (1970, 1972). For the purposes of this

report, tissue regeneration will refer to the minced muscle model wherein skeletal muscle follows a well-defined re- generative sequela leading to relatively rapid though im- perfect repair of the tissue. Epimorphic regeneration will refer to urodele limb regeneration, the most obvious and distinguishing feature of which is the formation of the blastema, an undifferentiated mass of cells arising locally from the stump. Once formed, the regeneration blastema has autonomous distal differentiation potential (Mettetal, 1939; Faber, 1960; Stocum, 1968; Faber and de Both, 1970). While it is now established that the auto- nomy of the blastema is developed early in its career, the explanted blastema nevertheless appears to remain sensitive for a time to morphogenetic influences emanating from het- erotopic graft beds which are competent in the transmission of such information. Previously published data on this topic from my laboratory (Dinsmore, 1974) is reconsidered in light of current concepts in pattern regulation and may subsequently lend insight to recent studies in which limb regeneration has been initiated on stumps recovering from extensive surgical trauma.

The enucleated orbit is a favorable graft site within which explanted wrist blastemas may differentiate to maxi- mum capacity (de Both, 1970; Dinsmore, 1974). If instead of a blastema, minced limb muscle is inserted into the orbit, it undergoes tissue regeneration (Dinsmore, 1974). However, if a wrist blastema is superimposed on the minced limb muscle, the epimorphic system provides it with a dom- inant organizing influence resulting in the differentiation of structures more proximal than the origin of the explant- ed blastema. The blastema differentiates providing distal structures while the epimorphic potential of the minced muscle appears to be unmasked by association with the blastema and subsequently produces appropriate, proximally contiguous elements. Another observation in this study, unexplained at the time, was the occasional occurrence of polydactyly (Figs. 1a-b). Evidence that skeletal muscle is one of the primary loci of morphogenetic activity in regenerating urodele limbs (Lheureux, 1972; Carlson, 1975b) and the demonstration that minced limb muscle in situ can also affect regenerate morphogenesis (Carlson, 1975a) suggests an explanation of the supernumerary structures. While the blastema initally pre- dominated by drawing cells from the tissue regeneration mode into the epimorphic system as described (Dinsmore, 1974), it is now clear that the relationship between modes of repair is

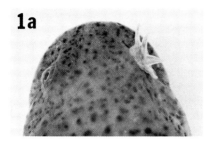
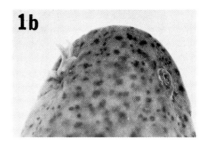

Fig. 1.a. An 8-week axolotl limb regenerate formed by an
intraorbital graft of minced limb muscle capped by a wrist
blastema. It forms an expanded 6-digit hand (five are seen
from this angle). b. A comparable 7-week regenerate with
9 presumptive digits in a relatively random array.

mutual in this case. Minced limb muscle once drawn into the
epimorphic system becomes activated in some as yet undefined
way and is then capable of compromising the otherwise autonomous morphogenetic expression of the explanted blastema.
        The extent of damage to local tissues in some experimental
procedures may therfore influence the epimorphic process.
This is an important and generally overlooked aspect of ex-
eriments in which surgical maniputation of limb tissues is
followed immediately or soon after by amputation. For example, when surgically created double-half limbs are amputated thirty or more days post-surgery, distal outgrowth is
much reduced or fails completely (Bryant, 1976; Tank, 1978).
Nevertheless, when the time between surgery and amputation
is less than thirty days, regeneration of more distally complete limbs, including a 4-digit regenerate, is achieved
(Tank and Holder, 1978).
        The predominance of the epimorphic response over tissue
regeneration (Dinsmore, 1974) and the ability of limb muscle
undergoing tissue regeneration to exert a morphogenetic effect
on an ajacent (ibid) or subsequently initiated (Carlson, 1975a)
epimorphic system have been demonstrated. It is thus unlikely
that healing time per se (Tank and Holder, 1978) is responsible for the observed differences in response. That morphogenetic information becomes stabilized by prolonged graft
healing (Tank 1979) is also speculative. It is equally likely
that there is a time-dependent partial loss in the ability of
tissue regeneration to participate morphogenetically in the
epimorphic response. Stabilization may produce a real or
virtual scar in the prospective pattern. This may also

àccount for the decline in regenerative success induced by
repeated amputation (Dearlove and Dresden, 1976). Nor does
it seem reasonable to assume that a longer healing time prior
to amputating a symmetrical stump allows freer cellular interaction across the midline (Tank and Holder 1981). The data
are more readily understood in the context of tissue regeneration induced by severe surgical trauma, and the time allowed for resolution of the tissue repair process prior to amputation.

Limb Regeneration following Skin Cuff Rotation
    Positional dislocation of a stump tissue produces an
atypical base which may or may not affect regenerate morphogenesis. Skin and skeletal muscle are the two limb tissues
which have been identified as active in regenerate pattern
regulation by this method (Lheureux, 1972, 1975; Carlson,
1975b). Cuffs of limb skin were positionally dislocated
and/or grafted to the contralateral limb to investigate differences in the morphogenetic strength represented by a particular axis (Lheureux, ibid.) The documented differences in
the rates of supernumerary limb formation demonstrate an
unequal circumferential distribution of morphogenetic strength
while the complexity of the supernumeraries initially defies
analysis of the regulatory mechanisms involved.

    Rotations of complete cuffs of limb skin as described
by Lheureux (1972) were recently repeated on Eastern red-
salamanders (Plethodon cinereus) and the limbs subsequently
amputated through the positionally dislocated cuffs. Although
the number and complexity of supernumerary regenerates was
low relative to earlier studies, each type of cuff rotation
was capable of inducing supernumerary structures (Figs. 2-4).
While confirming earlier observations that skin is morphogenetically active, the simple morphology of the supernumeraries in this study may provide insight into the primary interactions of tissues confronted with axial positional dishar-
monies. In its simplest form a skin cuff creating disharmony in, for example, the anteroposterior (a-p) axis would
induce only a-p expansion in the regenerate (Fig. 2). Simple
180° rotation of the skin cuff produces disharmony in both
anteroposterior and dorsoventral axes and the subsequent
regenerate may in its simplest form reflect this (Fig. 4).
Convergence and divergence of axial positional information
as well as the likely secondary mechanical effects so well
documented in embryogenesis (Smith, 1976) could then account
for many of the complex supernumerary (divergent) patterns

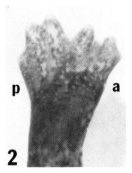 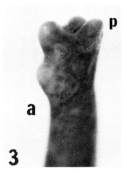 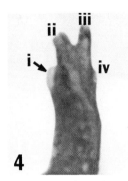

Fig. 2. A supernumerary regenerate showing expansion in only the anteroposterior axis of the hand. The humeral stump bore a cuff of limb skin the anteroposterior axis of which was reversed.

Fig. 3. A supernumerary regenerate demonstrating dorsoventral divergence. The humeral stump bore a grafted cuff of limb skin with a reversed dorsoventral axis only.

Fig. 4. Anteroventral view of a regenerate from a stump bearing 180° rotated limb skin. Digit 1 is duplicated dorsally (arrow) and ventrally while digit 2 bifurcates in the frontal plane.

seen in other urodele species. A positionally dislocated cuff of limb skin on the upper forelimb generally disrupts morphogenesis of only the hand (also noted by Carlson, 1975b). Although Lheureux (1972, 1975) found that positionally dislocated skin or skeletal muscle in the upper forelimb occasionally induced supernumerary skeletal elements in the regenerated zeugopod, it appears that these supernumerary elements are associated with the development of two or more hands. These data indicate that unless more than one primary center of regeneration is established, positional dislocation of a morphogenetically active tissue in the upper forelimb affects regenerate morphogenesis primarily at the level of the autopodium (wrist and hand/foot). This observation is likely to be important to our understanding of morphogenetic regulation along the proximodistal axis of the regenerating urodele limb.

Regeneration from Split Forelimbs

Weiss (1926), using Triton cirstatus, reflected skin from half of the lower fore- or hindlimb, the autopodium of which had been amputated. After splitting the zeugopodium, the complete circumference of skin was replaced about one half, while the other half was amputated more proximally.

TABLE 1.

Effects of Amputation Level and Suturing
on Split Limb Regeneration

| Amputation Level* | None | Type of Regenerate | | |
		Hypomorphic	4-digit	Hypermorphic
Lower forelimb unsutured	4	27	12	3
Upper forelimb sutured	0	2^+	21	0
unsutured	0	2^+	21	0

* limbs were amputated through the halved segment
+ 2 animals produced truncated regenerates bilaterally

His limb regenerates were of extremely variable morphology
even in the "Ganzregenerate" category. His partial and
mosaic regenerates are noteworthy, demonstrating that indeed
half stumps can produce half regenerates. Goss (1957) elab-
orated upon this experiment and proved that it is actually
the complete circumference of whole skin which allows an
otherwise half-stump to give rise to a 4-digit regenerate.
Recent studies on the regenerative ability of halved forelimbs have used the upper arm of Notophthalmus viridescens
(Bryant, 1976; Bryant and Baca, 1978) and while the authors
consider that their findings confirm the earlier studies,
with some exceptions, the studies, because of the different
levels of amputation, are not comparable.

The following investigation compares the results of
regeneration from upper and lower forelimbs in the same
urodele species. On the stage of a dissecting microscope
both forelimbs of anesthetized red-backed salamanders
(Plethodon cinereus) were split longitudinally between the
second and third digits proximally to the elbow. Postaxial halves containing the complete ulna and digits 3 and
4 were discarded. Forelimbs were amputated at mid-radius
2-10 days later and the regenerative responses followed for
at least 60 days.

In another series, the upper forelimbs were longitu-
dianlly halved by removing the anconeus (triceps), some adjacent muscle and the overlying skin. On the right limbs, the
skin was sutured closed over the ablated area while the left

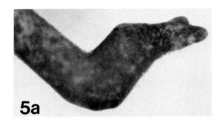
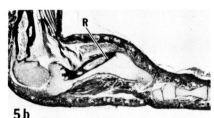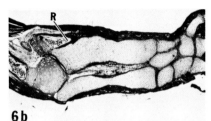

Fig. 5.a. A typical 2-digit regenerate produced on a split
lower forelimb. Digits 1-2 are present and, with time,
these regenerates usually strighten. b. Section of a
similar 2-digit regenerate showing typical cartilage formation along shaft of radius (R) with 1° growth originating
distolaterally.

Fig. 6. a. A normal, 4-digit regenerate produced on a split
lower forelimb. b. Frontal section of a similar 4-digit
regenerate showing extensive lateral cartilage deposition
along original radius (R) and a newly formed ulna.

limbs were allowed to undergo epidermal wound healing. Approximately 4-6 days later both limbs were amputated at mid-
humereus and the animals maintained for at least 60 days.
The results of both series are summarized in Table 1. The
data show that reduced distal regenerative expression on
halved lower forelimbs is the rule (Figs. 5-6) while halved
upper forelimbs regulate completely for a comparable defect.
The lower forelimbs regenerated an average of 2.43 digits per
hand which is comparable to Goss's average of 2.24 (op. cit.).
Halved upper forelimbs produced normal, 4-digit regenerates
which is consistent with earlier studies (Bryant, 1976; Bryant
and Baca, 1978). Moreover, they do so even when the skin is
sutured closed. An explanation for these level specific
variations is a circumferential intercalation of upper limb
positional values by migration of dedifferentiated cells with
the wound epithelium (Bryant and Baca, 1978). There is, however, no experimental evidence to support such a mechanism.
Furthermore, the present data from the suturing study precludes this explanation. Another mechanism which might explain the level-specific regulative behavior in the upper

forelimb is one suggested by Maden (1981). If one assumes
that growth and pattern formation are not strictly linked at
each stage of an epimorphic response, a half-cuff of upper
forelimb skin sutured closed to form a complete circumferential boundary about the upper forelimb may regulate its
contained positional information in a morphallactic (non-pro-
liferative) sense. That skin closure has no effect on upper
forelimb regeneration in this protocol (Table 1) argues
strongly for such an interpretation. Thus the split-limb
regenerate data suggest that the lower forelimb behaves as
a mosaic of at least two potential centers of regional
organization while the upper forelimb is regulative. Further
investigations of this point are in progress.

In summary, the effects that an atypical base may have
on urodele limb regeneration demonstrate several important
aspects of the stump-blastema relationship. A stump or
graft bed undergoing tissue regeneration will significantly
influence the outcome of the epimorphic response. A
positionally dislocated cuff of skin on the upper arm may
support normal forearm development while inducing a complex
supernumerary autopodium. And, the split limb model demonstrates level specific differences between upper and lower
forelimbs in the ways in which positional information may
by organized and/or governed.

These studies were supported in part by BRSG Grant
SO7 RR 05477 from the NIH. My sincere thanks to Ms. Laura
Pena for assistance in manuscript preparation.

de Both NJ (1970). The developmental potencies of the regen-
    eration blastema of the axolotl limb. Wilh Roux' Archiv
    165:242.
Bryant SV (1976). Regenerative failure of double half limbs
    in Notophthalmus viridescens. Nature 263:676.
Bryant SV, Baca BA (1978). Regenerative ability of double
    half and half upper arms in the newt, Notophthalmus viri-
    descens. J Exp Zool 204:307.
Carlson BM (1970). Relationship between tissue and epimorphic
    regeneration of muscles. Am Zool 10:175.
Carlson BM (1972). "The Regeneration of Minced Muscles."
    Monographs in Devel Biol Vol 4. New York: S. Karger,
Carlson BM (1975a). Multiple regeneration from axolotl limb
    stumps bearing cross-transplanted minced muscle regenerates
    Devel Biol 45:203.

Carlson BM (1975b). The effects of rotation and positional change of stump tissues upon morphogenesis of the regenerating axolotl limb. Devel Biol 47:269.

Dearlove GE, Dresden MH (1976). Regulative abnormalities in Notophthalmus viridescens induced by repeated amputations. J Exp Zool 196:251.

Dinsmore CE (1974). Morphogenetic interactions between minced limb muscle and transplanted blastemas in the axolotl. J Exp Zool 187:223.

Faber J (1960). An experimental analysis of regional organization in the regenerating forelimb of the axolotl (Ambystoma mexicanum). Arch Biol (Liège) 71:1.

Faber J (1971). Vertebrate limb ontogeny and limb regeneration: Morphogenetic parallels. In Abercrombie M, Brachet J, King TJ (eds): "Advances in Morphogenesis Vol 9," New York: Academic Press, p 127.

Faber J (1976). Positional information in the amphibian. Acta Biotheor 25:44.

Faber J, de Both NJ (1970). The role of innervation in the manifestation of digits in transplanted regeneration blastemas of the axolotl (Ambystoma mexicanum). Arch Biol (Liège) 81:215.

Goss RJ (1957). The relation of skin to defect regulation in regenerating half limbs. J Morphol 100:547.

Goss RJ (1961). Regeneration of vertebrate appendages. In Abercrombie M, Brachet J (eds): "Advances in Morphogensis," New York: Academic Press, p 103.

Goss RJ (1969). "Principles of Regeneration." New York: Academic Press.

Lheureux E (1972). Contribution à l étude du rôle de la peau et des tissues axiaux du membre dans le déclenchement de morphogenèses régénératrices anormales chez le triton, Pleurodeles waltlii Michah. Ann Embryol Morph 5:165.

Lheureux E (1975). Nouvelles données sur les rôles de la peau et des tissues internes dans la régénération du membre du triton, Pleurodeles waltlii Michah. (Amphibien Urodèle). Wilh Roux' Archiv 176:258.

Maden M (1981). Morphallaxis in an epimorphic system: size, growth control and pattern formation during amphibian limb regeneration. J Embryol Exp Morph 65:151

Mettetal C (1939). La régénération des membres chez la Salamandre et le Triton. Arch Anat Histol Embryol 28:1.

Smith DW (1976). "Recognizable Patterns of Human Malformation: Genetic, Embryologic and Clinical Aspects, 2nd ed," Philadelphia: WB Saunders, Co.

Stocum DL (1968). The urodele limb regeneration blastema:

A self-organizing system. II Morphogenesis and differentiation of autografted whole and fractional blastemas. Devel Biol 18:457.

Stocum DL (1975). Outgrowth and pattern formation during limb ontogeny and regeneration. Differentiation 3:167.

Tank PW (1978). The failure of doube-half forelimbs to undergo distal transformation following amputation in the axolotl, Ambystoma mexicanum. J Exp Zool 204:325.

Tank PW (1979). Positional information in the forelimb of the axolotl: Experiments with double-half tissues. Devel Biol 73:11.

Tank PW, Holder N (1978). The effect of healing time on the proximodistal organization of double-half forelimb regenerates in the axolotl, Ambystoma mexicanum. Devel Biol 66:72.

Tank PW, Holder N (1981). Pattern regulation in the regenerating limbs of urodele amphibians. Quart Rev Biol 56:113.

Thornton CS (1968). Amphibian limb regeneration. In Abercrombie M, Brachet J, King TJ (eds): "Advances in Morphogenesis Vol 7," New York; Academic Press, p 205.

Weiss P (1926). Ganzregenerate aus halbem Extremitätenquerschnitt. Wilh Roux' Archiv 107:1.

Wolpert L (1969). Positional information and the spatial pattern of cellular differentiation. J Theor Biol 25:1.

Limb Development and Regeneration
Part A, pages 587–596
© **1983 Alan R. Liss, Inc., 150 Fifth Avenue, New York, NY 10011**

THE ROLE OF ENDOGENOUS ELECTRICAL FIELDS IN LIMB
REGENERATION

Joseph W. Vanable, Jr., L. L. Hearson and M. E.
McGinnis

Dept. of Biological Sciences, Purdue University,
W. Lafayette, Indiana   47907, and Wabash College
Crawfordsville, Indiana   47933

For some time, we have been interested in learning the
extent to which electrical fields are a necessary component
of the process of limb regeneration.   There is evidence
suggesting that they are: When fields are imposed in frog
forelimb stumps by implanted battery-electrode assemblies,
distal negative, the stumps produce a growth consisting of an
extension of the radio-ulna, islands of cartilage, some muscle,
and a considerable quantity of nerve (Smith, 1974; Borgens et
al., 1977a).   When the field is imposed in the reversed direction, or no current is delivered to the implanted electrode,
no such growth occurs (Borgens et al., 1977a).   These effects
of imposed fields are seen even when current delivered to the
distal end of the stump is carried by a "liquid wire" salt
bridge to reduce the possibility of an effect by electrode
products (Borgens et al., 1977à).   Furthermore, when the
endogenous Na^+-dependent stump currents of newts and salamanders are reduced by blocking epidermal Na^+ channels with
amiloride, or by omitting Na^+ from the water in which the animals are immersed, regeneration is delayed considerably, and,
in a portion of the animals, it fails to occur (Borgens et
al., 1977b; 1979).

However, the notion that these skin-generated electrical
fields are important for normal amphibian limb regeneration has
met with some skepticism, perhaps properly so (e.g., Wallace,
1981).   In particular, Lassalle (1980) has questioned their
relevance in his experiments with axolotls, the neotenous
larvae of Ambystoma mexicanum.   In this work, he was unable to
detect proximo-distal potential differences along the skin
surface of their amputated stumps.   He proposed that this

indicated that these creatures lack the stump currents that
we have measured in newts and salamanders, and, since these
larvae regenerate exquisitely, he felt that this demonstrated
the irrelevance of electrical fields for regeneration.

We have found, however, with the use of a vibrating probe,
that axolotl limb stumps do in fact produce currents that are
similar to those we have measured in newts and salamanders,
and that associated with these currents are electrical fields
within the limb stumps. We now are beginning to explore, by
manipulating these endogenous fields, whether they are involved
in any way in axolotl limb regeneration.

AXOLOTL STUMP CURRENTS

Substantial currents leave the stump surface of axolotl
limbs after amputation (Fig. 1a). Approximately five hours
after amputation, currents averaging ca. 70 $\mu A/cm^2$ leave their
stumps. By the next day, currents from the stumps, now rather
well healed with wound epithelium, have fallen to 15 $\mu A/cm^2$,
and by the third day post-amputation, they have fallen to about
7 $\mu A/cm^2$. For the next several days, the average current density is approximately 5 $\mu A/cm^2$, and by day 13 after amputation, the outcurrents have fallen to approximately 2 $\mu A/cm^2$.
By this time blastemas are becoming well established. Typical
newt outcurrents (Fig. 1b) start at a lower average current
density than axolotls, but persist longer. Despite the differences, there seems to be a basic similarity between the two
species: In both cases, the currents (taken here as a flow of
positive charge) flow out of the stump, begin at rather high
current densities, taper off as wound epithelium becomes established, and fall to relatively low values as a blastema
becomes established.

Amiloride, which by blocking epithelial Na^+ channels
reduces newt stump currents significantly (Borgens et al.,
1977b; 1979), has no significant effect on axolotl stump
currents: Approximately one hour after amputation, stump
currents of 6 larvae in artificial pond water (APW; NaCl,
1.5 mM; KCl, 0.06 mM; $CaCl_2$, 1.0 mM) averaged 103 ± 7 (SEM)
$\mu A/cm^2$, while the currents from larvae immersed in 0.5 mM
amiloride averaged 86 ± 11 $\mu A/cm^2$ ($p > 0.1$, Student's t-test).
One day later, the control currents averaged 12 ± 2.4 $\mu A/cm^2$,
while the stump currents of the larvae immersed in amiloride
were 14 ± 3.6 $\mu A/cm^2$. By day 3, these values were 7.5 ± 2.0

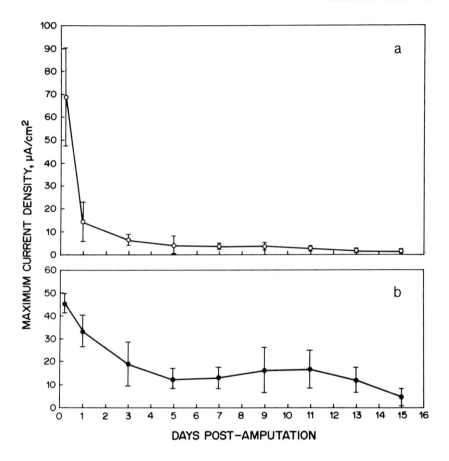

Fig. 1.   Stump currents as a function of time after amputation.
a, Axolotl; b, Newts. Axolotls (averaging 8.5 cm. snout-to-
vent) were anesthesized in 0.03% benzocaine (newts, 0.02%)
before amputation and each reading.   Maximum outcurrents were
measured with a vibrating probe constructed according to the
principle of the ultrasensitive vibrating probe (Jaffe,
Nuccitelli, 1974), as modified by Barker (Illingworth,
Barker, 1980).

$\mu A/cm^2$, and 7.2 ± 1.7 $\mu A/cm^2$, respectively.

The lack of effect of amiloride is made understandable by
the results of experiments in which the cation available for
current generation was varied (Fig. 2).   When K^+ (0.06 mM)

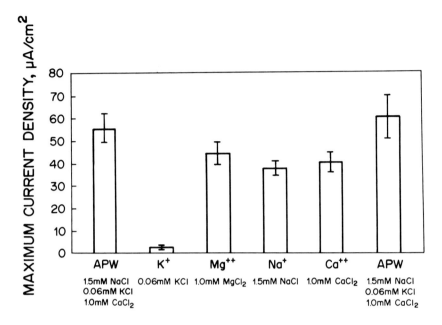

Fig. 2. Ion dependence of axolotl stump currents. Recently amputated axolotl larvae were anesthetized and rinsed thoroughly in deionized water. They then were immersed in the indicated solution for 5 minutes, and read with the vibrating probe while still immersed in that solution.

was the only cation present, quite low currents were generated (2.7 ± 0.6 $\mu A/cm^2$). With 1.5 mM Na^+ as the only cation, an average current of 37.2 ± 2.7 $\mu A/cm^2$ was generated, nearly 70% of the current produced in APW. When 1 mM Ca^{++} was the only cation available, approximately 70% of the current generated in APW was produced, also. With 1 mM Mg^{++}, the current was a bit higher: 44.2 ± 4.9 $\mu A/cm^2$, or 80% of the APW current. It is clear, therefore, that the divalent cation Ca^{++} in the APW can be used by the epidermis to generate substantial currents in the absence of Na^+. This makes it impractical to reduce stump currents in axolotls by blockage of Na^+ diffusion into the epidermis.

MANIPULATION OF INTERNAL LIMB FIELDS

Another way to test the relevance of stump currents to regeneration is to impose battery-driven currents through the stump, either in the same direction as the endogenous currents, increasing the endogenous fields, or in the reverse direction, diminishing them. In our earliest experiments with imposed currents in axolotls, we experienced difficulty in keeping the electrodes implanted, so there was considerable variability in the conditions for each animal. Nevertheless, we found in general that with current delivered so as to augment the normal endogenous fields, there was a slight acceleration of limb regeneration, whereas when the current was delivered so as to reduce the endogenous fields, there was a retardation of regeneration (Table 1).

Table 1. Effect of field manipulation on axolotl limb regeneration. Current was delivered to the distal end of the limb stump via a 15 cm "liquid wire" electrode.

#	Current Delivery Days*	Total Days	Result	Comparable Expected Stage	Δ
1	-13	15	No blastema	EB/MB	-1/-2
2	- 8	20	Early bud	MB/LB	-1/-2
9	- 4	20	Early bud	MB/LB	-1/-2
14	+5,- 7	12	No blastema	EB/MB	-1/-2
16	- 4	6	No blastema	EB	-1
5	+ 6	30	Digits	LB/Pal	+1/+2
6	+ 6	29	Digits	LB/Pal	+1/+2
8	+ 6	22	Early Palette	MB/LB	+1/+2
10	+ 4	20	Late Bud	MB/LB	0/+1

* - denotes current diminishing the endogenous fields
+ denotes current augmenting the endogenous fields
EB, early bud; MB, moderate bud; LB, late bud; Pal, Palette

A further complication of these earliest experiments is that, despite the use of a long salt bridge for the current delivery, there is a chance that in the case of the animals treated with a reversed current, some silver ions could have reached the end of the stump during the time of treatment. If this did occur, it would be possible that the silver ions alone could have retarded regeneration. We therefore began to use a very long salt bridge, with which there is virtually no chance that silver ions could reach the stump (Fig. 3a).

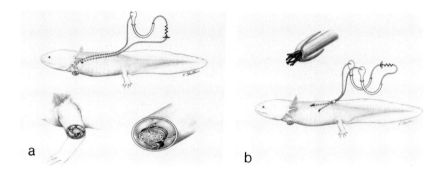

Fig. 3. Configuration of electrodes for the manipulation of internal limb stump fields. a. Electrode implanted at the end of the limb stump. b. External electrode. In both configurations, there is a Ringer-agar bridge 20 cm. long, 3 mm. i.d., separating the Ag-AgCl electrode from the Ringer reservoir to which the "liquid wire" that delivers the current to the animal is connected. The internal electrode in a. is a Teflon insulated stranded stainless steel wire; the "liquid wire" to the stump is a Ringer-soaked white cotton thread insulated with medical grade Silastic tubing. In b. both electrodes are Ringer-soaked white cotton thread insulated with Intramedic PE 50 tubing. The internal electrode is held in place by barbs fashioned from the PE tubing (inset).

Furthermore, we tied the electrode onto the end of the humerus quite firmly with a fine latex ligature, with the result that it was usually possible to keep the electrodes in place for a full two weeks. However, the continuous presence of the electrode interfered with the formation of a morphologically typical blastema, and made the histology difficult to interpret. Nevertheless, in these experiments, with 0.5 µA current delivered through the electrode such that the endogenous fields of the limb stump were reinforced, our impression is that there was a greater degree of muscle dedifferentiation than in the limbs in which the endogenous fields were reduced.

In view of the difficulties encountered with an electrode implanted in the limb stump, we have developed and have just begun to use an alternative means of current delivery. We realized that, since the larvae are

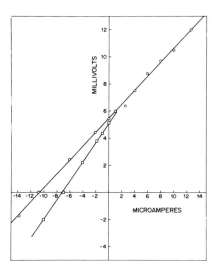

Fig. 4. The effect of imposed currents on the electric field measured in the axolotl limb stump. o, measuring electrodes 1 cm apart; ☐, measuring electrodes 1.5 cm apart. Current was delivered to the stump between an electrode implanted in trunk muscle near the base of the limb, and an external electrode close to the distal end of the stump. The animal's natural current and voltages are taken to be positive.

immersed in a conductive medium (APW) during these experiments, we could take advantage of the volume conduction properties of the system, and direct current through the limb stump from an electrode that lies outside the larva to one that is implanted in its trunk musculature (Fig. 3b). In addition to the advantage of not having an electrode in the stump whose physical presence interferes with regeneration, there is the further advantage that the current density through the limb stump would be more uniform than it had been with the electrodes implanted in the limb stumps. With the implanted electrodes, the current emanated from a source whose area was at most 15% of the limb's cross-sectional area, and the electrode path itself probably furnished a low resistance route for some of the current. Neither of these problems exists with supplying the current from an external electrode. That the internal field can indeed be manipulated in this fashion is shown in Fig. 4: When current is supplied so that it flows in the same direction as the endogenous current, the internal field is augmented; when it is supplied in the opposite direction, it can be reduced or even reversed.

We have just begun to use this approach with regenerating axolotls, and so we have only a few cases to report at the present. However, even with rather low current deliveries from the external electrode, there seem to be differences in the behavior of the stumps in which the endogenous

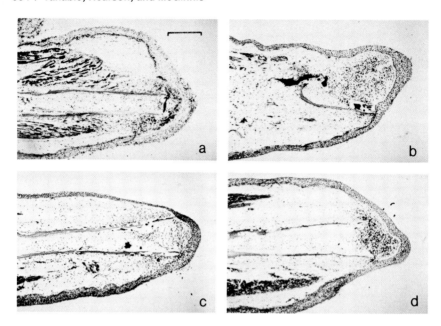

Fig. 5. Histology of axolotl limb stumps after field manipulation. a. Current diminishing endogenous field, 3 µA
for the first 10 days and 0.2 µA for the next 5 days after
amputation. Very little accumulation of blastema cells.
b. Current augmenting the endogenous field, same schedule
of delivery as in a. Medium-to-late bud blastema. c.
Current diminishing endogenous field, 1 µA for the first 10
days and 0.3 µA for the next 5 days after amputation. Early
bud. d. Current augmenting the endogenous field, same
schedule of delivery as c. Early-to-medium bud. Scale
marker indicates 0.5 mm.

currents were reinforced, and those in which they were diminished (Fig. 5). With 3 µA delivered for 10 days and then
0.2 µA for the next 5 days after amputation, very little
accumulation of blastema cells distal to the cut surface of
the humerus occurred when the direction of current was
opposite that of the natural currents, but with these currents in the same direction as the endogenous currents, a
medium-to-late bud blastema formed sooner than it normally
would have. When only 1 µA of reversed current was delivered for 10 days and then 0.3 µA for the next 5 days after
amputation, there was some blastema formation, but less than
when the same schedule of current was delivered so as to

reinforce the natural currents. This blastema, an early -
to-medium bud, is about at the stage that would be expected
if normal, unmanipulated regeneration were occurring.

DISCUSSION

It is quite clear that stump currents that are similar
to those we have measured in newts and salamanders flow from
the stumps of amputated axolotl limbs. Furthermore, the
evidence we have so far is consistent with the hypothesis
that the endogenous electrical fields associated with these
currents have a role in the process of limb regeneration:
When they are diminished, regeneration is retarded. This
result is clearly an effect of electric field manipulation,
rather than one of electrode products. With the present
design of current delivery, there is no reasonable chance
that electrode products could be affecting the outcome of
the experiments. It is likely, but not altogether clear,
that sufficiently augmenting the currents actually stimulates a more rapid regeneration than normal in axolotl limbs.
However, the limb of the axolotl regenerates quite rapidly
by itself; therefore, it probably is not the best object of
study for stimulation of regeneration. We plan to test the
question of whether augmenting natural electrical fields
accelerates normal regeneration with more favorable species.
Old axolotls and metamorphosed Pleurodeles waltl regenerate
more slowly than young axolotls, as does the newt,
Notophthalamus viridescens. Necturus maculosus is reported
not to regenerate at all (Scadding, 1977; Wallace, 1981),
and in this case, it is particularly interesting that this
species generates an unusually low transepithelial potential
across its skin (Bentley, 1975), and therefore should have
quite low stump currents.

We have not, however, exhausted the possibilities with
the younger larvae with which we have been working. The
natural currents are quite high at first, and then drop off
to low levels during the first few days after amputation.
It may be fruitful to attempt to mimic this pattern in our
current delivery schedules, rather than to hold at a constant
value for so long. The constant current approach has worked well with frogs (Borgens et al., 1977a), but it may not
be ideal for normally regenerating urodeles.

It is clear that a great deal more evidence is needed
before drawing firm conclusions regarding the relevance of

endogenous electrical fields to normal regeneration, par-
cularly in axolotls. The data are consistent with the hypothesis that they are, but the number of cases is still
small. However, the way now seems clear for ascertaining
clear-cut answers in the near future.

ACKNOWLEDGEMENTS

    We thank Richard Kontos for his excellent assistance
with the histology. Supported by NIH grants NS11545 and
NS17387, and NSF grant PCM104657.

REFERENCES

Bentley, PJ (1975). The electrical PD across the integument
    of some neotenous urodele amphibians. Comp Biochem
    Physiol 50A:639.
Borgens, RB, Vanable, JW, Jr, Jaffe, LF (1977a). Bioelec-
    tricity and regeneration I. Initiation of frog limb
    regeneration by minute currents. J Exp Zool 200:403.
Borgens, RB, Vanable, JW, Jr, Jaffe, LF (1977b). Bioelec-
    tricity and regeneration: Large currents leave the
    stumps of regenerating newt limbs. Proc Nat Acad Sci
    USA 74:4528.
Borgens, RB, Vanable, JW, Jr, Jaffe, LF (1979). Reduction
    of sodium dependent stump currents disturbs urodele limb
    regeneration. J Exp Zool 209:377.
Illingworth, CM, Barker, AT (1980). Measurement of elec-
    trical currents emerging during the regeneration of am-
    putated finger tips in children. Clin Phys Physiol Meas
    1:87.
Jaffe, LF, Nuccitelli, R (1974). An ultrasensitive vibrat-
    ing probe for measuring steady extracellular currents.
    J Cell Biol 63:614.
Lassalle, B (1980). Are surface potentials necessary for
    amphibian limb regeneration? Develop Biol 75:460.
Scadding, SR (1977). Phylogenetic distribution of limb
    regeneration capacity in adult amphibia. J Exp Zool
    202:57.
Smith, SD (1974). Effects of electrode placement on stimu-
    lation of adult frog limb regeneration. Ann NY Acad Sci
    238:500.
Wallace, H (1981). "Vertebrate Limb Regeneration."
    Chichester: Wiley, pp 2 and 115.

Limb Development and Regeneration
Part A, pages 597–608
© **1983 Alan R. Liss, Inc., 150 Fifth Avenue, New York, NY 10011**

THE ROLE OF IONIC CURRENT IN THE REGENERATION AND
DEVELOPMENT OF THE AMPHIBIAN LIMB

Richard B. Borgens, Ph.D.
Staff Scientist
Institute for Medical Research
2260 Clove Drive
San Jose, CA 95128

Hugh Wallace has called it "the limb bud paradigm."
The degree of likeness or difference between the developing
and regenerating amphibian limb has intrigued biologists
since the elegant experiments of Ross Harrison in the early
1900's. Today, one can still find those who stress the differences between limb regeneration and limb development.
However, the broad perspective suggests that these differences are really a matter of degree and probably not of
kind. For example, it has been suggested that regenerating
—but not developing—limbs possess absolute requirements
for nerves and certain hormones. Developing limbs are innervated about the same time the mesenchyme condensations
become buds in the flank. "Aneurogenic" limbs can form in a
larval urodele whose neural tube and cranial ganglia have
been surgically removed prior to limb development. (The
animal survives because it is parabiosed to another larval
host) (Piatt 1942). This supports the notion that limbs
need little or no innervation to develop fully - however,
such aneurogenic limbs also regenerate after amputation
(Yntema 1959). Limbs become addicted to the presence of
nerve during development, and can become experimentally re-
conditioned to an extremely low level of innervation and
still regenerate (Wallace et al. 1981; see also Wallace
1981). Since the role of most hormones believed to be critical to limb regeneration can also be viewed as primarily supporting the general health of the animal (having little direct action on limb regeneration) (reviewed by Wallace 1981)
there seems to be little room for an absolute comparison to
limb ontogeny where systemic hormones probably play little
direct role.

In recent years there has been a literal explosion in the interest in the development of biological pattern. This has led to a proliferation of models governing pattern formation. Many of these models have been supported and developed by ingenious grafting experiments using various limb systems (reviewed by Wallace 1981). Differences in the results of such experiments between developing and regenerating limbs have led some to state flatly that the actual mechanisms of pattern generation are very different during ontogeny and adult regeneration (Tickle 1981; see discussion by Maden 1982).

Maden (1982) has pointed out that surgically constructed double-half limbs (which regenerate hypomorphic structure) are not at all similar to limbs which develop these same abnormalities (for example, by limb bud rotation). The latter regenerate perfectly; thus, one can suggest that in double-half limbs, surgery has not produced a true alteration in the positional identity of cells within the stump. The abnormal or deficient structure maybe due to some curious effect of radical surgery. It is premature to suggest that there are indeed differences in the broad mechanisms of pattern generation between developing and regenerating limbs based on the results of such experimental surgery.

Developing limbs and regenerating limbs both share the ability to regulate for defects; both share a modest ability for autonomous development in foreign locations; both have many similarities in their physiology of development; and examples of the morphological similarities (for example, bud mesenchyme and its overlying apical ectodermal ridge, and limb blastema, with its overlying apical cap epitheleum) are well known. Although there are unreconciled questions that do not permit a convincing argument for, or against, the issue of homology—it is certainly probable that buds and blastemata share many of the same solutions to the problems they face in producing a limb.

Recently, using an ultrasensitive vibrating probe system, we have shown that a steady flow of ionic current leaves the stump end in regenerating salamanders (Borgens et al. 1977a). This current (on the order of 20–100 $\mu a/cm^2$) persists for about 10 days to 2 weeks, steadily declining in magnitude until the first overt signs of blastema formation. Recent electrical measurements by Kenneth Robinson's group at the University of Connecticut School of Medicine at Farmington demonstrate that a strong steady current emerges from developing hind limbs in Xenopus (personal communication). Louis DeLanney and I are studying similar developmental currents that predict hind limb formation in the axolotl. Thus, developing and regenerating amphibian limbs drive steady currents out the area of growth. First, I will briefly discuss the growing body of evidence suggesting that such currents are a necessary prerequisite for limb regeneration (for a detailed examination see Borgens 1982, see also Borgens et al. 1979a); secondly, I will examine the phenomenology of ionic currents associated with the generation of the amphibian limb; and lastly, I will speculate on what role current may play in both of these systems.

Ionic Current and Amphibian Regeneration

The first prerequisite for determining if the current leaving a salamander stump is relevant to its regeneration is to determine what the source of the current is, and develop ways to modify it in a predictable way. We discovered that the source of the current was the undamaged skin surrounding the limb stump, and possibly skin of the body as well (Borgens et al. 1977a). Amphibian skin is well known to produce a large potential difference across itself (40 to 80 mv, internally positive), and this potential is highly dependent on the presence of Na^+ in the pondwater or moisture in contact with the skin. When the limb is amputated, the skin battery drives charge through the stump and out its end. The circuit is completed via the surface moisture on the animal's skin, or the pondwater in more aquatic species.

Such stump currents have been reported in Triturus viridescens, (ibid.) Ambystoma mexicanum (Vanable et al. 1980), A. tigrinum (Borgens et al. 1979b), and four different species of plethodontid salamanders (Aneides lugubris, Desomagnathus quadrimaculatus, Batrachoseps attenuatus, and Ensatina eschscholtzi xanthoptica. (Borgens, unpublished measurements). In all cases, with the one exception of the axolotl (A. mexicanum), the currents are quite dependent on the concentration of Na^+ in the media.

By raising or lowering the Na^+ concentration in which electrical measurements are made, or by topically treating the skin with Na^+ blocking agents such as Amiloride or the methyl ester of lysine (MEL), we could modify the level of current driven out the stump in predictable ways.

In experiments using red-spotted newts and tiger sala-manders where the stump currents are greatly reduced by either Amiloride treatment or chronic immersions in a low Na^+ artificial pondwater, limb regeneration was in most cases either greatly retarded, caused to be abnormal, or completely inhibited (Borgens et al. 1979b). This supports the notion that such currents are indeed critical to the regeneration process.

However, adult ranids, who do not regenerate their limbs, would also be expected to produce similar currents, since leopard frogs are known to have a substantial skin battery. This paradox was resolved when we discovered that most of the skin generated current is shunted through the subdermal lymph spaces which are present in adult frogs, but not salamanders and newts (Borgens et al. 1979c). The result of this shunt is that little current traverses the core tissues of the frogs stump from where the blastema should arise.

When one artificially enhances the deficient level of current traversing this region in either <u>Xenopus</u> or <u>Rana</u> (by implanted batteries), limb regeneration can be initiated (in <u>Rana</u>) (Borgens et al. 1977b), or strikingly modified (in <u>Xenopus</u>) (Borgens et al. 1979d).

In summary, one can modify amphibian regeneration by a variety of means that, in common, all influence the charac-ter of the current driven out of the amputation surface. Thus, since an efflux of current appears to play a role in the <u>regeneration</u> of the amphibians limb, — what about its <u>generation</u>?

Steady Currents Predict Limb Formation in Amphibia

Kenneth Robinson has found that a substantial (1 - 10 $\mu a/cm^2$) current is driven out of developing hindlimbs of <u>Xenopus</u> embryos — apparently powered by an active uptake of charge at the gill. This current loop between the gill and limb forming area in the larvae predates the appearance of the bud itself, since this local outcurrent can be detected as early as stage 45 (Nieuwkoop and Faber). The direction

of current flow can be reversed in stage 49 embryos (possessing buds) with retinoic acid. In many cases, this causes a
regression of the limb bud as well.

In anurans, the hindlimb emerges first, in embryos and
very small larvae, and its development from a mesenchyme
condensation to a prominent limb bud can occur in 24 to 36
hours at room temperature. In urodeles, the hindlimb develops last, in large, well formed larvae. Moreover, the development of the hindlimb from a condensation to a bud in
the axolotl can take 3 or 4 days at room temperature—allowing greater temporal resolution of this process.

Using the vibrating probe, we have confirmed Robinson's
findings in larval axolotls (Borgens and DeLanney 1982).
Current densities on the order of 0.5 to 5 μa/cm^2 leave an
area of hindlimb formation as early as a week prior to the
first signs of a mesenchyme condensation. In general, the
current increases in magnitude until a tiny limb bud (less
than 0.1 mm) appears, and declines in intensity with the
growth of the bud. In large buds (over 0.4 mm in length),
the current usually reverses its polarity, now entering the
apex of the bud, while leaving its basal aspect and the
flank region on either side.

In the axolotl, this bud current is not driven by a
battery located in the gill. Rarely have we measured any
charge uptake at the gill, and when we have, it is minute in
comparison to gill currents in Robinson's Xenopus larvae
(0.5 μa/cm^2 compared to as much as 50 μa/cm^2). In these
large axolotl larvae (1 to 1.5 cm long) the site of active
charge uptake appears to be the skin. Small in-currents
(0.1 to 0.3 μa/cm^2) were found in most areas of the flank or
tail distant from the base of the cloaca where limbs will
appear. Thus far, we have been unsuccessful at producing a
consistent decrease in the bud currents by removing or
replacing Na$^+$, Ca^{++}, or K$^+$ in the media, or by attempting to
block the uptake of Na$^+$ (using MEL), or Ca^{++} (using Verapa-
mil). Unlike most other urodeles, the potential produced
across axolotl skin (at least of the larvae) is not strictly
dependent on Na$^+$ or Ca^{++} uptake. Thus we have not been able
to produce a consistent means of eliminating or reducing bud
currents. We are now approaching this problem in another
way—by designing experiments where current will be drawn
into the bud area from the culture media by means of an
implanted electrode in series with an external voltage

source and the media. In this way, we hope to test if these currents associated with limb development are indeed critical to the process. If so, it is reasonable to ask: what role do these currents play in development or regeneration?

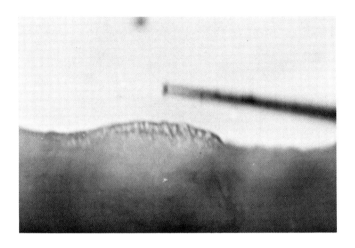

Fig. 1. Photograph of a thickened ridge of epidermal cells on the flank of an axolotl larvae. This marks the first overt sign of limb formation. The vibrating probe is seen vibrating at 400 Hz in a standard measuring position—about 100 μm from the surface of the bud.

What Are The Endogenous Fields Doing?

We have suggested that the currents leaving the limb stump in adult salamanders influence the amount, and the rate of growth of nerve into the critical area of blastema formation. In cases where we have implanted cathodes into the stumps of adult frogs, enhancing their regeneration, a striking enhancement of nerve in the terminal portion of the limb or regenerate was observed (Borgens et al. 1977a). This supported our assumption that the heightened regenerative response was due to an electrical hyperinnervation of stump tissues—producing a similar effect to Marcus Singer's well known surgical hyperinnervations of frog stumps. Moreover, in recent years the evidence demonstrating that weak applied electrical fields can indeed influence the rate, amount, even the directions of growth of a variety of ner-

vous tissue is incontrovertible (reviewed by Borgens 1982).
Modern experiments have demonstrated effects on the outgrowth of chick dorsal root ganglia in culture (Jaffe and
Poo 1979) and very striking effects on the direction of
growth of neurites emanating from differentiating Xenopus
neuroblasts in culture (Hinkle et al. 1981). In both of
these cases growth is enhanced, or directed, towards the
negative pole of an applied electric field. Peripheral
nerve regeneration has also been enhanced in vivo in two
species of frogs (Borgens et al. 1977b and 1979d), as well
as the linear growth of identifiable reticulospinal axons in
transected lamprey spinal cords (Borgens et al. 1981) —all
in response to a distally negative applied electric field.

It is not implausible that the electrical fields produced by current flowing through, and out of, limb buds may
help provide a coarse guide for innervation which is occurring simultaneously with bud development. Current flowing
in a proximal-distal direction within a bud (or regenerating
limb) and exiting the apex will produce a distally negative
electric field within the elongating bud (or stump). Such
polarity is consistent with the observations that all known
growth effects on nervous tissue are mediated by distally
negative electric fields. It is also consistent with the
observation that in limb buds and blastemata, the principle
innervation established early in these structures is a projection of nerve into the limb bud (or blastema) from the
body (or stump). Endogenous electrical fields may coarsely
influence the direction of cell movement other than nerve
during limb development, or the early stages of limb regeneration. Migrations of cells (such as the accumulation of
lateral plate mesoderm in limb development) and morphollac-
tic remodeling in the absence of cell division (in the early
stages of limb regeneration) may also be in part affected by
endogeneous electrical fields. We have no direct evidence
at present to support this view in limb primogenesis or regeneration; however, Barker et al. (1982) made a defensible
case for field effects on epidermal cell movement during
skin wound healing.

Considering the high degree of interest in modeling
rules for pattern formation, perhaps I would be remiss if I
did not discuss some possible interrelationship between endogeneous current flow and the generation of pattern in
limbs. First, it is well established that endogenous fields
are involved with the establishment of polarity in many sys-

tems. Professor Lionel Jaffe has spent most of his professional life performing experiments and producing elegant
discussions supporting this view—I direct the interested
reader to several of his discussions (Jaffe 1979, 1981;
Jaffe and Nuccitelli 1977). We have recently begun to dab-
ble in the notion that endogenous electrical fields may be
more involved in the developing and regenerating limb than
in just the initiation of these processes. A reconsideration
of several of our older observations has kindled our
interest:

Since the middle 1970's, we realized that soon after
amputation, current flowing out of the end of a salamander
stump is not homogeneous in its distribution. In 70% to 80%
of the individual measurements, a distinct peak in current
density can be found adjacent to the postaxial (posterior)
sector of the limb stump (Borgens et al. 1977a). We have
never been able to conclude why this should be so. There
are no obvious differences in morphology between anterior
and posterior hemispheres of the limb to account for this.
Implanted cathodes in adult frog stumps are most effective
in initiating regeneration of the limb if they are implanted
in the postaxial region of the limb (Smith 1974). This
finding also supports the idea that this area of the limb
stump is "special" in some respect. The biological importance of this posterior region during ontogeny is well known
to developmentalists. The posterior cells of the limb bud
(in chicks) were originally defined by Saunders and his co-
workers as an area intimately involved in the specification
of pattern, at least in the anterior-posterior axis, and
possibly providing wider morphogenetic influence during limb
development (reviewed by Hinchliffe and Johnson 1980).
Though this role for the "Zone of Polarizing Activity" has
lately been questioned (Iten and Murphy 1980) —and may be
modified in some respects—it is unquestioned that posterior
limb bud cells do exert some special general influence during the early development of the limb. As Wallace (1981)
has suggested, there is reason to hunt for an analogous posterior area with a special regional influence in regenerating limbs—however, at present there is no clear cut experimental evidence for one.

We have wondered if the peculiar electrical aspects of
the postaxial limb stump tissues may relate to this area of
developmental importance. Marie Rouleau, of our laboratory,
has constructed double-anterior and double-posterior limbs

in two species of plethodontid salamanders. After three weeks healing time, amputations were made through the graft. Curiously, in both double anterior and double-posterior constructions, peaks in current density had shifted to the pre-axial half of the limb. Control limbs (in which a hemisphere was removed and reimplanted) were normal with respect to their current profiles. This is a very preliminary observation made on 30 salamanders and needs more exploration; however, the point is that surgical intervention known to interfere with normal limb regeneration also shifts the locus of current exciting the stump. Just as there is no obvious explanation for the normally high current efflux from postaxial regions—there is no obvious explanation for this shift in density in response to surgery.

Another older observation suggests a relationship between current flow and the expressions of form in regenerating limbs. When we used implanted battery systems delivering about 0.1 - 0.2 μa of current to frog stump tissue - we achieved essentially two different types of responses depending on the species of anuran. In Rana, the regeneration achieved was a club shaped mass with one or two protuberances, very unlike a limb judging by its external appearance. However, a striking amount of histogenesis had occurred within this malformed regenerate (Borgens et al. 1977b).

Large, fully grown, Xenopus normally regenerate a long, thin, unbranched "spike" in response to amputations. The core of this spike is cartilage, with nerve, vascular tissue, etc., between it and its covering of skin. In these animals, the artificial application of current produced bifurcated regenerates with the cartilage core perforated throughout its length by nerve. However, several of these regenerates looked remarkably like limbs - in external appearance - and all developed through a paddle-shaped stage reminiscent of normal limb regeneration or development. However, even in the most "limb-like" regenerate there was no normal internal limb structure: Just an unbroken cartilage core bifurcated at its tip where two (sometimes three) "fingers" appeared to arise. No joints, muscle, phalanges, tendons or ligaments, nor even a hint of normal limb structure was observed within these regenerates (Borgens et al. 1979d).

How does such overall limb form arise? Under normal conditions of development, one would suggest that the overall <u>external</u> shape of an extremity is intimately connected with the developmental program for <u>internal</u> differentiation. However, this is clearly not the case here. These "pseudo-limbs", induced by current, challenge how we usually view the end product of differentiation. It is true that certain tissues have an innate program for form. Bone primordia, for example, differentiate in organ culture complete with the tuberosities normal for articulation and the attachment of muscle - of course, without the developing muscle to attach to or other bones to articulate with. However, where does the information reside that suggests external form in the complete absence of normally developing internal struc-ture? Finally, considering the <u>Xenopus</u> experiment, what role does current play in initiating the expression of this form? Answers to such questions will surely come with not only a better understanding of the genetics and cell biology of limb development; but with its electrophysiology as well.

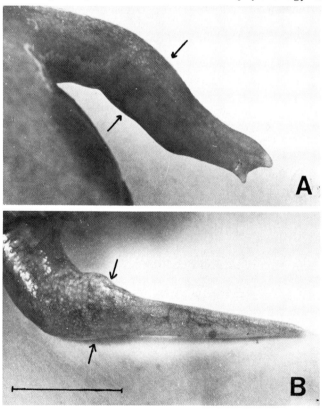

Fig. 2. External views of electrically treated and sham-treated <u>Xenopus</u> limbs 26 weeks after amputation. A) This limb-like structure developed in response to 0.1 μa of current pulled through stump tissues. B) The long tapering "spike" on this sham-treated animal is also typical of the usual response to amputation in adult <u>Xenopus</u>. Arrows indicate the approximate plane of amputation. Calibration line equals 1 cm. (From Borgens et al. 1979d.)

Acknowledgements. I thank Richmond Prehn and Louis DeLanney for a careful reading of this manuscript. This work is supported by Grant NS 18456-01 from the National Institutes of Health.

Barker AT, Jaffe LF, Vanable Jr JW (1982). The glabrous epidermis of cavies contains a powerful battery. Amer J Physiol 242:R358

Borgens RB (1982). What is the role of naturally produced electric current in vertebrate regeneration and healing? Int Rev of Cytol 76:245.

Borgens RB, Roederer E, Cohen MJ (1981). Enhanced spinal cord regeneration in lamprey by applied electric fields. Science 213:611.

Borgens RB, DeLanney LE (1982). An efflux of steady current predicts the place of hindlimb formation in the axolotl. (in preparation).

Borgens RB, Vanable Jr JW, Jaffe LF (1977a). Bioelectricity and regeneration: Large currents leave the stumps of regenerating newt limbs. Proc Nat Acad Sci USA 74:4528.

Borgens RB, Vanable Jr JW, Jaffe LF (1977b). Bioelectricity and regeneration I. Initiation of frog limb regeneration by minute currents. J Exp Zool 200:403.

Borgens, RB, Vanable Jr, JW, Jaffe, LF (1979a) Bioelectricity and regeneration. Bioscience 29:468.

Borgens RB, Vanable Jr JW, Jaffe LF (1979b). Reduction of sodium dependent stump currents disturbs urodele limb regeneration. J Exp Zool 209:377.

Borgens RB, Vanable Jr JW, Jaffe LF (1979c). Role of subdermal current shunts in the failure of frogs to regenerate. J Exp Zool 209:49.

Borgens RB, Vanable Jr JW, Jaffe LF (1979d). Small artificial currents enhance <u>Xenopus</u> limb regeneration. J Exp Zool 207:217.

Hinchliffe JR, Johnson DR (1980). Experimental embryology, pattern in the limb. In "The Development of the Vertebrate

Limb," Oxford: Clarendon Press, p 67, 109, 165.

Hinkle L, McCaig CD, Robinson KR (1981). The direction of growth of differentiating neurones and myoblasts from frog embryos in an applied electric field. Physiol (London) 314:121.

Iten LE, Murphy DJ (1980). Pattern regulation in the embryonic chick limb: Supernumerary limb formation with anterior (non-ZPA) limb bud tissue. Dev Biol 75:3773.

Jaffe LF (1979). Control of development by ionic currents. In Cone RA, Dowling JE (eds): "Membrane Transduction Mechanisms," New York: Raven Press, p 119.

Jaffe LF (1981). The role of ionic currents in establishing developmental pattern. Phil Trans R Sco London B 295:553.

Jaffe LF, Nuccitelli R (1977). Electrical controls of development. Ann Rev Biophys Bioeng 6:445.

Jaffe LF, Poo M-M (1979). Neurites grow faster towards the cathode than the anode in a steady field. J Exp Zool 209:115.

Maden M (1982). Supernumerary limbs in amphibians. Amer Zool 22:131.

Piatt J (1942). Transplantation of aneurogenic forelimbs in Amblystoma punctatum. J Exp Zool 91:79.

Smith SD (1974). Effects of electrode placement in stimulation of adult frog limb regeneration. Ann NY Acad Sci 238:500.

Tickle C (1981). Limb regeneration. The Amer Scient 69:639 See also discussion by Maden, 1982.

Vanable Jr JW, Hearson LL, Jaffe LF (1980). Currents leave amputated limb stumps of axolotl larvae. Amer Zool 20:739.

Wallace H (1981). Nervous control, hormonal influence, regional and axial determination, blastemal morphogenesis, and comments and speculations. In "Vertebrate Limb Regeneration" New York: John Wiley and Sons, p 22, 53, 156, 194 and 224.

Wallace H, Watson A, Egar M (1981). Regeneration of subnormally innervated axolotl arms. J Embryol Exp Morph 62:1.

Yntema CL (1959). Regeneration in sparsely innervated and aneurogenic forelimbs of amblystoma larvae. J Exp Zool 140:101.

Limb Development and Regeneration
Part A, pages 609–618
© 1983 Alan R. Liss, Inc., 150 Fifth Avenue, New York, NY 10011

BLASTEMA FORMATION IN REGULATING IMAGINAL DISC FRAGMENTS OF
DROSOPHILA MELANOGASTER

Gerold Schubiger and Gary H. Karpen*

Department of Zoology and *Department of Genetics
University of Washington
Seattle, Wa 98195

The development and organization of different cell
types depends on complex interactions between the genetic
program and the cellular environment. The heritable commitment of a cell to a particular developmental pathway is
known as determination. This process is believed to be
controlled in a stepwise manner by the activation or inact-
ivation of "selector genes" that steer the development of a
cell and its progeny into a particular direction (Garcia-
Bellido 1975; Lawrence, Morata 1976). Cells with the same
commitment form a pattern composed of different structures.
The particular structure within a pattern that a cell will
differentiate depends on its position (Wolpert 1969). Two
concepts have strongly influenced the field of pattern formation during the last ten years, the compartment hypothesis
and the polar-coordinate model.

The compartment hypothesis arose from studies on normal
development in Drosophila. Mitotic recombination can be
used to genetically mark a single cell that will continue
to divide and form a visible clone of mutant tissue (Stern
1936). Clones that are induced later in development produce
smaller mutant patches, because they are founded closer to
the time of differentiation. Garcia-Bellido et al. (1973),
in their classic analysis, were the first to show that
clones induced after specific developmental stages never
crossed certain boundaries in the adult cuticle. These
boundaries define compartments. At blastoderm, an early
stage of embryogenesis, the anterior and posterior compartments are formed, dividing each segment in half. As is best
shown in the wing, existing compartments become further

subdivided at later developmental stages. Once a compartment is established its cells and their progeny are
restricted to that identity. Compartments are an indication
of intradisc determination (Garcia-Bellido 1975).

Experiments on regulation led to the formulation of the
polar-coordinate model. French et al. (1976) proposed that
the position of each cell within a single morphogenetic
field is specified by radial and circumferential coordinates. If, for example, part of an imaginal disc is
removed, wound healing brings cells with disparate positional values together. The model states that this confrontation stimulates cell division. Newly formed cells intercalate missing positional values by the shortest route.
Thus, a fragment with less than half the number of values
will duplicate, whereas one with more than half of the
values will regenerate. We have used the imaginal discs of
Drosophila to test the validity of the two different models.
Our analysis of pattern regulation supports but also conflicts with both concepts.

THE BEHAVIOR OF COMPARTMENTS DURING REGULATION

The imaginal discs of Drosophila differentiate adult
cuticle structures. Fragments of imaginal discs will
differentiate when transplanted into metamorphosing larvae.
We have used this technique to construct a fate map of the
leg disc (Fig. 1). When the fragments are transplanted into
adult hosts the cells proliferate and will only differentiate when they are subsequently implanted into metamorphosing hosts (for review see Bryant 1978). This experimental design has allowed us to test the regulative capacity
of disc fragments. The 1/4UM fragment can regenerate a
complete leg whereas the complementary piece (3/4L+EK)
duplicates (Fig. 1d).

Steiner (1976) used clones induced at different times
during development to define the anterior-posterior compartment boundary and map it onto the leg disc (Fig. 1c). The
1/4UM fragment is composed entirely of anterior cells, yet
we know that it can regenerate an entire leg. Anterior
cells have the capacity to form posterior cells; therefore
their commitment to anterior determination is not binding.
We used clonal analysis to mark the progeny of single cells
during regeneration. As expected we found that clones

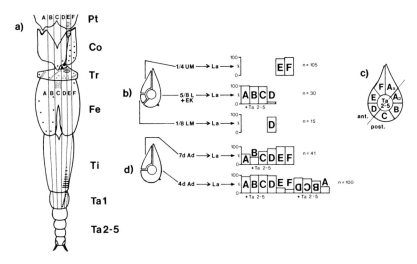

Fig. 1. Structures differentiated by fragments of the first leg disc. (a) An adult leg is split open longitudinally and the cuticle structures are divided into six sectors, A–F. Pt, prothorax; Co, coxa; Tr, trochanter; Fe, femur; Ti, tibia; Ta, tarsal segments. (b) Three different fragments (UM, upper medial; L, lateral; LM, lower medial; EK, end knob) were injected into larval hosts (La) which pupariated within 8 hr. Differentiated structures were analyzed and assigned to sectors A–F. The height of each column indicates the percentage of cases that differentiated structures of a particular sector. These data were used tó construct (c), a simplified fate map. ant., anterior; post., posterior compartment; Aa and Ap, anterior and posterior compartments of sector A, respectively. (d) Regulation of two complementary fragments. Fragments (1/4UM, 3/4L+EK) were injected into adult females. After culture in adults (Ad) implants were transplanted into larval hosts. Most of the 1/4UM fragments showed some regeneration (sectors D–A). The 3/4L+EK fragment predominantly duplicated (reversed letters) and expressed some regeneration (EF). Data from Abbott et al. (1981) and Karpen, Schubiger (1981).

induced at the beginning of regeneration differentiated structures of both compartments. Similarly, clones were induced in the duplicating 3/4L piece, which included cells of the anterior and posterior compartments (Fig. 1c,d), separated by the " old" boundary. Clones induced at the

time of fragmentation crossed the anterior-posterior boundary in the new or duplicated part, but crossing of clones in
the old part was not observed (Abbott et al. 1981). Thus,
existing boundaries are not broken.

Are compartments reestablished during the process of
regulation? This is expected to occur if compartments are
an integral part of pattern formation and regulation.
Clones were induced half-way through the culture period. In
the new part, clones now respected the same morphological
boundaries as observed in normal development (Abbott et al.
1981). In this regard the 1/4 UM and 3/4L behaved the same.
We conclude that at least the anterior and posterior compartments are established during regulation. Thus, it seems
that regulation recapitulates normal development. This
conclusion was also drawn by Szabad et al. (1979) and
Girton, Russell (1981).

THE CHARACTERISTICS OF REGULATING CELLS

What are the origins, proliferation dynamics and developmental capacities of regulating cells? In analyzing these
questions we made observations that conflict with the basic
tenets of the polar-coordinate model. Clonal analysis allowed us to infer the origin of regulated structures (Abbott
et al. 1981). We induced clones in regulating 3/4L fragments using mitotic recombination and found 18 clones that
encompassed structures from both the original and duplicated
parts. Fifteen of these clones included only sector D of
the original part. This demonstrates that cells closest to
the D-E cut (horizontal cut) were the major contributors to
pattern regulation. Further support for this conclusion
comes from an analysis of fragments with incomplete regulation. The appearance of new structures in both the 1/4 UM
and the 3/4L+EK fragments occurred in an orderly sequence.
We found that D was regulated more frequently than C and B
(Fig. 1d). This sequence also was observed in each case
that regulated. For example, when sector B was formed,
sectors C and D were always present. We also found
(Schubiger 1971; Abbott et al. 1981) that after prolonged
culture of the 1/4UM pieces, the frequency of regenerated
structures in sector A was higher than in B. This indicates
that regeneration can also occur from the F/A (vertical) cut
with the reversed sequence, but at a lower frequency than
that observed from the D/E cut.

Clonal analysis (Abbott et al. 1981) and direct observations of mitotic figures were used to determine the proliferation dynamics of cells in regulating fragments. In
collaboration with Louise Abbott we compared the mitotic
indices of cells near the cut with those of cells distant
from the cut. We subdivided the 3/4L+EK fragment into three
areas (W,Y,Z, Fig. 2). Discs isolated from larvae had equal
indices for these three regions. However, when the 3/4L+EK
fragments were first cultured for one day in adult females,
we observed a significant drop in the frequency of mitotic
figures for the middle area Y. A similar decrease in the
mitotic index was found in all areas of intact discs after
one day in adults (no regulation was observed for intact
discs). We conclude that cells close to the cut surface
(areas W and Z) maintain the proliferation dynamics that
discs cells have in the third instar. However, cells that
do not participate in regulation (area Y or all areas in
intact discs) show a significant drop in the mitotic index
during culture.

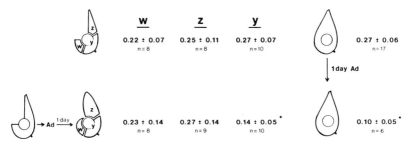

Fig. 2. Mitotic indices [(# of mitotic figures/# of cells)x
100, numbers presented are the mean + SD] calculated for
whole discs and different disc regions (W,Z,Y) of 3/4L+EK
fragments. Comparisons were made between cultured (Ad,
adult host) and uncultured material. * indicates significant
drop after 1 day in culture (p<0.001, student t-test).

We tested the developmental capacities of the cells
that were demonstrated to have higher mitotic activity.
3/4L+EK fragments were injected into adult females and were
isolated one day later, prior to fusion of the cut surfaces.
Fragment I was then separated from II (Fig. 3a). In control
experiments fragment I differentiated only structures of
sector D, whereas fragment II formed structures of sectors

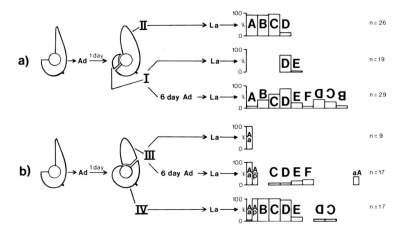

Fig. 3. Regulatory potential of two regions near the cut.
3/4L+EK fragments were cultured in adults (Ad) for 1 day.
The implants were then isolated and cut into (a) fragments I
and II; or (b) fragments III and IV. Differentiation of
fragments I–IV in larval hosts (La) was used to determine
their fates (control). Regulative capacities of fragments I
and III were tested with 6 more days of culture in vivo
followed by metamorphosis. Note that both fragments not
only duplicated (reversed letters) but also regenerated.
See legend of Fig. 1 for description of symbols. Fig. 3a
from Karpen, Schubiger (1981).

A,B,C and rarely D. Thus, during one day of adult culture
these areas retained the developmental fate predicted by the
fate map. When fragment I was cultured in adults to test
its regulative capacities, we found that in 39% of the cases
it duplicated. However, in 61% of the cases this fragment
regenerated and/or duplicated structures not expected from
the fate map (Karpen, Schubiger 1981). The structures appeared with the same sequence as observed for 3/4L+EK duplication. Cells near the vertical cut region (III) behaved in
a similar manner (Fig. 3b). Only 5 out of 17 cases (29%)
showed some regeneration. Four cases regenerated sector F
and the structures in the posterior compartment of sector A
(A_p). One case almost formed an entire leg. Note that
these fragments regenerated sector F,E,D and C with decreasing frequencies, indicating again that new structures
were formed in a sequence. The data in Fig. 3 show that
fragments I and III can regulate sequentially in either

direction. The experiments outlined in Fig. 3 were performed in a pairwise manner to exclude the possibility of cutting errors.

DISCUSSION AND CONCLUSIONS

Our analyses show that the cells near the cut (1) have a higher mitotic index, (2) form the regulated structures in a specific sequence, and (3) have extensive regulative capacities when isolated from the fragment. These characteristics have been used to define the blastema formed during vertebrate limb regeneration (Goss 1969; Stocum 1968). In vertebrates, regeneration via a blastema can be accounted for by the polar-coordinate model. However, in our experiments, the behavior of the blastema violates three aspects of this model. First, a single I or III fragment (Fig. 3) can both regenerate and duplicate, a behavior not predicted by any present model of positional information. Second, these fragments contain few positional values (Strub 1977), yet in our studies they are able to regenerate frequently. Third, a blastema with all three characteristics is formed prior to wound closure. The polar-coordinate model stipulates that the initiation of intercalation requires the juxtaposition of cells with non-adjacent positional values (Reinhardt, Bryant 1981).

We propose a major revision of the polar-coordinate model. After fragmentation a blastema is formed which has the information to regulate in a sequence. The cut surfaces independently begin to regulate, until the wound heals. If cells in the fused region differ in their positional values, regulation continues and only ceases when continuity of positional values is achieved (Lewis 1981). We have observed that isolation of the blastema permits expression of extensive regulatory capabilities; this suggests a release from an inhibition that normally restricts the blastema during regulation. We propose that the rules of the polar-coordinate model control the termination of regulation rather than the initiation.

A population of cells is subdivided morphallactically into two compartments during normal development. We have observed that anterior and posterior compartment identity is reestablished sometime during regulation. However, we have also found that regulated structures are formed by a blastema and appear in a sequence, indicating an epimorphic process. Nevertheless, we believe that the establishment of the compartment boundary at the same morphological position during development and regulation points to a common mechanism of compartment formation. We propose that the establishment of the compartment boundary is similar to the specification of any pattern element; it is established in response to positional information, and thus can be regenerated in the same sequence as morphological structures. On the other hand there is ample evidence that compartment identity influences pattern formation (Karlsson 1981). Cells of both the anterior and posterior compartments are required for distal regeneration (Schubiger, Schubiger 1978). We have observed that anterior cells regenerate posterior structures (e.g. 1/4UM), whereas posterior cells fail to regenerate the anterior compartment (also see Karlsson, Smith 1981). Additionally, posterior cells near a cut surface require more time than anterior cells to form a blastema and regulate new structures (unpublished observation). A recent model utilizes compartment-specific cell properties to explain distal regeneration (Meinhardt 1980).

Our studies have tested two different views of pattern formation in Drosophila, and we conclude that a fundamentalist interpretation of either model is incorrect. Compartment commitment is not binding, yet compartment identity plays a role in pattern formation. The initiation and direction of regulation is established independently of the juxtaposition of cells with disparate positional values. Therefore, it is time to unify these two views by choosing the valid components of each.

ACKNOWLEDGMENTS

This work was supported by NSF Grant PCM 8024535. We would like to express gratitude to Louise Abbott for her collaboration. Margrit Schubiger and Richard Fehon are acknowledged for their comments.

REFERENCES

Abbott LC, Karpen GH, Schubiger G (1981). Compartmental restrictions and blastema formation during pattern regulation in Drosophila imaginal leg discs. Devel Biol 87:64.

Bryant PJ (1978). Pattern formation in imaginal discs. In Ashburner M, Wright TRF (eds): "The Genetics and Biology of Drosophila, Vol 2c, London: Academic Press, p 229.

Garcia-Bellido A (1975). Genetic control of wing disc development in Drosophila. In "Cell Patterning", Ciba Foundation Symposium 29 Amsterdam: Elsevier, p 161.

Garcia-Bellido A, Ripoll P, Morata G (1973). Developmental compartmentalization of the wing disk of Drosophila. Nature New Biol 245:251.

French V, Bryant PJ, Bryant SV (1976). Pattern regulation in epimorphic fields. Science 193:969.

Girton JR, Russell MA (1981). An analysis of compartmentalization in pattern duplications induced by a cell-lethal mutation in Drosophila. Devel Biol 85:55.

Goss RJ (1969). "Principles of Regeneration". New York: Academic Press.

Karlsson J (1981). Sequence of regeneration in the Drosophila wing disc. In Gaze RM, French V, Snow M, Summerbell D (eds): "Growth and the development of pattern", J Embryol exp Morph, 65(suppl):37.

Karlsson J, Smith RJ (1981). Regeneration from duplicating fragments of the Drosophila wing disc. J Embryol exp Morph 66:117.

Karpen GH, Schubiger G (1981). Extensive regulatory capabilities of a Drosophila imaginal disk blastema. Nature 294:744.

Lawrence PA, Morata G (1976). The compartment hypothesis. In Lawrence PA (ed):" Insect Development", Royal Entom Soc London Symposium 8, Oxford: Blackwell, p132.

Lewis J (1981). Simpler rules for epimorphic regeneration: The polar-coordinate model without polar coordinates. J theor Biol 88:371.

Meinhardt H (1980). Cooperation of compartments for the generation of positional information. Z Naturf 35c:1086.

Reinhardt CA and Bryant PJ (1981). Wound healing in the imaginal discs of Drosophila. II. Transmission electron microscopy of normal and healing wing discs. J exp Zool 216:45.

Schubiger G (1971). Regeneration, duplication and transdetermination in fragments of the leg disc of Drosophila melanogaster. Devel Biol 26:277.

Schubiger G, Schubiger M (1978). Distal transformation in Drosophila leg imaginal disc fragments. Devel Biol 67:286.

Steiner E (1976). Establishment of compartments in the developing leg imaginal discs of Drosophila melanogaster. Wilhelm Roux's Arch Dev Biol 180:9.

Stern C (1936). Somatic crossing—over and segregation in Drosophila melanogaster. Genetics 21:625.

Stocum DL (1968). The urodele limb regeneration blastema: A self—organizing system I. Differentiation in vitro. Devel Biol 18:441.

Strub S (1977). Pattern regulation and transdetermination in Drosophila imaginal leg disk reaggregates. Nature 269:688.

Szabad J, Simpson P, Nothiger R (1979). Regeneration and compartments in Drosophila. J Embryol exp Morph 49:229.

Wolpert L (1969). Positional information and the spatial pattern of cellular differentiation. J theor Biol 25:1.

Limb Development and Regeneration
Part A, pages 619–628
© **1983 Alan R. Liss, Inc., 150 Fifth Avenue, New York, NY 10011**

THE SHAPING OF TRIPLICATED TIPS IN CRAYFISH LEGS

Jay E. Mittenthal[1], Ph.D., and Rachel Warga[2]
[1]Department of Anatomical Sciences, SBMS, University of Illinois, Urbana, IL 61801; [2]Department of Biology, University of Oregon, Eugene, OR 97403

How do cells generate the form of a multicellular organism? Steinberg and his associates have suggested that within an aggregate of cells, in tissue culture or in an embryo, cells rearrange themselves to minimize the free energy of their adhesive interactions (Steinberg 1963). Here we show how this concept can be extended to understand the shaping of limbs in arthropods. In these animals an epithelium, the hypodermis, secretes the exoskeleton and so determines the external form. Thus arthropods provide a simplified system for studying morphogenesis of limbs, and more generally of epithelia, without the added complexities of interactions between epithelium and mesenchyme which are so important in vertebrate morphogenesis.

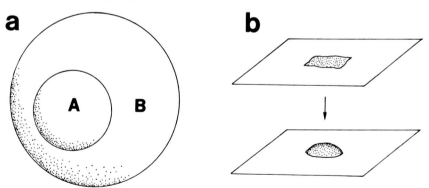

Fig. 1. Cells of one adhesive affinity enclose an aggregate of cells of a different affinity: (a) in a three-dimensional mass of cells; (b) in an epithelium. The enclosed patch of epithelial cells (stippling) tends to round and dome.

If the cells in a three dimensional aggregate are motile and adhesive, the aggregate behaves as a liquid droplet. Over a range of adhesive affinities, cells of two types will tend to sort out into spheres nested inside each other (Steinberg, 1963). In Fig. 1a the A cells of the inner sphere have a greater work of cohesion than do the B cells which enclose them. The spherical shape of the A aggregate minimizes the area of contact between A and B cells.

Cells will form a two-dimensional aggregate -- an epithelium -- if each cell has apical and basal ends which adhere poorly to other cells. If the cells can move they may sort out into patches. The cells in a patch have similar adhesive affinities, and differ in affinity from nearby cells surrounding the patch (Fig. 1b). Alternatively, clusters of cells with similar adhesive properties may develop directly, without appreciable rearrangement of cells. The perimeter of a patch will decrease to reduce the area of contact between cells of the patch and the surrounding cells, if the adhesive affinities of patch and surrounding cells differ appreciably. The area of contact between patch and surround does decrease if a patch of integument is grafted to an abnormal host site on the body surface of an insect (Nardi, Kafatos 1976; Nübler-Jung 1977). The perimeter of the graft becomes more circular, the graft thickens, and it may evaginate or invaginate. These changes in form apparently manifest a difference in the intercellular affinities of host and graft cells.

If the sheet contains cells of suitably different adhesive types, the most stable distribution of cells is a bull's-eye in which all the cells of one type form an annular band, as in Fig. 2a (Goel et al. 1970). The planar sheet containing this bull's-eye will tend to deform, with rearrangement of cells, into a blunt banded cone because the area of contact between unlike cells in adjacent bands is thereby

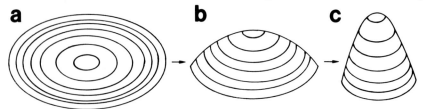

Fig. 2. Deformation of a planar bull's-eye (a) with low strain and high adhesive disparity into a banded blunt cone (b,c) in which strain balances adhesion.

reduced. However, the epithelium is stiff; it resists deformation, as does an elastic sheet. This stiffness tends to
keep the bull's-eye more nearly planar.

Mittenthal and Mazo (1982) have proposed that such an
arrangement of cells represents an arthropod limb. In the
model the shape of the limb represents a compromise between
an elongate shape, which minimizes contact between bands and
thereby maximizes the total work of adhesion among cells, and
a short broad shape which minimizes the energy of mechanical
strain in the epithelium. For cylindrical leg segments with
radius R and length L, this strain-adhesion model implies
that the curvature 1/R is nearly proportional to the shape
factor R/L. Leg segments of <u>Drosophila</u> have proportions
consistent with this scaling relation.

In the present work we have asked what pattern of
adhesion might determine the shape of the terminal segment,
the dactyl, in a crayfish leg. Crayfish have five pairs of
legs. The anterior three pairs (legs 1 - 3) terminate in a

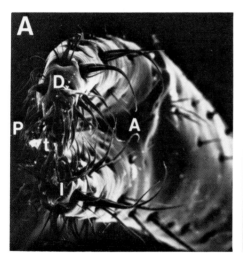
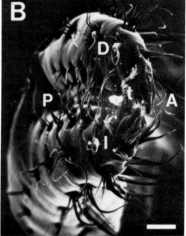

Fig. 3. Distal views of a normal dactyl (D) and index (I) of
2nd leg. A, anterior; and P, posterior, here and in subsequent
figures. The leg is tilted toward the posterior (A) or the
anterior (B) to show that the posterior side is flatter than
the anterior side. t, large distal teeth; a row of teeth is
proximal to the distal tooth on the dactyl and on the index.
Scale bar: 150 μm.

pincer, formed by the dactyl and an extension, the index, from the subterminal segment (Fig. 3). The dactyl and index have facing rows of fine teeth; each row terminates in a larger tooth distally. Tufts of setae form proximo-distally oriented rows around the rest of the circumference.

The surface is slightly deformed near the teeth and setae. In applying the strain-adhesion model we have neglected these local deformations, focusing rather on the large-scale curvatures of the surface. The dactyl and index deviate from antero-posterior bilateral symmetry in two ways: The posterior surface is less curved than the anterior surface; and the proximo-distal axis is skewed toward the anterior (Fig. 3). (We shall neglect dorso-ventral asymmetries.)

The dactyl resembles the blunt cone in Fig. 4a. To aid us in finding a pattern of adhesion which would shape this cone to resemble a dactyl, we assume that cells differ in adhesive affinity around the circumference of the cone, as well as along its length. (In support of this assumption, it

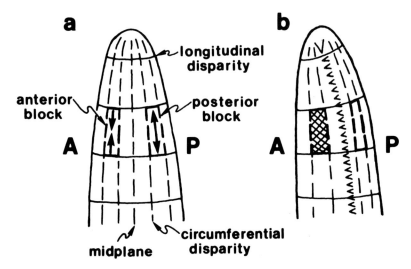

Fig. 4. A possible pattern of adhesive disparities on the distal part of a leg; ventral views. (a) An axially symmetric pattern of disparities shapes an axially symmetric blunt cone. (b) Greater longitudinal disparity on the posterior side skews the cone toward the anterior, as in a crayfish dactyl. Arrows in (a) represent the direction of longitudinal distortion of the block.

seems likely that cells with different positional information
have different adhesive properties. Grafting operations have
demonstrated that cells at different places on the circumference of the leg have different positional information (French
1980).) We refer to the boundary between two groups of cells
differing in adhesive properties as an adhesive disparity.
Thus, we are assuming that the epithelium forming the cone
contains longitudinal and circumferential adhesive disparities. Two circumferential disparities -- one mid-dorsal, the
other mid-ventral -- define a mid-plane which separates anterior and posterior sides of the cone. We regard the tooth
row, which occurs in a mid-ventral position, as lying on the
ventral mid-plane disparity.

The disparities partition the epithelium into blocks.
We assume that all cells with the same adhesive properties
are together in one block. The size of a block is unknown;
if every cell has distinct adhesive properties, each cell is
a block. The pattern of blocks is mirror-symmetric about the
mid-plane; that is, a posterior block corresponds to every
anterior block. However, the areas or shapes of corresponding blocks must differ to make the dactyl asymmetric. The
following argument supposes that corresponding anterior and
posterior blocks have nearly the same area, and that the
epithelium has uniform stiffness. Increasing the posterior
longitudinal disparities relative to the corresponding anterior disparities will cause a posterior block to be longer
and more slender than the corresponding anterior block (Fig.
4b). In a more slender posterior block fewer cells will
contact the longitudinally adjacent blocks, with which contact now produces higher adhesive disparity and hence higher
energy. Because the posterior blocks are longer and more
slender, they will contribute less to the circumference, and
more to the proximo-distal length, than anterior blocks.
Thus this difference in block shape will flatten the posterior side and skew the proximo-distal axis toward the anterior, as in the dactyl.

We are proposing that a skewed bull's-eye pattern of
longitudinal adhesive disparities underlies the shape of the
normal dactyl. If this hypothesis is correct it should also
predict the shape of abnormal limbs, containing supernumerary dactyls, which regenerate after surgery. In many animals, interchanging the left and right distal ends of limbs
consistently elicits the formation of two supernumerary ends,
each with host symmetry, from opposite sides of the host-

graft junction (e.g. cockroach: Bohn 1965; urodele: Iten and Bryant 1976). This result is predicted by the hypothesis that pattern regulation restores the continuity of positional information (French et al. 1976; review: Mittenthal 1981b).

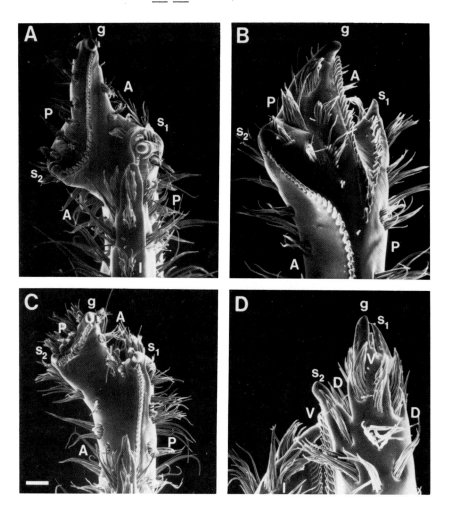

Fig. 5. Triplicated ends developed after grafting the end of a right dactyl into the stump of a left dactyl. (A)-(C) ventral views; (D) posterior view. I, index; D, dorsal; V, ventral; g, graft; s_1, s_2, supernumerary ends, here and in following figures. s_1 is better aligned with the host than s_2. Scale bar: (A) 220 μm; (B) 270 μm; (C,D) 165 μm.

To look for such supernumeraries we grafted the distal 25% of the dactyl of leg 3 into the proximal 10% of the contralateral dactyl of leg 2, matching dorsal to dorsal and ventral to ventral margins, using the methods of Mittenthal 1981a. Of 19 operated legs, 4 regenerated a normal dactyl, 4 went unscored through loss or death of the animal, 1 regenerated a single supernumerary, and 10 regenerated two supernumeraries. In the latter 10 legs, the two supernumeraries were dorsal and ventral to the graft in 2 legs, and anterior and posterior to the graft in the other 8 legs (Fig. 5).

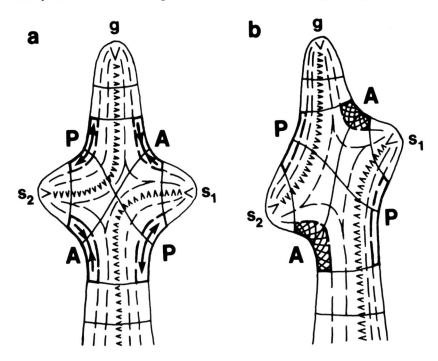

Fig. 6. A possible pattern of adhesive disparities in a triplicated end. (a) If the normal end has an axisymmetric pattern of adhesive disparities, intercalation can produce two supernumeraries which are roughly mirror images. (Ignoring shortening of s_1 and s_2, the pattern of disparities is symmetric about planes between g and s_1, and between g and s_2, even if the normal leg is asymmetric (Bateson, 1894). However, the pattern need not be mirror symmetric about the mid-plane of the host.) Greater longitudinal disparity on the posterior sides (arrows) could skew the triplicated end, more nearly aligning s_1 with the host (b).

The legs with two supernumeraries showed two consistent characteristics. First, the supernumeraries were invariably shorter than the graft. This feature has been noted after left-right interchange in other arthropods also, but not in urodeles. Second, in those legs having the supernumeraries anterior and posterior to the graft, the posterior supernumerary was well aligned with the host; the angle between their proximo-distal axes was small. However, the angle between the axes of the anterior supernumerary and the host was appreciably larger.

To interpret these features with the strain-adhesion model, we first need to define a pattern of adhesive disparities in which pattern regulation has restored the continuity of positional information. Fig. 6a shows one such pattern. If the graft and host are axially symmetric, the two supernumeraries will roughly be mirror images. Now suppose we bias the adhesive disparities as in the normal dactyl, making the longitudinal disparities greater on the posterior than on the anterior surface. Posterior blocks elongate and anterior blocks shorten, skewing the three tips toward the anterior side (Fig. 6b). The angles between posterior faces become larger, and the angles between anterior faces become smaller, as observed.

In the strain-adhesion model the shape having minimum energy depends on the strain energy, as well as on the adhesive disparities. What effect does the asymmetry of disparity have on the strain energy? The strain energy depends on the principal curvatures K_1 and K_2 at each point on the surface of the epithelium (Fig. 7a). $1/K_1$ and $1/K_2$ are the radii of curvature of two circles, in perpendicular planes, tangent to the surface at a point. Roughly, in a normal leg the plane for a circle with curvature K_2 transects the leg transversely, and the plane for a circle with curvature K_1 passes through the proximo-distal axis. By convention $K_2 < 0$ for a tube such as a limb.

The strain energy is the integral, over the surface of the epithelium, of the strain energy density:

$$\text{strain energy density} \quad f = K_1{}^2 + K_2{}^2 + K_1 K_2$$

$$\text{strain energy} \quad E = \int f \, da.$$

The integral is over elements of surface area, da, of the

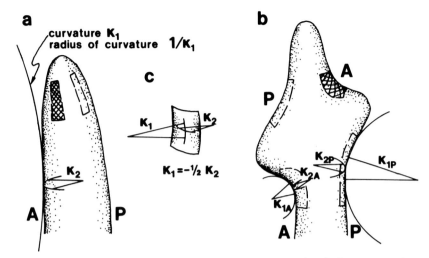

Fig. 7. Principal curvatures, K_1 and K_2, in (a) a normal dactyl; (b) a triplicated dactyl; (c) a saddle-shaped patch with relative curvatures that minimize strain energy.

epithelium (Mittenthal and Mazo 1982). The equation for the strain energy density assumes that the epithelium is a fluid shell -- a sheet of incompressible fluid in small elastic-sheathed packets, the cells. In response to stress cells can rearrange, but the sheet exhibits elastic strain in bending.

What is the shape of a block which minimizes its strain energy? Given K_2, f has a minimum when the block is saddle-shaped, with the two curvatures matched so that $K_1 = -(1/2)K_2$ (Fig. 7c). Development of this saddle shape contributes more to minimizing the total strain energy, the larger is $|K_2|$, because $f_{min} = (3/4)K_2^2$.

If we compare two saddle-shaped blocks in the triplicated dactyl (Fig. 7b), both K_1 and $|K_2|$ are smaller in the posterior block than in the anterior block. That is, the asymmetry of the triplicated dactyl matches K_1 and K_2 in the way which tends to minimize the strain energy. Hence the strain pattern, as well as the pattern of adhesive disparity, favors development of the observed asymmetry.

The strain-adhesion model can interpret, not only the angles at which supernumeraries emerge from the limb, but also the finding that the supernumeraries are shorter than the graft. This model, extended to deal with adhesive disparities in a system with an epithelium and an underlying mesenchyme, may eventually provide insight into the early shaping of the vertebrate limb bud.

We thank Chuck Kimmel for providing facilities at the University of Oregon, and Jim Nardi for helpful comments. N.S.F. grant BNS 7920260, P.H.S. grant HD 16577, and funds from the University of Illinois supported this work.

Bateson W (1894). "Materials for the Study of Variation." London: MacMillan.

Bohn H (1965). Analyse der Regenerationsfahigkeit der Insektenextremitat durch Amputations- und Transplantationsversuche an Larven der Afrikanischen Schabe Leucophaea maderae Fabr. (Blattaria). II. Achsendetermination. W Roux Arch EntwMech Org 156:449.

Bryant SV, Iten LI (1976). Supernumerary limbs in amphibians: Experimental production in Notophthalmus viridescens and a new interpretation of their formation. Develop Biol 50:212.

French V (1980). Positional information around the segments of the cockroach leg. J Embryol exp Morph 59:281.

French V, Bryant PJ, Bryant SV (1976). Pattern regulation in epimorphic fields. Science 193:969.

Goel N, Campbell RD, Gordon R, Rosen R, Martinez H, Ycas M (1970). Self-sorting of isotropic cells. J Theor Biol 28:423.

Mittenthal JE (1981a). Intercalary regeneration in legs of crayfish: Distal segments. Develop Biol 88:1.

Mittenthal JE (1981b). The rule of normal neighbors: A hypothesis for morphogenetic pattern regulation. Develop Biol. 88:15.

Mittenthal JE, Mazo RM (1982). A model for shape generation by strain and cell-cell adhesion in the epithelium of an arthropod leg segment. J Theor Biol (submitted).

Nardi JB, Kafatos FC (1976). Polarity and gradients in lepidopteran wing epidermis. I. Changes in graft polarity, form, and cell density accompanying transpositions and reorientations. J Embryol exp Morph 36:469.

Nübler-Jung K (1977). Pattern stability in the insect segment. I. Pattern reconstitution by intercalary regeneration and cell sorting in Dysdercus intermedius Dist. W Roux Arch Develop Biol 183:17.

Steinberg MS (1963). Reconstruction of tissues by dissociated cells. Science 141:401.

INDEX